Introduction to Computers for Engineering and Technology

Kenneth C. Mansfield Jr.
James L. Antonakos

Broome Community College

Prentice Hall
Upper Saddle River, New Jersey   Columbus, Ohio

Library of Congress Cataloging-in-Publication Data

Mansfield, Kenneth C.
    Introduction to computers for engineering and technology / Kenneth C. Mansfield Jr., James L. Antonakos
      p. cm.
    Includes index.
    ISBN 0-13-227786-7
    1. Computers.   2. Microcomputers.   I. Antonakos, James L.
II. Title.
QA76.5.A557   2000
004.16—dc21
                                                                                                                99-27117
                                                                                                                         CIP

Publisher: Charles E. Stewart, Jr.
Production Editor: Alexandrina Benedicto Wolf
Production Coordination: Custom Editorial Productions, Inc.
Cover Design Coordinator: Karrie Converse-Jones
Cover Art: SuperStock, Inc.
Cover Designer: Linda Fares
Production Manager: Deidra M. Schwartz
Marketing Manager: Ben Leonard

This book was set in Times Roman by Custom Editorial Productions, Inc., and was printed and bound by The Banta Company. The cover was printed by Phoenix Color Corp.

© 2000 by Prentice-Hall, Inc.
Pearson Education
Upper Saddle River, New Jersey 07458

All rights reserved. No part of this book may be reproduced, in any form or by any means, without permission in writing from the publisher.

Notice to the Reader: All product names listed herein are trademarks and/or registered trademarks of their respective manufacturer.

    The publisher and the authors do not warrant or guarantee any of the products and/or equipment described herein, nor has the publisher or the authors made any independent analysis in connection with any of the products, equipment, or information used herein. The reader is directed to the manufacturer for any warranty or guarantee for any claim, loss, damages, costs, or expense arising out of or incurred by the reader in connection with the use or operation of the products or equipment.

    The reader is expressly advised to adopt all safety precautions that might be indicated by the activities and experiments described herein. The reader assumes all risks in connection with such instructions.

Printed in the United States of America

10 9 8 7 6 5 4 3 2 1

ISBN: 0-13-227786-7

Prentice-Hall International (UK) Limited, *London*
Prentice-Hall of Australia Pty. Limited, *Sydney*
Prentice-Hall Canada Inc., *Toronto*
Prentice-Hall Hispanoamericana, S. A., *Mexico*
Prentice-Hall of India Private Limited, *New Delhi*
Prentice-Hall of Japan, Inc., *Tokyo*
Prentice-Hall (Singapore) Pte. Ltd., *Singapore*
Editora Prentice-Hall do Brasil, Ltda., *Rio de Janeiro*

Preface

Students graduating from college today must be equipped with increasingly sophisticated skills. In addition to the fundamental core of knowledge, students should know how to use the proper tools to enhance their job performance. Within the last fifteen years, the ability to use a personal computer has become one of the most important additions to a graduate's portfolio.

It is our belief as educators that being comfortable with the personal computer, its peripherals, the DOS and Windows operating systems, and application software is required for students graduating from two- and four-year science, technology, and engineering programs. Students who have never used a computer before, and students who have a working knowledge of the personal computer but want more details, will find this textbook a valuable resource. One student may ask, "What is a hard drive?" whereas another might ask, "How can I replace my 840MB hard drive with an 8GB drive?" There are exercises designed to answer both of these questions and more.

After using this text, the students will be able to explain the internal operation of the personal computer, including its microprocessor, memory, and input/output systems. The DOS and Windows operating systems, including Windows 95/98 and NT, are covered in detail. Students will be able to perform various tasks with each operating system, such as word processing, preparing new disks, spreadsheets, printing, installing and upgrading software, troubleshooting, and programming, to name a few. There are numerous exercises designed to provide hands-on experience in these areas.

The social aspects of using computers and their technical challenges are illustrated through the exploits of Joe Tekk, a fictitious computer specialist working at a fictitious software company called RWA Software. Joe encounters the successes and failures commonly associated with computers and their operation, and also interacts with many different individuals regarding their computer experiences. Students are encouraged to consider the social implications of computers whenever possible.

OUTLINE OF COVERAGE

The textbook is divided into seven major units.

UNIT I: INTRODUCTION

This unit sets the stage for the remaining units by describing the presentation of the material and through coverage of electrical and mechanical safety. The microcomputer and its various components are also introduced.

UNIT II: THE DISK OPERATING SYSTEM

In this unit students are exposed to the features of the first operating system designed for the personal computer. All versions of DOS, up through version 6.x, are covered. Students will learn skills such as handling floppy disks (formatting, copying, scanning for viruses), navigating through the DOS file system, customizing their personal computer via CONFIG.SYS and AUTOEXEC.BAT, and the basics of memory management.

UNIT III: THE WINDOWS OPERATING SYSTEM

The Windows 3.x, Windows 95/98, and Windows NT operating systems are all presented, with numerous screen shots provided to illustrate exactly what students should see as they try new commands and procedures. They are shown how to manage the graphical user interface provided by Windows, how to install new software and upgrade existing software, how to use the Control Panel to customize Windows, and much more. Coverage of NetWare is included for contrast.

UNIT IV: COMPUTER NETWORKS

From local area networks to the global communication services provided by the Internet, computer networks and their associated hardware and software are explained in detail. Topics include cabling (thinwire, 10-base T, fiber optic), topology (star, ring, bus), TCP/IP, Ethernet, token ring, ISO-OSI hierarchy, e-mail, FTP, and Web browsers.

UNIT V: MICROCOMPUTER SYSTEMS

This unit contains exercises designed to familiarize students with all major hardware aspects of the personal computer. Some of the hardware examined includes the motherboard (microprocessor, memory, I/O, and expansion buses), floppy and hard drives, CD-ROMs, display adapters, sound cards, and modems.

UNIT VI: APPLICATION SOFTWARE

This unit covers a number of popular software packages, showing the student how to use word processors, spreadsheets, databases, and a host of technology-specific applications. Demonstration versions of selected applications are available on the companion CD.

UNIT VII: SELECTED TOPICS

This unit covers a wide range of microcomputer-related topics, from computer languages to processor architecture and computer viruses. These exercises are designed to introduce students to many advanced topics so that they are more familiar with the complete world of microcomputer systems.

THE COMPANION CD

The companion CD included with the textbook contains useful example programs and files designed to aid students in developing and understanding the concepts presented in each unit.
The demonstration software includes:

- Electronics Workbench®
- HoTMetaL PRO™

- LanExplorer®
- Paint Shop Pro®

In addition, there are comprehensive study guide materials designed to provide the necessary exposure to the material prior to actual testing. View the README document (text, Microsoft Word, and HTML formats) to get a detailed description of the companion CD contents.

A comprehensive set of ancillaries is also available:

- **Instructor's Manual:** This manual includes solutions to selected text problems, teaching suggestions, and sample syllabi.
- **PowerPoint® Slides:** Figures from the text are designed to help instructors with classroom/lecture presentations. The slides are contained on a CD packaged with the Instructor's Manual.
- **Test Item File:** Test bank containing sample test questions.
- **Test Manager:** This computerized testing system allows instructors to build tests from an electronic database of questions. This program also has on-line testing capabilities.
- **News Updating Service:** Adopters of this book will be able to access a news updating service for timely information on technology, which will greatly enhance lectures.
- **Companion Web Site — www.prenhall.com/mansfield:** The companion Web site complements the text as an on-line study guide. Review questions help students understand the topics presented in each exercise. In addition, students will find links to many sites in the field of microcomputers. Using the Syllabus Builder™ feature, instructors can post on the Internet syllabi specifically designed for their classes.

ACKNOWLEDGMENTS

We would like to thank our editor, Charles Stewart, for his encouragement and assistance during the development of this project. Thanks also go to our copyeditor, Cindy Lanning, and our production supervisor at Custom Editorial Productions, Kevin Walzer.

Many different companies provided their software (for student use) on the companion CD. These companies and the following individuals were especially helpful to us regarding permission requests and demonstration versions, and we deeply thank them:

- Naomi Bulock of The MathWorks, Inc., for allowing screen shots of MATLAB®. MATLAB is a registered trademark of The MathWorks, Inc.
- Michele Burgoyne of Visio Corporation, for allowing screen shots of Visio®. © 1997 Visio Corporation. Visio is a registered trademark of Visio Corporation in the United States and/or other countries.
- Scott Duncan of Interactive Image Technologies, Ltd. for providing Electronics Workbench. Contact Interactive at www.interactiv.com.
- Nancy Johnson of Jasc Software, Inc., for providing Paint Shop Pro.
- Cami Layton of Netscape Communications Corporation, for allowing screen shots of Netscape Navigator. Screen shots copyright Netscape Communications Corporation, 1999. All rights reserved. Netscape, Netscape Navigator, and the Netscape N Logo are registered trademarks of Netscape in the United States and other countries.
- Ken Jowe of Intellimax Systems, Inc., for providing LanExplorer. LanExplorer is a registered trademark of Intellimax Systems, Inc.
- Microsoft® Corporation for allowing screen shots of their operating systems and software applications, particularly Microsoft Office. Screen shots reprinted by permission of Microsoft Corporation.
- Network Associates, Inc. and McAfee.com Corp., for allowing screen shots of VShield and VirusScan.
- Jennifer Yanez-Pastor of Caere Corporation, for allowing screen shots of OmniPage Pro.
- Carol Parcels of Hewlett-Packard Company for allowing screen shots of Colorado Backup.
- Jerry Gowen of Power Quest Corp., for allowing screen shots of PartitionMagic.
- Jordannah Shuster of SoftQuad™ Software, Inc., for providing of HoTMetaL PRO™.

- Elizabeth Whiteley of Symantec Corp., for allowing screen shots of Norton Utilities and pc ANYWHERE. © 1991, 1995, 1997 Symantec Corporation, 10201 Torre Avenue, Cupertino, CA 95014 USA. All rights reserved.

We also wish to thank our reviewers, who provided many useful and constructive suggestions: Bill Iverson, Renton Technical College; George Kannel, New Jersey Institute of Technology and LUCENT; Yehuda Nishli, Bramson ORT Technology Institute; Tom Young, Rochester Institute of Technology; and Mengchu Zhou, New Jersey Institute of Technology.

Kenneth C. Mansfield Jr.
mansfield_k@sunybroome.edu
http://www.sunybroome.edu/~mansfield_k

James L. Antonakos
antonakos_j@sunybroome.edu
http://www.sunybroome.edu/~antonakos_j

Brief Contents

UNIT I	Introduction 1
EXERCISE 1	Using the Instructional System 3
EXERCISE 2	Electrical and Mechanical Safety 7
EXERCISE 3	Microcomputer Familiarization 11

UNIT II	The Disk Operating System 17
EXERCISE 4	Using Floppy Disks 19
EXERCISE 5	Introduction to DOS 33
EXERCISE 6	DOS Commands and File Management 47
EXERCISE 7	Configuring DOS with CONFIG.SYS and AUTOEXEC.BAT 67
EXERCISE 8	DOS Utilities 87

UNIT III	The Windows Operating System 105
EXERCISE 9	An Overview of Windows 3.x 107
EXERCISE 10	An Introduction to Windows 95 and Windows 98 115
EXERCISE 11	An Introduction to Windows NT 129
EXERCISE 12	The Desktop 139
EXERCISE 13	The Control Panel 165
EXERCISE 14	Windows Explorer 179
EXERCISE 15	Managing Printers 197
EXERCISE 16	Accessories 209
EXERCISE 17	An Introduction to Networking with Windows 227
EXERCISE 18	Installing New Software 235
EXERCISE 19	Installing New Hardware 245
EXERCISE 20	Other Network Operating Systems 255

UNIT IV	Computer Networks 263
EXERCISE 21	What Is a Computer Network? 265
EXERCISE 22	Network Topology 273
EXERCISE 23	Networking Hardware 281
EXERCISE 24	Networking Protocols 299
EXERCISE 25	Network Applications 313
EXERCISE 26	The Internet 321
EXERCISE 27	Windows NT Domains 333
EXERCISE 28	An Introduction to Telecommunications 345

UNIT V	Microcomputer Systems 353
EXERCISE 29	Computer Environments 355
EXERCISE 30	System Teardown and Assembly 371
EXERCISE 31	Power Supplies 377
EXERCISE 32	Floppy Disk Drives 385
EXERCISE 33	The Motherboard Microprocessor and Coprocessor 399
EXERCISE 34	The Motherboard Memory 409
EXERCISE 35	Motherboard Expansion Slots 423
EXERCISE 36	Power-On Self-Test (POST) 437
EXERCISE 37	Motherboard Replacement and Setup 445
EXERCISE 38	Hard Disk Fundamentals 451
EXERCISE 39	Hard Drive Backup 471
EXERCISE 40	Hard Disk Replacement and File Recovery 489
EXERCISE 41	Video Monitors and Video Adapters 497
EXERCISE 42	The Computer Printer 513

EXERCISE 43	Keyboards and Mice 527	
EXERCISE 44	Telephone Modems 535	
EXERCISE 45	CD-ROM and Sound Card Operation 549	
EXERCISE 46	Multimedia Devices 559	

UNIT VI Application Software 567

EXERCISE 47	Word Processors 569
EXERCISE 48	Spreadsheets 579
EXERCISE 49	Databases 593
EXERCISE 50	Presentation Software 607
EXERCISE 51	Web Development 619
EXERCISE 52	Science and Technology 637

UNIT VII Selected Topics 657

EXERCISE 53	An Introduction to Intel Microprocessor Architecture 659
EXERCISE 54	Computer Languages 673
EXERCISE 55	Hardware and Software Interrupts 687
EXERCISE 56	The Advanced Intel Microprocessors 697
EXERCISE 57	A Detailed Look at the System BIOS 709
EXERCISE 58	Windows Internal Architecture 715
EXERCISE 59	Computer Viruses 721
EXERCISE 60	A Typical Computer Center 731
APPENDIX A	ASCII Character Set 737

Answers to Odd-Numbered Self-Test Questions 739

Index   745

Contents

UNIT I	**Introduction** 1	

EXERCISE 1 Using the Instructional System 3
Introduction 3
Performance Objectives 3
Background Information 3
Self-Test 4
Familiarization Activity 4
Questions/Activities 4
Review Quiz 4
Answers to Self-Test 5

EXERCISE 2 Electrical and Mechanical Safety 7
Introduction 7
Performance Objectives 7
Background Information 7
Troubleshooting Techniques 8
Self-Test 8
Familiarization Activity 9
Questions/Activities 9
Review Quiz 9

EXERCISE 3 Microcomputer Familiarization 11
Introduction 11
Performance Objectives 11
Background Information 11
Troubleshooting Techniques 14
Self-Test 14
Familiarization Activity 15
Questions/Activities 16
Review Quiz 16

UNIT II **The Disk Operating System** 17

EXERCISE 4 Using Floppy Disks 19
Introduction 19
Performance Objectives 19
Background Information 19

*Discussion   Basic Computer System   Disk Drives
Floppy Disks   Labeling Floppy Disks   Write-Protecting
Disks   Storage Capabilities   Formatting a Disk
Booting DOS   Required DOS Files   FORMAT Revisited
Volume Labels   Displaying the Volume Label*

Troubleshooting Techniques 29
Self-Test 30
Familiarization Activity 31
Questions/Activities 31
Review Quiz 31

EXERCISE 5 Introduction to DOS 33
Introduction 33
Performance Objectives 33
Background Information 33

*Introduction   Booting Methods   Setting the Date and
Time   The DOS Prompt   Changing the Date and Time
Resetting the Computer   Copying a Disk   Internal and
External DOS Commands   Definition of DOS Files
How DOS Names Files   Displaying DOS Files*

Troubleshooting Techniques 43
Self-Test 43
Familiarization Activity 44
Questions/Activities 44
Review Quiz 45

EXERCISE 6 DOS Commands and File Management 47
Introduction 47
Performance Objectives 47
Background Information 48

*Introduction   Copying the Screen to the Printer
Transferring a DOS File to Another Disk   Displaying
One File   DOS Wild Card Characters   Copying a
Group of Files   Displaying File Contents   Redirecting
Files   The F1 Key   Creating a DOS File   Using EDIT
Starting EDIT   Editing an Existing File   Other EDIT
Features   Renaming Files   Erasing Files   Organizing
Your Files   Creating Directories   Getting into
Directories   Getting out of Directories   Some Directory
Examples   Creating Other Directories (Pathnames)
Removing Directories   Deleting a Nonempty Directory
Transferring Files*

Troubleshooting Techniques 61
Self-Test 61
Familiarization Activity 62
Questions/Activities 65
Review Quiz 66

EXERCISE 7	Configuring DOS with CONFIG.SYS and AUTOEXEC.BAT 67 Introduction 67 Performance Objectives 67 Background Information 67		Troubleshooting Techniques 125 Self-Test 125 Familiarization Activity 126 Questions/Activities 127 Review Quiz 127

Introduction Setting the Path System Configuration DRIVEPARM FILES LASTDRIVE Stacks Definition of Batch Processing Batch Files Batches Within Batches The REM Command The ECHO Command The PAUSE Command Batch Parameters The IF Command The NOT Command The GOTO Command Combining the IF and GOTO Commands AUTOEXEC.BAT Files Changing the DOS Prompt

		EXERCISE 11	An Introduction to Windows NT 129 Introduction 129 Performance Objectives 129 Background Information 129

Windows NT Operating System Logon Windows NT Security Menu Windows NT Desktop, Taskbar and Start Menu Windows NT Control Panel Windows NT Domains

Troubleshooting Techniques 81
Self-Test 82
Familiarization Activity 84
Questions/Activities 85
Review Quiz 85

Troubleshooting Techniques 136
Self-Test 137
Familiarization Activity 137
Questions/Activities 138
Review Quiz 138

EXERCISE 8	DOS Utilities 87 Introduction 87 Performance Objectives 87 Background Information 87	EXERCISE 12	The Desktop 139 Introduction 139 Performance Objectives 139 Background Information 140

Standard Input and Output DOS Redirection Operators DOS Output Redirection The SORT Command The MORE Command The FIND Command Setting File Attributes Using ATTRIB Attribute of a File Read-Only Files Archive Files Hidden Files System Files Checking the File Status The XCOPY Command Copying a Set of Files from a Hard Disk The MODE Command Controlling the Video Display with the MODE Command Setting Printer Characteristics with MODE Serial Port Configuration with MODE

The Start Button The Taskbar The Background The Systems Settings Area Creating and Using Shortcuts Changing Other Desktop Properties Managing the Desktop Contents My Computer The Network Neighborhood The Recycle Bin My Briefcase The Inbox Getting Back to the Desktop from DOS Windows 98 Help Windows 98 Settings Menu Windows 98 Explorer Windows 98 Web-Style Desktop Windows NT Desktop Evolution

Troubleshooting Techniques 100
Self-Test 100
Familiarization Activity 102
Questions/Activities 103
Review Quiz 104

Troubleshooting Techniques 160
Self-Test 161
Familiarization Activity 162
Questions/Activities 163
Review Quiz 163

UNIT III	**The Windows Operating System 105**	EXERCISE 13	The Control Panel 165 Introduction 165 Performance Objectives 165 Background Information 165
EXERCISE 9	An Overview of Windows 3.x 107 Introduction 107 Performance Objectives 107 Background Information 107		

Accessibility Add New Hardware Add/Remove Programs Date/Time Display Keyboard Modems Mouse Multimedia Network Passwords Printers Regional Settings Sounds System

Starting Windows Leaving Windows Opening a Program Group Window Moving a Window Scrolling Through a Window Resizing a Window Running a DOS Application from Inside Windows

Troubleshooting Techniques 176
Self-Test 177
Familiarization Activity 177
Questions/Activities 178
Review Quiz 178

Troubleshooting Techniques 113
Self-Test 113
Familiarization Activity 114
Questions/Activities 114
Review Quiz 114

		EXERCISE 14	Windows Explorer 179 Introduction 179 Performance Objectives 179 Background Information 179
EXERCISE 10	An Introduction to Windows 95 and Windows 98 115 Introduction 115 Performance Objectives 115 Background Information 115		

Changing the View Creating New Folders Deleting a Folder Creating Shortcuts Checking/Setting Properties Editing Features Finding Things Working with Network Drives Using Go To Help Windows 98 Explorer Windows NT

The Desktop Long File Names Context-Sensitive Menus Improved Help Facility Windows Explorer The Registry Networking DOS Preemptive Multitasking Recycle Bin Other New Features Windows 95 Version B The Windows 98 Desktop What Else Is New in Windows 98? Windows 98 Is Easier to Use Windows 98 Is Enhanced for the Web Windows 98 Is Faster and More Reliable Windows 98 Is More Entertaining

Troubleshooting Techniques 195
Self-Test 195
Familiarization Activity 196
Questions/Activities 196
Review Quiz 196

EXERCISE 15	Managing Printers 197 Introduction 197

Performance Objectives 197
Background Information 197

Printer Properties   Adding a New Printer   Checking the Printer Status   Pausing a Print Job   Resuming a Print Job   Deleting a Print Job   Network Printing   Windows NT Printer Security

Troubleshooting Techniques 206
Self-Test 207
Familiarization Activity 208
Questions/Activities 208
Review Quiz 208

EXERCISE 16   Accessories 209
Introduction 209
Performance Objectives 209
Background Information 209

Games   Internet Tools   Multimedia   System Tools   Calculator   Calendar   Cardfile   Character Map   Clipboard Viewer   Missing Shortcuts   Dial-Up Networking   Direct Cable Connection   HyperTerminal   Notepad   Paint   Phone Dialer   WordPad   Windows NT Administrative Tools

Troubleshooting Techniques 224
Self-Test 224
Familiarization Activity 225
Questions/Activities 226
Review Quiz 226

EXERCISE 17   An Introduction to Networking with Windows 227
Introduction 227
Performance Objectives 227
Background Information 227

Microsoft Networking   The Network Neighborhood   Network Printing   Sharing Files over a Network   Dial-Up Networking   Connecting to the Internet

Troubleshooting Techniques 231
Self-Test 231
Familiarization Activity 232
Questions/Activities 233
Review Quiz 233

EXERCISE 18   Installing New Software 235
Introduction 235
Performance Objectives 235
Background Information 235

Existing Windows 3.x Software   Installing New Software from the Windows CD-ROM   Getting the Latest Updates from Microsoft   Installing Third-Party Software   Removing Application Software

Troubleshooting Techniques 242
Self-Test 242
Familiarization Activity 243
Questions/Activities 243
Review Quiz 243

EXERCISE 19   Installing New Hardware 245
Introduction 245
Performance Objectives 245
Background Information 245

Starting Off   Detecting New Hardware   The Results   Finishing Up   General Comments   Installing Hardware in Windows NT

Troubleshooting Techniques 252
Self-Test 252
Familiarization Activity 253
Questions/Activities 253
Review Quiz 253

EXERCISE 20   Other Network Operating Systems 255
Introduction 255
Performance Objectives 255
Background Information 255

NetWare   Installing/Upgrading NetWare   NetWare-Windows Time Line   NDS   HCSS   Network Protocols Management   Print Services   The Linux Operating System

Troubleshooting Techniques 261
Self-Test 261
Familiarization Activity 262
Questions/Activities 262
Review Quiz 262

UNIT IV   **Computer Networks 263**

EXERCISE 21   What Is a Computer Network? 265
Introduction 265
Performance Objectives 265
Background Information 265

Computer Network Topology   Representing Digital Data   Communication Protocols   Ethernet   Token-Ring Networks

Troubleshooting Techniques 270
Self-Test 270
Questions/Activities 271
Review Quiz 271

EXERCISE 22   Network Topology 273
Introduction 273
Performance Objectives 273
Background Information 273

Physical Topology Versus Logical Topology   Fully Connected Networks   Star Networks   Bus Networks   Ring Networks   Hybrid Networks   Network Hierarchy   Subnets

Troubleshooting Techniques 277
Self-Test 278
Familiarization Activity 279
Questions/Activities 279
Review Quiz 279

EXERCISE 23   Networking Hardware 281
Introduction 281
Performance Objectives 281
Background Information 281

Ethernet Cabling   The NIC   Token Ring   Repeaters   Transceivers   Hubs   Bridges/Switches   Routers   Cable Modems   Satellite Network System

Troubleshooting Techniques 295
Self-Test 296
Familiarization Activity 297
Questions/Activities 297
Review Quiz 297

EXERCISE 24   Networking Protocols 299
Introduction 299
Performance Objectives 299
Background Information 299

The NetBEUI Protocol   The IPX/SPX Protocols   The TCP/IP Protocol Suite   RFCS   IP   IP Version 6   TCP   UDP   DNS   ARP and RARP   ICMP   FTP   SMTP   Telnet   SLIP and PPP   SNMP   Routing Protocols   Protocol Analyzers

Troubleshooting Techniques 309
Self-Test 310
Familiarization Activity 311
Questions/Activities 311
Review Quiz 311

xi

| EXERCISE 25 | Network Applications 313
Introduction 313
Performance Objectives 313
Background Information 313
PING TRACERT WINIPCFG Echo Servers and Chat Servers pcANYWHERE
Troubleshooting Techniques 318
Self-Test 319
Familiarization Activity 319
Questions/Activities 320
Review Quiz 320 |

| EXERCISE 26 | The Internet 321
Introduction 321
Performance Objectives 321
Background Information 321
The Organization of the Internet World Wide Web HTML CGI Java Related Sites
Troubleshooting Techniques 329
Self-Test 330
Familiarization Activity 330
Questions/Activities 331
Review Quiz 331 |

| EXERCISE 27 | Windows NT Domains 333
Introduction 333
Performance Objectives 333
Background Information 333
Domains Domain Clients Logging onto a Network Running a Network Server User Profiles Security
Troubleshooting Techniques 341
Self-Test 342
Familiarization Activity 343
Questions/Activities 343
Review Quiz 343 |

| EXERCISE 28 | An Introduction to Telecommunications 345
Introduction 345
Performance Objectives 345
Background Information 345
TDM Circuit Switching Packet Switching Frame Relay ATM ISDN SONET FDDI Mobile Communication
Troubleshooting Techniques 351
Self-Test 351
Familiarization Activity 352
Questions/Activities 352
Review Quiz 352 |

UNIT V   Microcomputer Systems   353

| EXERCISE 29 | Computer Environments 355
Introduction 355
Performance Objectives 355
Background Information 356
Physical Environment Electrical Environment
Troubleshooting Techniques 367
Self-Test 367
Familiarization Activity 368
Questions/Activities 369
Review Quiz 369 |

| EXERCISE 30 | System Teardown and Assembly 371
Introduction 371
Performance Objectives 371
Background Information 371
System Overview Disassembly Procedures Assembly Procedures Other Considerations
Troubleshooting Techniques 374
Self-Test 374
Familiarization Activity 375
Questions/Activities 375
Review Quiz 376 |

| EXERCISE 31 | Power Supplies 377
Introduction 377
Performance Objectives 377
Background Information 377
Power Supply Characteristics
Troubleshooting Techniques 381
Self-Test 381
Familiarization Activity 382
Questions/Activities 382
Review Quiz 383 |

| EXERCISE 32 | Floppy Disk Drives 385
Introduction 385
Performance Objectives 385
Background Information 385
Overview How a Floppy Disk Drive Works Operating Sequence Disk Drive Support System
Troubleshooting Techniques 392
System Overview Troubleshooting Logic Troubleshooting Steps
Self-Test 394
Familiarization Activity 396
Questions/Activities 397
Review Quiz 397 |

| EXERCISE 33 | The Motherboard Microprocessor and Coprocessor 399
Introduction 399
Performance Objectives 399
Background Information 399
Definition of the Motherboard Contents of the Motherboard The Microprocessor The Coprocessor
Troubleshooting Techniques 405
Self-Test 406
Familiarization Activity 407
Questions/Activities 407
Review Quiz 407 |

| EXERCISE 34 | The Motherboard Memory 409
Introduction 409
Performance Objectives 409
Background Information 409
Computer Memory Primary Storage—RAM and ROM Bits, Bytes, and Words SIMM DIMM SDRAM EDO DRAM VRAM Level-2 Cache Chip Speed How Memory Is Organized How Memory Is Used Breaking the DOS Barrier Virtual Memory Memory Usage in Windows
Troubleshooting Techniques 419
Self-Test 420
Familiarization Activity 421
Questions/Activities 422
Review Quiz 422 |

| EXERCISE 35 | Motherboard Expansion Slots 423
Introduction 423
Performance Objectives 423
Background Information 423
Overview Makeup of an Expansion Slot Expansion Slot Designs The Micro-Channel Expansion Slot The Local Bus The PCI Bus The PCMCIA Bus AGP |

Troubleshooting Techniques 434
Self-Test 434
Familiarization Activity 435
Questions/Activities 435
Review Quiz 436

EXERCISE 36 Power-On Self-Test (POST) 437
Introduction 437
Performance Objectives 437
Background Information 437

*Definition of POST   POST Error Messages
Understanding POST Error Codes*

Troubleshooting Techniques 441
Self-Test 441
Familiarization Activity 442
Questions/Activities 443
Review Quiz 443

EXERCISE 37 Motherboard Replacement and Setup 445
Introduction 445
Performance Objectives 445
Background Information 445

*Overall Safety   Why Replace the Motherboard?
Motherboard Form Factors   Chip Sets   BIOS Upgrades
Removal Process   New Motherboard Installation*

Troubleshooting Techniques 448
Self-Test 448
Familiarization Activity 449
Questions/Activities 450
Review Quiz 450

EXERCISE 38 Hard Disk Fundamentals 451
Introduction 451
Performance Objectives 451
Background Information 451

*Partitions   Windows NT Disk Administrator
LBA   FAT32   NTFS   Hard Drive Interfaces
Data Storage   Disk Caching   Disk Structure
Disk Storage Capacity   Disk Fragmentation
File Allocation   Optimizing Disk Performance
Using the PATH Command*

Troubleshooting Techniques 469
Self-Test 469
Familiarization Activity 470
Questions/Activities 470
Review Quiz 470

EXERCISE 39 Hard Drive Backup 471
Introduction 471
Performance Objectives 471
Background Information 471

*Disk Backup   Backup Schedules   Windows 95/98
Backup   Restoring Files   Windows NT Backup
Disk Backup Using DOS   Restoring Files Using DOS
Using PKZIP and PKUNZIP for Backups   Preparation
of a New Hard Disk*

Troubleshooting Techniques 484
Self-Test 485
Familiarization Activity 486
Questions/Activities 487
Review Quiz 487

EXERCISE 40 Hard Disk Replacement and File Recovery 489
Introduction 489
Performance Objectives 489
Background Information 489

Physical Considerations   The DOS RECOVER Command

Troubleshooting Techniques 493

Self-Test 494
Familiarization Activity 495
Questions/Activities 495
Review Quiz 496

EXERCISE 41 Video Monitors and Video Adapters 497
Introduction 497
Performance Objectives 497
Background Information 497

*Overview   Monitor Servicing   Monitor Fundamentals
Monochrome and Color Monitors   Energy Efficiency
Video Controls   Pixels and Aspect Ratio   Monitor
Modes   Types of Monitors   Display Adapters   VESA
Graphics Accelerator Adapters*

Troubleshooting Techniques 509
Self-Test 510
Familiarization Activity 512
Questions/Activities 512
Review Quiz 512

EXERCISE 42 The Computer Printer 513
Introduction 513
Performance Objectives 513
Background Information 514

*Impact Printers   Nonimpact Printers   Technical
Considerations   Testing Printers   System Software
Extended ASCII Codes   Other Printer Features
Printer Escape Codes   Multifunction Print Devices
Energy Efficiency*

Troubleshooting Techniques 523
Self-Test 524
Familiarization Activity 525
Questions/Activities 525
Review Quiz 525

EXERCISE 43 Keyboards and Mice 527
Introduction 527
Performance Objectives 527
Background Information 527

The Keyboard   The Mouse   Trackballs

Troubleshooting Techniques 531
Self-Test 534
Familiarization Activity 534
Questions/Activities 534
Review Quiz 534

EXERCISE 44 Telephone Modems 535
Introduction 535
Performance Objectives 535
Background Information 535

*The Modem   The RS-232 Standard   Telephone Modem
Setup   Windows Modem Software   Telephone Modem
Terminology   Modulation Methods   MNP Standards
CCITT Standards   ISDN Modems   Cable Modems
Fax/Data Modems   Protocols*

Troubleshooting Techniques 545
Self-Test 546
Familiarization Activity 547
Questions/Activities 547
Review Quiz 547

EXERCISE 45 CD-ROM and Sound Card Operation 549
Introduction 549
Performance Objectives 549
Background Information 549

*CD-ROM Operation   Physical Layout of a Compact
Disk   The High Sierra Format   CD-ROM Standards
Photo CD   ATAPI   CD-ROM Installation   CD-ROM
Properties in Windows   Sound Card Operation   MIDI
Sound Card Installation*

xiii

Troubleshooting Techniques 556
Self-Test 556
Familiarization Activity 558
Questions/Activities 558
Review Quiz 558

EXERCISE 46 Multimedia Devices 559
Introduction 559
Performance Objectives 559
Background Information 559

*Cameras   Scanners   Television Cards   DVD
Bar Code Readers   Interfacing Multimedia Devices
Device Drivers   Other Multimedia Applications*

Troubleshooting Techniques 564
Self-Test 564
Familiarization Activity 565
Questions/Activities 566
Review Quiz 566

UNIT VI Application Software 567

EXERCISE 47 Word Processors 569
Introduction 569
Performance Objectives 569
Background Information 569

*Creating a New Document   Exiting from Word
Pull-Down Menus and Toolbars   Moving the Window
Resizing the Window   Scrolling Through a Document
Saving the Document   Opening an Existing Document
Printing Documents   Other Word-Processing Applications*

Troubleshooting Techniques 576
Self-Test 577
Familiarization Activity 577
Questions/Activities 578
Review Quiz 578

EXERCISE 48 Spreadsheets 579
Introduction 579
Performance Objectives 579
Background Information 579

*Spreadsheet Planning   Building a Spreadsheet   Viewing
the Output from a Spreadsheet   Formulas   Spreadsheet
Appearance   Other Spreadsheet Software*

Troubleshooting Techniques 590
Self-Test 591
Familiarization Activity 591
Questions/Activities 592
Review Quiz 592

EXERCISE 49 Databases 593
Introduction 593
Performance Objectives 593
Background Information 593

*Database Design   Using an Existing Database
Relationships Between Data Elements   Database
Internal Structure   Database Forms   Creating a
Database   Database Reporting and Queries
Other Database Software Applications*

Troubleshooting Techniques 604
Self-Test 604
Familiarization Activity 605
Questions/Activities 605
Review Quiz 605

EXERCISE 50 Presentation Software 607
Introduction 607
Performance Objectives 607
Background Information 607

*Microsoft PowerPoint   Opening an Existing Presentation
Viewing a Presentation   Adding Slides to a Presentation
Inserting Objects into a Slide   Saving a Presentation
Printing a Presentation   Creating a Presentation
Other Software*

Troubleshooting Techniques 616
Self-Test 616
Familiarization Activity 617
Questions/Activities 617
Review Quiz 617

EXERCISE 51 Web Development 619
Introduction 619
Performance Objectives 619
Background Information 619

*Planning the Page   Images   Creating Images   HTML
Creating a Web Page   HoTMetaL PRO   Adding an
Image   Adding Tables   Adding Links   The Final Web
Page   Maintaining a Web Page   Other Software*

Troubleshooting Techniques 635
Self-Test 635
Familiarization Activity 636
Questions/Activities 636
Review Quiz 636

EXERCISE 52 Science and Technology 637
Introduction 637
Performance Objectives 637
Background Information 637

*LanExplorer   Electronics Workbench   Visio
Visual BASIC   Visual C++   Other Software*

Troubleshooting Techniques 654
Self-Test 654
Familiarization Activity 655
Questions/Activities 656
Review Quiz 656

UNIT VII Selected Topics 657

EXERCISE 53 An Introduction to Intel Microprocessor Architecture 659
Introduction 659
Performance Objectives 659
Background Information 659

*Binary Numbers   Hexadecimal Numbers   The Real-
Mode Software Model of the 80x86   80x86 Processor
Registers   80x86 Data Organization   80x86 Instruction
Types   80x86 Addressing Modes*

Troubleshooting Techniques 670
Self-Test 670
Familiarization Activity 671
Questions/Activities 671
Review Quiz 671

EXERCISE 54 Computer Languages 673
Introduction 673
Performance Objectives 673
Background Information 674

*Machine Language Versus Assembly Language   The
NUMOFF Program   High-Level Languages   Simple
BASIC Program   Object-Oriented Programming*

Troubleshooting Techniques 683
Self-Test 685
Familiarization Activity 685
Questions/Activities 686
Review Quiz 686

EXERCISE 55 Hardware and Software Interrupts 687
Introduction 687
Performance Objectives 687
Background Information 687

The Interrupt Vector Table   Viewing the Interrupt Vector Table with DEBUG   Hardware Interrupt Assignments

Troubleshooting Techniques   693
Self-Test   694
Familiarization Activity   695
Questions/Activities   695
Review Quiz   696

EXERCISE 56   The Advanced Intel Microprocessors   697
Introduction   697
Performance Objectives   697
Background Information   697

A Summary of the 80186   A Summary of the 80286 A Summary of the 80386   A Summary of the 80486 Pipelining   A Summary of the Pentium

Troubleshooting Techniques   705
Self-Test   705
Familiarization Activity   706
Questions/Activities   707
Review Quiz   707

EXERCISE 57   A Detailed Look at the System BIOS   709
Introduction   709
Performance Objectives   709
Background Information   709

Getting into the System BIOS   The CMOS RAM Standard CMOS Setup   Advanced CMOS Setup Advanced Chip Set Setup   Auto Configuration with BIOS Defaults   Auto Configuration with Power-On Defaults Change Password   Hard Disk Utilities   Write to CMOS and Exit   Do Not Write to CMOS and Exit   Examining the BIOS Data Area

Troubleshooting Techniques   713
Self-Test   713
Familiarization Activity   713
Questions/Activities   714
Review Quiz   714

EXERCISE 58   Windows Internal Architecture   715
Introduction   715
Performance Objectives   715
Background Information   715

A Look Inside Windows 95/98   The Windows API Layer The System Virtual Machine   The MS-DOS Virtual Machine   The Base System   Windows NT Architecture

Troubleshooting Techniques   718
Self-Test   718
Familiarization Activity   719
Questions/Activities   719
Review Quiz   719

EXERCISE 59   Computer Viruses   721
Introduction   721
Performance Objectives   721
Background Information   721

Operation of a Typical File-Infecting Virus   Getting the First Infection   The Anatomy of a Virus   Boot Sector Viruses   Worms   Trojan Horses   Macro Viruses Virus Detection   Virus Detection in Windows   Virus Prevention

Troubleshooting Techniques   729
Self-Test   729
Familiarization Activity   729
Questions/Activities   730
Review Quiz   730

EXERCISE 60   A Typical Computer Center   731
Introduction   731
Performance Objectives   731
Background Information   731

Network Equipment   Personnel   Supplies   Maintenance Agreements   Other Expenses   Services Provided

Troubleshooting Techniques   735
Self-Test   735
Familiarization Activity   736
Questions/Activities   736
Review Quiz   736

APPENDIX A   ASCII Character Set   737

Answers to Odd-Numbered Self-Test Questions   739

Index   745

To our Editor,
Charles E. Stewart, Jr.

Thank you for everything Charles,
you are the greatest!

UNIT 1 Introduction

EXERCISE 1 Using the Instructional System
EXERCISE 2 Electrical and Mechanical Safety
EXERCISE 3 Microcomputer Familiarization

1 Using the Instructional System

INTRODUCTION

This exercise serves as an introduction to the entire book. Here you will see how all the remaining exercises are laid out and how best to use them in order to gain the maximum benefit.

Each exercise begins with a short example of why the exercise is applicable to microcomputer use. A fictitious employee is typically used to convey the problem or situation. Here is the first example:

Joe Tekk has just been hired by RWA Software as its new computer specialist. His manager was impressed during Joe's job interview by the fact that Joe had read RWA's company literature before the interview. When asked why, Joe said, "I like to be prepared, to know in advance what to expect of a situation."

PERFORMANCE OBJECTIVES

Every exercise starts with **performance objectives.** The performance objectives let you know what new skills and knowledge you will learn in the course of completing the exercise. The performance objectives also tell you how you can check to see if you have acquired these new skills and what degree of accuracy you can expect to attain.

Your instructor will usually administer the requirements of the performance objectives. You can think of these objectives as a test. However, unlike most tests, you know exactly what the test will be. Thus, you can practice the performance objectives as many times as you want before attempting them with your instructor. By making sure you can satisfy the objectives before being tested on them, you can ensure your success. And that is the whole idea behind performance objectives—making sure you know exactly how to apply your new skills and knowledge.

BACKGROUND INFORMATION

The **background information** section presents all the information you need in order to perform the exercise and pass the review quiz. The background information section, which is usually the longest section in the exercise, contains important and detailed information. Because of this, you may want to read through the section more than once. A good rule is to read through this section first, just to get an overview of what it covers. Then read through it again, this time much more slowly. Use a highlighter or marking pen to point out key areas for yourself. You may even want to jot notes in the margins. Remember, as a computer user you will be using this book as a constant reference. The more meaningful the notes you place in it, the more useful it will become for you. Also, keeping notes in this lab manual will help you prepare for job interviews.

The background information section also contains a **troubleshooting** area. Tips, techniques, and actual problems and their solutions are presented.

SELF-TEST

True/False

Multiple Choice

Completion

Open-Ended

The next section you will encounter in each exercise is the **self-test.** The self-test is there to help you check your understanding of the material you just covered in the background information section.

As you can see, the self-test is divided into several types of test questions. This is done to make the test more interesting and more reflective of what you may need to review, and to help you get used to the different types of questions you may be asked during job interviews. In order to help you check your progress, answers to odd-numbered self-test questions are given at the end of the book. It's a good idea to try the self-test before going on to the next section of the book, where you will *apply* the information you learned in the background information section.

FAMILIARIZATION ACTIVITY

The next section in each exercise is the **familiarization activity.** It is here that you get "hands-on" applications of what you have just learned.

The familiarization activity is just that—an activity or series of activities whose purpose is to familiarize you with the applications of what you have just learned. It is here that you will use the equipment and other supplies listed in the required materials section of the exercise.

You will usually perform the familiarization activity in the lab. However, there may be some exercises that your instructor will assign as outside work. This is usually the case for exercises on software. In these exercises, the familiarization activity usually consists of a series of software interactions with a computer. These interactions can be performed as a homework assignment or done in some other place such as a computer room that provides access to computers for all students. You can then complete the review quiz at another time under the direction of your instructor.

QUESTIONS/ACTIVITIES

The next section you will find in all the exercises is the **questions/activities** section. Here you will find questions about the familiarization activity you just completed. These questions are designed to help reinforce important concepts you should have picked up when you were doing the familiarization activity. There may be times when other activities are suggested, and your instructor may or may not assign them. These other activities usually include outside assignments and are selected to give you the opportunity to broaden your understanding of the subject of the exercise.

REVIEW QUIZ

The last section, the **review quiz,** restates the performance objectives. Thus you start the exercise knowing what you should learn from it and you end at the point at which you should be ready to quiz yourself. At this point in the exercise, you should check with your instructor to see when and how you will be quizzed on the stated performance objectives.

ANSWERS TO SELF-TEST

The **answers to the odd-numbered self-test questions** are given at the end of the book. They are placed there to keep you from getting distracted while you are trying the self-test. Keep in mind that the self-test is designed to help you. It is a personal self-check. The best way to benefit from it is to try all the questions first, writing down your answers as you go. Then, after completing the entire test, check your answers. Taking the test in this manner prevents you from seeing the answer to the next question, which will happen if you look at the answer each time you complete a question. The answers to the even-numbered self-test questions are provided in the instructor's manual.

You now have an overview of how each exercise is set up. You should also have an understanding of the purpose of each section in the exercise. Keep in mind that all the sections of each exercise are important. They are designed to be used as an integrated whole. By using them in the manner in which they were designed to be used, you can gain the skills needed to become a successful computer user.

2 Electrical and Mechanical Safety

INTRODUCTION

At the end of his first week at RWA Software, Joe Tekk felt very comfortable in the computer lab. His first real troubleshooting experience had gone well. A customer's completely dead PC turned out to have a defective component in the power supply. Joe found the bad part and replaced it. The computer was still partially disassembled on a workbench, but it was plugged in and operational. The other technicians took Joe out to lunch to celebrate.

When Joe came back from lunch, he casually tossed his car keys onto the shelf above the workbench. The keys fell off and landed inside the exposed power supply unit of the customer's PC, causing a direct short between the AC power line and ground. The short drew so much current it blew the etched power line trace off the circuit board and tripped the circuit breaker on the workbench.

It took another week to repair the power supply again.

PERFORMANCE OBJECTIVES

Upon completion of this exercise, you will be able to

1. State verbally the safety rules of the microcomputer lab.
2. Discuss the methods for handling unexpected situations during lab sessions.

BACKGROUND INFORMATION

Safety in the microcomputer laboratory requires observance of the commonsense safety rules that you should follow in any situation when working with or on electrical and mechanical equipment. The following safety rules should be observed at all times.

1. *Do not allow "horseplay" in the lab.* Many lab injuries are caused by students playing jokes or "booby-trapping" equipment. This practice can cause serious permanent injuries to yourself and others. This kind of behavior should not be tolerated in any laboratory situation.
2. *Always get instructor approval.* Your instructor is there to help. Always ask for instructor approval before starting any new task. Doing this can save valuable lab time and also help prevent injuries to yourself and/or damage to equipment.
3. *Report any injuries immediately.* Always report any injury to your lab instructor. You should do this no matter how small the injury. What may appear to be a small cut, mild shock, or minor bruise could lead to serious complications if not properly treated.

4. *Use safety glasses.* When any mechanical or electrical equipment is used, there is always the chance of sparks or particles being ejected. This can happen in an electrical circuit when a part such as an electrolytic capacitor has been installed incorrectly. Remember that it takes only a very small particle to cause permanent eye damage.
5. *Use tools correctly.* The improper use of tools can result in injuries to yourself or others as well as permanent damage to the tools. Never attempt to use a tool that is damaged. Never use a tool that you do not know how to use. Never use a tool for a purpose other than that for which it was designed.
6. *Use equipment correctly.* What applies to tools applies equally well to electrical or mechanical equipment. If you are not sure about the operation of any piece of equipment, ask your instructor before attempting to use it.
7. *Do not distract others.* Do not talk to or otherwise distract someone who is in the process of using electrical or mechanical equipment. Doing so could lead to personal injury and/or damage to the equipment.
8. *Use correct lifting techniques.* Always use the proper method for lifting or pushing heavy objects. Ask for help in lifting very heavy objects. If you cannot lift or move an object for any reason, let your instructor know.
9. *Remove jewelry.* Remove all rings, watches, chains, and other jewelry, which are all capable of conducting electricity and causing a shock.
10. *Avoid static discharge.* Follow the appropriate measures to avoid static damage of sensitive equipment and devices.

TROUBLESHOOTING TECHNIQUES

In addition to the physical kind of professionalism required of you in lab, you must also prepare yourself mentally for the lab environment. You are in laboratory to observe, to listen, to use all of your senses as an active participant. If you are required to write a report about the lab exercise, everything that occurs in lab from the moment you walk in the door is important. Spend a few minutes trying to be extra observant of what you see around you. You may find that there is a lot going on that you simply filter out without even thinking about it. Practice being a good observer. Your lab work will show the results.

SELF-TEST

The following self-test is designed to test your knowledge of the safety requirements of your microcomputer lab. Answers to odd-numbered questions are given at the end of the book.

Multiple Choice

Select the best answer.

1. When in doubt about operating a piece of equipment, you should
    a. Ask your lab partner for a demonstration.
    b. Try it yourself first so you don't appear stupid.
    c. Ask your instructor.
    d. Simply proceed. You are expected to know how to work all the equipment before starting any lab assignment.
2. If you accidentally cut your finger while using a small hand tool, you should
    a. Report the accident immediately to your instructor.
    b. Wait until after class and then report the accident to your instructor.
    c. Quietly leave the lab in order not to disturb anyone, and seek first aid.
    d. Ignore the incident, and continue with the lab experiment.
3. If you find that the tool you are using in your lab experiment is damaged, you should
    a. Try to repair it to save the school money.
    b. Not use it and let your instructor know that the tool is damaged.
    c. Use it so that you don't waste time in the lab.
    d. Put the damaged tool aside and borrow a similar tool from the lab group next to you.
4. When your lab partner is performing a complex measurement on a piece of equipment, it's best for you to
    a. Start a conversation to ease his or her nerves.
    b. Hum or whistle a soft tune.

c. Shout at others to keep quiet so that he or she is not disturbed.
   d. Not talk or otherwise distract your partner.
5. If you are asked to lift something and you feel that it is too heavy for you to lift, you should
   a. Put your back into it and do the best you can.
   b. Let your instructor know and ask for assistance.
   c. Try a "test lift" first.
   d. Wait for someone else to do it.
6. If you find that you need a screwdriver but your lab group has not checked one out, you should
   a. Check out a screwdriver.
   b. Use a pocket knife to save lab time.
   c. Omit that part of the experiment and come back to it in the next lab session, when someone will remember to check out a screwdriver.
   d. Borrow a screwdriver from another lab group.
7. If a piece of lab equipment appears to be operating incorrectly, you should
   a. Get the service manual for that piece of equipment and attempt to repair it.
   b. Use it anyway to become accustomed to the kind of equipment you may find in the field.
   c. Wait until the lab group next to you is finished with their assignment and then use their piece of equipment.
   d. Immediately report the problem to your instructor.
8. Unknown to you, your lab partner, as a "practical joke," rigs a piece of equipment in the lab in such a manner that it sparks when you attempt to use it.
   a. This kind of behavior shows the great creativity of your lab partner.
   b. Such behavior is very dangerous and is not allowed in any lab situation.
   c. This kind of behavior may cause your lab partner to be expelled from the lab.
   d. Both b and c are correct.

Open-Ended

Answer the following questions.

9. State the ten safety rules presented in this exercise.
10. When should safety glasses be worn in the lab?

FAMILIARIZATION ACTIVITY

Your lab instructor may give a safety demonstration or lecture. After completing the self-test, check your answers against those at the end of the book. Your instructor may have an open class discussion concerning the answers to this self-test.

QUESTIONS/ACTIVITIES

Answer the following questions.

1. Why is safety needed in the computer lab?
2. Who is responsible for safety in the computer lab?
3. What should you do when you are not sure of an assignment?
4. Name the potential safety hazards of the microcomputer lab.
5. List the nine safety rules presented in this exercise.
6. What are the special safety precautions that should be observed in your lab situation?

REVIEW QUIZ

Under the supervision of your instructor, within 15 minutes and with 100% accuracy,

1. State the potential safety hazards in the microcomputer lab.
2. Outline the nine safety rules presented in this exercise.
3. Explain any special safety precautions that should be observed in your lab situation.

3 Microcomputer Familiarization

INTRODUCTION

Don, Joe Tekk's manager at RWA Software, asked Joe if he would come with him to pick up some new computer equipment. They drove to a local computer store and loaded the company van with several large boxes.

When they returned, Don asked Joe to unpack everything and set up the new computer. Thirty minutes later, Joe turned power on and the new computer booted up. He showed it to Don with admiration. "It's a real nice system, Don. It has everything: 450 MHz Pentium III, 64 megabytes of RAM, 3-D hardware acceleration, a 12-gigabyte hard drive, and lots of other goodies."

Don smiled at Joe. "I'm glad you like it. It's your new computer."

PERFORMANCE OBJECTIVES

Upon completion of this exercise, you will be able to

1. Identify the major parts of a microcomputer system.
2. Explain the purpose of each of these major parts.

BACKGROUND INFORMATION

Figure 3.1 illustrates the major parts of a microcomputer system. Table 3.1 lists the major parts shown in Figure 3.1, as well as the purpose of each part.

Each of these major parts is called a **peripheral device** because the device (such as the printer) is separate from the microcomputer.

Figure 3.2 shows the relationship of each peripheral device to the microcomputer. As shown in Figure 3.2, some of the peripheral devices serve only as input devices. An **input device** is one that can only input information to the microcomputer.

Other devices serve as output devices. An **output device** is one that can only get information from the microcomputer.

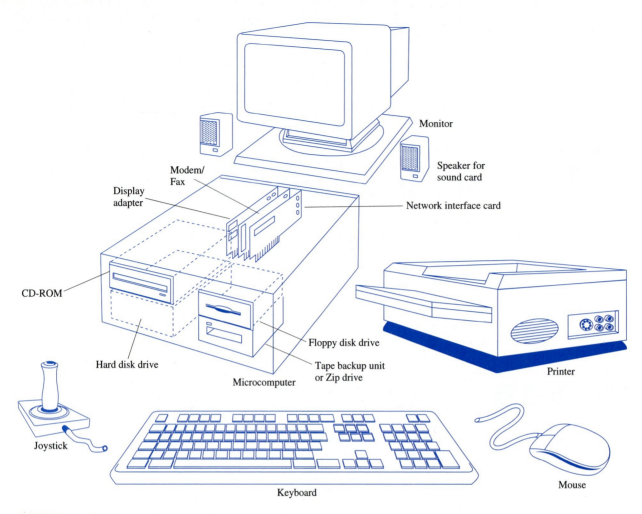

FIGURE 3.1   A microcomputer system

TABLE 3.1   Major parts of a microcomputer system

Part	Purpose
Microcomputer	Central component of the system. Performs all the calculations and logic functions. Also called the CPU (central processing unit).
Keyboard	Consists of miniature switches with alphanumeric and other labels. Allows the program user to enter information directly into the computer.
Monitor	Contains a viewing screen. Gives the program user temporary information useful in the operation of the microcomputer. Requires a display adapter card, such as a graphics accelerator.
Hard disk drive	Serves as a storage place for information. Consists of one or more rigid magnetic disks used to store programs and other items useful to the user. These disks cannot be changed by the user.
Tape backup unit or Zip drive	Used to back up files to/from the hard drive. Tapes and cartridges are removable.
Floppy disk drive	Will copy information from or place information on small disks consisting of magnetic material. These disks are an easy and quick way of getting information into the microcomputer and can be changed by the user.

TABLE 3.1 *continued*

Part	Purpose
CD-ROM	Provides very large storage capability. Reads compact disks provided by the user. Rewritable CD-ROMs are also available.
Printer	Consists of a printing head and paper mechanism for the purpose of making permanent copies of useful information contained in the microcomputer.
Mouse	A small device moved by hand across a smooth surface. Used with information on the screen to control the microcomputer quickly and easily.
Joystick	Used for quick interaction with the monitor. Usually used for interacting with computer games.
Speakers	Provide left and right audio output from a soundboard.
Telephone modem	A device for transferring information between computers by use of telephone lines.
Network interface card	Used to connect the computer to a network.

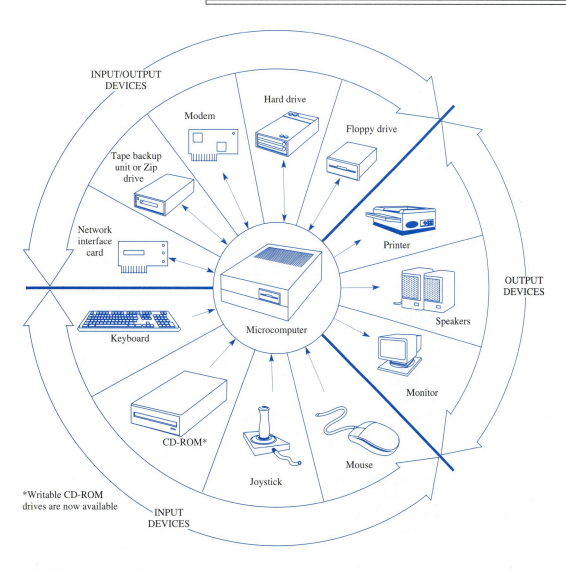

FIGURE 3.2 **Relationships of peripheral devices to the microcomputer**

The third type of peripheral device is the kind that serves as both an input and an output device. These devices are capable of putting information into the microcomputer as well as getting information from the microcomputer. Peripheral devices that are capable of both inputting information and getting information from the microcomputer are called **I/O (input/output) devices.**

TROUBLESHOOTING TECHNIQUES

A brand-new computer fresh out of the box should assemble and boot up with a minimum of fuss. Many of the external connectors (keyboard, printer, mouse) only allow one way to plug in the appropriate cable, so there is no need to worry about accidentally plugging the printer cable into the mouse port.

If all external connections are correct and the machine will not boot, it could be the result of vibration damage that could have occurred during shipment. For example, an already-loose peripheral card may have simply popped out of its slot during transit. It is good to take a look inside the chassis (even before powering on for the first time, if necessary). A good visual inspection might turn up the problem.

SELF-TEST

This self-test is designed to help you check your understanding of the background information presented in this exercise.

True/False

Answer *true* or *false*.

1. A complete microcomputer system consists of more than just the microcomputer itself.
2. All the calculations and logic functions are performed by the microcomputer.
3. The keyboard can be viewed as an example of an output device.
4. The monitor screen can be thought of as an input device.
5. The hard disk drive is a storage place for information.

Multiple Choice

Select the best answer.

6. An output device
   a. Copies information from the microcomputer.
   b. Puts information into the microcomputer.
   c. Can copy information from or put information into the microcomputer.
   d. Is the keyboard.
7. An input device
   a. Copies information from the microcomputer.
   b. Puts information into the microcomputer.
   c. Can copy information from or put information into the microcomputer.
   d. Is the monitor.
8. An I/O device
   a. Copies information from the microcomputer.
   b. Puts information into the microcomputer.
   c. Can copy information from or put information into the microcomputer.
   d. Is the printer.
9. A joystick is classified as an
   a. Output device.
   b. Input device.
   c. I/O device.
   d. None of the above.

10. The monitor is used
    a. For storing permanent information.
    b. When the printer fails.
    c. For putting information into the computer.
    d. As a temporary storage for immediately useful information.

Matching

Match each phrase on the left with the correct peripheral device or devices on the right.

11. Copies information from or places information into the microcomputer
12. Used with the monitor for quick control of the microcomputer
13. Can transfer information between computer systems
14. Makes a permanent copy of information in the microcomputer
15. Makes a temporary copy of information for immediate use

a. Monitor
b. Keyboard
c. Printer
d. Mouse
e. Joystick
f. Modem
g. Microcomputer
h. Hard drive
i. Floppy drive
j. CD-ROM

Completion

Fill in the blanks with the best answers.

16. Any device that copies information into the microcomputer is called a(n) _____ device.
17. Any device that copies information from the microcomputer is called a(n) _____ device.
18. I/O devices are capable of copying information _____ the microcomputer as well as putting information into it.
19. A(n) _____ is a small device moved by the hand across a smooth surface.
20. The disk drive that can have its disks changed by the user is called the _____ disk drive.

FAMILIARIZATION ACTIVITY

Your instructor may give a laboratory demonstration showing a complete microcomputer system, or your lab station may be equipped with a complete microcomputer system. In either case, you should know how to identify the following components of your system:

1. Microcomputer
2. Floppy, hard, and CD-ROM disk drives
3. Keyboard
4. Monitor
5. Printer
6. Modem
7. Network interface card
8. Mouse
9. Joystick
10. Tape backup unit or Zip drive

As an aid in familiarizing yourself with each of these peripheral devices, answer the following questions as they apply to the system used in the demonstration or at your lab station:

1. Who is the manufacturer of the microcomputer?
2. How many disk drives does the system contain?
3. Describe the monitor used in this system.
4. Who is the manufacturer of the printer?
5. How many keys are contained on the keyboard?
6. What is the capacity of the tape backup?

QUESTIONS/ACTIVITIES

Using a current computer magazine, list two manufacturers of each of the following:

1. Microcomputer
2. Modem
3. Printer
4. Monitor
5. Mouse/joystick
6. Disk drive
7. CD-ROM
8. Tape backup unit

Using a current computer catalog, list the prices of each of the following:

1. Telephone modem
2. Printer
3. Monitor
4. Disk drive
5. Mouse/joystick
6. CD-ROM
7. Network interface card

REVIEW QUIZ

Under the supervision of your instructor, within 10 minutes and with 100% accuracy,

1. Identify the major parts of a microcomputer system.
2. Explain the purpose of each of these major parts.

UNIT II The Disk Operating System

EXERCISE 4 Using Floppy Disks
EXERCISE 5 Introduction to DOS
EXERCISE 6 DOS Commands and File Management
EXERCISE 7 Configuring DOS with CONFIG.SYS and AUTOEXEC.BAT
EXERCISE 8 DOS Utilities

4 Using Floppy Disks

INTRODUCTION

Joe Tekk opened the cabinet where all the software packages were kept. Several floppy disk storage cases were on one shelf. Joe estimated that they contained more than 200 floppy disks. Unfortunately, many of the disks did not have labels. Joe did not know if they were blank floppies or if they contained important data.

For the next three hours, Joe sat at a computer, placing disks one by one into the A drive and viewing their directories. For disks that contained information, Joe wrote a brief description on the label. He formatted the blank disks and put blank labels on them.

Don, Joe's manager, walked by when Joe was finishing and remarked, "That's what I like about you, Joe, you are so organized."

Joe shrugged. "How long does it take to put a label on a disk? How did these disks get so disorganized?"

Don told him to ask the former computer specialist.

PERFORMANCE OBJECTIVES

Upon completion of this exercise, you will be able to

1. Identify all installed floppy disk drives by physical size and storage capacity within 3 minutes and with 100% accuracy.
2. Properly apply and inscribe a label with your last name on one floppy disk within 2 minutes and with 100% accuracy.
3. Properly write-protect one floppy disk within 1 minute, with 100% accuracy.

BACKGROUND INFORMATION

DISCUSSION

One of the most important tools used with the microcomputer is the **floppy disk.** The floppy disk contains programs that make the computer the useful device it is. Many of the problems a user has with a personal computer involve floppy disks, either because the user does not have a complete understanding of their use or because of actual physical damage to the disks. This experiment helps prepare you for the next series of exercises, which concentrates on this vital area of the personal computer.

BASIC COMPUTER SYSTEM

Figure 4.1 shows the major parts of a basic microcomputer system. The main unit contains all the essential electrical circuits that make up the microcomputer, including the motherboard, power supply, and all storage devices. The **keyboard** acts as one of the main input devices, and the **monitor** acts as one of the main output devices. Another device for inputting information into the computer is the **mouse;** another output device is the **printer.**

DISK DRIVES

Disk drives act as both input and output devices, which means that they are used for getting information into the computer as well as storing information from the computer. This capability makes them necessary for any microcomputer.

There are many different kinds of disk drives popularly used in microcomputers:

- The hard (fixed) disk drive
- The CD-ROM drive
- The $5^{1}/_{4}''$ floppy disk drive (rarely seen today)
- The $3^{1}/_{2}''$ floppy disk drive

Figure 4.2 illustrates each of these different disk drives.

The hard disk and CD-ROM drives are located inside the main unit of the microcomputer. The hard disk itself is not directly accessible to the computer user. The other types of drives are usually housed in the main unit of the microcomputer, but the user may freely insert and remove the disks used by these drives. Newer drives, called **Zip drives,** operate like hard drives but use removable cartridges.

Many systems use one or more hard disk and floppy disk drives. In order to distinguish one disk drive from the other, they are named using letters of the English alphabet. The hard disk drive is normally referred to as the C: drive. (It is conventional to place a colon after the drive letter. Doing this emphasizes that you are talking about a disk drive for the computer. You will also find other important reasons for doing this, so it is a good habit to begin now.) A second hard drive is usually drive D:.

If one floppy disk drive is used, it is called the A: drive; if a second floppy disk drive is used, it is called the B: drive.

FIGURE 4.1 Major parts of a basic microcomputer system

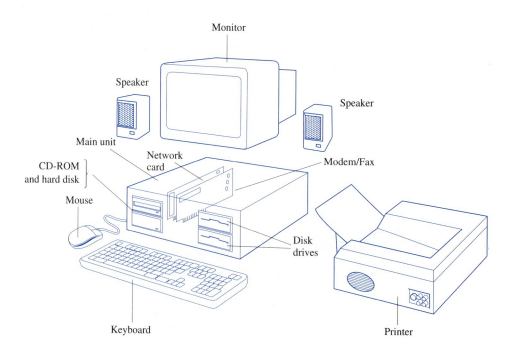

FIGURE 4.2 Four types of disk drives

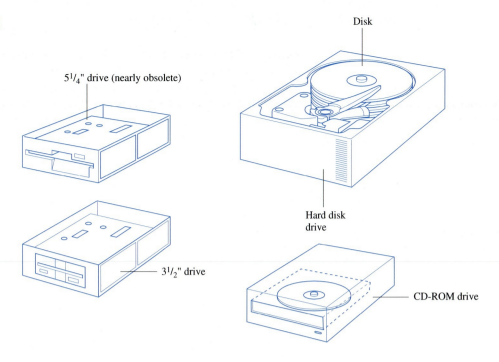

FLOPPY DISKS

Figure 4.3 shows the two most popular types of floppy disks: the $3^1/_2''$ disk and the $5^1/_4''$ disk. Figure 4.4 shows how the $3^1/_2''$ disk is constructed. Surprisingly, you will find that the smaller $3^1/_2''$ floppy can generally store more information than its larger $5^1/_4''$ counterpart. This is because of the higher quality of the magnetic surface on the $3^1/_2''$ disk.

There are important precautions that you should exercise when handling floppy disks. These precautions are illustrated in Figure 4.5.

LABELING FLOPPY DISKS

It's important to know what is contained on floppy disks, and there is a correct way of labeling them. There are special labels made for both the $5^1/_4''$ and $3^1/_2''$ disks. These labels are designed to stick to the surfaces of these disks. Figure 4.6 illustrates how to apply the labels.

FIGURE 4.3 The two most popular types of floppy disks

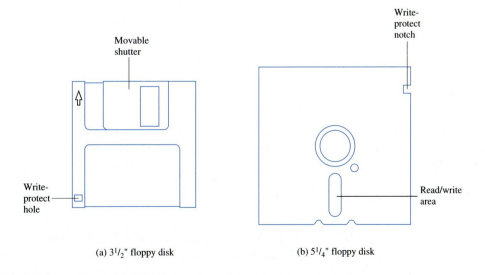

FIGURE 4.4 Construction of a 3½" floppy disk

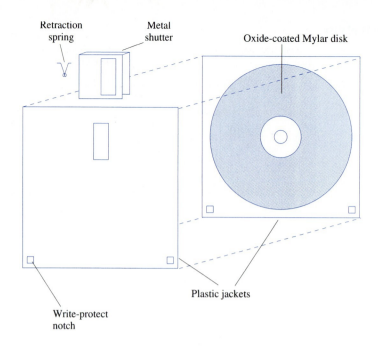

FIGURE 4.5 Precautions for handling floppy disks

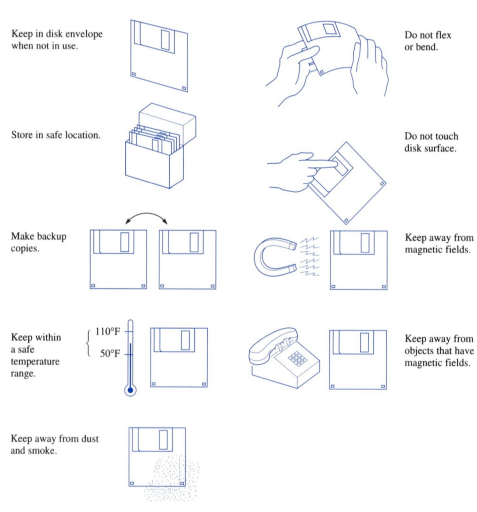

These labels are made so that they will not get stuck in a disk drive when you remove a disk. Also, the glue used does not cause damage to the disk surface. For these reasons, you should use labels made especially for this purpose. Figure 4.7 shows the correct method for writing information on the disk labels.

FIGURE 4.6 Applying labels to floppy disks

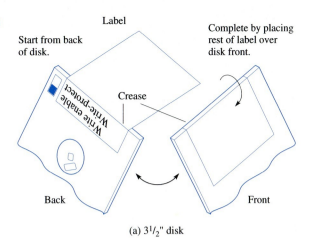

FIGURE 4.7 Writing information on disk labels

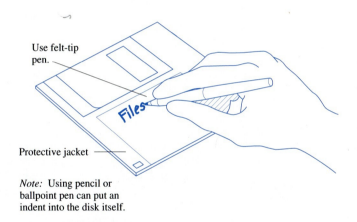

WRITE-PROTECTING DISKS

When information is placed on a floppy disk, it can sometimes replace existing information. To ensure that this does not happen, the disk can be write-protected. Write-protecting a disk also ensures that the user cannot accidentally erase information from the disk when it is being used in the system. Figure 4.8 illustrates how to write-protect both types of floppy disks.

STORAGE CAPABILITIES

Both the 5¼″ and 3½″ floppy disks have different storage capacities. These capacities are shown in Table 4.1.

Recall that in the binary number system, a byte is 8 bits (1s and 0s).

A disk that is capable of storing 737,280 bytes is theoretically capable of storing 737,280 characters. This number is more commonly referred to as 720KB, where 1KB equals 1024 bytes. You will see that other information must also be placed on the disk. Just to give you some idea, a single-spaced typed page contains about 4000 characters (including all the spaces and punctuation marks).

FIGURE 4.8 Write-protecting floppy disks

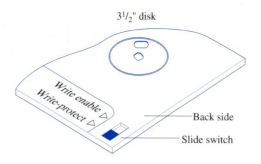

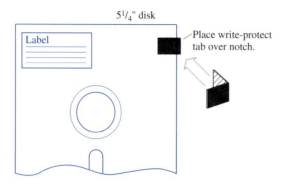

TABLE 4.1 Storage capacities of floppy disks

Disk Type	Storage Capacity	Storage Capacity in Kilobytes*
5¼″ disk		
Double-density, single-sided	184,320 bytes	180
Double-density, double-sided	368,640 bytes	360
Quad-density, double-sided	737,280 bytes	720
High-density	1,228,800 bytes	1200
3½″ disk		
720KB* disk, double-density (DD)	737,280 bytes	720
1.44MB** disk, high-density (HD)	1,474,560 bytes	1440
2.88MB disk, extra high-density (ED)	2,949,120 bytes	2880

*One kilobyte equals 1024 bytes.
**One megabyte equals 1024^2 or 1,048,576 bytes.

Besides having a disk capable of storing large amounts of information (the 2.88MB disk, for instance), your system must also have a disk drive capable of reading and writing with this amount of information. A disk drive that is capable of reading and writing only a 720KB disk will not work with the higher-density 1.44MB disk.

Of course the advantage of the larger-capacity disks is that they can store more information. The disadvantage is that these disks cost more, as do the corresponding disk drives.

FORMATTING A DISK

When you buy a new floppy disk, there is absolutely no software on it. The manufacturer of the disk has no way of knowing what brand of computer you will be using the disk for. **Formatting** a disk does the following to your disk:

- Erases any and all programs that may be on the disk.
- Places invisible (to you) information on the disk to get it ready to store other programs.
- Puts some of its own programs on the disk for use by DOS.

Read carefully what formatting a disk does. Note that if you format a disk that already has files on it, you will lose all those files *forever.* However, a new disk must be formatted before your system can use it. The basic DOS command for formatting a disk in drive A: is

FORMAT A:

where A: is the drive containing the disk you want to format. The display screen will look similar to this during the formatting process:

```
Insert new diskette for drive A:
and press ENTER when ready...

Checking existing disk format.
Verifying 1.44M
    1 percent completed.
```

The "1" will gradually increase all the way to "100," and then DOS will ask for a volume label (optional):

```
    100 percent completed.
Format complete.

Volume label (11 characters, ENTER for none)?
```

Then the results of the formatting are displayed:

```
    1,457,664 bytes total disk space
    1,457,664 bytes available on disk

        512 bytes in each allocation unit.
      2,847 allocation units available on disk.

Volume Serial Number is 2633-1ED4

Format another (Y/N)?
```

Another disk may be formatted if necessary.

DOS offers you many different options to use when formatting a disk. For example, if you want DOS to make your new formatted disk in drive A: bootable, you must enter the following:

```
FORMAT A:/S
```

The /S extension tells DOS to place system programs on your disk (including DOS) so that the newly formatted disk can be used to boot your system. You will discover other qualifiers for the FORMAT command as you learn more about DOS in the next few exercises.

BOOTING DOS

A strict process occurs when your computer system is first turned on. For the purpose of this discussion, consider your computer system as consisting of three main parts:

1. *ROM.* ROM stands for read-only memory. It contains programs installed permanently at the factory. One of the programs in ROM is a program that loads the first part of DOS (called the BOOTSTRAP program) every time your computer is turned on. It is a program in ROM that turns on drive A:, looks for the BOOTSTRAP program there, and goes to the hard drive if it doesn't find the program in A:.
2. *RAM.* RAM stands for random-access memory. However, usage has changed its meaning to read/write memory. This memory can take in new programs (from you, the disk, or other input devices). This is where the BOOTSTRAP program starts loading necessary DOS files used to continue the booting process of your computer. However, unlike ROM, anything stored in RAM is lost once you turn your computer off.

FIGURE 4.9 The DOS booting process

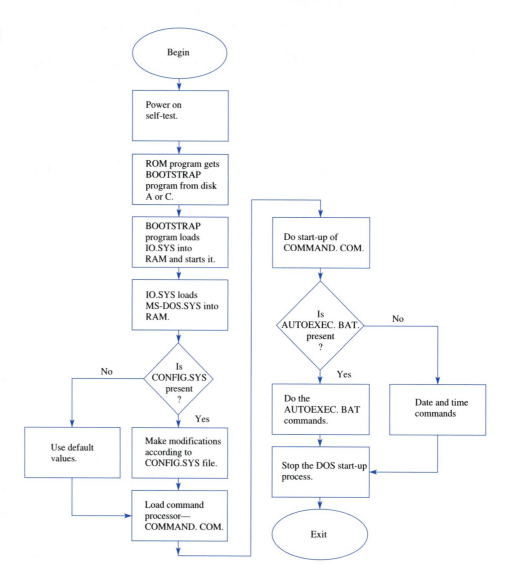

3. *DOS files.* There are three files on a bootable DOS disk that are needed to complete the booting process. Two of these files are hidden (so that they cannot be accidentally erased) and the other is called COMMAND.COM. If, for any reason, any one of these files is missing, the booting process will not be able to continue.

The DOS booting process is illustrated in Figure 4.9.

REQUIRED DOS FILES

The files required by DOS to perform the booting process are listed in Table 4.2. As previously mentioned, if any of these files is missing, the booting process will not be able to continue. The hidden files do not show up in your directory, only COMMAND.COM does. You can test for the presence of the hidden files by issuing the DOS command ATTRIB.

FORMAT REVISITED

The FORMAT command can be used to make a floppy disk bootable.
 The syntax that is used for the DOS FORMAT command in this exercise is:

FORMAT[drive:][/S][/V][/B]

TABLE 4.2 DOS programs required for booting

File Name	Type	Use
BOOTSTRAP program	On the outer track of the disk in the boot sector	This file contains information about the disk, the disk size, how information is placed on the disk, how many disk read/write heads are used, and how the disk may boot.
IBMDOS.COM (PC DOS) or MSDOS.SYS (MS-DOS)	Hidden system, read-only	This file is responsible for creating the DOS kernel. Here, the file creation, directory management, and applications to DOS interface services are installed in RAM.
IBMBIOS.COM (PC DOS) or IO.SYS (MS-DOS)	Hidden system, read-only	This file, along with the built-in programs in ROM, build up an area in RAM that is responsible for input and output.
COMMAND.COM	Normal	This is the DOS command processor. It is responsible for performing your DOS commands. It also looks for and executes the AUTOEXEC.BAT file if it exists. It also contains all the internal DOS commands.

where: drive: = the drive containing the disk to be formatted
/S = information causing the DOS system files to be placed on the disk so that the system may be booted from this disk. (In DOS 4.0 and later, this command will work only on a 1.2MB disk or one of higher density.)
/V = allows placement of a volume label on the disk after it is formatted
/B = leaves space on the disk to place an operating system to make the disk bootable but does not actually place the system on the disk

As an example, if you want the disk in drive A: to be made into a bootable DOS disk (meaning that you can now leave this disk in drive A: when first turning the system on and the computer will use this disk to boot from, ignoring the hard disk), use

```
FORMAT A: /S
```

When the formatting is completed, the message

```
System Transferred
```

appears, meaning that all the required system files (the two hidden files and COMMAND.COM) have been copied to the disk to be formatted.

Warning: Remember that all existing files are lost when a disk is formatted. Later you will discover the correct process for formatting a hard drive, but you should not do this here.

If you issue the command:

```
FORMAT A: /V
```

the disk will be formatted without the system files, and after the formatting process is completed, you will be prompted for the name of a volume label.

Entering

```
FORMAT A: /B
```

again formats the disk, this time leaving space for the system files but not actually transferring them. If you later want to transfer the system files, you issue the DOS command

```
SYS A:
```

and the system files are transferred.

VOLUME LABELS

When you do a directory listing (DIR) of a disk, the top of the screen presents information about a **volume,** as shown in Figure 4.10. As shown in the figure, there is no volume name for this particular disk. You can give a disk a volume name of up to 11 characters. Volume names are important in some program applications in which many different disks are used and the program must ensure that the correct disk has been inserted. For example, a payroll department may be using a payroll program that helps keep records of employees' pay. It may be required by the program that a different floppy disk be used for each month. The program tells if the correct disk is inserted by checking the volume name of the disk.

Another reason for giving the disk a volume name is simply to help you or the customer keep disks organized. Because you can use up to 11 characters for the volume name, you have a great deal of flexibility in naming your disks.

There are two methods for creating a disk volume name. One method is used when the disk is first formatted. In this case, you enter an extension to the format command:

```
FORMAT A:/V
```

This instructs DOS to format the disk in drive A:. After the disk is formatted, you are given an opportunity to enter the volume name. This process is shown in Figure 4.11.

The other method of creating a volume name is by using the DOS command

LABEL[d:][Name]

For this DOS command:

   d: = drive containing the disk that is to have the new label name
Name = new label name

For example, if the disk to have a new label name is in drive A:, then the following command will place the label MYDISK on the disk in the A: drive:

```
LABEL A:MYDISK
```

Once your disk is labeled, a directory listing will display the label name, as shown in Figure 4.12.

DISPLAYING THE VOLUME LABEL

DOS also has a command that will simply display the volume label of a disk:

VOL[drive:]

where: drive = the drive having its label examined. (If you do not specify a drive, DOS will use the active drive.)

FIGURE 4.10 Display of disk volume name information

```
A:\> DIR

Volume in drive A has no label
Directory of A:\

File not found

A:\>
```

FIGURE 4.11 Creating a volume name during the formatting process

```
A:\> FORMAT A:/V

Insert new diskette for drive A:
and strike ENTER when ready

Format complete

Volume label (11 Characters, ENTER for none)?
```

FIGURE 4.12 Display of label name from a directory

```
A:\> DIR

Volume in drive A is MYDISK
Directory of A:\

File not found

A:\>
```

FIGURE 4.13 Volume name display from DOS command

```
A:\> VOL

Volume in drive A is MYDISK

A:\>
```

For example, to display the label of the disk in drive A: (assuming A: is the active drive), simply enter

A:\> VOL

An example of the display is shown in Figure 4.13.

TROUBLESHOOTING TECHNIQUES

The DOS utility program ScanDisk performs an important job: analyzing and repairing files on a floppy or hard disk. Imagine pulling a floppy out of the drive while it is being written to, or accidentally dropping it, or having your children touch the oxide coating of the floppy while playing with it. All of these situations, and many more, provide the means for file and directory entries stored on the floppy to be damaged. ScanDisk will automatically find and repair any errors it can, and generate a report of the results.

To scan the floppy in drive A, use:

SCANDISK A:

The log file saved by ScanDisk might look like this for a floppy with typical problems:

```
ScanDisk checked drive A for problems, with the following results:

Directory structure

        Directory entry A:\AUTOEXEC.BAK had an incorrect size.
        ScanDisk successfully corrected the directory entry.

        Directory entry A:\DBLSPACE.BIN had an incorrect size.
        ScanDisk successfully corrected the directory entry.

File allocation table

        Directory entry A:\AUTOEXEC.BAK had an invalid FAT chain.
        ScanDisk corrected the FAT chain by truncating it.

        Directory entry A:\DBLSPACE.BIN had an invalid FAT chain.
        ScanDisk corrected the FAT chain by truncating it.

File system

        ScanDisk did not find any problems.
```

```
Surface scan
              Data could not be read from cluster 847.
              There is no file using cluster 847.
              ScanDisk patched the cluster successfully.
```

ScanDisk replaces the older CHKDSK utility. To see a list of its options, use:

```
SCANDISK /?
```

SELF-TEST

This self-test is designed to help you check your understanding of the background information presented in this exercise.

True/False

Answer *true* or *false*.

1. Many of the problems that users have with microcomputers are associated in some way with floppy disks.
2. A floppy disk contains information that the computer uses to operate itself.
3. A hard disk drive uses hard disks, which the user can easily insert and remove to produce new programs for the computer.
4. One disk may be identified from another through the use of a volume label.
5. A disk volume label may not have more than eight characters.

Multiple Choice

Select the best answer.

6. Disk drives are
   a. Used as input devices.
   b. Used as output devices.
   c. Devices that can only store information.
   d. Both a and b.
7. Disk drives are
   a. Labeled using letters of the alphabet.
   b. Called either the primary or secondary drives.
   c. Given names such as the left and right or top and bottom.
   d. Labeled according to the manufacturer.
8. It is common practice to follow the disk drive identification letter by a
   a. Semicolon (;).
   b. Period (.).
   c. Colon (:).
   d. Space.
9. The DOS command to display the volume label is
   a. DISVOL.
   b. VOL.
   c. VOLDIS.
   d. None of the above.
10. The ScanDisk utility will
    a. Display all the errors on the disk.
    b. Fix all the disk errors and put them in a file.
    c. Show you the contents of all the disk files.
    d. None of the above.

Matching

Match each phrase on the left with one or more items on the right.

11. Will damage a disk
12. Proper care for a disk
13. Cannot be done with a disk

a. Keep in protective jacket.
b. Keep in direct sunlight.
c. Keep away from magnetic fields.
d. Look at the disk to see if it contains data.

Completion

Fill in the blanks with the best answers.

14. There are _____ major types of disks used with microcomputers.
15. A quad-density double-sided floppy disk can hold _____ bytes of information.
16. A disk drive that can read a $3^1/2''$ 1.44MB disk can also read a $3^1/2''$ _____ disk.
17. To make a disk bootable, issue _____ to format the disk.
18. Formatting a disk _____ any and all existing files on the disk.

FAMILIARIZATION ACTIVITY

1. Using the available computers, identify all of their floppy disk drives. Look for the following:
   - Which are $5^1/4''$ drives and which are $3^1/2''$ drives?
   - How are the drives designated: which is the A: drive, which is the B: drive? Is there a D: drive?
2. With your lab partners, practice stating the rules for handling floppy disks. Be sure you know them.
3. Working with your lab partners, practice labeling both kinds of floppy disks.
4. Practice writing neat labels on disks using a felt-tip pen. Be sure to include all necessary information.
5. Be sure you and your lab partners know how to write-protect both kinds of disks.
6. Working with your lab partners, practice stating the storage capacities of the different kinds of disks.
7. Run ScanDisk on your hard drive and on a floppy. Comment on any errors that occur.

QUESTIONS/ACTIVITIES

1. State the following information about the computer or each of the computers available during this lab exercise:
   a. The number of floppy disk drives
   b. The type of disk drives
   c. Label designation for each drive
   d. Storage capacity of each drive
2. Determine the number of single-spaced typewritten pages that could theoretically be contained on each of the following types of floppy disks, assuming that the disk did not need any other information on it. Use 4000 characters per page in your calculations.
   a. Double-density, double-sided $5^1/4''$ disk
   b. 1.44MB microfloppy $3^1/2''$ disk
3. What does the VOL command do? Why is it useful?

REVIEW QUIZ

Given a PC and the necessary supplies and equipment,

1. Identify all installed floppy disk drives by physical size and storage capacity within 3 minutes and with 100% accuracy.
2. Properly apply and inscribe a label with your last name on one floppy disk within 2 minutes and with 100% accuracy.
3. Properly write-protect one floppy disk within 1 minute and with 100% accuracy.

5 Introduction to DOS

INTRODUCTION

During his lunch hour, Joe Tekk sometimes made house calls to friends who were having trouble with their computers. One friend was a professor at a local community college who has a laboratory set up to study small animals, such as mice and rats. An old 386 PC running DOS was used to record sensor data from each of the animal cages, and turn lights on and off at precise intervals to simulate day/night intervals. The control program was written in BASIC years ago and never updated.

Joe's friend the professor asked him how he still knew so much about DOS when Windows 98 seemed to be the standard. Joe replied, "Even Windows 98 still supports DOS operations. I know the commands I need to navigate around directories, edit and copy files, and perform simple DOS operations. This allows me to help people like you, who have older equipment that might not benefit by upgrading to Windows 98. Since the machines are still out there, I don't think I should abandon DOS completely."

PERFORMANCE OBJECTIVES

Upon completion of this exercise, given a PC and the necessary supplies and equipment, you will be able to

1. Perform a cold boot on the computer within 3 minutes and with 100% accuracy.
2. Perform a warm boot on the computer within 3 minutes and with 100% accuracy.
3. Enter the current date and time into the system time-of-day clock within 2 minutes and with 100% accuracy.
4. Format a floppy disk as a system disk within 5 minutes and with 100% accuracy.
5. Display the directory listing of a floppy disk using any two forms of the DOS DIR command within 2 minutes and with 100% accuracy.

BACKGROUND INFORMATION

INTRODUCTION

This is an important exercise for getting you started in the technical side of software. Here is an opportunity to gain skills with one of the most powerful and popular operating systems used in personal computers. DOS, as you learned in the previous exercise, is system software; DOS stands for "disk operating system." Recall that DOS is simply a program that helps you and your computer, with its peripheral devices, to work together.

BOOTING METHODS

In the previous exercise, you learned the meaning of booting; in this exercise you will learn how to boot your computer. There are basically two methods of booting your computer (loading it with DOS). One method is to place a floppy disk containing DOS into drive A: and turn the power on. When you turn on the computer, it will go through the sequence shown in Figure 5.1.

The other method for getting DOS into your computer is to have a hard disk inside your computer and a copy of DOS on the disk. A more detailed explanation of how DOS begins execution is shown in Figure 5.2.

SETTING THE DATE AND TIME

Once DOS is loaded into your computer, you will see the message shown in Figure 5.3. Some computers have a system clock whose time is maintained by a small battery (much the same as your digital watch). When this is available, the date and time will be put in automatically. DOS itself cannot remember the date. The message shown in Figure 5.3 is your opportunity to put the correct date into the computer. An important rule is always to make sure your computer has the correct date when working with DOS. You will see the reason for this shortly.

To enter the date, use any one of the following formats:

mm-dd-yy
mm/dd/yy
mm.dd.yy

where mm is the month number (1 through 12), dd is the day number (1 through 31), and yy is the last two digits of the year (80 through 99). After the year 2000, you must enter a four-digit year (2000 to 2099) to get the date command to accept your input. Entering a "yy" value between 80 and 99 assumes the year is 1980 through 1999, so all four digits must be used to set the date in the twenty-first century.

Figure 5.4 shows what happens if you enter an invalid date. As you can see from the figure, DOS knows there are only 12 months in a year.

After correctly entering the date, press the ENTER (or RETURN) key. (All DOS commands require you to press ENTER, or RETURN, before the computer takes any action.) To correct a mistake, simply use the BACKSPACE key to "rub out" the incorrect input and start over again.

FIGURE 5.1 Computer turn-on sequence

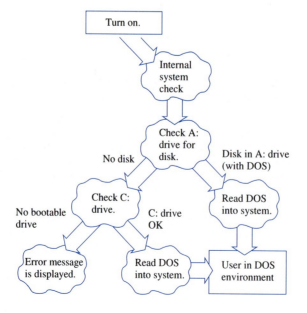

FIGURE 5.2 Detailed MS-DOS boot sequence

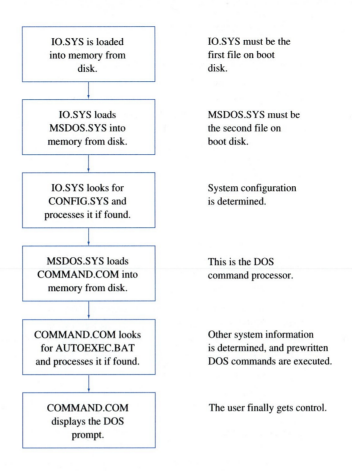

FIGURE 5.3 First DOS message

```
Current date is Sat 11-07-1998
Enter new date (mm-dd-yy):
```

FIGURE 5.4 Entering an invalid date in DOS

```
Current date is Sat 11-07-1998
Enter new date (mm-dd-yy): 16-08-98

Invalid date
Enter new date (mm-dd-yy):
```

After the date has been entered, DOS will ask for the correct time, as shown in Figure 5.5. The format for entering the time is

HH:MM:SS:hh

where HH is hours (expressed in military time, in which 1:00 P.M. is written as 13:00), MM is minutes, SS is seconds, and hh is hundredths of a second. You need enter only the hour and minute and then press the ENTER key. As an example, if it is 3:15 P.M. you would enter

15:15

FIGURE 5.5 Second DOS message

```
Current time is 04:12:28.25a
Enter new time:
```

FIGURE 5.6 DOS message and prompt

```
Microsoft (R) MS-DOS (R) Version 6.20
(C) Copyright Microsoft Corp 1981-1993.

C>
```

Note that when you press the ENTER key, it means that you are returning control of the computer to the computer. Newer versions of DOS allow you to enter "a" or "p" after the time to specify A.M. or P.M.

THE DOS PROMPT

After correctly entering the date and time, you will see a message similar to the one shown in Figure 5.6. As shown, you will see the DOS prompt

C>

The C means that disk drive C: is active (remember that the colon is used to indicate a disk drive). The "greater than" sign (>) is the symbol used by DOS to indicate that the control of the computer has been turned over to you. To change the DOS prompt—that is, to make a different disk drive active—simply type the drive letter followed by the colon:

B:

This will cause drive B: to be the active drive, and the DOS prompt will change to B:. Similarly, typing in A: will change the active drive to the A: drive. This assumes that these drives are on your system.

CHANGING THE DATE AND TIME

Table 5.1 shows two DOS commands for changing the date and the time. Figure 5.7 shows the actions of these two DOS commands.

RESETTING THE COMPUTER

Resetting the computer is the process of restarting DOS in its initial state. There are two methods of doing this. One is called a **cold boot;** the other, a **warm boot.** A cold boot

TABLE 5.1 DOS date and time commands

Command	Action
DATE	Allows you to change the date.
TIME	Allows you to change the time.

FIGURE 5.7 Actions of DATE and TIME commands

```
Current date is Sat 11-07-98
Enter new date (mm-dd-yy):
```

(a) DATE

```
Current time is 4:27:36.14p
Enter new time:
```

(b) TIME

occurs when you turn the system off, wait a few seconds, and then turn the system on again. A warm boot occurs when you start with the system already on and press the CTRL, ALT, and DEL keys on your keyboard all at the same time (or the RESET button on the computer's front panel).

COPYING A DISK

There are many times when you will find it necessary to make backup copies of programs on a disk. Remember that purchased software is copyrighted, and the legal owner of the software is usually allowed to make one backup copy to keep in case the original is damaged. It is illegal to make copies of purchased software for any other reason.

To make a copy of a disk on a single-drive system, follow the procedure shown in Figure 5.8.

The DOS DISKCOPY command is used to make backup copies of original disks. You should not use original disks on a daily basis; instead, you should run your programs from the backup copies.

Note: It is not possible to use DISKCOPY to copy a $5^{1}/_{4}''$ disk to a $3^{1}/_{2}''$ disk, or vice versa.

INTERNAL AND EXTERNAL DOS COMMANDS

When you boot DOS into your system, certain commands, such as DATE and TIME, are placed within your computer. Thus, when you enter the command DATE, your computer, through its internal DOS, will immediately respond to the command. Because DOS has placed this command inside your computer memory (and it will stay there until you turn

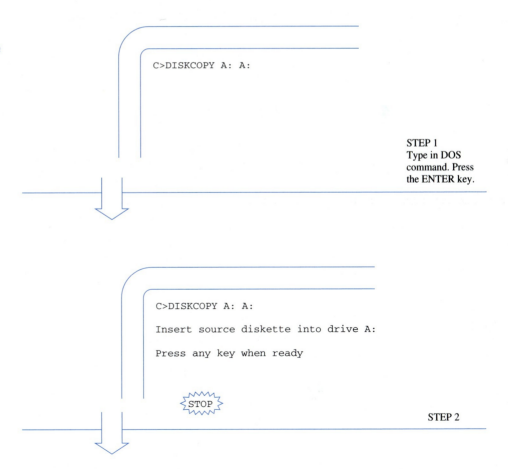

FIGURE 5.8 Procedure for copying a disk on a single-drive system

FIGURE 5.8 *continued*

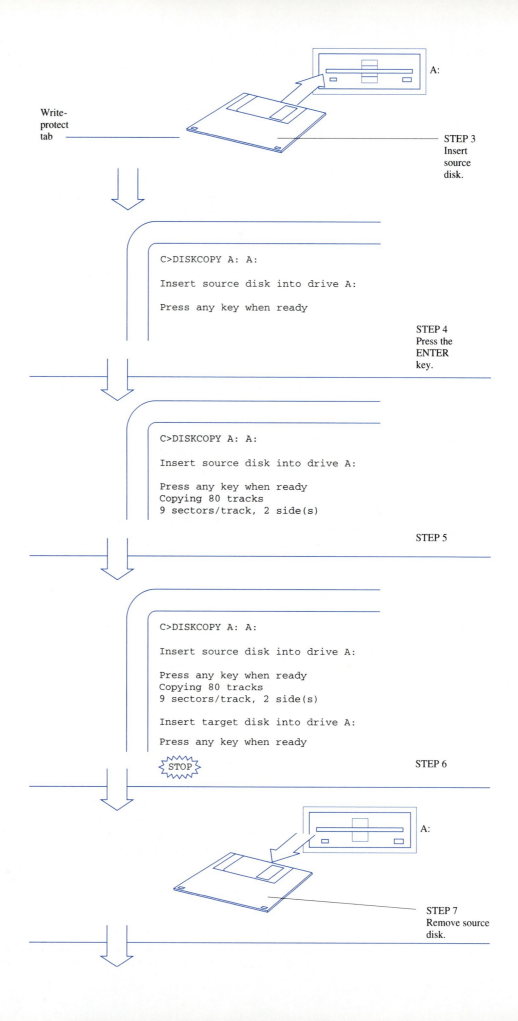

FIGURE 5.8 *continued*

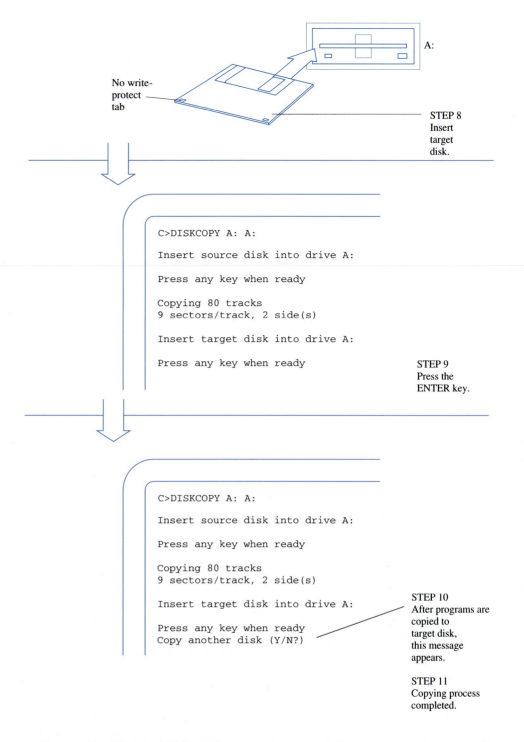

your computer off), it is called an **internal DOS command.** What this means is that you do not need the DOS disk in any of your disk drives in order to use it.

The FORMAT command is not an internal DOS command. This command is stored as a separate DOS program in the DOS directory. As a result, the FORMAT command is called an **external DOS command.** What this means is that the DOS directory must be contained on your C: drive in order for the command to be executed.

DEFINITION OF DOS FILES

The purpose of a disk is for storing programs. When you store information on a disk, DOS places this information in what is called a *file,* much as you might place any information you have in a filing cabinet. As you will see in later exercises with DOS, you can change the contents of disk files, give them names, change their names, transfer them to other places within the computer system (to other disks or to the printer or monitor), or erase them completely. To start with, you should know how DOS names a file. Every DOS file has a name. The DOS file name consists of two important parts: the file name and the extension.

FILENAME.EXT

The following rules apply:

FILENAME: Must contain from one to eight characters, starting with a number or letter of the alphabet and then any letter or number, including the following symbols: ~, {, }, _, -, @, #, %, ^, &. No blanks are allowed.

EXT: Optional; may contain up to three characters or numbers.

As an example, the following are all legal DOS file names:

```
COMMAND.COM
PLAYIT.EXE
DO_12
WOW.
FILE.05
12345.678
```

The following are not legal DOS file names and will not be accepted by DOS:

TOOMANYLETTERS	More than eight characters
.WOW	No file name
BIG.EXTENSION	More than three letters in the extension
HEY YOU.	Blank used in file name

As you will see later, the file extension can be a very powerful tool for keeping, maintaining, and using files.

DOS has some file names that it reserves for its own special use. You will see the purposes of these file names in forthcoming DOS exercises. They are listed here so that you won't use them for now:

```
AUX   CON   COM1  COM2  NUL
LPT1  LPT2  LPT3  PRN   CLOCK$
```

Note: Windows 95/98/NT allow the use of long file names, such as "My Documents," that appear in truncated form in DOS (MYDOCU~1). See Unit III for more details.

HOW DOS NAMES FILES

Table 5.2 lists the file-naming conventions used by DOS. These conventions consist of file extensions.

In addition, special programs (such as business application programs) use their own extensions. You need to consult the application program in order to know what these extensions are.

DISPLAYING DOS FILES

DOS places files on your disk in a list called a **directory.** You will learn more about directories, but for now you should know how to display a list of file names on the computer screen. The command for displaying the list of file names is

```
DIR
```

This command causes DOS to list the disk files, as shown in Figure 5.9.

TABLE 5.2 Common DOS file extensions

Extension	Meaning
.COM	DOS commands or programs you can directly execute. To execute the program, simply enter the file name of the program (without the .COM extension), press ENTER, and the program will be loaded into your computer and will begin running. Application programs may have this extension.
.BAT	DOS batch files. This extension provides a way of storing several commands in a file and having them done automatically just by entering the name of the file.
.DAT	Data files used by programs.
.CPI	Code page information containing foreign character sets.
.EXE	Executable programs. To execute the program, simply enter the file name of the program (without the .EXE extension) and press ENTER, and the program will be loaded into your computer and will begin running. Some application programs may have this extension.
.SYS	For installing device drivers. These are programs that allow DOS to communicate with hardware devices attached to your computer, such as the keyboard, printer, and monitor.
.TXT or .DOC	Files used for word processing.
.BAS	File is a BASIC program.
.ASM	File is an assembly language source program.
.C or .CPP	File is a C/C++ source program.

Note: The main difference between the .COM and .EXE extensions is in how the program uses memory when loaded into the computer.

Note from Figure 5.9 that the volume name of the disk is given at the top of the screen. The drive the disk is in and a list of the files is also given. At the bottom of the screen is a summary of how many files there are and how many bytes are left on the disk.

Each file is displayed as follows:

FILENAME.EXT   SIZE   DATE/TIME

One example is

MYFILE.TXT   1230   11-22-98   5:07a

You see the name of the file with its optional extension; this is followed by the size of the file in bytes. The date and time tell you when the file was created or last modified. This is why it is so important to set the date and the time correctly in your computer.

To clear the screen on your monitor, use the internal DOS command

CLS

and your computer screen will be cleared, with the DOS prompt appearing at the top left corner of the screen. You may wish to clear the screen before displaying a directory.

There are several ways that you can use the DIR command. Sometimes there are so many files on the disk that they will simply scroll up to the top of the screen until the last file is displayed at the bottom of the screen. In this case you may not see some of the beginning files on the monitor. To prevent this, you can use the DOS DIR modifier:

DIR/P

The /P stands for *pause;* it causes the displaying of files to stop once the screen is full. You may then press any key on the keyboard to see the rest of the files. Another way of stopping

FIGURE 5.9 Typical DOS directory

```
Volume in drive C is FIREBALLXL5
Volume Serial Number is 3729-11D6
Directory of C:\DOS

.            <DIR>         01-18-95   10:22a
..           <DIR>         01-18-95   10:22a
COMMAND  COM         47,845 11-11-91    5:00a
FORMAT   COM         33,087 11-11-91    5:00a
COUNTRY  SYS         17,069 11-11-91    5:00a
KEYB     COM         14,986 11-11-91    5:00a
KEYBOARD SYS         34,697 11-11-91    5:00a
ANSI     SYS          9,029 11-11-91    5:00a
ATTRIB   EXE         15,796 11-11-91    5:00a
CHKDSK   EXE         16,200 11-11-91    5:00a
EDIT     COM            413 11-11-91    5:00a
MORE     COM          2,618 11-11-91    5:00a
SYS      COM         13,440 11-11-91    5:00a
DEBUG    EXE         20,634 11-11-91    5:00a
FDISK    EXE         57,224 11-11-91    5:00a
MODE     COM         23,537 11-11-91    5:00a
EGA      CPI         58,873 11-11-91    5:00a
EGA      SYS          4,885 11-11-91    5:00a
HIMEM    SYS         11,616 11-11-91    5:00a
XCOPY    EXE         15,820 11-11-91    5:00a
EDIT     HLP         17,898 11-11-91    5:00a
HELP     EXE         11,473 11-11-91    5:00a
PRINT    EXE         15,656 11-11-91    5:00a
SETVER   EXE         12,015 11-11-91    5:00a
APPEND   EXE         10,774 11-11-91    5:00a
DISKCOPY COM         11,879 11-11-91    5:00a
GRAPHICS COM         19,694 11-11-91    5:00a
LABEL    EXE          9,390 11-11-91    5:00a
SORT     EXE          6,938 11-11-91    5:00a
TREE     COM          6,901 11-11-91    5:00a
DOSKEY   COM          5,883 11-11-91    5:00a
       31 file(s)       526,270 bytes
                     22,773,760 bytes free
```

the files from scrolling off the screen is to press the CTRL key at the same time you press the S key (for stop/start). Press the same keys again to allow scrolling to continue.

Another useful modifier is

DIR/W

Think of the /W as standing for **wide.** The results of issuing this command are shown in Figure 5.10. As shown in the figure, the DIR/W command allows the display of more files

FIGURE 5.10 Result of the DIR/W command

```
Volume in drive C is FIREBALLXL5
Volume Serial Number is 3729-11D6
Directory of C:\DOS

[.]            [..]            COMMAND.COM     FORMAT.COM      COUNTRY.SYS
KEYB.COM       KEYBOARD.SYS    ANSI.SYS        ATTRIB.EXE      CHKDSK.EXE
EDIT.COM       MORE.COM        SYS.COM         DEBUG.EXE       FDISK.EXE
MODE.COM       EGA.CPI         EGA.SYS         HIMEM.SYS       XCOPY.EXE
EDIT.HLP       HELP.EXE        PRINT.EXE       SETVER.EXE      APPEND.EXE
DISKCOPY.COM   GRAPHICS.COM    LABEL.EXE       SORT.EXE        TREE.COM
DOSKEY.COM
       31 file(s)       526,270 bytes
                     22,773,760 bytes free
```

at the same time. What is lost here is the individual file size and the creation or modification date and time.

TROUBLESHOOTING TECHNIQUES

Much has been said about the "Year 2000" (or Y2K) problem, where the simplified storage format used by many software packages (and early versions of BIOS) to represent the year breaks down after midnight December 31, 1999. Estimated costs of the resulting software crisis (including litigation) is in the hundreds of billions of dollars.

Numerous programs are available on the Web that are designed to help cope with the Y2K problem. Many are designed to automatically correct the date on older machines. Others scan for anything resembling a date in a document, and make the necessary modifications.

No one ever thought the original hardware and software architecture of the PC would still exist 20 years after its creation. The exact impact of the Y2K problem on computers and the world in general will not be known for quite some time.

SELF-TEST

This self-test is designed to help you check your understanding of the background information presented in this exercise.

True/False

Answer *true* or *false*.

1. The process of loading DOS into your computer is called booting.
2. The easiest method of getting DOS into your computer is by programming it through the keyboard.
3. A cold boot is achieved by starting with your computer turned off.
4. Booting DOS when your computer is already turned on is called a warm boot.
5. If there is a hard disk in your computer, you must load DOS using a floppy disk.

Multiple Choice

Select the best answer.

6. To enter the correct date, you use
    a. mm-dd-yy.
    b. mm/dd/yy.
    c. mm.dd.yy.
    d. Any of the above.
7. To enter the correct time, you use
    a. HH:MM:SS:hh.
    b. HH\MM\SS\hh.
    c. HH.MM.SS.hh.
    d. Any of the above.
8. When entering the time into the computer, you
    a. Must enter the hours, minutes, seconds, and hundredths of a second.
    b. Need enter only the hour.
    c. Need enter only the hour and minutes.
    d. You cannot enter the time into your computer.
9. The DOS prompt tells you
    a. The letter of the active drive.
    b. That you now have control of the computer.
    c. What is going to happen next.
    d. Both a and b.
10. The DOS command for clearing the screen is
    a. CLEAR.
    b. CLS.

    c. CLEAR SCREEN.
    d. None of the above.

Matching

Match each definition on the left with one or more DOS extensions on the right.

11. Programs that you can directly execute	a. .SYS
12. Programs that contain other DOS commands that can be automatically executed	b. .TXT
	c. .EXE
13. Contains foreign characters	d. .COM
14. Files used in word processing	e. .BAT
15. Programs that allow DOS to communicate with hardware devices	f. .CPI
	g. .DAT

Completion

Fill in the blanks with the best answers.

16. A DOS file displayed on the directory contains the elements of _____ _____ _____ _____ _____.
17. The DOS command for displaying files is _____.
18. The DOS command DIR/P will cause the display of files to _____ once the screen is full.
19. The DOS command _____ causes files to be displayed across the screen rather than just down the left edge.
20. The DOS command DIR is a(n) _____ DOS command.

FAMILIARIZATION ACTIVITY

1. With your lab partners, practice loading DOS into your computer first by using a cold boot and then by using a warm boot.
2. Try changing the DOS prompt to another disk drive. If you have only a one-drive system, this won't be possible. Be sure to reset the active drive back to drive A:.
3. With your lab partners, practice setting the system date and time.
4. Insert your new floppy disk (the one to be formatted), and format this disk as a DOS system disk. Use the FORMAT /S command. Follow the instructions presented in the background information.
5. Boot your system using your newly formatted disk to ensure that it works as a system disk.
6. With your lab partners, practice viewing DOS files on the DOS disk.
7. Make a copy of the disk containing public domain software to your newly formatted disk. If your system has a dual-disk drive, practice doing this using both drives.
8. With your lab partners, practice displaying the files of the public domain disk you copied. Use the /P as well as the /W extension and the DOS CLS command.
9. With your lab partners, be sure you understand the information displayed on the screen after you issue the DIR command.

QUESTIONS/ACTIVITIES

1. How do you change the active drive to the B: drive?
2. Explain, in your own words, the difference between an internal and an external DOS command.
3. State exactly what happens when you format a disk. Why is this process necessary?
4. How do you format a disk to make it a bootable disk?
5. What is the function of the following keystrokes?

    CTRL-ALT-DEL

    CTRL-S

6. State what you must do in order to see the files contained on a disk.
7. Explain the different methods of displaying disk files.

8. State the meaning of all the information displayed on the screen after executing the DOS DIR command.
9. Give the common DOS extensions and the meaning of each.
10. Give three different examples of illegal DOS file names and state what is wrong with them.

REVIEW QUIZ

Given a PC and the necessary supplies and equipment,

1. Perform a cold boot on the computer within 3 minutes and with 100% accuracy.
2. Perform a warm boot on the computer within 3 minutes and with 100% accuracy.
3. Enter the current date and time into the system time-of-day clock within 2 minutes and with 100% accuracy.
4. Format a floppy disk as a system disk within 5 minutes and with 100% accuracy.
5. Display the directory listing of a floppy disk using any two forms of the DOS DIR command within 2 minutes and with 100% accuracy.

6 DOS Commands and File Management

INTRODUCTION

Don pulled a floppy disk out of his shirt pocket. He placed it into the A: drive on his office machine and viewed its directory contents. Joe Tekk watched Don from the doorway. "Why do you keep a floppy disk in your pocket?" Joe asked.

Don finished printing out a file that was on the disk. "I call this my 'transit' disk. Any small files that I want to bring back and forth between here and my home, I place on this floppy." Don motioned at the display screen. "Look at the contents. See how I've got everything arranged by subdirectory?"

Joe looked at the screen. Several subdirectories were listed.

"There are more than 300 files on that disk," Don said. "But all we see in the root directory are six subdirectories."

"Why bother with subdirectories? Why not just put all of the files in the root directory?"

"Putting files into subdirectories allows me to easily copy or delete entire groups of files at once, without having to search through a huge list of files. Plus, there are limits to how many files a single directory can store."

Joe got the message. "That's really neat, Don. I think I need to do some housecleaning on my floppies."

PERFORMANCE OBJECTIVES

Upon completion of this exercise, given a PC, a printer, and the necessary supplies and equipment, you will be able to

1. Display a listing of a given group of files, using the appropriate command and wild cards, within 3 minutes and with 100% accuracy.
2. Use the proper command to copy the information displayed on the monitor to the printer device (make a hard copy) within 2 minutes and with 100% accuracy.
3. Create a disk file with the name MYFILE.TXT that contains

   Your name
   Your social security number
   Today's date

   within 5 minutes and with 100% accuracy.

BACKGROUND INFORMATION

INTRODUCTION

This exercise gives you the opportunity to become very familiar with DOS. Here you will discover how to work with DOS files. You will gain many new and important skills. Take your time, and be sure you understand each section before moving on to the next one.

COPYING THE SCREEN TO THE PRINTER

Your PC makes it easy to copy the contents of the screen to the printer. This means that whatever text is displayed on the monitor can be copied to your printer, giving you a hard copy of what is on the screen. To do this, use the sequence given in Table 6.1.

TRANSFERRING A DOS FILE TO ANOTHER DISK

In the preceding exercise you learned how to make a backup copy of a disk. There are times when you will need to transfer just a single file from one disk to another. The DOS command that lets you do this is

$$\text{COPY source\_file target\_file [/V][/A][/B]}$$

where: source_file = the complete file name of the file to be copied.
target_file = the name of the destination file. *Note:* If a file with the same name exists, it will be overwritten by the new copy.
/V = disk verification. It does a double check to make sure that the file was copied exactly the same as the original. It takes a little more copying time to do this.
/A = a modifier that lets the COPY command know that the file is an ASCII file (a text file)
/B = a modifier that lets the COPY command know that the file is a binary image file (a file that contains computer code instead of text)

Table 6.2 shows three examples of the use of the DOS COPY command. In all the examples, the active drive is drive A:.

A special form of the COPY command allows you to copy several files and merge them all into one single file. As an example, suppose you have three text files on the disk, as shown:

```
MYFILE1.TXT
MYFILE2.TXT
MYFILE3.TXT
```

TABLE 6.1   Copying the screen to the printer

Step	Process
1	On the monitor, view the text that you want to copy to the printer.
2	Make sure the printer is attached to the computer, the printer is on, and paper is in the printer.
3	Hold down the SHIFT key and press the key marked <Print Scrn> at the same time. (The cursor will now move across the screen from top to bottom and left to right, transferring the screen's contents to the printer.)
4	Remove the copy from the printer, and turn the printer off.

TABLE 6.2 DOS copy command examples

DOS Command	Resulting Action
`A>COPY MYFILE.TXT B:`	Copies the file MYFILE.TXT from the A: drive to the disk on the B: drive and uses the same name.
`A>COPY MYFILE.TXT MYFILE.BAK`	Copies the file MYFILE.TXT into a new file on the same drive called MYFILE.BAK (the extension .BAK is normally used to indicate a backup copy of a file).
`A>COPY MYFILE.TXT B:NEW.TXT`	Copies the file MYFILE.TXT from the A: drive to the disk on the B: drive and uses the name NEW.TXT, which will now be the file's name on the B: drive.

You can now use the COPY command to copy all three files into one large file called ALLFILES.TXT, as follows:

`A> COPY MYFILE1.TXT + MYFILE2.TXT + MYFILE3.TXT ALLFILES.TXT`

DISPLAYING ONE FILE

Recall from the preceding exercise that you used the DOS command DIR to display the files contained on the active drive. There may be times when you want to find out if a particular file is on the disk at all. One way to do this is, of course, to use the DIR command and then look through all the files on the screen. Another, more professional way is to use the name of the file with the DIR command. Suppose you want to see if a file by the name of THISONE.TXT is in your disk directory. You can enter the DIR command as follows:

`DIR THISONE.TXT`

If the file is there, it will be the only one displayed. If it isn't there, DOS will present a message telling you so.

DOS WILD CARD CHARACTERS

There may be times when you want to display a listing of all the files on a disk that have the extension .EXE and none of the others. For example, suppose your disk contains the following files:

```
TEXT1.TXT     GAME1.EXE     TEXT2.TXT     WONKERS.EXE
CONFIG.SYS    TEXT1.BAK     WONKERS.BAK   WONKINS.EXE
```

You can use the DOS wild card symbol * and enter the following DOS command:

`DIR *.EXE`

The only files that will be displayed are

`GAME1.EXE     WONKERS.EXE     WONKINS.EXE`

The * symbol tells DOS to use any group of characters in its search for a file. As an example, if you instead enter the DOS command

`DIR TEXT1.*`

then the only files you will get are

`TEXT1.TXT     TEXT1.BAK`

Of course, if you enter

`DIR *.*`

you will get all the files.

Another wild card DOS character is the question mark, ?. The ? allows substitution for only one letter. For example, if you enter the DOS command

`DIR WONK???.EXE`

you will get the files

`WONKERS.EXE    WONKINS.EXE`

COPYING A GROUP OF FILES

The DOS wild card characters are useful when you want to copy a group of files. For example, suppose you want to copy all the .TXT extension files from the hard drive (drive C:) to the A: drive. To do this (assuming that your active drive is the C: drive), all you need to enter is

`COPY *.TXT A:`

All the files with the .TXT extension will be copied to the A: drive.

DISPLAYING FILE CONTENTS

DOS allows you to display the contents of a text file. Actually, you can display the contents of any file, but only the text file will make any sense. To do this, use the DOS internal command

> TYPE file_specification

where: file_specification = the complete name of the file

To display the contents of the text file MYTEXT.TXT, make sure that the file is on the active drive and enter

`TYPE MYTEXT.TXT`

and the contents of the file will be displayed on the monitor. Again, you can use the CTRL-S combination to stop and start the text display just in case it is so large that it starts scrolling off the top of the screen. A better method involves the use of a text editor (such as EDIT) to view the file.

Many software vendors have a file on their disk called README.TXT. This file usually contains important last-minute information about the software on the disk. In order to see the contents of this file, you simply enter

`TYPE README.TXT`

REDIRECTING FILES

DOS provides a way of redirecting the information in a file. For example, if you want the contents of a file to be displayed on the printer (meaning printed on the paper in the printer) rather than on the screen, you can issue the following command:

`TYPE README.TXT > PRN`

This command sends the text of the file README.TXT directly to the printer. The > symbol is the DOS redirection operator, and it means that the file is to be directed elsewhere than the monitor. PRN is the DOS device name for the printer. So, to transfer the contents of any text file to the printer, use

`TYPE FILENAME.EXTENSION > PRN`

Using the DOS redirection operator, you can also place a copy of the directory listing into a file:

`DIR > DIRFILE.TXT`

This command instructs DOS to copy the contents of the directory into a file on the active drive and name the file DIRFILE.TXT. You can then send the contents of that file to the printer:

```
TYPE DIRFILE.TXT > PRN
```

You can also transfer the disk directory to the printer with

```
DIR/W > PRN
```

or

```
DIR > PRN
```

In either case, the directory output is redirected from the monitor to the printer.

THE F1 KEY

DOS provides an easy way to enter repeated information. Whatever you type in the DOS environment is retained in a section of computer memory called the *keyboard buffer*. If you enter

```
DIR A:
```

then, for example, after the directory is displayed, simply pressing the F1 key repeatedly will print out

```
DIR A:
```

on the monitor, just as if it were being typed. Now, if you want to see the directory of the C: drive, press the F1 key enough times to have

```
DIR
```

displayed and then type in the C:.

The entire command can be retrieved by pressing F3 once.

CREATING A DOS FILE

You can create a DOS text file while in the DOS environment by using the DOS command

```
COPY CON FILENAME.EXT
```

This tells DOS to copy what is coming from the keyboard (the console). The DOS device CON stands for the keyboard console in much the same way as PRN stands for printer. When you do this, DOS will let you type in lines of text. When you are finished, you let DOS know by placing the end-of-file marker CTRL-Z. You achieve this by holding down the CTRL key and at the same time pressing the Z key. DOS will respond by showing the CTRL-Z character ^Z on the screen and saving what you have typed into a text file of the name of your choice. You may also use the F6 key in place of CTRL-Z. The F6 key will produce the ^Z file marker to signal DOS that the file has ended.

For example, suppose you want to enter the text

> This is a test of your
> computer's ability to
> work with text files.

Also suppose you want to store this text in a file called TEST.TXT. Here is how you do it:

```
COPY CON TEST.TXT
This is a test of your
computer's ability to
work with text files.
^Z
```

The text you just typed is now copied into the disk on the active drive with the name TEST.TXT. You can now treat this file exactly as any other text file, by redirecting it to the printer or another disk or displaying the contents on the monitor.

USING EDIT

EDIT is an external DOS command that allows *screen* editing. EDIT allows you to use arrow keys to move the cursor around on the screen. This makes it much easier to correct typos, enter new text, and select other text for deletion.

EDIT is available only on systems using DOS version 5.0 or later.

STARTING EDIT

Since EDIT is actually a program (EDIT.COM), it must reside either on your active DOS disk or in a subdirectory accessible by DOS. EDIT allows you to specify the name of the text file you wish to edit, as in:

```
A> EDIT MYBATCH.BAT
```

If the MYBATCH.BAT file does not exist, EDIT creates it. If the file does exist, EDIT opens the file and shows you the initial screen of file text. Figure 6.1 shows what the EDIT start-up screen looks like when the file does not exist. At the top of the screen are *pull-down menu items* File, Edit, Search, Options, and Help. The Help menu is especially helpful for new users. The large box with arrows on the right side and bottom is called the *editing window* and is where your text file is displayed. Horizontal and vertical scroll bars provide a graphic display of your current position within the text file. The four arrow keys can be used to navigate around inside the editing window.

At the lower right corner are two numbers that display the current line and column position of the cursor in the text file. The cursor controls where text is inserted and deleted.

Since MYBATCH.BAT does not yet exist, EDIT creates the file and waits for you to begin entering text. Enter the following three lines right now:

```
CLS
VOL
VER
```

Those three lines should now appear in the upper left-hand corner of the editing window.

FIGURE 6.1 Initial EDIT screen

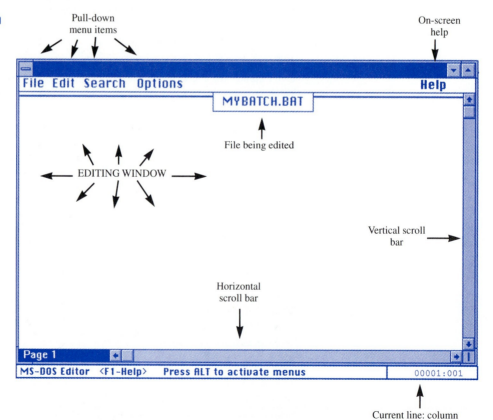

FIGURE 6.2 The EDIT File menu

```
New
Open...
Save
Save As...

Print...

Exit
```

To save the new text file, it is necessary to access the File menu. This is accomplished by pressing ALT and then the F key. When you do this, EDIT displays the File menu, which is shown in Figure 6.2. Choosing a command from the File menu can be done in several ways:

- Enter the letter of the command you want to use (N for New, O for open, etc.).
- Use the up and down arrows to highlight the desired command, then press ENTER.
- Use the mouse to select a command and click the left mouse button.

To save the file, use the SAVE command from the File menu (press S). Then exit EDIT by choosing the Exit command (ALT-F-X). EDIT writes the new text into MYBATCH.BAT and exits to DOS.

EDITING AN EXISTING FILE

Now that we have a text file to work with, let us edit it again. First, enter

```
A> EDIT MYBATCH.BAT
```

EDIT opens the file and displays the three lines of text. The cursor will be at line 1, column 1. Use the arrow keys to position the cursor so that it is underneath the V in the VER command on line 3. Now press the DELETE button on your keyboard. EDIT erases the V character and shifts the remaining two characters (ER) to the left. Press DELETE again. Now only the R from VER should remain. Enter the characters D and I now so that the third line reads DIR. This is one way we can modify an existing file. The BACKSPACE key can also be used to delete single characters, and there are other EDIT commands that allow entire blocks of text to be moved, deleted, and copied.

OTHER EDIT FEATURES

Large text files can be navigated quickly through the use of the Page-Down and Page-Up buttons, which replace the entire editing window with a new set of lines. CTRL-END takes you to the last page of the text file, and CTRL-HOME takes you to the beginning.

Also, EDIT allows the use of user-selectable screen colors. This is a great advantage over monochrome (black and white) editors, because it allows the screen colors to be adjusted for a visually pleasing effect. And, as stated before, EDIT can be controlled with the mouse, adding another dimension of ease to the task of editing.

Finally, choices within the Search menu allow the user to search for a word, phrase, or group of symbols, and to replace it or them with a different character or set of characters, if necessary. This is very useful for programmers, writers, and individuals who handle large data files.

RENAMING FILES

DOS provides a command for renaming any file (or group of files using wild card characters). The internal DOS command is

```
RENAME file_name newfilename[.ext]
```

or

```
REN file_name newfilename[.ext]
```

For example, if you want to rename the file MYFILE.TXT to YOURFILE.TXT, type

```
RENAME MYFILE.TXT YOURFILE.TXT
```

and the name of MYFILE.TXT will be changed to YOURFILE.TXT.

Entering REN HELLO.* BYE.* uses the wild card on the extension to rename all files beginning with HELLO to BYE.

ERASING FILES

Because you can create DOS files, you can also get rid of them; however, use this command with caution. The DOS command is

DEL filename

For example, if you want to delete the file OLDSTUF.TXT, you enter

```
DEL OLDSTUF.TXT
```

and the file is deleted. You can also use wild card characters to delete files. Be very cautious when you do this so you don't delete more files than you intend to delete.

ORGANIZING YOUR FILES

An important part of using a computer system is understanding the different methods by which files are kept on the disk. Up to this point, when you needed to store information on a disk, you simply gave this information a file name and used the proper DOS command to store it on the disk. Storing files in this way is a rather simple and direct approach. However, you will find that this storage method is not the most practical one actually used in the field. Figure 6.3 shows the direct approach to storing disk files.

The problem with this storage method is that any time you want to access a file or get information about it, you must be in the environment of all the other files on the disk. In practical applications, there are usually stored files that contain related kinds of information or have some other kind of unique relationship, as illustrated in Figure 6.4.

It would be nice to be able to deal with just one file type at a time. One method of doing this is to use a separate disk for each type of file. This, however, may not be practical, especially if a single hard disk is used to store a variety of disk files.

FIGURE 6.3 Direct approach to storing disk files

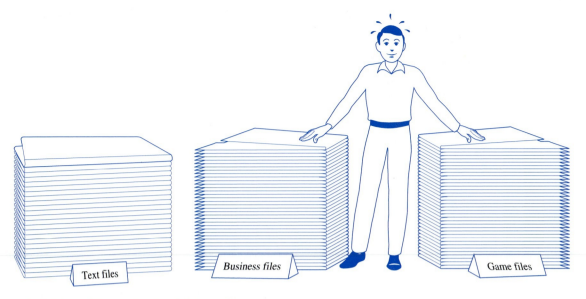

FIGURE 6.4 Files with related information

FIGURE 6.5 Breaking a disk into smaller parts

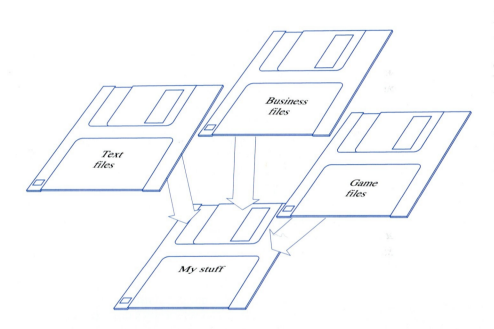

Fortunately, DOS provides a way of making a single disk appear as if it were many smaller, separate disks, as shown in Figure 6.5. DOS does this by creating **directories.** A directory is nothing more than a part of a disk that looks like a smaller disk, while the rest of the disk is invisible to the user. This important concept is illustrated in Figure 6.6.

Theoretically, you can break your disk into as many smaller disks as you want. You can even break up your directories into even smaller parts, called **subdirectories.** A block diagram representing the use of subdirectories is shown in Figure 6.7.

CREATING DIRECTORIES

The DOS command for creating a directory on your disk is

MKDIR[drive:]directory

where: drive = the name of the drive (optional)
directory = the name you want to give the directory

FIGURE 6.6 Concept of DOS directories

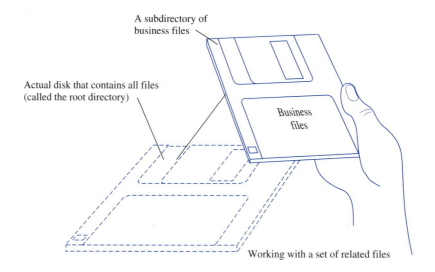

FIGURE 6.7 Block diagram representing the use of subdirectories

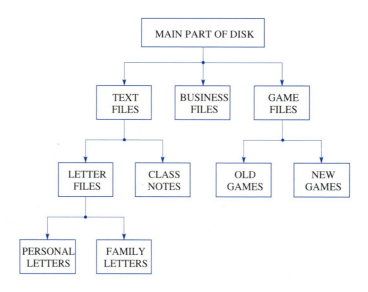

A directory must be named following the same DOS rules used for naming a file. This means that each name can consist of from one to eight characters, with a three-character extension. If you do not specify a drive, DOS will use the active drive.

For example,

```
MKDIR DIR_1
```

will cause the directory DIR_1 to be created on the disk in the active drive.

When you use the DIR command to list the disk directory, if there is nothing else on the disk, your output will be as shown in Figure 6.8.

Because the MKDIR command is frequently used in DOS, the abbreviated form MD is provided. For example, using the same disk, you could create two other directories:

```
MD DIR_2
MD DIR_3
```

You have effectively created three new disks called DIR_1, DIR_2, and DIR_3. You still have the main disk as well. The main disk is called the *root directory,* meaning that it is the one with which you always start.

Now that you know how to create a directory, the next step is to discover how to get inside one.

FIGURE 6.8 Disk listing of a directory

```
A:\> DIR

Volume in drive A is MYDISK
Directory of A:\

DIR_1         <DIR>         12-23-98   2:32a  dir_1
    0 file(s)        0 bytes
    1 dir(s)     1,456,640 bytes free

A:\>
```

GETTING INTO DIRECTORIES

The DOS command for getting into a directory is

CHDIR[drive:]directory

where: drive = the name of the drive that contains the directory
    directory = the name of the directory

If you don't use a drive name, DOS will look for the directory on the active drive. If it isn't there, DOS will respond with an error message, telling you that the directory cannot be found.

As an example, to get inside the DIR_1 directory, the command

`CHDIR DIR_1`

will activate the DIR_1 directory. If you now do a listing with the DOS DIR command, you will see what is shown in Figure 6.9. As you can see from the figure, all the other information on the disk (such as the other directories) is no longer displayed, just as if you were now on another disk.

Note that the top part of the directory display shows the drive the disk is in and the directory:

`Directory of A:\DIR_1`

As shown in Figure 6.9, there are two other listings:

```
.     <DIR>   12-23-98   2:33a   .
..    <DIR>   12-23-98   2:33a   ..
```

The single period is the abbreviation for the current directory. The double period is the abbreviation for the directory immediately above the current directory. For example, entering DIR . will produce exactly what is shown in Figure 6.9, whereas entering DIR .. will produce the directory immediately above DIR_1, which is, in this case, the root directory.

Since the CHDIR command is so common, DOS allows you to abbreviate it CD. Thus, CD DIR_1 works the same as CHDIR DIR_1.

Figure 6.9 Inside a DOS directory

```
A:\> CHDIR DIR_1

A:\DIR_1> DIR

Volume in drive A is MYDISK
Directory of A:\DIR_1

  .           <DIR>         12-23-98     2:33a.
  ..          <DIR>         12-23-98     2:33a..
       0 file(s)        0 bytes
       2 dir(s)     1,456,640 bytes free
A:\DIR_1>
```

GETTING OUT OF DIRECTORIES

To get back to the root directory (the one you start with when the system first boots up), use

CD \

This will automatically put you back into the root directory. The directory that is just above a current directory is called the **parent directory.** Thus, the root directory is the parent directory for DIR_1. If there were a directory named SUB_1 inside DIR_1, then DIR_1 would be the parent directory for SUB_1. Using CD, you move up one directory to the parent directory of the current directory.

SOME DIRECTORY EXAMPLES

Suppose you had a disk with the directory structure shown in Figure 6.10. Note from the figure that there are two subdirectories (the root directory is the main directory). These subdirectories are called DIR_1 and DIR_2. Also note that there are three files, each called MYFILE.01. Each of these files, even though they all have the same name (MYFILE.01), can be entirely different *because they are in different directories.* Remember to think of a directory as if it were a separate minidisk. The directory structure of Figure 6.10 is similar to having three different disks. Of course, you could have three entirely different files using exactly the same name on three different disks.

Normally you wouldn't give the files inside different directories the same name; we did this here just to illustrate that the files inside directories are indeed isolated from files inside other directories. Figure 6.11 shows another example of two directories, each with some files contained in them.

Table 6.3 shows some of the various combinations that you could achieve with the directory structure of Figure 6.11.

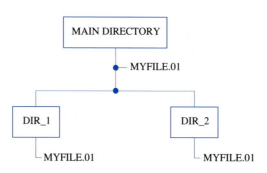

FIGURE 6.10 Example of directory structure

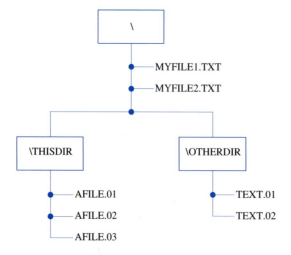

FIGURE 6.11 Directories with different files

TABLE 6.3 Various directory combinations from Figure 6.11

Active Directory	Directory Listing Display				Comments
\(root)	MYFILE1.TXT	128	3-5-98	9:30a	Active files: MYFILE1.TXT, MYFILE2.TXT. Files not accessible: all the files in the other two subdirectories.
	MYFILE2.TXT	250	3-6-98	10:05a	
	THISDIR	<DIR>	3-6-98	10:15a	
	OTHERDIR	<DIR>	3-7-98	9:03a	
Change to another directory with CD\THISDIR					
\THISDIR	AFILE.01	356	3-9-98	1:08p	Active files: AFILE.01, AFILE.02, AFILE.03. Files not accessible: all the files in the other directories (the root directory and OTHERDIR).
	AFILE.02	550	3-9-98	1:15p	
	AFILE.03	128	3-9-98	1:25p	
Change to the other directory with CD\OTHERDIR					
\OTHERDIR	TEXT.01	256	3-7-98	9:20a	Active files: TEXT.01, TEXT.02. Files not accessible: all the files in the other directories (the root directory and THISDIR).
	TEXT.02	256	3-8-98	11:46a	

CREATING OTHER DIRECTORIES (PATHNAMES)

As a technician you will encounter directories contained inside directories. Directories inside directories are usually found on hard disk drives (because hard drives can hold so much information), but they are possible on floppy disks as well. Figure 6.12 shows an example of what you might encounter on a hard disk drive directory used by a small business.

As shown in Figure 6.12, there are three main subdirectories, called TEXT, GAMES, and ACCOUNTS. Each of these subdirectories has subdirectories within it. For example, the subdirectory TEXT contains the subdirectories LETTERS and MAILINGS. If you start at the root directory and you want to play the game ZAP.EXE, you first enter

CD\GAMES\ZAPPER

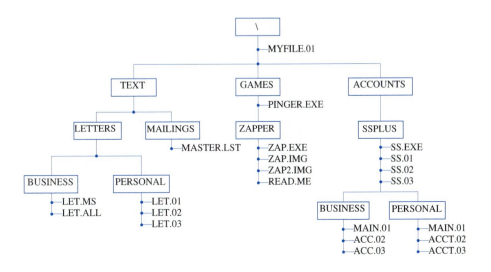

FIGURE 6.12 Typical directory tree used by a small business

This DOS command puts you into the ZAPPER subdirectory, where you can access the game ZAP.EXE.

In DOS, a **pathname** is simply a series of subdirectory names DOS must combine to give you access to the files you want. Thus, in the DOS command CD\GAMES\ZAPPER, \GAMES\ZAPPER is the pathname.

To get to the game PINGER.EXE, you enter CD\GAMES (the pathname here is \GAMES). This allows you to access the game PINGER.EXE, because you are in the GAMES directory.

Table 6.4 gives some examples of how to access various files in the directory structure given in Figure 6.12.

Table 6.5 summarizes the DOS directory rules.

REMOVING DIRECTORIES

The DOS command for removing a directory is

$$RMDIR[drive:]pathname$$

or

$$RD[drive:]pathname$$

where: drive = the drive containing the subdirectory. (If no drive is specified, DOS will look for the subdirectory in the active drive.)
pathname = the subdirectory names required to get to the directory

TABLE 6.4 DOS pathname examples for Figure 6.12

If the Current Directory Is	To Gain Access to File	You Must Use DOS Command (Pathname)
\	LET.MS	CD\TEXT\LETTERS\BUSINESS
TEXT\LETTERS\BUSINESS	LET.01	CD\TEXT\LETTERS\PERSONAL
TEXT\LETTERS\PERSONAL	ACC.02	CD\ACCOUNTS\SSPLUS\BUSINESS
ACCOUNTS\SSPLUS\BUSINESS	SS.01	CD..
ACCOUNTS\SSPLUS	MYFILE.01	CD\

TABLE 6.5 Summary of DOS directory rules

Main Point	Comments
Directory names conform to DOS file name standards.	A file name can consist of eight characters with a three-character extension.
Do not create a subdirectory called \DEV.	DOS uses a *hidden* internal subdirectory called \DEV to perform I/O operations to devices such as the printer.
DOS allows the creation of any number of subdirectories according to the amount of disk space.	Disk Space Maximum Subdirectories 160/180KB 64 320/360KB 112 1.2MB 224 (The more disk space, the more subdirectories the disk can contain.)
DOS pathnames cannot exceed 63 characters.	

For example, in order to remove the directory called MYGRADES, use RMDIR MYGRADES or RD MYGRADES, assuming that this directory is on the active drive. To remove the subdirectory LETTERS in the subdirectory TEXT, use RMDIR\TEXT\LETTERS or RD\TEXT\LETTERS.

A directory must be empty (not contain any files or other directories) before it can be removed. Thus, if you want to remove a directory that contains files or other subdirectories, you must first get rid of all the files and subdirectories.

DELETING A NONEMPTY DIRECTORY

The RMDIR command requires an empty directory for the command to proceed. Beginning with DOS 6.0, the DELTREE command can be used to delete entire directories (and any subdirectories) quickly and easily. For example, the "punchin" directory is deleted as follows:

```
C:\> deltree punchin
Delete directory "punchin" and all its subdirectories? [yn] y
Deleting punchin...
```

Wild cards are allowed in the directory name, which allows multiple directories to be deleted with a single command. Use the DELTREE command with caution, especially when using wild cards.

TRANSFERRING FILES

There are times when you need to copy a file from one directory to another. To do this, you use the DOS COPY command just as before, except now you include the pathname. Thus, to copy the file MYFILE.01 in the root directory to a subdirectory called TEXTFILE, you enter

```
COPY MYFILE.01 \TEXTFILE
```

In this example both the root directory and the subdirectory are on the same disk in the same active drive.

If you want to copy a file from one directory to another on a different drive (assuming the active drive is A:), you use

```
COPY MYFILE.01 B:\TEXTFILE
```

TROUBLESHOOTING TECHNIQUES

Keeping your files organized into subdirectories reduces the time required to search for an important file, keeps the directory listing small and manageable, and provides an easy way to group files together for copying.

If you are using a floppy disk to carry important files, bear in mind that each subdirectory you create reserves a number of sectors on the disk. A large number of subdirectories reduces the amount of available storage space on the floppy.

SELF-TEST

This self-test is designed to help you check your understanding of the background information presented in this exercise.

True/False

Answer *true* or *false*.

1. A printed output of a file is called a hard copy.
2. You can make a copy of a whole disk but not just a part of it.

3. You can make a direct copy of the screen to the printer by pressing just two keys on the keyboard.
4. In practical applications, disk files are sorted so that every one can be accessed at the same time.
5. The DOS root directory is the main directory contained by all disks.

Multiple Choice

Select the best answer.

6. The DOS command

    ```
    A> COPY THIS.TXT B:
    ```

    a. Copies one file from drive A: to drive B:.
    b. Makes another file with a new name.
    c. Creates a file on drive B: and deletes the one on drive A:.
    d. All of the above.
7. From the root directory, the DOS command

    ```
    MKDIR NEW_1
    ```

    will cause
    a. A new file by the name NEW_1 to be created.
    b. A new subdirectory by the name of NEW_1 to be created.
    c. Nothing to happen because you are already in the root directory.
    d. None of the above.
8. To change from any subdirectory to the root directory, use the DOS command
    a. CD\.
    b. CHDIR ROOT.
    c. ROOT DIR.
    d. None of the above.

Matching

Match each phrase on the left with the proper command on the right.

9. Incorrect DOS directory name
10. Means the previous directory
11. Means the current directory

    a. .<DIR>
    b. MYDIR.TXT
    c. .. <DIR>
    d. None of these

Completion

Fill in the blanks with the best answers.

12. The DOS command _____ is used to change the name of a file.
13. The _____ is a series of subdirectory names DOS must combine to give you access to the files you want.
14. Before you can _____ a directory, the directory must be empty.
15. There are two DOS commands for changing a directory, _____ and CHDIR.

FAMILIARIZATION ACTIVITY

This activity is designed to help you get acquainted with DOS file management. Be sure to start this activity with a newly formatted disk (no existing files or subdirectories on the disk). You will be creating subdirectories and text files to store in these directories. You will then remove one of the directories.

1. Boot up your computer and make sure DOS is resident.
2. With your newly formatted disk, list its directory. It should not show any files or subdirectories.

3. Create a subdirectory called DIRECT.ONE.
4. Do a directory listing of your disk. You should see only one directory listed (called DIRECT ONE <DIR>).
5. Create a second subdirectory called DIRECT.TWO.
6. Do a directory listing again. You should now see two subdirectories listed:

```
Directory of A:\
 DIRECT ONE   <DIR>
 DIRECT TWO   <DIR>
```

7. Change directory to DIRECT.TWO.
8. Do a directory listing; you should see

```
Directory of A:\DIRECT.TWO
 .    <DIR>
 ..   <DIR>
```

9. Make another subdirectory called SUBDIR.
10. Again display the listing; you should see

```
Directory of A:\DIRECT.TWO
 .          <DIR>
 ..         <DIR>
 SUBDIR     <DIR>
```

11. Go back to the root directory.
12. A listing of the root directory should display

```
Directory of A:\
 DIRECT ONE   <DIR>
 DIRECT TWO   <DIR>
```

13. Using the COPY CON command or the EDIT utility, create three text files with the following names and text:

   First file: `COPY CON FILE.01`
                 `This is in file one.`
                 `^Z`
   Second file: `COPY CON FILE.02`
                 `This is in file two.`
                 `^Z`
   Third file: `COPY CON FILE.03`
                 `This is in file three.`
                 `^Z`

14. Again, do a directory listing of the root directory. You should now see

```
Directory of A:\
 DIRECT   ONE    <DIR>
 DIRECT   TWO    <DIR>
 FILE     01
 FILE     02
 FILE     03
```

15. Copy both FILE.01 and FILE.02 into the directory DIRECT.ONE. Use

    `copy file.01 \DIRECT.ONE and COPY FILE.02 \DIRECT.ONE.`

16. Copy FILE.03 into the directory DIRECT.TWO. Use

    `COPY FILE.03 \DIRECT.TWO.`

17. Now copy FILE.01 into the subdirectory SUBDIR. Use

    `COPY FILE.01 \DIRECT.TWO\ SUBDIR.`

18. Next, delete all the files in the root directory. Use

    ```
    DEL FILE.*
    ```

19. A directory listing of the root directory should now look like this:

    ```
    Directory of A:\
      DIRECT  ONE    <DIR>
      DIRECT  TWO    <DIR>
    ```

20. Now change to the directory DIRECT ONE.
21. A directory listing of DIRECT.ONE should be

    ```
    Directory of A:\DIRECT ONE
      .            <DIR>
      ..           <DIR>
      FILE    01
      FILE    02
    ```

22. Change directory to DIRECT.TWO.
23. A directory listing of DIRECT.TWO should be

    ```
    Directory of A:\DIRECT TWO
      .            <DIR>
      ..           <DIR>
      SUBDIR       <DIR>
      FILE    03
    ```

24. Now change directory to SUBDIR.
25. A listing of SUBDIR should produce

    ```
    Directory of A:\DIRECT TWO\SUBDIR
      .            <DIR>
      ..           <DIR>
      FILE    01
    ```

26. Return to the root directory. You will now remove the directory called **DIRECT.ONE**, but recall that you must first make sure the directory is empty.
27. First try to remove the directory DIRECT.ONE while it still contains files. What DOS message do you get?
28. Change the directory to DIRECT.ONE.
29. A directory listing should produce

    ```
    Directory of A:\DIRECT ONE
      .            <DIR>
      ..           <DIR>
      FILE    01
      FILE    02
    ```

30. Now, remove the two files in this directory.
31. A listing should now produce

    ```
    Directory of A:\DIRECT ONE
      .            <DIR>
      ..           <DIR>
    ```

32. Return to the root directory and remove the subdirectory **DIRECT.ONE** with **RMDIR DIRECT.ONE**.
33. A listing of the root directory should now produce

    ```
    Directory of A:\
      DIRECT TWO    <DIR>
    ```

34. Using EDLIN or EDIT, create a batch file called **MYBATCH.BAT**.

35. Enter the following lines of code into the file:

```
ECHO OFF
CLS
ECHO        *************************
ECHO        *       BATCH FILE       *
ECHO        *     Created by:        *
ECHO        *    [Your name here]    *
ECHO        *       Date:            *
ECHO        *  [Today's date here]   *
ECHO        *************************
PAUSE
DIR
```

36. End the file-editing session.
37. Check the directory of your disk and verify that there is now a text file named MYBATCH.BAT.
38. List the text file on your monitor by using the DOS TYPE command:

    `TYPE MYBATCH.BAT`

    Verify that the text in this file is exactly as you entered it.
39. Activate the batch file by typing **MYBATCH** from the DOS prompt. The monitor screen should clear, your title should be displayed, a pause requiring you to press any key should occur, and then the directory of your disk should be displayed.
40. Now reopen MYBATCH.BAT for editing.
41. Remove your name and replace it with your lab partner's name.
42. Insert a new line after the date and enter the name of your instructor.
43. End the editing session, and make sure you are returned to DOS.
44. Use the DOS TYPE command to view the contents of the MYBATCH.BAT file.
45. Execute your new batch file to verify that you get the new display (with your lab partner's and instructor's names).

QUESTIONS/ACTIVITIES

To answer the following questions, assume you are working with a $3\frac{1}{2}''$ disk in the A: drive and a C: drive that contains the directory structure shown in Figure 6.13.

1. Assume you need to add a file called FILE.03 from drive A: to the subdirectory TX99. What DOS command would you use to do this? (Assume your active drive is A:.)
2. If you needed to copy all the files from \PAYROLL\WEEK to a disk in drive A:, what DOS command would you use? (Assume your active drive is C:.)
3. Give the DOS command you would use to transfer the file FILE.02 in the \TAXES\TX97 subdirectory to the \PAYROLL\YEAR subdirectory. (Assume your active drive is C:.)
4. State the DOS commands you would use to create a new directory called OLDFILE, with the root directory as the parent directory, and transfer all the files in the \TAXES\TX97 subdirectory to the OLDFILE directory.
5. State the DOS commands you would use to remove the subdirectory \TAXES\TX97.

FIGURE 6.13 Directory structure for questions/activities

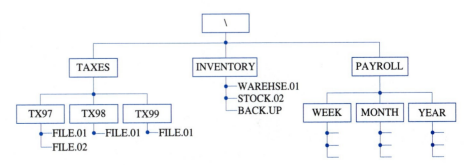

REVIEW QUIZ

Given a PC, a printer, and the necessary supplies and equipment, you will be able to

1. Display a listing of a given group of files, using the appropriate command and wild cards, within 3 minutes and with 100% accuracy.
2. Use the proper command to copy the information displayed on the monitor to the printer device (make a hard copy) within 2 minutes and with 100% accuracy.
3. Create a disk file with the name MYFILE.TXT. that contains

    Your name
    Your social security number
    Today's date

    within 5 minutes and with 100% accuracy.

7 Configuring DOS with CONFIG.SYS and AUTOEXEC.BAT

INTRODUCTION

Joe Tekk led Don into RWA Software's corporate training room, which contains 16 computers, several different laser printers, and a color scanner. "I've rewritten all the CONFIG.SYS and AUTOEXEC.BAT files on these machines so that there is more memory available for DOS, and each machine knows what laser printer it is connected to. The user is allowed to choose from a menu of printers at boot time."

Don was very satisfied. "That's great, Joe. We can use your files in the training classes to show them how to make menus, too."

PERFORMANCE OBJECTIVES

Upon completion of this exercise, within 10 minutes and with 100% accuracy, you will be able to

1. Set up a DOS path.
2. Make a CONFIG.SYS file that creates a number of buffers and files as assigned by your instructor.
3. Create a batch file that will do one of the following:
   - Clear the screen and display a message selected by your instructor.
   - Clear the screen and display the current directory.
   - Clear the screen and display the contents of a user-selected directory.
   - Clear the screen and direct the contents of a user-selected file to the printer.
4. Create an AUTOEXEC.BAT file that will do one of the following:
   - Clear the screen and display a message selected by your instructor.
   - Clear the screen and change the DOS prompt to display either the date or current directory, as selected by your instructor.
   - Clear the screen and call the batch file you created in item 3.

BACKGROUND INFORMATION

INTRODUCTION

The skills presented in this exercise will help you customize DOS and build on the DOS concepts you have learned to this point.

SETTING THE PATH

Figure 7.1 shows the structure of the program in a typical hard drive configuration (usually drive C:). As you can see from the figure, a typical hard drive configuration has many subdirectories, including a subdirectory for DOS files. DOS provides a command that makes it easy to tell DOS where to look for external commands. For example, every time you execute an external command, DOS first searches the current directory for the command file. If this file exists (such as a .COM, .EXE, or .BAT file), it is executed. Otherwise, DOS checks to see if you have defined a **path** or other subdirectories or disk drives to use in searching for the command.

The DOS command used for this purpose is PATH; it has the form

PATH[drive:][pathname][;drive][pathname] . . .

where: drive = the drive with a disk that may contain a required file
pathname = the name of the path that may contain a required file

For example, the command

PATH = C:\;C:\DOS;C:\MYSTUFF;C:\WORDS;C:\GAMES

tells DOS to search these five directories for external commands. The DOS path command is normally placed in the AUTOEXEC.BAT file so that it is automatically activated every time the system is booted.

SYSTEM CONFIGURATION

The file called **CONFIG.SYS** is a special file for which DOS looks every time the system is booted. CONFIG.SYS is a **system configuration file.** If the file exists, DOS reads its contents and configures itself according to the entries provided. If the file does not exist, DOS leaves the system in the default condition. Table 7.1 lists the DOS configuration parameters.

There will be times when you will have to create a CONFIG.SYS file to meet certain specifications. Some of the more common configuration requirements are covered in this exercise.

BREAK Command

The BREAK command has the form

BREAK = Condition

where: Condition = the word **on** or the word **off**

Normally the CTRL-C key combination can be used to stop operation of most programs. In normal operation, DOS checks for a CTRL-C only when it is reading from the keyboard or writing to the printer or screen. If you want DOS to check for a CTRL-C at other times (such as during a disk read or write), you must set BREAK ON in the CONFIG.SYS file:

BREAK = ON

FIGURE 7.1   Typical hard drive configuration

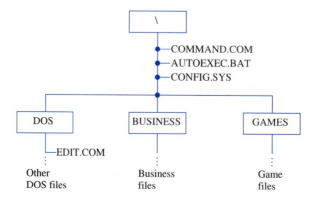

TABLE 7.1 DOS configuration parameters

Parameter	Meaning
BREAK	May be set ON or OFF. Determines how often DOS checks for a CTRL-C.
BUFFERS	Allows you to set the number of buffers (memory locations) used to hold data when it reads from or writes to a disk.
COUNTRY	Lets you change specific items (such as the currency format) used by certain countries.
DEVICE	Allows you to let DOS control specific devices such as plotters or create a spot in memory that looks like an extra disk drive.
DRIVEPARM	Sets parameters for floppy disk drives.
FCBS	Used with early versions of DOS for file control blocks.
FILES	Introduced with DOS 3.0. Specifies the number of files that DOS can open at a time.
INSTALL	Invokes a special **fastopen** command from processing of the CONFIG.SYS file. Available with DOS 4.01 and later.
LASTDRIVE	Lets you specify the last disk drive to which DOS can refer.
REM	Allows you to add remarks to the CONFIG.SYS file. Available with DOS 4.01 and later.
SHELL	Lets you specify an alternate command process other than COMMAND.COM.
STACKS	Designates how many places in memory DOS can use to store what it was doing while servicing an interrupt.

To set BREAK back to its normal default condition, use

```
BREAK = OFF
```

If BREAK is not specified in the CONFIG.SYS file, DOS uses the default condition, BREAK = OFF.

The disadvantage of setting BREAK = ON is that it may slow down the system somewhat because DOS has to check for the CTRL-C more often.

BUFFERS Command

The BUFFERS command has the form

$$BUFFERS = n[,m][/x]$$

where: n = the number of disk buffers, from 1 to 99
m = the maximum number of sectors that can be read or written in one I/O operation (1–8); the default setting is 1
/x = a qualifier that, when used, allows the maximum number of disk buffers to be 10,000 or the largest number of buffers to fit into computer memory, whichever value is less

A **buffer** can be thought of as a special storage place in memory that holds data when it is being written to or read from a disk. The size of a buffer is 528 bytes. Whenever an application program is required to read data from a disk, it first checks the buffers to see if the data is there. If the required data is not in any of the buffers, the program performs a read operation from the disk. This operation is illustrated in Figure 7.2.

Some application programs (such as database programs) access the disk many times to get required information. Doing this can slow down the system. The system can be speeded

FIGURE 7.2 Example of reading buffers

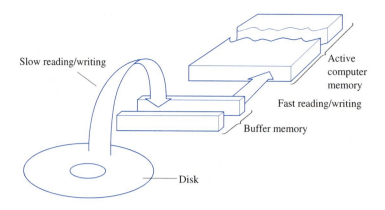

up by increasing the number of buffers, normally from 10 to 25. As an example, to increase the number of buffers to 20, you enter the following in the CONFIG.SYS file:

BUFFERS = 20

The use of too many buffers can result in a waste of computer memory, with a resulting slowdown in system operation. The slowdown occurs because every time DOS performs a disk read, it must first search through all the buffers. If you have allocated 60 or more buffers, for instance, this reading operation can take as much time as a simple disk I/O.

COUNTRY Command

DOS gives you the opportunity to use international time, date, and currency conversions as well as folding formats and decimal separators. These are specialty items not normally used in the United States. You should refer to your DOS manual for any changes that you need to make here. What is important for now is that these changes are implemented with the COUNTRY command used in the CONFIG.SYS file.

DEVICE Command

Your computer needs to know what **devices** it is required to operate. By default, DOS supplies **device drivers** (programs), which instruct it on how to read the keyboard (standard input), place information on the monitor (standard output), and use the printer, the system clock, and the disk drives. However, if you add devices such as a mouse or CD-ROM to the system, the computer needs to know this. This information is placed again in the CONFIG.SYS file.

The format for the DEVICE command is

DEVICE = [drive]:[path]filename[argument]

where: drive = the drive containing the device driver
         path = the path of the device driver
   filename = the device driver file name
 argument = any information needed with the file name

Two device drivers that you can install are ANSI.SYS and VDISK.SYS. The ANSI.SYS driver gives your system some enhanced control characteristics. VDISK.SYS (some versions of DOS use RAMDRIVE.SYS) is a driver that makes a section of the computer memory look like a disk drive. This has the advantage of allowing very quick reading and writing of files. However, the disadvantage is that when you turn off your system, all the information saved to this *virtual* drive is lost.

The syntax for such a device driver installation is

DEVICE = VDISK.SYS[disksize][sectorsize][entries][/E]

or

DEVICE = RAMDRIVE.SYS[disksize]a[sectorsize][entries][/E][/A]

where: VDISK.SYS or RAMDRIVE.SYS = a DOS device driver
                       disksize = disk size in kilobytes. (Default is 64KB;
                                  minimum size is 16KB.)

sectorsize = sector size in bytes. (Default size is 512 bytes; values of 128, 256, 512, and 1024 are available.)

entries = the number of root directory entries (the number of files or subdirectories). (Default value is 64, with a minimum of 4 to a maximum of 1024).

/E = a qualifier that forces the system to use extended memory (if available) above 1MB. (If the /E qualifier is used, the /A qualifier cannot be used.)

/A = a qualifier that forces the system to use a memory board that meets the Lotus/Intel/Microsoft (LIM) Expanded Memory Specification for a RAM drive (if the system has such a board installed). (If you use the /A qualifier, you cannot use the /E qualifier.)

As an example, to install a RAM drive, you would have, in the CONFIG.SYS file,

DEVICE = VDISK.SYS

This gives you a virtual disk drive with the default value of 64KB and with 128-byte sectors and 64 file entries. You can now treat this as a typical disk drive. DOS will assign it a drive name. For example, if you have drives A:, B:, and C:, DOS will assign your RAM drive as D:. You may then use it as a disk drive with the understanding that if you want to save any files, you must transfer them to a disk located on a real drive before turning off the system.

There may also be other requirements for the DOS DEVICE settings. Some of these are shown in Figure 7.3. DOS 5.0 and later versions allow the use of the DEVICEHIGH command to load a device driver into high memory. This may be necessary if RAM below 640KB must be conserved.

DRIVEPARM

There are times when you will encounter computer systems that use an external floppy disk drive or CD-ROM. When such a system is encountered, it is necessary to let DOS know that the drive exists. This is accomplished in the CONFIG.SYS file with the DOS DEVICE command

DEVICE = DRIVER.SYS[/D:driveNumber][/C][/F:formFactor][/H:heads][/N] [/S:sectors] [/T:tracks]

where: DRIVER.SYS = a DOS device driver

/D:driveNumber = the physical number of the drive. (This value ranges from 0 to 127. The first physical floppy disk is drive 0, which is drive A:.)

/C = a qualifier to make sure DOS checks to ensure that the disk drive door is closed before proceeding

FIGURE 7.3 Some other devices requiring the DEVICE COMMAND

```
DEVICE=C:\HIMEM.SYS
DEVICE=C:\DOS\ANSI.SYS
DEVICE=C:\CCARD\CCDRIVER.SYS
DEVICE=C:\MOUSE\MOUSE.SYS
DEVICE=C:\WINDOWS\SMARTDRV.SYS 2048
```

/F:formFactor = a qualifier that tells DOS what type of disk drive is being used. It takes on the following values:
- 0 = 180KB or 360KB drives
- 1 = 1.2MB drive
- 2 = 720KB, $3^1/2''$ drive
- 5 = fixed disk
- 6 = tape drive
- 7 = 1.44MB, $3^1/2''$ drive
- 8 = CD-ROM
- 9 = 2.88MB, $3^1/2''$ drive

/H:heads = a qualifier that tells DOS the maximum number of read/write heads on the disk. (This value ranges from 1 to 99, with a default value of 2.)

/N = a qualifier that tells DOS that the device is not removable

/S:sectors = a qualifier that gives the number of sectors per track. (This value ranges from 1 to 99, with a default value of 9.)

/T:tracks = the number of tracks per side. (This value ranges from 1 to 999, with a default value of 9.)

As an example, if you install a 1.44MB, $3^1/2''$ drive, you would place the following in your CONFIG.SYS file:

```
DEVICE = DRIVER.SYS /F:2
```

What is important is that you check the documentation that comes with the disk drive to see exactly how you should set the correct DRIVER.SYS entry.

FILES

The FILES entry in the CONFIG.SYS file lets DOS know the number of disk files that DOS can have open at the same time. Normally DOS allows up to eight files to be opened at the same time. However, some application programs require more than eight files to be opened. When this is the case, you need to use the DOS FILES entry in the CONFIG.SYS file. The syntax is:

FILES = NumberFiles

where: NumberFiles = the number of files to be opened. (The minimum value is 8—the default value—and the maximum is 255.)

For example, if you have a database program that requires a maximum of 20 files to be opened, you add

```
FILES = 20
```

to your CONFIG.SYS file.

LASTDRIVE

The DOS LASTDRIVE entry lets you tell DOS the name of the last drive to be used in the system. The default value for this is E (for the E: drive). The syntax is

LASTDRIVE = DriveLetter

where: DriveLetter = a letter from A to Z

The LASTDRIVE designation must be equal to or greater than the actual number of drives in the system. For example, an entry in the CONFIG.SYS file of

```
LASTDRIVE = E
```

makes the last drive in the system the E: drive.

STACKS

Some computer systems must operate many different devices, and some computer programs have a lot of interaction with these devices. When this happens, DOS is interrupted many times. Each time DOS is interrupted, it must find a place in memory to store what it was doing, so that when it completes its interrupt it can retrieve this information and continue. Memory spaces used in this fashion are referred to as **stacks.** If there are too many interrupts, DOS may run out of stack space, and the following error message will appear:

```
Fatal: Internal Stack Failure, System Halted
```

When this message appears, you need to increase the number of stacks in your CONFIG.SYS file. The syntax is

STACKS = Number,Size

where: Number = the number of stacks to be used. (The values range from 0 to 64.)
Size = the size of each stack. (The values range from 0 to 512 bytes.)

For example, the entry

```
STACKS = 8,512
```

in your CONFIG.SYS file results in eight stacks of 512 bytes each.

DOS 6.0 and later versions allow function keys F5 and F8 to skip or single-step through CONFIG.SYS and AUTOEXEC.BAT. This is very helpful when you are troubleshooting a computer, because the specific hardware or software problem can be detected quickly through the process of elimination.

DEFINITION OF BATCH PROCESSING

Up to this point, when you entered DOS commands, you did so by entering one command at a time, pressing the RETURN key, waiting for the command to act, and following up with an action of your own. This is called **interactive processing.** Simply stated, interactive processing consists of the user typing a command at the system prompt and then waiting for the command to complete before proceeding.

DOS allows you to develop a file that will contain all the DOS commands you need. All you have to do is activate the file (by typing its name); the file will then process each DOS command you entered into it, just as if you had entered these commands from the keyboard. For example, you could have a file that you named DOIT that would automatically clear the screen and display the directory. This type of processing, where the commands are obtained from a file rather than from the user at the keyboard, is called **batch processing.** The difference between interactive and batch processing is shown in Figure 7.4.

FIGURE 7.4 Difference between interactive and batch processing

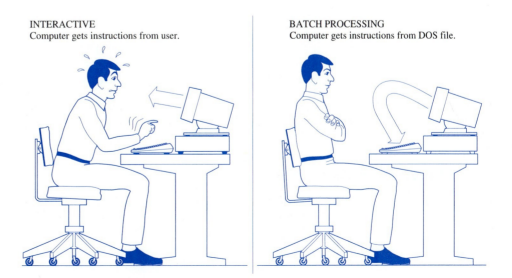

INTERACTIVE
Computer gets instructions from user.

BATCH PROCESSING
Computer gets instructions from DOS file.

BATCH FILES

A DOS batch file allows you to group DOS commands in a file. This file must have the extension .BAT (meaning "batch"). When you type the name of this file (without the extension), DOS automatically opens the file and executes each DOS command just as if it were entered by you from the keyboard. As an example, you could create a batch file called TEST.BAT that would clear the screen and display the current directory, as follows:

1. First, create the batch file using your preferred editor. Include these two commands in TEST.BAT:

   ```
   CLS
   DIR
   ```

2. The new file now exists on your disk:

   ```
   TEST.BAT
   ```

3. To activate the file, simply enter the file name (without the .BAT extension):

   ```
   A> TEST
   ```

4. DOS automatically opens the file, looks at the first DOS command (CLS), executes it, looks for the next DOS command (DIR), and executes it.

Effectively, you have created a new external DOS command. That is, every time you enter the word TEST, the screen will automatically be cleared and the directory will be displayed. This DOS feature gives you tremendous power to customize any computer system for a variety of different needs.

Figure 7.5 shows the execution of a batch file that clears the screen and changes the DOS prompt to display the current time. Note from the figure that, by default, DOS displays each of the commands it uses. Do you know why the CLS command is not shown?

BATCHES WITHIN BATCHES

You can evoke a batch file from another batch file. For example, suppose you have a batch file called FIRST.BAT that contains the following:

```
TIME
DATE
```

This batch file displays the time (TIME) and the date (DATE). You can create another batch file, called DOIT.BAT, that first clears the screen, displays the directory, and then calls the batch file FIRST.BAT. DOIT.BAT should contain the following commands:

```
CLS
DIR
FIRST
```

If the second batch file is not the last DOS command, you must use a special DOS command to activate it. For example, if you want the file DOIT.BAT to activate FIRST.BAT before showing the directory, you have to write DOIT.BAT as follows:

```
CLS
CALL FIRST
DIR
```

FIGURE 7.5 Example of DOS batch file execution

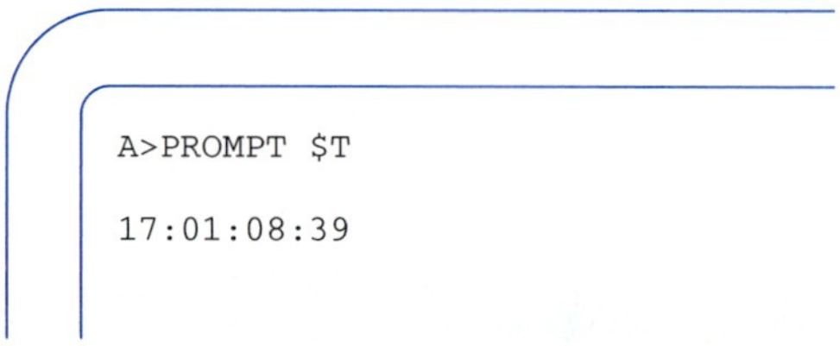

```
A>PROMPT $T

17:01:08:39
```

The CALL command allows the original batch file (DOIT) to resume processing when the second batch file (FIRST) completes.

THE REM COMMAND

As you will soon discover, you can create very powerful batch files that can perform many complex tasks. Because of this, you may want to get into the habit of using the REM command to put *remarks,* or comments, in your batch files that explain what the file is to do. You will find this useful, especially if you come back to a file after a time and have forgotten your original intention in creating it. This command can also be used to display messages to the screen. The DOS command for doing this is

$$\text{REM [message]}$$

where: message = an optional character string that can contain up to 123 characters

A batch file that activates several .EXE files may use remarks like this:

```
CLS
REM Activating ACCOUNTS
ACCOUNTS.EXE
REM Activating PROCESS
PROCESS.EXE
```

When creating a batch file, it is a good habit to state what the file is to do, the name of the person who made the batch file, and when it was created. Look at the contents of the batch file called MYBATCH.BAT:

```
REM MYBATCH.BAT
REM
REM File created to activate accounting process.
REM
REM Developed by A. Student, Feb. 24, 1999
REM
REM Activating ACCOUNTS
ACCOUNTS.EXE
REM Activating PROCESS
PROCESS.EXE
```

As always, good documentation is essential. REM commands provide this important capability.

THE ECHO COMMAND

DOS provides a command that will turn on or turn off characters being displayed on the screen as a batch file is processed. The command has the form

$$\text{ECHO[ON/OFF/message]}$$

where: ON = turn on screen displays
        OFF = turn off screen displays
   message = an optional string of characters that will appear on the screen

Note: ECHO is ON by default.

The ECHO command may also be used to display information directly to the screen. For example, the following batch program displays information about the program developer.

```
ECHO OFF
CLS
ECHO           ***************************
ECHO           *      BATCH PROGRAM      *
ECHO           *   Brad's Repair Service *
ECHO           *      Phone: 339-3115    *
ECHO           ***************************
```

FIGURE 7.6 Using the ECHO message feature

```
***************************
*      BATCH PROGRAM      *
*   Brad's Repair Service *
*      Phone: 339-3115    *
***************************

A>
```

Execution of this program results in the output shown in Figure 7.6.

DOS 3.3 and later versions allow you to use the @ before a command to suppress its display during execution. Thus, a batch file that contains

```
@ACCOUNTS.EXE
@PROCESS.EXE
```

would not display these file names during processing.

THE PAUSE COMMAND

There may be times when it is necessary to halt the batch process temporarily. A halt may be needed to give the user a chance to insert or change a disk or to remind the user that something is about to happen. This feature is provided for by the DOS PAUSE command:

PAUSE [message]

where: message = an optional string of characters to be displayed on the screen

An example of the use of the DOS PAUSE command in a batch file (called DOACCT.BAT) is as follows:

```
CLS
REM Activating ACCOUNTS
ACCOUNTS.EXE
PAUSE Be sure to insert disk in drive B:
REM Activating PROCESS
PROCESS.EXE
```

Activation of the program produces the output shown in Figure 7.7.

FIGURE 7.7 Using the PAUSE command

```
A>REM Activating ACCOUNTS

A>PAUSE Be sure to insert disk in drive B:
Strike a key when ready...
```

(Press the RETURN key.)

```
A>REM Activating ACCOUNTS

A>PAUSE Be sure to insert disk in drive B:
Strike a key when ready...

A>REM Activating PROCESS

A>
```

The user may continue batch processing by pressing any key or not continue it by a CTRL-BREAK (holding down the CTRL key at the same time as the BREAK key). When you do a CTRL-BREAK, the following message will appear:

```
Terminate batch job (Y/N)
```

To terminate the batch process, press Y; otherwise, press N.

BATCH PARAMETERS

DOS provides a way of allowing the user to decide how the batch file will operate without having to know much about DOS or anything about the batch process. This is done by making a batch file to which a user can pass instructions. To give you an idea of how this works, consider a batch file called WATCH.BAT, which displays the user's instructions:

```
A> WATCH Testword
```

The batch program would now display

```
This is an example of
a batch parameter.
The parameter you entered
was Testword for the
batch file called WATCH.
```

The actual contents of the batch file that does this are as follows:

```
ECHO OFF
ECHO This is an example of
ECHO a batch parameter.
ECHO The parameter you entered
ECHO was %1 for the
ECHO batch file called %0.
```

A **parameter** is nothing more than something that is passed on to the batch program itself. You may pass up to three parameters (the %0 parameter will always contain the name of the batch file). These parameters are identified by DOS as %1, %2, and %3. Consider the following batch file, called ALLTHREE.BAT:

```
ECHO OFF
ECHO The first parameter is %1
ECHO The second parameter is %2
ECHO The third parameter is %3
```

If you activated this batch file as

```
A> ALLTHREE Look at this
```

the result would be

```
The first parameter is Look
The second parameter is at
The third parameter is this
```

The ability to pass parameters to a batch file is a very powerful feature, as you will soon see.

THE IF COMMAND

DOS provides a method of allowing decision making in batch files by the use of the IF command. For example, one of the common uses of the IF command is in the following form:

IF EXIST file.ext DOS Command

where: file.ext = the name and extension of a file
DOS Command = a legal DOS command

As an example of the usefulness of this command, suppose you want to have the batch file (called CHECK.BAT) check to see if a specified program (called MYTEXT.TXT) is available in the current directory and, if it is, to have its contents displayed on the screen. The batch file CHECK.BAT would contain

```
ECHO OFF
REM Batch file to check for program
REM MYTEXT.TXT and display its
REM contents to the screen if it
REM exists.
IF EXIST MYTEXT.TXT TYPE MYTEXT.TXT
```

When executed, the program displays the contents of the file MYTEXT.TXT. If the file does not exist, the contents are not displayed. You will learn more about the DOS IF command as other commands are presented.

THE NOT COMMAND

You can modify a DOS command to do the reverse of an IF condition by using the NOT command. As an example, suppose you have a batch file called REDO.BAT that checks to see if a file exists and, if it does exist, renames it. The batch file REDO.BAT would contain

```
ECHO OFF
REM Batch file to check if a
REM program exists and, if it
REM does, to rename it.
IF NOT EXIST %1 ECHO File %1 does not exist.
IF EXIST %1 RENAME %1 %2
```

To use the batch file, proceed as follows. Suppose the user is in the root directory and wants to change the name of a file called OLD.TXT to NEW.TXT. All that is necessary is

```
A> REDO OLD.TXT NEW.TXT
```

The program first checks whether the file OLD.TXT exists and then automatically renames it NEW.TXT, all without the user ever having to change the directory.

THE GOTO COMMAND

DOS provides a method of allowing a batch file to take a different set of instructions or to repeat a set of instructions. The action of the DOS GOTO command is shown in Figure 7.8.

The form of the GOTO command is

>GOTO Labelname

where: Labelname = the place where commands will continue. (It is distinguished from other DOS commands by starting with a colon.)

An example using labels is shown here for the batch file FINDIT.BAT. This batch program looks for the existence of a file and lets the user know if the file exists.

```
ECHO OFF
REM File Locator Program
IF EXIST %1 GOTO OK
IF NOT EXIST %1 GOTO NOGOOD
:OK
ECHO File %1 exists.
GOTO EXIT
:NOGOOD
ECHO File %1 does not exist.
:EXIT
```

Figure 7.9 illustrates the action of this batch program.

FIGURE 7.8 Action of the GOTO command

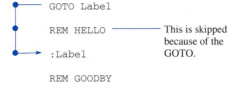

FIGURE 7.9 Action of FINDIT.BAT

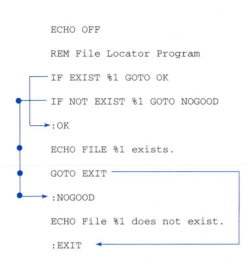

COMBINING THE IF AND GOTO COMMANDS

Using the DOS GOTO makes the DOS IF even more powerful. The DOS IF can be used to compare two strings, as shown:

IF String1 == String2 DOS Command

Note: Be sure to use the double equal sign (==) for this command.

What this means is that if String1 is identical to String2, the DOS command (DOS Command) that follows will be executed. As an example, consider the batch file called YOURID.BAT that checks for a user ID. If the user gives the correct ID, the contents of a file are displayed; if not, the screen is cleared.

```
ECHO OFF
REM ID check program,
IF '%1' == 'HOWDY' GOTO YES
CLS
ECHO You are an imposter!
GOTO EXIT
:YES
TYPE SECRET.TXT
:EXIT
```

Observe that strings (a *string* is a sequence of keyboard characters *strung* together) must be enclosed by single quotation marks. To implement this program, enter

```
A> YOURID HOWDY
```

This command causes the contents of the text file SECRET.TXT to be displayed on the screen (assuming that the file exists).

AUTOEXEC.BAT FILES

DOS provides a method of automatically activating a batch file when the system is first booted. What happens is that each time the system boots, DOS examines the root directory for a file named

```
AUTOEXEC.BAT
```

If a file by the name of AUTOEXEC.BAT is found, DOS executes all the commands the file contains. If it does not find the file, DOS then issues the DATE and TIME commands.

An AUTOEXEC.BAT file is exactly the same as any other batch file, with the exception that it is automatically executed each time the system is booted. All the rules you have discovered about batch files apply to the AUTOEXEC.BAT file as well.

An AUTOEXEC.BAT file might clear the screen, change the DOS prompt so it displays the root directory, and then display a directory listing. You could create such a file with the following commands:

```
ECHO OFF
REM Batch file for creating
REM the DOS Prompt and listing
REM the directory.
PROMPT $N$G
CLS
DIR
```

Frequently users want particular application programs to be activated automatically every time they turn on their computers. For example, a user may want an accounting program, a word-processing program, or a listing of computer games to appear. This is the place to use an AUTOEXEC.BAT file. The following AUTOEXEC.BAT file sets the DOS prompt to the current date and activates an accounting program in the file ACCOUNT.EXE.

```
ECHO OFF
PROMPT $D
ACCOUNT.EXE
```

DOS 6.0 and later versions allow the use of two function keys to control what happens during boot time. If F5 is pressed at the beginning of the boot process, DOS skips of the AUTOEXEC.BAT file entirely (as well as CONFIG.SYS). Booting up with the F5 key is useful for times when a new software package must be installed into a machine that does not contain any *extras* (such as a CD-ROM or sound card), and for certain diagnostic programs.

Another feature DOS 6.0 provides at boot time is activated when the F8 key is pressed. This feature allows the user to *single-step* through the CONFIG.SYS and AUTOEXEC.BAT files one line at a time. DOS prompts the user with

```
[Y/N]
```

after each line, allowing custom boot sequences depending on how the user answers each prompt.

Furthermore, DOS 6.0 provides a mechanism that allows the CONFIG.SYS and AUTOEXEC.BAT files to be written in such a way that a **menu** of boot options is provided at boot time. This allows the user to pick a particular machine configuration (for example, CD-ROM only, Network only, or CD-ROM and Network) as needed, without having to constantly modify AUTOEXEC.BAT for each configuration.

CHANGING THE DOS PROMPT

Normally, the DOS prompt is just a small flashing line. It indicates the position on the screen where the next keyboard entry will appear. However, DOS allows you to modify the prompt to display useful information. To change the DOS prompt, use the command

PROMPT PromptString

where the meaning of PromptString is given in Table 7.2. (*Note:* The PromptString letters may be either upper- or lowercase.)

Using the command

```
A:\> PROMPT $P
```

TABLE 7.2 Meanings of PromptString letters

PromptString Letter	Causes the DOS Prompt to Display
$b	:
$d	The current date
$e	The ESC character
$h	Backspace
$g	>
$l	<
$n	Current disk drive
$p	Current directory
$q	=
$t	Current time
$v	DOS version number
$$	$
$_	Carriage-return line-feed combination

results in the following DOS prompt (assuming you are in the directory called MYDIRECT):

`A:\MYDIRECT`

The command

`A:\> PROMPT [$P]`

causes the prompt to display

`[A:\MYDIRECT]`

Entering

`A:\> PROMPT $P $G`

produces

`A:\MYDIRECT>`

To restore the prompt back to its original condition, simply enter

`PROMPT`

This produces the *standard* DOS prompt

`A>`

It is common to have the DOS prompt modified in order to display the current directory (as used throughout this exercise).

TROUBLESHOOTING TECHNIQUES

It is convenient to create small batch files that help make life a little easier inside the DOS world. For example, here are the contents of a batch file called TY.BAT:

`TYPE %1 | MORE`

This one-line batch file uses an input parameter to represent the name of a file that gets TYPEd out one page at a time by piping the output of the TYPE command into the MORE command.

The TY batch file is used like this:

`TY README.TXT`

The advantage for the user is never having to worry about a large file zipping by on the display faster than it can be read.

Another time-saving batch file is DW.BAT, which also contains a single statement:

```
DIR %1 /W /P
```

This displays a wide directory listing for the supplied parameter, pausing if necessary.

Any DOS command you use often is a candidate for use in a simple batch file. Try creating your own time-saving batch file.

SELF-TEST

The following self-test is designed to help you check your understanding of the background material presented in this exercise.

True/False

Answer *true* or *false*.

1. When DOS searches for a file, it starts with the current active directory.
2. DOS will search other directories, but it must be instructed to do so in the AUTOEXEC.BAT file.
3. The DOS PATH entry is used to direct DOS to other directories when searching for a file.
4. Interactive processing means entering a command, waiting for the results, and then entering another command.
5. Batch processing means having a set of commands performed automatically by the computer without human interaction.
6. Interactive processing and batch processing are really the same thing.

Multiple Choice

Select the best answer.

7. The function of the CONFIG.SYS file is
   a. To let DOS know the maximum number of disk drives.
   b. To let DOS know whether there are other unique devices attached to the computer.
   c. To set floppy disk drive conditions.
   d. All of the above.
8. If a customer requires that CTRL-C be active for as many computer operations as possible, it is necessary for you to add which of the following to the CONFIG.SYS file?
   a. CTRLC
   b. BREAKON
   c. BREAK = ON
   d. CTRLC = ON
9. Some application programs require that the number of buffers in the system be increased. From this you can conclude that
   a. The more buffers allocated, the better.
   b. You should increase the number of buffers only to meet the needs of the application program.
   c. The number of buffers should never be increased.
   d. You cannot change the number of buffers in the system.
10. A DOS batch file must
    a. Have the extension .BAT.
    b. Use legal DOS commands.
    c. Be activated by typing in its file name without the extension.
    d. All of the above.
11. The batch file

    ```
    CLS
    CALL QP
    DIR
    ```

    will

a. Use a program called COMMAND.QP.
   b. Not execute properly.
   c. Execute another batch file QP.BAT.
   d. None of the above.
12. The ECHO command can be used to
    a. Prevent screen displays.
    b. Cause screen displays.
    c. Place messages on the screen.
    d. All of the above.

Matching

Match each command on the right with each stated action on the left.

13. DEVICE = VDISK.SYS
14. DEVICE = DRIVER.SYS /F:2
15. DEVICE = PLOTTER.SYS
16. Jump to a different place in the file.
17. The name of the batch file.
18. Hold the batch process until the user presses a key.

a. Prepares system for operating a plotter
b. Creates a virtual disk in the computer's memory
c. Sets up the system to operate a disk drive
d. None of the above.
e. PAUSE
f. :JUMP
g. GOTO
h. %0
i. %1

Completion

Fill in the blanks with the best answers.

19. The DOS _____ entry in the CONFIG.SYS file tells DOS the name of the last drive in the system.
20. A REM entry is allowed in a CONFIG.SYS file in DOS _____ and later versions.
21. The DOS PATH is set in the _____ file.
22. A(n) _____ is a value that is passed on to the batch program.
23. The DOS _____ command is used for making decisions.
24. A(n) _____ file is automatically activated each time the system is booted.

Open-Ended

Answer the following questions as they apply to the following batch file:

```
ECHO OFF
CLS
ECHO        *****************************
ECHO        *     MY SPECIAL BAT FILE    *
ECHO        *             By             *
ECHO        *      Howard's PC Service   *
ECHO        *         Phone 428-5712     *
ECHO        *****************************
REM Special batch file developed by Howard.
IF '%1' = '' GOTO PLACE1
TYPE %1 > PRN
GOTO EXIT
:PLACE1
ECHO A file by that name does not exist.
GOTO EXIT
:EXIT
```

25. Briefly describe what this batch file does.

26. Are there any lines (outside of ECHO strings) that could be omitted from the program without affecting its operation? If so, which one(s)?
27. What feature is missing from the batch file?

FAMILIARIZATION ACTIVITY

1. Boot your system, making sure DOS is loaded.
2. Insert your blank formatted disk in drive A: and create a text file that is called MYTEXT.TXT in the root directory.
3. Create a directory called PLACEIT. Do not place any files in this directory yet.
4. Create the following batch file, called TRANSFER.BAT, in your root directory.

```
ECHO OFF
CLS
ECHO.
ECHO.
ECHO            *******************************
ECHO            *    FILE REDIRECTION PROGRAM  *
ECHO            *       Brad's Repair Service  *
ECHO            *         Phone 339-3115       *
ECHO            *******************************
ECHO.
ECHO.
REM Batch file, copy a program to another directory.
REM Created 3/18/99
REM By: A student.
IF '%1' == '' GOTO MESSAGE1
IF '%2' == '' GOTO MESSAGE1
IF EXIST %2 GOTO MESSAGE2
ECHO Copy [ %1 ] to [ %2 ] and delete [ %1 ].
COPY %1 %2
IF NOT EXIST %2 GOTO ERROR
REM
ECHO ON
PAUSE About to delete source file [ %1 ].
ECHO OFF
DEL %1
GOTO EXIT
:ERROR
ECHO Output file name [ %2 ] not correct.
ECHO Program terminated...
GOTO EXIT
:MESSAGE1
ECHO Must enter the name of a source and target file.
ECHO Program terminated...
GOTO EXIT
:MESSAGE2
ECHO Target file already exists.
ECHO Program terminated...
:EXIT
PAUSE
CLS
DIR
```

The purpose of this batch file is to allow the user to copy a file from any directory to another directory and erase the original file.

5. To test the operation of the batch file, enter the following:

```
TRANSFER MYTEXT.TXT \PLACEIT\NEWTEXT.TXT
```

6. When the batch process has been completed, the file MYTEXT.TXT should be gone from the root directory and should be found inside the directory called PLACEIT under the new name NEWTEXT.TXT. Verify that this is true. If not, check your batch file carefully for any errors you may have made when entering the program.
7. Create another text file in the root directory called MYTEXT.01.
8. Make an AUTOEXEC.BAT file that contains the following:

   ```
   ECHO OFF
   CLS
   PROMPT $D
   DIR
   ```

9. With your new AUTOEXEC.BAT file in the root directory, reboot your system. The screen should automatically become cleared, the cursor should display the current date, and the directory of the disk should be displayed. If this doesn't happen, check your AUTOEXEC.BAT file for any errors you may have made when entering the program.

QUESTIONS/ACTIVITIES

Questions

1. What is the purpose of the CONFIG.SYS file?
2. Explain how batch processing and interactive processing differ.
3. How does DOS distinguish a batch file from others?
4. What is the purpose of the ECHO command?
5. State how the @ is used.
6. Explain what a *parameter* is. Give an example.

Activities

1. Make a batch file that will allow the user to set the date and the time.
2. Make a batch file that will clear the screen and display only the batch files in the root directory.
3. Create an AUTOEXEC.BAT file that clears the screen, displays a message, changes the DOS prompt to display the current directory, and invokes CHKDSK/V to display directories and files.

REVIEW QUIZ

Within 10 minutes and with 100% accuracy,

1. Set up a DOS path.
2. Make a CONFIG.SYS file that creates a number of buffers and files as assigned by your instructor.
3. Create a batch file that will do one of the following:
   a. Clear the screen and display a message selected by your instructor.
   b. Clear the screen and display the current directory.
   c. Clear the screen and display the contents of a user-selected directory.
   d. Clear the screen and direct the contents of a user-selected file to the printer.
4. Create an AUTOEXEC.BAT file that will do one of the following:
   a. Clear the screen and display a message selected by your instructor.
   b. Clear the screen and change the DOS prompt to display either the date or current directory, as selected by your instructor.
   c. Clear the screen and call one batch file you created in item 3.

8 DOS Utilities

INTRODUCTION

The phone rang in Joe Tekk's office. It was his mother. She was running an old accounting program on her computer when it suddenly crashed and left the screen filled with "large blocky letters, much larger than normal, and no color."

Joe asked her what happened if she pressed ENTER a few times. She said that each time she presses ENTER a new DOS prompt appears but with larger-than-normal letters.

Joe told her to enter "MODE CO80" and describe what happens. She told him, "That fixed the problem. The letters are all back to normal. How did you know what it was?"

Joe explained that the problem sounded as if the display mode had changed to 40-character/line black-and-white mode. The "MODE CO80" command set her display back to 80-character/line color mode.

Later, Joe wondered how the program crash actually caused the display to enter the different display mode. After some thought, his mind wandered back to what he was doing before his mother called—reading about the SORT command in an old DOS manual.

PERFORMANCE OBJECTIVES

Upon completion of this exercise, within 10 minutes, you will be able to

1. Demonstrate the use of the redirection operators by
    a. Redirecting standard output to a file or printer.
    b. Appending standard output to an existing file.
    c. Redirecting standard input from a file.
    d. Redirecting standard input and output of commands.
2. Demonstrate the use of the following commands:
    a. SORT
    b. MORE
    c. FIND

BACKGROUND INFORMATION

Knowledge of software has become a major component of computer usage. It's important to know how to use the various DOS commands, and even more important to know the capabilities of these commands and the concepts they represent. As time passes, the specific commands may change, but the concepts will stay the same.

The material in this exercise introduces some of the DOS **utilities.** You can think of a utility as a program or computer function that assists the user in some operation of the computer.

STANDARD INPUT AND OUTPUT

Figure 8.1 shows a computer system and what is referred to as **standard input** and **standard output.** As shown in the figure, the computer normally gets its input from the keyboard and normally sends its output to the monitor. You can cause the computer to change its standard input and get its instructions from some place other than the keyboard. You can also instruct the computer to direct its output to places other than the monitor. You have already done this with disk files, as shown in Figure 8.2.

DOS comes with built-in **redirection operators.** You have already used one of them (>) to redirect output to the printer.

DOS REDIRECTION OPERATORS

A redirection operator allows you to tell DOS to change its standard input and/or standard output to something else. The redirection operators used by DOS are listed in Table 8.1.

Figure 8.3 shows some of the uses of the redirection operators.

DOS OUTPUT REDIRECTION

As indicated in Table 8.1, the DOS redirect output operator > instructs DOS to redirect its output from the monitor to somewhere else. As an example, you can use the > operator to direct output to a DOS file rather than to the monitor. You can do this by entering

```
A> DIR > DIRFILE.TXT
```

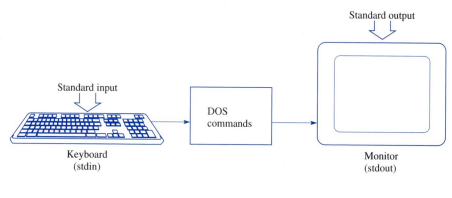

FIGURE 8.1 Concept of standard input and standard output

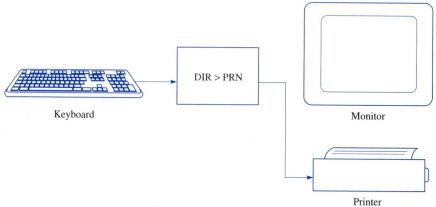

FIGURE 8.2 Redirecting input and output with DOS files

TABLE 8.1 Standard redirection operators

Name	DOS Symbol	Action
Redirect Output	>	Redirects output from monitor to somewhere else.
Append Output	>>	Redirects output from monitor and appends it to an existing file.
Redirect Input	<	Redirects input from the keyboard to input from somewhere else.
DOS Pipe	|	Redirects both output and input at the same time.

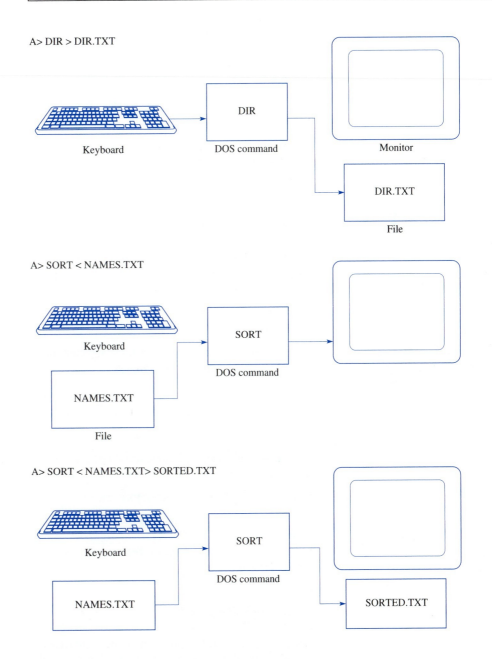

FIGURE 8.3 Some applications of the redirection operators

Recall that you can also use the redirect output operator to send output to the printer:

```
A> DIR > PRN
```

This causes the directory to be printed on the printer instead of being displayed on the monitor. PRN is a standard DOS device name meaning *printer*.

FIGURE 8.4 Difference between the > and >> operators

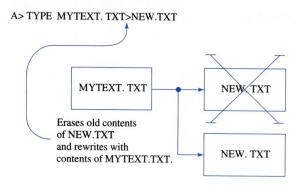

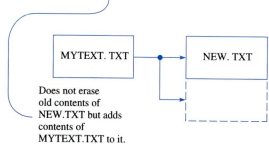

By using the DOS append output operator >>, you can append information to an existing file. Figure 8.4 shows the difference between the DOS > and >> redirection operators.

As an example of the use of the DOS >> operator, suppose you need to merge a group of separate text files into one text file. Given that the original group of text files is

TEXT.01
TEXT.02
TEXT.03

you can place all the text in these three separate files into one text file as follows:

```
A> TYPE TEXT.01 > ALLTEXT.TXT
```

This command creates a new file called ALLTEXT.TXT and copies the contents of TEXT.01 into it. Next, to merge the contents of TEXT.02, enter

```
A> TYPE TEXT.02 >> ALLTEXT.TXT
```

This command places the contents of TEXT.02 into ALLTEXT.TXT without removing the previous contents there from TEXT.01. Next issue the following:

```
AA> TYPE TEXT.03 >> ALLTEXT.TXT
```

Now the contents of all three text files, TEXT.01, TEXT.02, and TEXT.03, are contained in one file called ALLTEXT.TXT.

To show an application of the redirect input operator <, the DOS utility program SORT.EXE is presented.

THE SORT COMMAND

DOS provides a utility program called SORT.EXE. This utility allows you to sort redirected input from the highest to the lowest or the other way around. You can sort information based on a specific column. The syntax of the SORT command is

SORT[/+N][/R]

where: /+N = the column number that the sort is to use
   /R = the indicator that DOS is to perform the sort in descending order

As an example of the DOS SORT, suppose you have a file called LIST.01, which contains the names of students and their grades.

```
Tom       Baker       98
Jim       Rober       89
Mary      Avalar      98
Jerry     Commings    87
Alice     Cooper      72
Tom       Anderson    68
Jerry     Brown       83
Harry     Downs       90
John      Evens       95
```

The first letters of the first names are in column 1 of the list. The first letters of the last names are in column 10, and the first numbers of the grades are in column 21.

To sort the list by the first letter of the first name, you enter

```
A> SORT < LIST.01
```

Note the use of the redirect input operator, <. This tells DOS to direct the input to the SORT utility from the DOS file called LIST.01. The result of the sort is shown in Figure 8.5.

As shown in Figure 8.5, the sort has been done on the first names. This is because the default column for the sort is column 1. To sort the list by the last names, use

```
A> SORT /+10 < LIST.01
```

The modifier /+10 tells DOS to start the sort at column 10 of the list (which is where the first letters of the last names begin). The resulting sort is shown in Figure 8.6.

To sort the list by grades, simply start the sort on column 21 (the column of the first numbers of the grades):

```
A> SORT /+21 < LIST.01
```

The results are shown in Figure 8.7.

Note from Figure 8.7 that the grades are listed in ascending order. To change this, issue

```
A> SORT /+21 /R < LIST.01
```

FIGURE 8.5 Result of the SORT < LIST.01 command

```
A:\>SORT < LIST.01
Alice     Cooper      72
Harry     Downs       90
Jerry     Brown       83
Jerry     Commings    87
Jim       Rober       89
John      Evens       95
Mary      Avalar      98
Tom       Anderson    68
Tom       Baker       98

A:\>
```

FIGURE 8.6 Result of the SORT /+10 < LIST.01 command

```
A:\>SORT /+10 < LIST.01
Tom       Anderson    68
Mary      Avalar      98
Tom       Baker       98
Jerry     Brown       83
Jerry     Commings    87
Alice     Cooper      72
Harry     Downs       90
John      Evens       95
Jim       Rober       89

A:\>
```

FIGURE 8.7 Result of the SORT /+21 < LIST.01 command

```
A:\>SORT /+21 < LIST.01
Tom       Anderson    68
Alice     Cooper      72
Jerry     Brown       83
Jerry     Commings    87
Jim       Rober       89
Harry     Downs       90
John      Evens       95
Tom       Baker       98
Mary      Avalar      98

A:\>
```

It's important to note that the SORT utility sorts according to the ASCII value of the number (refer to the ASCII Table in Appendix A) and not the numerical value of the number. This means that a value of 100 is treated as a *smaller* value than 99 because the ASCII value of 1 is smaller than the ASCII value of 9.

THE MORE COMMAND

The MORE command gets input from standard input and displays it, one screen at a time, to standard output, as shown in Figure 8.8. This DOS utility is a program called MORE.COM. The syntax of the MORE command is

$$\text{MORE} < \text{filename}$$

or

$$\text{DIR | MORE}$$

where: filename = the name of the file to be viewed one screen at a time
        DIR | MORE = the directive allowing the directory to be viewed one screen at a time. This routes the output of the directory to the input of the MORE command through a "pipe" (indicted by the | symbol).

An alternative way of viewing a text file with MORE is to use this command:

$$\text{A> TYPE FILENAME | MORE}$$

which causes the output of the TYPE command to be piped to the MORE utility.

FIGURE 8.8 Concept of the MORE command

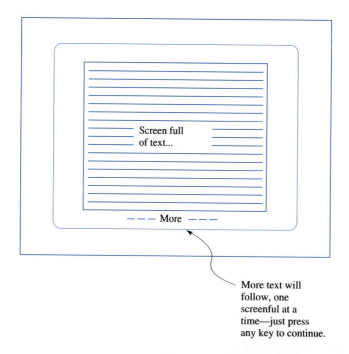

THE FIND COMMAND

The FIND command locates and displays each occurrence of a string within a file. This useful command will help you find in which file a particular string of text is located. Suppose, for example, you have several different text files, each containing a list of employee names. Suppose also that you need information about a particular employee and don't remember in which file the employee's name is located. You could search through each file yourself, or you could let the MORE command do it for you automatically.

The syntax of the FIND command is

$$FIND[/C][/N][/V]\text{"String"}Filespec[\ldots]$$

where: /C = information telling DOS to display the number of occurrences of the search string in the file(s)
/N = information telling DOS to display the number of each line containing the search string
/V = information telling DOS to display all of the lines that do not contain the search string
"String" = the search string for which you are looking

Figure 8.9 demonstrates the action of the FIND command.

SETTING FILE ATTRIBUTES USING ATTRIB

There are several important characteristics of a DOS file. These characteristics have applications when it comes to write-protecting files. Write-protecting a file means fixing it so that its contents may be read but not changed. Another characteristic of a DOS file is its ability to remember whether it has been changed since a backup copy of it was last made. All of these file characteristics have important applications in troubleshooting microcomputer systems. As a computer user, you will be the one responsible for knowing what these special file characteristics mean and how to apply them. This exercise will give you the necessary background to help you meet that responsibility.

ATTRIBUTE OF A FILE

An **attribute** of a DOS file is a software assignment to that file that instructs it to behave in a specified manner. Table 8.2 lists the attributes of a DOS file that will be presented in this exercise.

You can set the attribute of a file from the following DOS command:

$$ATTRIB\ [\pm R][\pm A][\pm H][\pm S][drive:]pathname[/S]$$

where: +R = sets the file to read only
−R = removes the read-only status of the file
+A = sets the file to archive status
−A = removes the archive status of the file
+H = sets the file to hidden
−H = removes the hidden status of the file
+S = sets the system attribute of the file
−S = removes the system attribute of the file
/S = directs ATTRIB to process all files that reside in a given directory

FIGURE 8.9 Action of the FIND command

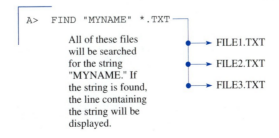

TABLE 8.2 DOS file attributes

Attribute	Meaning
Read Only	The contents of the file can be read but cannot be modified, nor can the file be erased from the disk.
Archive	The file has not been backed up since it was last modified.
Hidden	The file is hidden and will not show up in a DIR listing.
System	The file is a system file. System files are hidden by default.

READ-ONLY FILES

Suppose you have an important text file that you want to protect from modification or erasure. Using the DOS ATTRIB command, you can change the file to a **read-only file.** If the name of the file is MYFILE.TXT, you enter

```
A> ATTRIB +R MYFILE.TXT
```

The file MYFILE.TXT will now appear exactly the same on the directory as before. You can look at its contents with the TYPE command, but you cannot change its contents with the EDIT or any other text processor; you also cannot erase the file. Figure 8.10 illustrates the characteristics of write-protected and non-write-protected files.

If later you decide that you want to change the file contents (or erase the file), you remove its read-only status:

```
A> ATTRIB -R MYFILE.TXT
```

ARCHIVE FILES

When a file is first created, DOS automatically labels it an **archive** file. Anytime a file is modified, DOS also automatically labels it an archive file. When a file is copied from one place to another, the copied file is automatically labeled an archive file by DOS. This is done as an aid in backing up files from one disk to another and backing up only those files on the original disk that either are new or have been recently modified.

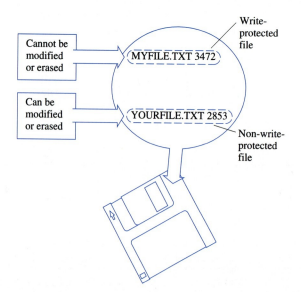

FIGURE 8.10 Characteristics of write-protected and non-write-protected files

Later in this exercise, you will be introduced to a DOS command that selects files (such as only those that are archive files) to be copied from one place to another. For now it is important that you know how to set and undo the archive status of a file.

To set a file (say, MYFILE.TXT) to be an archive file, use

`A> ATTRIB +A MYFILE.TXT`

The file is now an archive file. It still appears the same on your disk directory, but—as you will later see—DOS knows its status.

To remove the archive status of an archive file (again, MYFILE.TXT), use

`A> ATTRIB -A MYFILE.TXT`

Now the file no longer has the status of an archive file.

HIDDEN FILES

Often, for reasons of security or privacy, you may want to prevent a file or files from showing up in the output of a DIR command. DOS allows you to protect files in this manner by setting its **hidden attribute.** To turn MYFILE.TXT into a hidden file, you use

`A> ATTRIB +H MYFILE.TXT`

After the hidden attribute is set, the file will no longer show up in directory listings. It is also impossible to copy or delete a hidden file. This is a valuable feature to keep in mind, and helps provide a measure of security to DOS's file system.

To change a hidden file back to its normal, visible status, use the command

`A> ATTRIB -H MYFILE.TXT`

Note that a directory may also be hidden through the use of ATTRIB.

SYSTEM FILES

The last file attribute is reserved for **system files.** A system file is a file used exclusively by DOS, such as IO.SYS and MSDOS.SYS, and is hidden by default. System files cannot be copied or deleted. Normally, you have no reason to set the system attribute of a file, but you can do so by means of the command

`A> ATTRIB +S MYFILE.TXT`

CHECKING THE FILE STATUS

You can check the status of any DOS file by issuing the following command:

`A> ATTRIB [filename]`

As an example, to check the attribute of the file MYFILE.TXT, you would enter, from the DOS prompt,

`A> ATTRIB MYFILE.TXT`

Figure 8.11 illustrates the different results from this command and their meanings. As you can see, the attributes of a file can easily be determined.

THE XCOPY COMMAND

The XCOPY command is a powerful copy command. Unlike the COPY command or the DISKCOPY command, XCOPY has many special features. The syntax for this command is

XCOPY[Filespec1][Filespec2][/A][/D:mm-dd-yy][/E][/M][/P][/S][/V][/W]

FIGURE 8.11 Possible results of the ATTRIB command

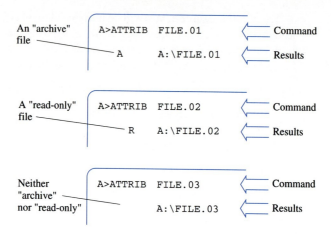

where: Filespec1 = the source file(s) to copy
Filespec2 = the destination file name(s) for the copied file(s)
/A = copies only archive files. (Does not change the archive status of the original files.)
/D:mm-dd-yy = copies only files that have the date mm-dd-yy or later
/E = creates subdirectories on the target disk that have the same contents as those on the source disk
/M = copies only archive files and changes the status of the copied files. (This modifier allows the XCOPY command to be used as a file backup command.)
/P = causes XCOPY to present a "(Y/N?)" prompt, letting you decide whether you want to create each target file
/S = causes XCOPY to copy directories and lower-level subdirectories. (When this modifier is omitted, only the contents of the current directory are accessed.)
/V = forces XCOPY to verify that each file that is copied is an exact match of the original file
/W = makes XCOPY wait before starting the copying process. The following message is

Press any key when ready to start copying files.
(This gives you a chance to change disks if necessary.)

Figure 8.12 illustrates the effects of the XCOPY command with the use of various modifiers.

COPYING A SET OF FILES FROM A HARD DISK

The XCOPY command is a convenient command to use for copying the contents of the hard disk (C: drive) or a large subdirectory to a floppy disk. To do this, modify the attributes of the files on your hard drive with

```
A> ATTRIB +A C:*.* /S
```

Next, use the XCOPY command:

```
A> XCOPY C:*.* B: /M
```

The files will then be copied from the hard drive to the floppy disk in drive B:. When the floppy disk becomes full, the following message is displayed:

```
Insufficient disk space
        nn files copied
A>
```

FIGURE 8.12 XCOPY command examples

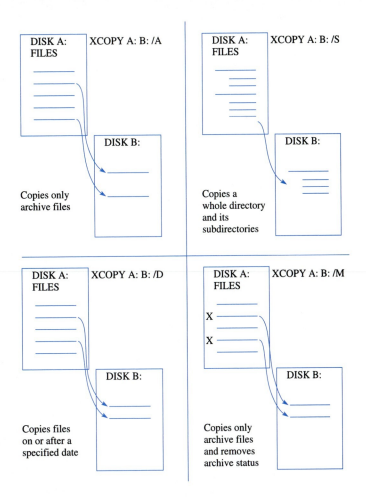

Now all you need do is remove the disk from drive B:, insert another formatted disk, and issue the same command:

```
A> XCOPY C:*.* B: /M
```

XCOPY will now pick up where it left off. It will copy only those files that it has not yet completely copied to the previous backup disk (in drive B:). These are the files that still have the archive attribute. Those files already copied have had their archive attributes turned off (because of the /M modifier used in XCOPY).

THE MODE COMMAND

DOS provides a command that allows you to set the characteristics of communications adapters—the color/graphics adapter as well as the system printer. In terms of your video display, the MODE command gives you the option of displaying either 40 or 80 characters across the width of the screen on the monitor. It also allows you to enable or disable the ability of your monitor to display color as well as set character alignment on the screen. It is important to note that the MODE command works only on color graphics or extended graphics adapter systems.

As you will see, the MODE command also allows you to change the default character sizes on your system printer as well as how printer "timeouts" are handled. You can also use the MODE command to set various values of your computer's data communication requirements. What follows are the various forms of the MODE command:

MODE [device][status]

where: device = name of specific device whose status is to be displayed
status = a switch required only for requesting the status of a redirected parallel printer

This command, by itself, displays the status of all the devices installed in your computer system.

```
C> MODE

Status for device LPT1:
----------------------
LPT1: not rerouted
Retry=NONE

Code page operation not supported on this device

Status for device LPT2:
----------------------
LPT2: not rerouted

Status for device LPT3:
----------------------
LPT3: not rerouted

Status for device CON:
----------------------
Columns=80
Lines=25

Code page operation not supported on this device

Status for device COM1:
----------------------
Retry=NONE

Status for device COM2:
----------------------
Retry=NONE
```

A **code page** is simply a table that defines the character set you are using. For example, a character set can be thought of as the set of all letters, numbers, and symbols—including accent marks. These characters are translated from the code page table and then are used by the keyboard, screen, and printer attached to the computer system. There are actually two types of code pages. One is a hardware code page, which is usually a ROM chip built into the device. As an example, a printer for use in Germany has a hardware code page built into it for the particular letters, symbols, and accent marks used in the German language. The other is a software code page, which is provided by DOS. The message "Code page operation not supported on this device" simply means that the code page cannot be used with the indicated device.

CONTROLLING THE VIDEO DISPLAY WITH THE MODE COMMAND

The MODE command can be used to affect your monitor display. The form of the MODE command is:

MODE[N],Shift[,T]

where: N = the video display mode (see Table 8.3)
Shift = a command that specifies the direction you want to move the display on the monitor: L for left; R for right
T = a command that lets MODE display a test pattern across the top of the screen to help you in character alignment

TABLE 8.3 Video display types

Value of N	Meaning
40	Sets the video display to 40 columns per line.
80	Sets the video display to 80 columns per line.
BW40	Sets the video display to 40 columns per line while disabling color. (You can think of the BW as black and white.)
BW80	Sets the video display to 80 columns per line while disabling color.
CO40	Sets the video display to 40 columns per line with color enabled.
CO80	Sets the video display to 80 columns per line with color enabled.
MONO	Sets the output for a monochrome monitor.

Table 8.3 lists the various video display types that can be used with the MODE command. For example, the DOS command

```
A> MODE BW40
```

causes the monitor screen to display 40 columns in monochrome.

SETTING PRINTER CHARACTERISTICS WITH MODE

You can also set the characteristics of your printer using the MODE command. The form of the MODE command for printer control is

MODE LPTn[:][C][[,[L]][,R]]

You can also use

MODE LPTn[COLS = C][LINES = L][RETRY = R]

where: LPTn = the printer port whose characteristics are to be set (such as LPT1, LPT2, or LPT3, where more than one printer may be attached to the system)
C = the number of characters per line. (You may have either 80 or 132 per line.)
L = the amount of vertical spacing, expressed as a number of lines per inch. (The two values are 6 and 8. The default DOS value is six lines per vertical inch.)
R = a command that specifies that the MODE command is using the COM port for a serial printer and will retry if necessary. (The following values are possible:
E: Returns a status check error of a busy port
B: Returns busy from the status check if the printer port is found to be busy
R: Returns a ready from the status check if the port is busy)

An example of using the MODE command is

```
A> MODE LPT2: 80,8,B
```

This MODE command directs the printer connected to LPT2 to print in 80 columns with eight lines per inch and to keep trying to print the file until the printer is ready to print. With this command, to stop retrying you simply press CTRL-BREAK (holding down both keys at the same time). Another way is to type the MODE printer command without the R option.

The following MODE printer command changes the number of characters per line from the default value of 80 to 132:

```
A> MODE LPT1: COLS = 132
```

You should note that not all the MODE printer commands are supported by all printer systems. Just because the printer MODE command does not work with your system does not necessarily mean there is a problem with the system. It could mean that the electronics of your printer system do not support the MODE command. The printer MODE command does not work with laser-type printers.

SERIAL PORT CONFIGURATION WITH MODE

DOS allows you to configure a serial communications adapter with the MODE command. The form of doing this is

$$\text{MODE COMn[:][B[,P[,D[,S[,R]]]]]}$$

where: n = the asynchronous communications (COM) port number
- B = the first two digits of the transmission rate (baud rate). (Acceptable values of the transmission rate are 110, 150, 300, 600, 1200, 2400, 4800, 9600, and 19,200.)
- P = a command that tells DOS what parity to use. (Acceptable values are N [for none], O [for odd parity], E [for even parity], M [for a mark], and S [for space]. E is the default value.)
- D = a command that tells DOS the number of data bits. (Acceptable values are 5, 6, 7, and 8.)
- S = a command that tells DOS the number of stop bits. (Acceptable values are 1, 1.5, and 2. Note that if the baud rate is 110, then the default value for stop bits is 2. If the baud rate is any other value, the default value is 1.)
- R = a command that tells DOS what kind of action to take on a retry of transmission. If the communications port is being used as part of a computer network, do not use this option for continuous retry. Acceptable options for this switch are:
   - E: Gives an error from a status check for a busy port; the default value
   - B: Gives a busy from a busy port status check
   - R: Gives a ready from a busy port status check
   - NONE: Do not take any retry action

The DOS default for this command is COM1, even parity, and 7 data bits.

CHANGING SERIAL AND PARALLEL OUTPUTS

You can use the MODE command to redirect a parallel output to a serial port. For example, if your system has only a serial printer, you can use the MODE command to direct the normally directed parallel output to the asynchronous port. This is done as follows:

```
A> MODE LPT1 := COM1
```

TROUBLESHOOTING TECHNIQUES

In the ever-growing world of Windows 95/98/NT, the need for DOS applications is growing smaller every day. The material presented in this exercise, and the preceding ones, is intended to help those users who are still comfortable using DOS and those who use DOS occasionally from the DOS Prompt icon in Windows 95/98/NT.

There is almost no need for an AUTOEXEC.BAT file anymore. Windows 95/98/NT does not require one, but will process the AUTOEXEC.BAT file found in the root of the C: drive. AUTOEXEC.BAT should be used primarily to load any real-mode device drivers required by older hardware.

Any special path settings or other environment-related variables can be set by adjusting the properties of the DOS Prompt icon in Windows 95/98/NT.

SELF-TEST

This self-test is designed to help you check your understanding of the background information presented in this exercise.

True/False

Answer *true* or *false*.

1. By default, DOS gets standard input from the keyboard.
2. By default, DOS directs standard output to the monitor.
3. Standard input and standard output cannot be changed by software alone.
4. DOS files can always be modified by simply editing them.
5. DOS has no way of knowing whether a file has been modified.

Multiple Choice

Select the best answer.

6. The command

    A> DIR > DIR.01

    causes
    a. A text file to be made of the directory.
    b. The contents of one file to be placed in another.
    c. The directory to be displayed as a binary file.
    d. None of the above.

7. The command

    A> SORT < NAMES.TXT

    causes
    a. The names in the file NAMES.TXT to be sorted.
    b. A display of the contents of the file NAMES.TXT sorted on the first column of the file.
    c. The contents of the file NAMES.TXT to be placed in a file called SORT.
    d. This is not a valid DOS operation.

8. The command

    A> DIR | FIND "<DIR>"

    causes
    a. The contents of DIR to be placed in a file called FIND with the extension <DIR>.
    b. The name of each subdirectory in the current directory to be displayed on the screen.
    c. A directory listing of each directory on the disk to be displayed.
    d. This is not a valid DOS command.

9. The *attribute* of a DOS file is a software assignment that
    a. Instructs it to behave in a specified manner.
    b. Can prevent it from being erased.
    c. Prevents it from being modified.
    d. All of the above.

10. The DOS command

    A> ATTRIB +R THISFILE.TXT

    sets up the file THISFILE.TXT so it can
    a. Not be copied.
    b. Not be modified.
    c. Be modified.
    d. Be erased.

Matching

Match a function on the right with each DOS redirection operator on the left.

11. \>   a. Redirecting the output of a command to a file or a device
12. <   b. Redirecting the source of input for a command to a file or device
13. | c. Redirecting the output of one command to become the input of a second command
           d. None of these

Completion

Fill in the blanks with the best answers.

14. The DOS command _____ presents one full screen of information at a time, going on to the next screen after a key is pressed.
15. The DOS _____ command allows information in a text file to be arranged in a particular order.
16. The DOS _____ command lets you determine which text file(s) may contain a particular string of text.

FAMILIARIZATION ACTIVITY

1. Make sure you have the following on your DOS disk:

   EDIT.COM
   MORE.COM
   FIND.EXE
   SORT.EXE

2. Using your blank formatted disk, create the directory structure shown in Figure 8.13.

   Note: The files FILE.01 to FILE.02 may all be copies of the same single-line text file. To make multiple copies easily, first create a single-line text file called FILE.01. Next use the COPY command:

   A> COPY FILE.01 FILE.02

   After the file is created, simply press the F3 key on your keyboard. This will flush the keyboard buffer and display the last command you entered:

   ```
           A> COPY FILE.01 FILE.02
   <F3>    A> COPY FILE.01 FILE.02_
   ```

   Now use the BACKSPACE key to erase the 2 in FILE.02 and replace it with a 3, making the entry look like

   A> COPY FILE.01 FILE.03

   When FILE.03 is created, repeat the process by pressing the F3 key again. Do this to create the 20 files.

3. Make two text files called LIST.01 and LIST.02, with contents as shown in Figure 8.14.

The following steps in the activity familiarize you with the DOS utilities presented in the background information section of this exercise.

4. Use the DOS redirection operator > to direct output to a new file. Go to the root directory of your disk and enter the following:

   A> DIR > DIR.TXT

FIGURE 8.13 Directory structure for familiarization activity

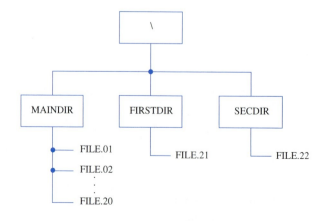

FIGURE 8.14 Contents of LIST.01 and LIST.02

```
         LIST.01
Tom      Baker       98
Jim      Rober       89
Mary     Avalar      98
Jerry    Commings    87
Alice    Cooper      72
Tom      Anderson    68
Jerry    Brown       83
Harry    Downs       90
John     Evens       95
```

```
         LIST.02
Alice    Baker       87
James    Todder      72
Harry    Zepper      58
Walter   Reason      91
Janet    Dempster    89
William  Farmer      96
Debbie   Castor      72
Thomas   Adams       62
```

5. Do a directory listing and verify that there is now a file called DIR.TXT.
6. Display the contents of this new file (DIR.TXT). You now should see the directory listing of your disk as a text file.
7. Next, you will have an opportunity to see the difference between the DOS redirection operator > and merge operation >>. Create a copy of the file LIST.01 in a new file called LIST.TOT. While in your root directory, enter

   ```
   A> TYPE LIST.01 > LIST.TOT
   ```

8. Check your directory to ensure that the new file, LIST.TOT, now exists. Use the TYPE command to verify that the contents of this new file (LIST.TOT) are the same as the contents of LIST.01.
9. Now, issue the following command:

   ```
   A> TYPE LIST.02 > LIST.TOT
   ```

10. Again check the contents of the file LIST.TOT. Note that this time the contents are identical to those of the file LIST.02 and the old contents have been written over.
11. To append the contents of LIST.01 to the current contents of LIST.TOT, issue the following command:

    ```
    A> TYPE LIST.01 >> LIST.TOT
    ```

12. Now look at the contents of LIST.TOT. What you should see is the text of both LIST.02 and LIST.01 together. The use of >> simply adds onto what is already in the file, whereas > removes the old contents before adding the new ones.
13. Next, you will have an opportunity to see the effect of the MORE command. Get into the directory MAINDIR:

    ```
    A> CD\MAINDIR
    ```

14. Now do a directory list of the contents of MAINDIR:

    ```
    A> DIR
    ```

    Observe that the information at the top of the screen scrolls off the screen. To prevent this, use the MORE command.
15. Enter the following:

    ```
    A> DIR | MORE
    ```

    Now notice that only a screenful of information at a time appears and the message

    —More—

    appears at the bottom of the screen. You need only to press any key to continue with more information.

QUESTIONS/ACTIVITIES

1. For each of the following commands, state the standard input and the standard output:
   a. A> SORT < NAMES.LST
   b. A> FIND "<DIR>" > DIRLST.TXT < OTHER.LST
   c. A> DIR | MORE

2. State the function of the following commands:
   a. A> TYPE NAME.LST | MORE
   b. A> DIR | SORT /R > SORT.DIR
   c. A> DIR | FIND "<DIR>"
3. An individual wants to be able to check the contents of text files for the occurrence of a particular name and, when the name is found, print this information. Create a batch file that will do this.
4. Create a batch file that will sort on a given column from the largest value to the smallest, place the results into a file named by the user, and then display the results on the screen using the MORE command. When the batch file is completed, it should require only the name of the file to be sorted, the column number, and name of the file to be created (remember the parameters such as %1 in batch files).

REVIEW QUIZ

Within 10 minutes,

1. Demonstrate the use of the DOS redirection operators by
   a. Redirecting standard output to a DOS file or printer.
   b. Appending standard output to an existing DOS file.
   c. Redirecting standard input from a DOS file.
   d. Redirecting standard input and output of DOS commands.
2. Demonstrate the use of the following DOS commands:
   a. SORT
   b. MORE
   c. FIND

UNIT III The Windows Operating System

EXERCISE 9	An Overview of Windows 3.x
EXERCISE 10	An Introduction to Windows 95 and Windows 98
EXERCISE 11	An Introduction to Windows NT
EXERCISE 12	The Desktop
EXERCISE 13	The Control Panel
EXERCISE 14	Windows Explorer
EXERCISE 15	Managing Printers
EXERCISE 16	Accessories
EXERCISE 17	An Introduction to Networking with Windows
EXERCISE 18	Installing New Software
EXERCISE 19	Installing New Hardware
EXERCISE 20	Other Network Operating Systems

9 An Overview of Windows 3.x

INTRODUCTION

Joe Tekk was working on a computer with a Windows problem. He opened and closed applications, resized several application windows, switched between different tasks, and tested a few screen savers. Then he just let the computer sit and run for a few hours.

Don, Joe's manager, saw the computer later in the day and asked Joe what he was doing with it.

"It's an old 25-megahertz 80386 system one of our clients uses at his office. He says that Windows periodically locks up on him, usually when a screen saver is running."

"You mean Windows 3.1? Not Windows 98?"

"Yes," Joe answered, "The machine has only 4 Meg of RAM and a 100-Meg hard drive. Windows 98 would probably run very poorly on it, if it ran at all."

"Is it Windows 3.1 or Windows for Workgroups?"

"Windows for Workgroups. All I know so far is that the screen saver doesn't lock up when I turn off the network stuff."

Don was curious. "What does the network have to do with the screen saver?"

"Nothing, I think," Joe replied. "It might be a memory problem instead. I'm going to check lots of different things before I send it back."

PERFORMANCE OBJECTIVES

Upon completion of this exercise, within 8 minutes and with 100% accuracy, you will be able to

1. Enter and leave Windows 3.x.
2. Select icons with the mouse.
3. Resize and move windows on the desktop.
4. Start up a Windows application.
5. Start up a DOS shell and run a DOS application from inside Windows.

BACKGROUND INFORMATION

The Windows operating system allows you to control the computer in a totally different way than you are used to with DOS. For example, DOS commands (such as DIR and FORMAT) must all be entered from the keyboard, on a line-by-line basis. If your typing is

FIGURE 9.1 Typical Windows 3.x desktop

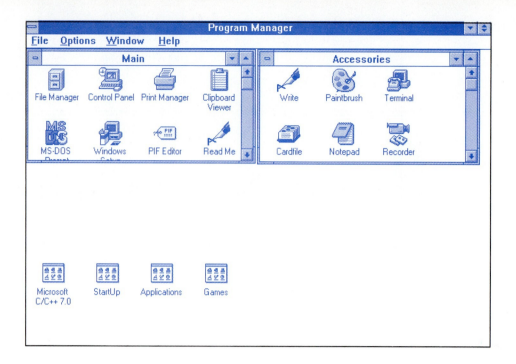

not up to speed, it may be very inconvenient for you to continually enter DOS commands while servicing a computer or working on a project.

Windows practically eliminates the need to use the keyboard, getting all of its commands from the mouse. Clicking the mouse button once, or twice quickly, is all that is needed to generate a command. This is possible through the use of a *graphical interface*. In Windows, graphical symbols are used to represent application programs and groups of programs, as well as commands. This is illustrated in Figure 9.1, which shows a typical Windows 3.x start-up screen (called the **desktop**).

The four square icons at the bottom of the screen in Figure 9.1 represent *program groups*. Each program group may contain one or more programs, as indicated by the open windows for the Main and Accessories groups. To open up the Games window, all you need to do is position the mouse cursor over any portion of the Games icon and press the left mouse button twice quickly. This is roughly equivalent to changing directories in DOS (as in CD\GAMES).

In the next few sections, you will learn how to perform many useful functions inside Windows 3.x.

STARTING WINDOWS

All that is necessary to start Windows 3.x from the DOS prompt is the command

`C:\> WIN`

Windows does all the rest, eventually bringing up the Program Manager window shown in Figure 9.1. Depending on your processor's speed, total system RAM size, and hard disk access time, Windows will take different amounts of time to start up and perform other operations. Increasing the amount of RAM (from 4MB to 8MB) is the easiest way to make Windows run a little faster.

Windows is known as a *protected-mode* operating system, whereas DOS is a *real-mode* operating system. These two modes of operation are found on the 80386, 80486, and Pentium processors from Intel. All of these processors start up in the real mode, acting like really fast 8086 (or 8088) machines, the first machines DOS ran on. Windows switches the processor into the protected mode, where the full 32-bit power of the processor is available.

FIGURE 9.2 File menu

FIGURE 9.3 Exit Windows dialog box

LEAVING WINDOWS

At the end of a Windows 3.x session, you must follow a specific sequence to exit to DOS. Simply turning the power off may have drastic consequences for your Windows environment because of the way Windows uses the hard disk to support virtual memory.

There are two ways to leave Windows 3.x. The first involves the use of the File menu. Examine Figure 9.1 again. Do you see the menu bar near the top of the screen? The four menu items are File, Options, Window, and Help. To choose the File menu, either press the ALT-F key on the keyboard or position the mouse cursor over the menu name (File) and press the left mouse button. You will get the File menu shown in Figure 9.2.

To select Exit Windows from the File menu, either press X on the keyboard or position the mouse cursor over the Exit Windows line and click the left mouse button again. This will cause the Exit Windows *dialog box* to appear, as shown in Figure 9.3. A dialog box usually contains a brief description of what Windows is doing, and one or more *buttons* to choose from. Notice that the two buttons in the Exit Windows dialog box provide for two choices. If you have changed your mind and are not yet ready to exit, use the Cancel button. Otherwise, to exit Windows simply press ENTER on the keyboard, or position the mouse cursor over the OK button and click the left mouse button. Windows will perform its shutdown sequence and return you to the DOS prompt.

The second way to exit Windows 3.x involves the use of the Program Manager window **Close** button. In all windows, there is a button in the upper left-hand corner that looks like the front of a file cabinet drawer. Clicking twice quickly (double-clicking) on this button in the Program Manager window brings up the Exit Windows dialog box. Clicking on this button in any other window simply closes the window.

OPENING A PROGRAM GROUP WINDOW

Recall from Figure 9.1 that there are four icons at the bottom of the Windows start-up screen that look identical except for their different names. These icons represent program groups. A program group window is opened by double-clicking the mouse pointer on the desired icon. For example, double-clicking on the Games icon brings up the Games window shown in Figure 9.4.

Notice that there are seven game applications inside the Games window. Any of these game programs may be executed by double-clicking on the associated icon. Also notice that when the Games window opens, it overwrites portions of the Main and Accessories

FIGURE 9.4 Open Games window

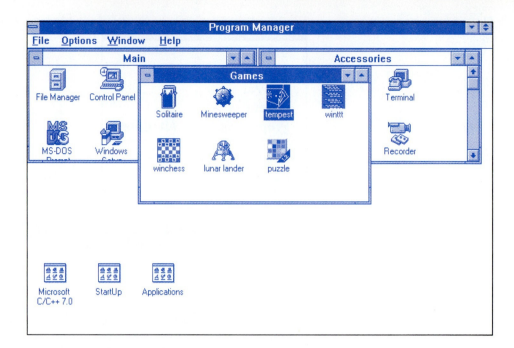

FIGURE 9.5 Reselecting the Main window

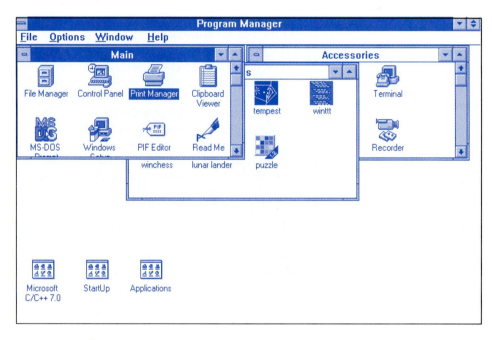

windows. This is similar to what happens when we place more than one folder on a desktop. Sometimes other folders get covered up. It is easy to view the covered (hidden) window again: simply click the mouse pointer anywhere inside the visible portion of the covered window. Figure 9.5 shows what happens when the mouse is clicked inside the Main window.

Windows provides an easy way for you to tell which window is the active window. Compare Figures 9.4 and 9.5. The bar containing the name of the Games window is highlighted in Figure 9.4, and in Figure 9.5 the bar for the Main window is highlighted.

MOVING A WINDOW

Refer again to Figure 9.1. There is a large empty portion on the lower right of the desktop. In order to see all three windows that are open (Main, Accessories, and Games), you can move one of the windows into this area. To move a window,

1. Place the mouse cursor inside the name bar of the window.
2. Press *and hold* the left mouse button.

FIGURE 9.6 Moving the Games window

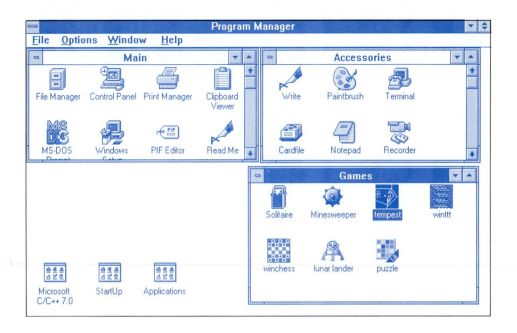

3. Drag the window to its new place on the desktop. The window will be outlined as it is being dragged.
4. Release the mouse button.

If you use these four steps to move the Games window, you end up with the desktop shown in Figure 9.6.

SCROLLING THROUGH A WINDOW

In Figure 9.6, there are differences between the Games window and the Main and Accessories windows. Notice the scroll bars at the right sides of the Main and Accessories windows. The scroll bar is present whenever there are more icons inside the window than the current window size can display. In the Main window, we cannot fully read the names of the first two icons in the bottom row. In the Accessories window, the scroll bar indicates that there are other icons present that are not shown at all.

One way to view hidden icons in a window is to use the scroll bar to move up/down (or left/right with a horizontal scroll bar) within the window. The up and down arrows shown in Figure 9.6 are used for this purpose. Clicking the down arrow inside the Main window results in the new window shown in Figure 9.7. Now the names of all four bottom-row icons are completely visible.

RESIZING A WINDOW

Instead of using the scroll bar, we can graphically resize the window, making it as large or as small as Windows will allow. This is accomplished in the following way:

1. Place the mouse cursor on an edge or corner of the window.
2. Press *and hold* the left mouse button.
3. Drag the window edge (or corner) until the window is the desired size.
4. Release the mouse button.

Using these steps on the Accessories window results in the desktop shown in Figure 9.8.

Since the scroll bar is still present in the window, there must be more icons that are still hidden. To make the window as large as the entire screen (and, hopefully, see the entire contents), simply double-click on the upward-pointing triangle at the top right-hand corner of the Accessories window. The result of this operation is shown in Figure 9.9. To get back to the original desktop, click on the button at the far right end of the menu bar (two triangles pointing in different directions).

FIGURE 9.7 Seeing more of the Main window

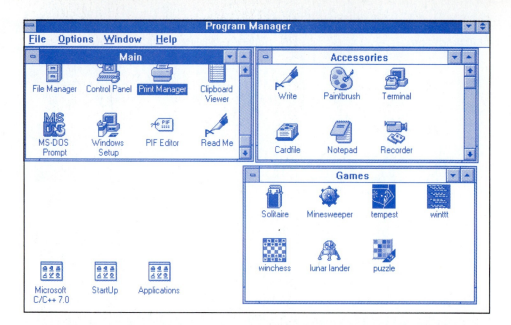

FIGURE 9.8 Resizing the Accessories window

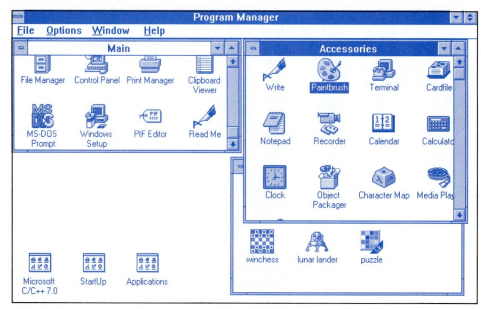

FIGURE 9.9 Resizing the Accessories window to full screen

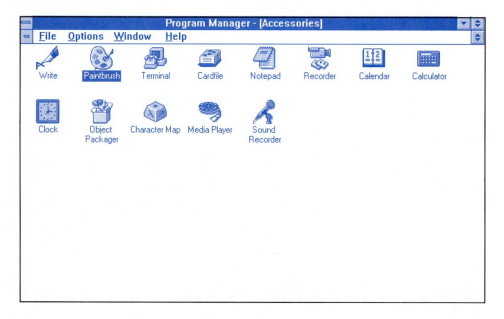

RUNNING A DOS APPLICATION FROM INSIDE WINDOWS

Refer again to Figure 9.8. The MS-DOS Prompt icon in the Main window is used to start up a new DOS *shell,* a copy of the DOS environment maintained by Windows. Once inside a DOS shell, you can issue DOS commands as you normally would (DIR, for example). Depending on the settings for the MS-DOS Prompt icon, your DOS shell may run in a small window or may cover the full screen.

Windows requires that you enter EXIT when you are through using the DOS shell. EXIT returns you to the Windows desktop.

TROUBLESHOOTING TECHNIQUES

Windows 3.1 users unlucky enough to get a general protection fault understand the frustration of having the operating system *lock up* during an unexpected problem. The advanced Windows operating systems (95, 98, and NT), as you will see in the remaining exercises of this unit, are much more *robust,* handling even unexpected errors gracefully, rarely getting hung up by a single application. For example, we can always press Ctrl-Alt-Del in Windows 98 to get the operating system's attention, but would only do so in Windows 3.1 under extreme circumstances.

This is just one of the major differences between Windows 3.1 and the advanced Windows operating systems. Read on to discover the others.

SELF-TEST

This self-test is designed to help you check your understanding of the background information presented in this exercise.

True/False

Answer *true* or *false.*

1. Windows and DOS are both real-mode operating systems.
2. Windows uses a graphical interface.
3. All icons in a window are displayed when the scroll bar is absent.
4. To exit Windows, you simply turn the computer off.

Multiple Choice

Select the best answer.

5. To open a program group window, position the mouse cursor over the icon and
   a. Single-click the mouse button.
   b. Double-click the mouse button.
   c. Press and hold the mouse button.
   d. Press ENTER on the keyboard.
6. To move a program group window, position the mouse cursor inside the name bar and
   a. Single-click the mouse button and drag.
   b. Double-click the mouse button and drag.
   c. Press and hold the mouse button and drag.
   d. Press ENTER on the keyboard.

Completion

Fill in the blanks with the best answers.

7. Windows is a(n) _____-mode operating system.
8. The main screen used by windows is called the _____.
9. When Windows is exited, a(n) _____ _____ appears with a message and two buttons.
10. An application program is executed by _____ _____ on its icon.

FAMILIARIZATION ACTIVITY

Be sure you have a floppy disk with some files on it.

1. Start Windows.
2. Open the Main program group by double-clicking on its icon.
3. Double-click on the File Manager icon.
4. Make sure drive A: is empty.
5. Click on the drive icon for drive A:. A dialog box should appear saying "There is no disk in drive A."
6. Place your disk in drive A: and click the Retry button. You should get a directory listing of your disk.
7. Click the drive C: icon. Your hard drive directory should now be displayed.
8. Leave File Manager.
9. Open the Accessories program group.
10. Double-click on the Clock icon.
11. Choose Settings from the Clock menu.
12. Change the clock setting to Analog (if not already set). What does the clock look like?
13. Change the clock setting to Digital. Now what does it look like?
14. Resize the clock to an appropriate size.
15. Exit Windows without closing the Clock window.
16. Restart Windows. Is the Clock still visible?
17. Double-click on the MS-DOS Prompt icon.
18. Enter the DOS command DIR.
19. When the DIR command completes, use EXIT to return to Windows.
20. Leave Windows by double-clicking the Program Manager Close button.

QUESTIONS/ACTIVITIES

1. What are the three main factors affecting the speed of Windows?
2. List some of the differences between the DOS environment and the Windows environment.
3. Explain why a program group is similar to a DOS directory.
4. Examine the files in your \WINDOWS directory. How many files are there? What are the more common extensions? Do you notice any relationship between the file names and the graphical screen information?

REVIEW QUIZ

Within 8 minutes and with 100% accuracy,

1. Enter and leave Windows 3.x.
2. Select icons with the mouse.
3. Resize and move windows on the desktop.
4. Start up a Windows application.
5. Start up a DOS shell and run a DOS application from inside Windows.

10 An Introduction to Windows 95 and Windows 98

INTRODUCTION

Joe Tekk was just about to leave for his lunch break when he was stopped by Don, his manager. "Hey, Joe, what happened with the laser printer?"

Joe replied that he had no luck trying to install the old drivers on the printer installation disks, so "I downloaded the new driver from the Web and installed it. Now the printer works fine."

"You did what?" Don asked, seeking more of an explanation.

"I brought up Netscape and used Yahoo to search for the printer manufacturer. Yahoo gave me a link to their home page. I went to it and found new drivers on their support page. I just had to click on the right one to download a copy to my hard drive."

Don was impressed. "Joe, you fixed a problem the customer has had for months with that laser printer. Good job!"

After Joe left, Don smiled to himself. "I would have searched with Alta Vista."

PERFORMANCE OBJECTIVES

Upon completion of this exercise, within 10 minutes and with 100% accuracy, you will be able to

1. Explain the various items contained on the Windows 95 desktop.
2. Discuss the differences between Windows 3.x and Windows 95.
3. Briefly list the new features of Windows 95.
4. Identify the differences between Windows 95 and Windows 98.

BACKGROUND INFORMATION

The Windows operating system has gone through many changes since it first appeared in the mid-1980s. It has evolved from a simple add-on to DOS to a multitasking, network-ready, object-oriented, user-friendly operating system. As the power of the underlying CPU running Windows has grown (from the initial 8086 and 8088 microprocessors through the Pentium III, as well as other microprocessors), so too have the features of the Windows operating system. For users familiar with Windows 3.x, the good news is that many of the operating system features are still there. For example, a left double-click is still used to launch an application. The purpose of this exercise is to familiarize you with many of the features that are new in Windows 95. Where possible, comparisons will be made to Windows 3.x to help you gain an appreciation for how things have changed. In addition, many of the new features of Windows 98 will be introduced.

FIGURE 10.1 Windows 95 desktop

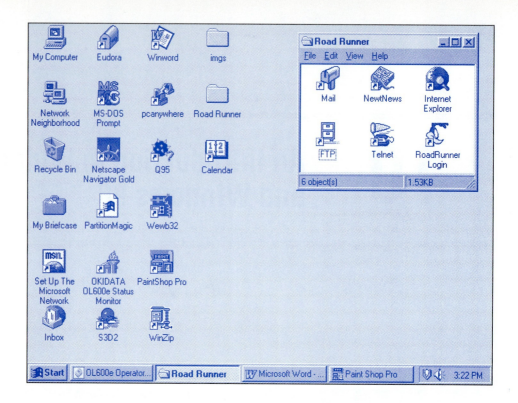

THE DESKTOP

Once Windows 95 has completely booted up, you may see a display screen similar to that shown in Figure 10.1. This graphical display is called the *desktop,* because it resembles the desktop in an office environment. The desktop may contain various *folders* and *icons,* a *taskbar,* the *current time,* and possibly *open folders* containing other folders and icons. A folder is more than just a subdirectory. A folder can be shared across a network, and *cut* and *pasted* just like any other object. You can even e-mail a folder if you want.

Typically, the bottom of the display will contain the *taskbar,* which contains the Start button, icons for all applications currently running or suspended, open desktop folders, and the current time. You can hide, resize, or move the taskbar to adjust the display area for applications. Simply left-clicking on an application's icon in the taskbar makes it the current application.

A new application may be launched by left double-clicking its desktop icon. The desktop may contain a picture, centered or tiled, called the *background image.* The desktop itself is an object that has its own set of properties. For example, you can control how many colors are available to display the desktop. Everything is controlled through the use of easily navigated pop-up menus. The desktop is the subject of Exercise 12.

LONG FILE NAMES

File names in DOS were limited to eight characters with a three-letter extension (commonly called 8.3 notation). Because Windows 3.x ran on top of DOS, it, too, was limited to file names of the 8.3 variety, even though Program Manager allowed longer descriptive names on the program icons.

Windows 95 eliminates the short file name limitation by allowing up to 255 characters for a file name. As shown in Figure 10.2, a *long file name* has two representations. One is compatible with older DOS applications (the old 8.3 notation using all uppercase characters). The other, longer representation is stored exactly as it was entered, with uppercase and lowercase letters preserved. To be compatible with older DOS applications, Windows

FIGURE 10.2   Two examples of long file names

```
Volume in drive D is FIREBALLXL5
Volume Serial Number is 245F-15E6
Directory of D:\repair3e\e15

.              <DIR>        12-31-97  12:58a .
..             <DIR>        12-31-97  12:58a ..
LIST      TXT            0  01-06-98   2:01p list.txt
E15       DOC       31,232  01-06-98   2:01p e15.doc
WOWTHI~1  DOC       20,480  01-06-98   1:38p Wow This is a LONG filename.doc
WOWTHI~2  DOC       20,992  01-06-98   2:01p Wow This is LONG too.doc
         4 file(s)        72,704 bytes
         2 dir(s)     23,674,880 bytes free
```

the first six characters of a long file name, followed in most cases by ~1. When two or more long file names appear in the same directory, Windows 95 will enumerate them (~1, ~2, etc.), as you can see in the directory listing of Figure 10.2. To specify a long file name in a DOS command, use the abbreviated 8.3 notation, or enter the entire long file name surrounded by double quotes. For example, both of these DOS commands are identical in operation:

```
TYPE    WOWTHI~1.DOC
TYPE    "Wow This is a LONG filename.doc"
```

The great advantage of long file names is their ability to describe the contents of a file, without having to resort to cryptic abbreviations.

CONTEXT-SENSITIVE MENUS

In many instances, right-clicking on an object (a program icon, a random location on the desktop, the taskbar) will produce a *context-sensitive menu* for the item. For example, right-clicking on a blank portion of the desktop produces the menu shown in Figure 10.3. Right-clicking on the time in the lower corner of the desktop generates a different menu, as indicated by Figure 10.4. Note that the two example menus are different. This is what the "context-sensitive" term is all about. Windows 95 provides a menu tailored to the object you right-click on. This is a great improvement over Windows 3.x, which rarely did anything after right-clicking.

FIGURE 10.3   Context-sensitive desktop menu

FIGURE 10.4   Context-sensitive time/date menu

FIGURE 10.5   Help menu

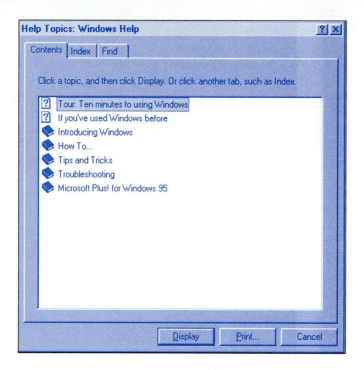

IMPROVED HELP FACILITY

The built-in help available with Windows 95 is significantly different from that provided by Windows 3.x. To get help, go to the Start menu. Figure 10.5 shows the Help Topics window that is displayed when Help is selected.

Three tabs appear at the top of the display. The Index tab allows the user to enter keywords that might be found by looking in an index. As each letter is entered, the display is updated to show all matching items. Help for the highlighted selection (left-click to choose a different help topic) is then displayed by left-clicking the Display button or by double-clicking on the help topic.

The two other tabs, Contents and Find, provide additional support in the form of a guided tour of Windows 95, troubleshooting methods, tips, and alternate ways of finding specific help topics.

WINDOWS EXPLORER

Windows 3.x provided two main applications that made life bearable: Program Manager and File Manager. In Windows 95, the services provided by these two applications, as well as many new features, are found in the new Windows Explorer program. Figure 10.6 shows a typical Explorer window. Although we will cover Windows Explorer in detail in Exercise 14, it is worth taking a quick look at now. The small box in the upper left corner containing W95 (C:) indicates the current folder selected. Clicking the down arrow produces a list of folders to choose from. The two larger windows display, respectively, a directory tree of drive C: (folders only) and the contents of the currently selected folder (which also happens to be drive C:). Note the different icons associated with the files shown. Windows Explorer allows you to change the icon, or associate it with a different file. In general, as with Windows 3.x, double-clicking on a file or its icon opens the application associated with it.

Windows Explorer also lets you map network drives, search for a file or folder, and create new folders, among other things. It is truly one of the more important features of Windows 95.

THE REGISTRY

The Registry is the Windows 95 replacement for the SYSTEM.INI and WIN.INI configuration files used by Windows 3.x. The Registry is an internal operating system file maintained

FIGURE 10.6 Windows Explorer

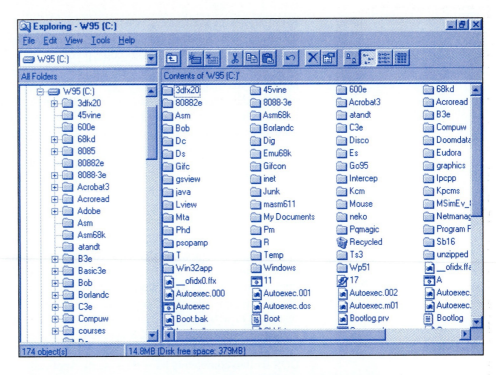

FIGURE 10.7 REGEDIT window

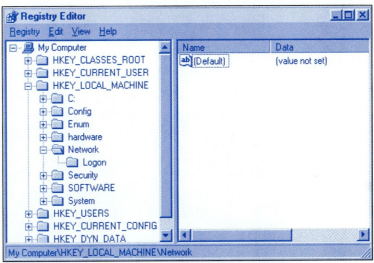

by Windows 95. As each application is installed, the installation program makes "calls" to the Registry to add configuration information, storing similar information to what was previously stored in the .INI files. In this way, the Registry is protected and therefore is harder to corrupt. The Registry is accessed using the REGEDIT program as illustrated in Figure 10.7. The Registry is nothing to fool around with simply because you feel like experimenting. A corrupt Registry can prevent Windows 95 from booting and could possibly require a complete reinstallation of Windows 95.

The Registry contains all the information Windows 95 knows about both the hardware and software installed on the computer.

NETWORKING

Windows 95 offers a major improvement in networking capabilities. Windows 3.1 had no built-in networking support and required network software to be loaded and maintained by DOS. This situation was improved slightly with the release of Windows for Workgroups 3.11, which provided limited networking via a network protocol called **NetBEUI** (NetBIOS Extended User Interface), which allows e-mail and file sharing in small peer-to-peer networks.

NetBEUI is still available in Windows 95 (providing the Network Neighborhood feature, network drives, shared printers), along with other additions. Two important protocols have been added, PPP (point-to-point protocol) and TCP/IP (transmission control protocol/Internet protocol). PPP is used with a serial connection (such as a modem) and is the basis for the dial-up networking provided by Windows 95. TCP/IP is the protocol used by the Internet. You can use TCP/IP applications such as Netscape or Internet Explorer to browse the World Wide Web, connect to a remote computer and share files, send and receive e-mail, and much more. Typically, TCP/IP is used in conjunction with a network interface card for fast data transmission, although it can also be used over a modem connection (by encapsulating it inside the PPP).

DOS

Yes, DOS is still a part of Windows. However, the role of DOS in Windows 95 is different in many ways from what was required under Windows 3.x. For example, Windows 3.x relied on the file system set up and maintained by DOS. This is commonly referred to as running "on-top" of DOS. Windows 95 does things differently, providing its own improved file system (long file names) and a *window* to run your DOS application in. You can open several DOS windows at the same time if necessary, and run different applications in each. Figure 10.8 shows a typical DOS window.

Furthermore, in Windows 3.x, it is possible for a wayward application running in a DOS shell to completely hang the system. Applications are not allowed that much control in Windows 95. If a DOS application (or any other, for that matter) hangs up, all you need to do is press Ctrl-Alt-Del (oddly enough) to bring up the Close window. Figure 10.9 illustrates the Close window.

FIGURE 10.8 DOS window

FIGURE 10.9 Close window

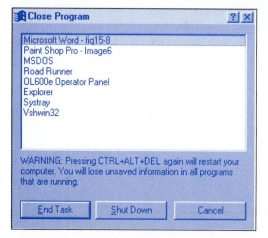

It is important to note that Windows 95 is always in control somewhere in the background, monitoring everything. A DOS application cannot hang the system like its Windows 3.x counterpart.

PREEMPTIVE MULTITASKING

A significant change in the way multiple programs are executed was made when Windows 95 was developed. Windows 3.x used a technique called *cooperative multitasking* to run more than one 16-bit application at a time. Each application would get a slice of the processor's computation time, with periodic switches between applications. The problem with this form of multitasking is that one task can take over the system (by requesting all available memory or other resource) and prevent the other tasks from running.

Windows 95 uses a method called *preemptive multitasking* to run multiple 32-bit applications while still running the older 16-bit Windows 3.x applications in cooperative mode (in a single shared block of memory). Preemptive multitasking means that the current application can be interrupted and another application started or switched to. This provides a degree of *fairness* to the set of applications competing for processor time. For example, referring back to Figure 10.1, it is possible to quickly double-click on several icons (Netscape, Eudora, Winword) before the first application has completely come up. All applications clicked will start up, with the first one clicked getting the initial shot at the desktop.

In Windows 3.x, when you double-clicked an application to start it, you had to wait until the hourglass went away before you could do anything else. In Windows 95, the hourglass still indicates that the operating system is busy with a chore, but it may be preempted to begin a new task at any time. Once again, we see that both the user and the operating system have more control under Windows 95 than was possible with Windows 3.x.

RECYCLE BIN

The Recycle Bin shown in Figure 10.10(a) is a holding place for anything that is deleted in Windows 95. The nice thing about the Recycle Bin is that *you can get your files back* if you want to, using the Undelete option. Double-clicking on the Recycle Bin icon brings up the window shown in Figure 10.10(b). Any or all of the files shown in the window may be recovered. *Warning:* If you delete files while in DOS, they are not deposited in the Recycle Bin, and may be impossible to recover at a later time (with the old UNDELETE command).

FIGURE 10.10 Recycle Bin

(a) Program icon

Name	Original Location	Date Deleted	Type	Size
~$fig	D:\repair3e\e15	1/9/98 3:44 PM	Microsoft Word Doc...	1KB
~$list	D:\repair3e\e15	1/9/98 3:44 PM	Microsoft Word Doc...	1KB
ENGLISH.RES	C:\GRAVIS\GRAVU...	1/4/98 12:06 AM	Intermediate File	102KB
GRAVUTIL	C:\GRAVIS\GRAVU...	1/4/98 12:06 AM	Application	365KB
GRAVUTIL	C:\GRAVIS\GRAVU...	1/4/98 12:06 AM	Help File	9KB
GRAVUTIL	C:\GRAVIS\GRAVU...	1/4/98 12:06 AM	Shortcut to MS-DOS...	1KB
OKIDATA OL600e...	C:\WINDOWS\Des...	1/7/98 9:24 PM	Shortcut	1KB

7 object(s) — 475KB

(b) Window

OTHER NEW FEATURES

A summary of several other new features is included here to help you get a good idea of how much of an improvement Windows 95 is over Windows 3.x.

- *Remote Procedure Calls*   RPCs allow computers in a network to share their processing capabilities. For example, a 486 machine could issue an RPC to a Pentium-based machine to execute code *on that machine* and send the results back to the 486.
- *Support for OLE 2*   Object Linking and Embedding 2 (OLE 2) provides powerful features supporting a dynamic application environment. Cutting and pasting text, images, and other types of objects are just the beginning of OLE 2. Windows 95 provides a large set of enhanced OLE 2 features, such as drag-and-drop, nested objects, and optimized object storage.
- *Accessibility Options*   Many new features have been added to allow differently abled users to customize their computers to meet individual needs. The keyboard, display, sounds, and mouse can be set to a variety of combinations to suit most needs.
- *Microsoft Exchange*   E-mail is now a standard feature on the Windows 95 operating system.

WINDOWS 95 VERSION B

A number of bugs encountered in version A (the first release) of Windows 95 can be fixed by downloading Service Pack 1 from Microsoft's Web site (http://www.microsoft.com). The service pack updates Windows 95 by modifying various portions of the operating system (such as the kernel), and provides additional drivers and other improvements.

Version B of Windows 95 (called OEM Service Release 2) offers additional improvements, such as FAT32 (a better way of organizing files on your hard drive to reduce lost storage space), many new drivers, and other enhancements. Unfortunately, you cannot update Windows 95 A to Windows 95 B.

THE WINDOWS 98 DESKTOP

Many changes in Windows 98 are visible right on the desktop. A *channel bar* provides one-click access to your favorite or often-used Internet links. A channel is a connection to a Web site that allows you to schedule automatic updates of the information you need. The desktop can be organized and operated as a Web page if you desire, even displaying a specific Web page as its background.

Figure 10.11 shows a sample Windows 98 desktop. The channel bar is to the right, and contains several preassigned Internet links. The taskbar has a new area that holds often-used shortcuts. This is a nice feature because desktop shortcuts are hidden from view in Windows 95 whenever a window uses the full screen.

WHAT ELSE IS NEW IN WINDOWS 98?

The Welcome to Windows tour (Programs, Accessories, System Tools) contains details for users of Windows 3.x, Windows 95, and new users. All the new features of Windows 98 are highlighted and presented in an easy-to-view multimedia format. Figure 10.12 shows the main feature screen. Clicking on a feature opens up a short slide show with narration and a quick run-through of the menus required to use the feature. Let us take a brief look at each feature group.

WINDOWS 98 IS EASIER TO USE

In addition to the Web-style features we will cover shortly, many other improvements are found in Windows 98. Two or more monitors (with associated display adapter cards) can

FIGURE 10.11   Windows 98 desktop

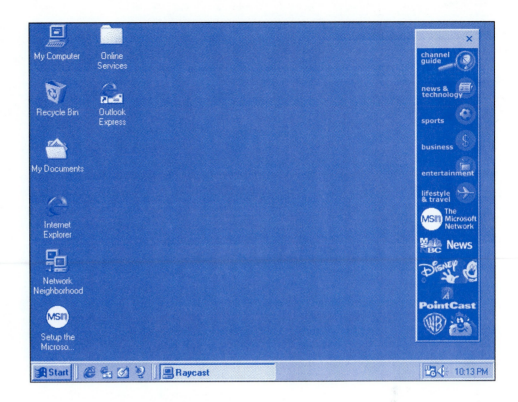

Figure 10.12   New features in Windows 98

be used at the same time. Special OnNow hardware can be managed to reduce power use. Advanced Plug & Play includes support for the Universal Serial Bus. A new utility called Microsoft Magnifier uses a portion of the desktop as a magnifying glass. The portion of the desktop closest to the mouse is displayed with an adjustable level of magnification. In addition, a substantial amount of online help is provided.

FIGURE 10.13 Web-style desktop

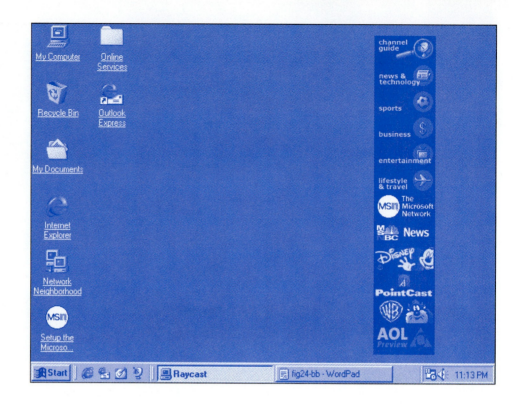

WINDOWS 98 IS ENHANCED FOR THE WEB

Windows 98 comes with Internet Explorer 4.0 included, a Connection Wizard to help make your connection to the Internet a smooth process, desktop Web page capabilities, Channels (favorite Internet links), Outlook Express, an all-purpose e-mail application, NetMeeting (use a digital whiteboard to sketch proposals, use Chat to confer with group members), and even FrontPage Express for creating your own Web pages. Figure 10.13 shows the Windows 98 desktop setup for Web-style appearance. If a Web page is used as the background, you can elect to hide the desktop icons to get a true browser appearance.

WINDOWS 98 IS FASTER AND MORE RELIABLE

Microsoft has added a number of features to Windows 98 to enhance its operating speed and increase reliability. The Hard Disk Optimizer will convert FAT16 file systems into the newer FAT32 file system, which allows for larger drives and more efficient use of drive space. In addition, applications that are used frequently can be relocated to a special portion of the disk to allow quicker launching.

Windows 98 can be updated automatically over the Web. For reliability, a System File Checker application checks the integrity of the files required by the operating system. At boot time, ScanDisk runs automatically if the previous shutdown was not completed correctly.

WINDOWS 98 IS MORE ENTERTAINING

A number of multimedia improvements are found in Windows 98. With special hardware, Windows 98 allows you to view broadcast television from the airwaves or gather and present information from the Internet (using NetShow). The Universal Serial Bus is also supported; this allows devices such as DVD (digital video disk) players and *force-feedback* joysticks to be connected.

DirectX support is built-in, enabling real-time high-quality 3-D graphics for supported graphics adapters. There are also many new screen savers (which take advantage of DirectX for stunning visual effects).

Overall, Windows 98 is another big step forward, as Windows 95 was compared to Windows 3.x.

TROUBLESHOOTING TECHNIQUES

There are so many new features in Windows 95/98 it is easy to lose track of them and not use one or more of them to make your life easier. For instance, rather than getting out a large, heavy Windows reference book, simply search the topics in Help for the answers you need.

Also, remember that it is usually OK to right-click on *anything*, any object (file, program icon, taskbar, display), and a context-sensitive menu will typically pop up. Sometimes you might get a What's This? button, which provides a short description of whatever you right-clicked on.

SELF-TEST

This self-test is designed to help you check your understanding of the background information presented in this exercise.

True/False

Answer *true* or *false*.

1. Long file names are limited to 500 characters.
2. NetBEUI is a networking protocol.
3. The Registry is just a database of .INI files.
4. You can open only one DOS window at a time.
5. Once a file enters the Recycle Bin it is gone forever.
6. FAT32 is available on all Windows 95 computers.
7. Windows 98 contains a channel bar used to watch television.
8. The Windows 98 desktop can be organized and operated as a Web page.
9. The Windows 98 operating system requires an 80486 processor or better.

Multiple Choice

Select the best answer.

10. The desktop contains
    a. The taskbar, application icons, and folders.
    b. The taskbar, running applications, and a communications console.
    c. A set of folders for each hard drive.
    d. All of the above.
11. A context-sensitive menu is displayed when
    a. Double-clicking on a program icon.
    b. Pressing both mouse buttons at the same time.
    c. Right-clicking almost anything.
    d. None of the above.
12. Windows 95 supports
    a. 16-bit applications.
    b. 32-bit applications.
    c. Both a and b.
    d. DOS applications only.
13. Windows Explorer
    a. Replaces Program Manager and File Manager.
    b. Explores the Internet.
    c. Searches for viruses on the hard drive.
    d. None of the above.

14. Windows 98 includes Outlook Express, which provides:
    a. A digital whiteboard and a chat feature.
    b. An HTML editor to create custom Web pages.
    c. An all-purpose e-mail application.
    d. Reduced system power consumption during peak system activity.
15. The new Windows 98 utility program Microsoft Magnifier is used to:
    a. Show all channels on the desktop.
    b. Show a portion of the display as a magnifying glass.
    c. Send e-mail messages on the Internet.
    d. Access the devices on a Universal Serial Bus.
16. FAT32 is better than FAT16 because:
    a. Hard drive space is used more efficiently.
    b. Memory resources are reduced and the system runs faster.
    c. It cannot be used on small disk drives.
    d. 32 is higher than 16 and it is a multiple of 2.

Completion

Fill in the blank or blanks with the best answers.

17. Two new Windows 95 protocols are PPP and _____.
18. Windows 95 runs 32-bit applications using _____ multitasking.
19. The _____ contains folders, icons, and the taskbar.
20. Windows Explorer can be used to map a(n) _____ drive.
21. A centralized location storing all configuration information about a Windows 95 system is called the _____.
22. Windows 98 can use multiple _____ at the same time.
23. ScanDisk runs automatically on Windows 98 if the previous _____ was not successful.

FAMILIARIZATION ACTIVITY

1. Boot up Windows 95/98.
2. On the taskbar, left-click the Start button.
3. Move the mouse pointer up to the Help icon and left-click it.
4. In the Help Topics window, left-click the Index tab.
5. Enter the word "network" in the text box (left-click inside the box if the cursor is not visible).
6. From the list of network topics, chose Dial-Up Networking by left-clicking on it.
7. Left-click the Display button.
8. Left-click the Cancel button.
9. Double left-click the Dial-Up Network topic.
10. Left-click the Cancel button.
11. Left-click the Cancel button to return to the desktop.
12. Right-click on an empty portion of the desktop.
13. Move the mouse pointer to the New menu selection.
14. Move the mouse pointer to the Shortcut selection on the submenu.
15. Left-click on the Shortcut item.
16. Click the Browse button.
17. Locate the Windows folder icon (using the scroll bar).
18. Double left-click on the Windows folder icon.
19. Locate the Calendar icon (using the scroll bar).
20. Left-click on the Calendar icon.
21. Left-click the Open button.
22. Left-click on the Next button.
23. Click on the Finish button.
24. Double left-click on the Calendar icon.
25. Close the Calendar application.

QUESTIONS/ACTIVITIES

1. Does Windows 95 ever lose control of the system?
2. How are long file names backward compatible with the older DOS 8.3 notation?
3. What types of networking does Windows 95 provide?
4. Name five features found in Windows 98 that are not found in Windows 95.

REVIEW QUIZ

Within 10 minutes and with 100% accuracy,

1. Explain the various items contained on the Windows 95 desktop.
2. Discuss the differences between Windows 3.x and Windows 95.
3. Briefly list the new features of Windows 95.
4. Identify the differences between Windows 95 and Windows 98.

11 An Introduction to Windows NT

INTRODUCTION

Joe Tekk was examining the computer book section of a local bookstore. Shelf after shelf contained books about the Windows operating system. There were books about Windows 95, Windows 98, Windows NT, and Windows 2000.

Joe thought about where the future of the Windows operating system is headed. Just skimming through the new features of Windows 98 showed him that a major push toward integrating the Web into the operating system was undertaken.

Next, Joe turned his attention to the Windows NT operating system. With support for multiple processors, improved security and system management tools, and full 32-bit code utilization, Windows NT has significant differences from Windows 95/98, and is well suited for network server operation.

A voice from behind broke into Joe's thoughts. "I remember when the entire operating system for the PC fit into 32KB of RAM."

Joe turned to see who had spoken to him. It was an old man, well into his 80s, hunched over and standing with the help of a cane. Joe noticed that the cane had a microchip encased in clear plastic on the handle.

"Now you need 16MB just to run the install program," Joe replied. He spent the next two hours talking to the old man about operating systems, and learned a great deal from him.

PERFORMANCE OBJECTIVES

Upon completion of this exercise, within 10 minutes and with 100% accuracy, you will be able to

1. Identify the differences between Windows 95/98 and Windows NT.
2. Identify the key features of Windows NT.
3. Identify any common features of Windows 95/98 and Windows NT.

BACKGROUND INFORMATION

Windows NT is another operating system developed by Microsoft. It was developed to create a large, distributed, and secure network of computers for deployment in a large organization, company, or enterprise. Windows NT actually consists of two products: Windows NT Server and Windows NT Workstation. The server product is used as the server in the client-server environment. Usually a server will contain more hardware than the regular desktop-type computer, such as extra disks and memory. The workstation product is designed to run on a regular desktop computer (consisting of an 80486 processor or better). Windows NT provides users a more stable and secure environment, offering many features not available in Windows 95/98 such as NTFS, a more advanced file system than FAT 16/32. The newest version of Windows NT, called Windows 2000, is in beta testing at the time of this writing.

We will use the Windows NT Server product to illustrate the user interface into the Windows NT environment. Let us begin by looking at the Windows NT login process.

WINDOWS NT OPERATING SYSTEM LOGON

One of the first things a new user will notice about the Windows NT environment is the method used to log in. The only way to initiate a logon is to press the Ctrl-Alt-Delete keys simultaneously as shown in Figure 11.1. This, of course, is the method used to reboot a computer running DOS or Windows 95/98. Using Windows NT, the Ctrl-Alt-Delete keys will no longer cause the computer to reboot, although it will get Windows NT's attention.

If the computer is not logged in, Windows NT displays the logon screen, requesting a user name and password. During the Windows NT installation process, the Administrator account is created. If the computer (Windows NT Server, Windows NT Workstation, Windows 95/98, or Windows for Workgroups 3.11) is configured to run on a network, the logon screen also requests the domain information. After a valid user name, such as Administrator, and the correct password is entered, the Windows NT desktop is displayed, as shown in Figure 11.2.

WINDOWS NT SECURITY MENU

After a Windows NT Server or Workstation is logged in, pressing the Ctrl-Alt-Delete keys simultaneously results in the Windows NT Security menu being displayed as illustrated in

FIGURE 11.1 Windows NT Begin Logon window

FIGURE 11.2 Windows NT desktop

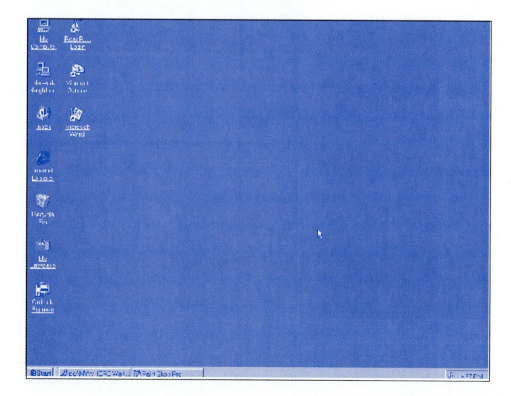

Figure 11.3. From the Windows NT Security menu it is possible for the operator to select from several different options, including Cancel to return to Windows NT.

The Lock Workstation option is used to put the Windows NT Server or Workstation computer in a *locked* state. The locked state is usually used when the computer is left unattended, such as during lunch, dinner, nights, and weekends. When a computer is locked, the desktop is hidden and all applications continue to run. The display either enters the screen saver mode or displays a window requesting the password used to unlock the computer. The password is the same one used to log in.

The Logoff option is used to log off from the Windows NT computer. The logoff procedure can also be accessed from the shutdown menu by selecting the appropriate setting. The logoff procedure terminates all tasks associated with the user but continues running all system tasks. The system returns to the logon screen shown in Figure 11.1. The Shutdown option will cause the computer to perform a system reset operation and restart the operating system. After the reboot is complete, the system returns to the logon screen shown in Figure 11.1.

The Change Password option is used to change the password of the currently logged-in user.

The Task Manager option causes the Windows NT Task Manager window to be displayed. The Task Manager is responsible for running all the system applications, as indicated by Figure 11.4. Notice that individual applications may be created, selected, and ended or switched by using the appropriate buttons. It is sometimes necessary to end tasks that are not functioning properly for some reason or another. In these cases, the status of the application is usually "not responding."

FIGURE 11.3 Windows NT Security menu

FIGURE 11.4 Task Manager applications

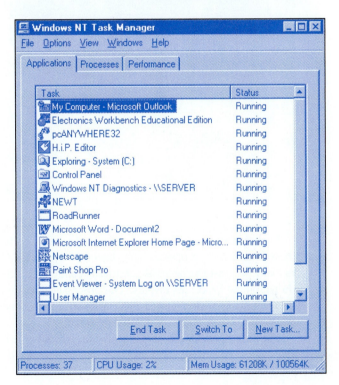

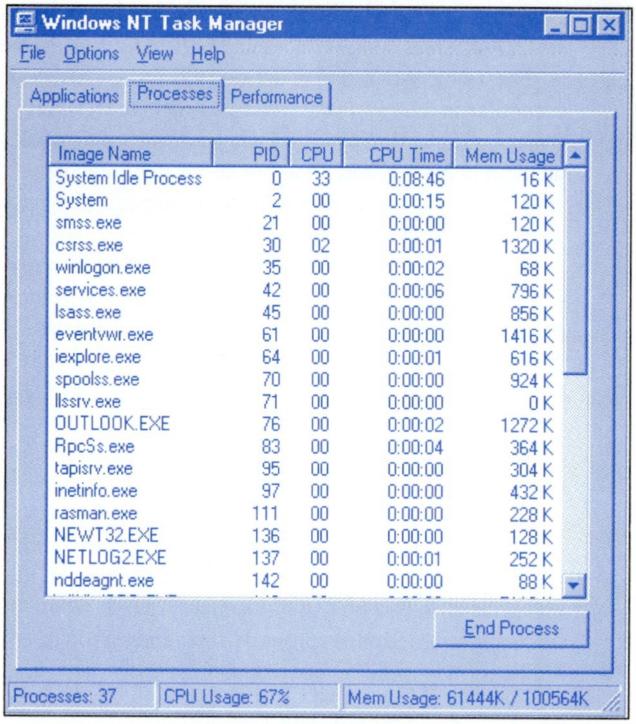

FIGURE 11.5 Windows NT processes

FIGURE 11.6 Task Manager display

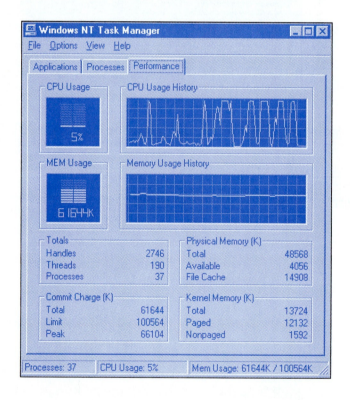

Each application controls processes that actually perform the required tasks. Figure 11.5 shows a number of processes being executed by the Task Manager. Applications may create as many processes as necessary. Extreme caution must be exercised when ending a process shown on the Processes display. The processes that are used to control Windows NT can also be ended, causing the computer to be left in an unknown state. If processes must be terminated, it is best to use the Applications tab.

The Task Manager can also display the system performance. Figure 11.6 shows a graphical display of current CPU and memory utilization. It also shows a numeric display of other critical information.

FIGURE 11.7 Windows NT Start menu

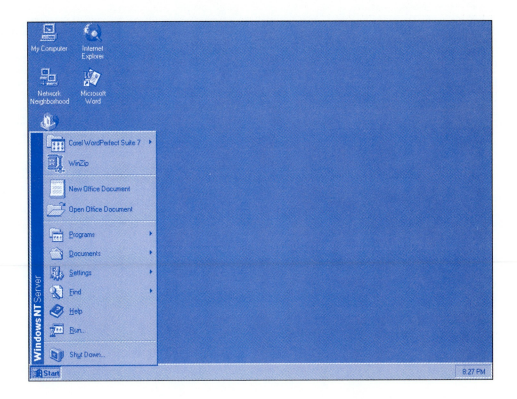

WINDOWS NT DESKTOP, TASKBAR, AND START MENU

The look and the feel of the Windows 95/98 desktop have been incorporated into the Windows NT desktop. At first glance, it might be hard to tell the difference between the two operating systems. Figure 11.7 shows the Start menu of a typical Windows NT desktop. Aside from the Windows NT text display along the left margin of the Start menu, it contains some of the same categories found on Windows 95/98 computers. As you will see, there are many similarities to Windows 95/98 (such as the desktop, Start menu, and the taskbar), and many differences.

WINDOWS NT CONTROL PANEL

Let us continue our investigation of Windows NT by looking in the Control Panel as illustrated in Figure 11.8. Within the Control Panel, there are several changes. For example, the System icon in Windows NT brings up a completely different display than Windows 95/98, as shown in Figure 11.9. This is no longer the place to go when examining system hardware components. The Devices icon provides that function in Windows NT. This is illustrated in Figure 11.10. Each individual device may be configured from this window.

The Services icon is used to control the software configuration on Windows NT. Usually, a service is started when the computer is booted, but it can also be started or stopped when a user logs in or out of Windows NT. Note that the Administrator account must be used when making changes to the Windows NT service configuration. A typical Services window is shown in Figure 11.11. You are encouraged to explore the different icons in the Windows NT Control Panel.

WINDOWS NT DOMAINS

Windows NT computers usually belong to a computer network called a *domain*. The domain will collectively contain most of the resources available to members of the domain. Computers running Windows NT Server software offer their resources to the network clients (Windows NT workstations, Windows 95/98, and Windows for Workgroups 3.11).

FIGURE 11.8 Items in the Windows NT Control Panel

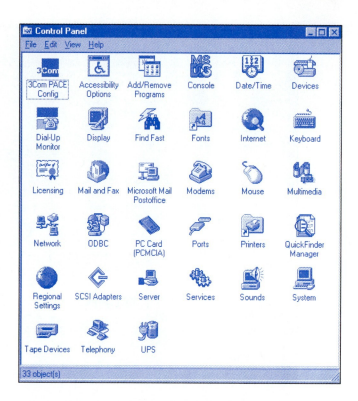

FIGURE 11.9 Startup and shutdown system properties

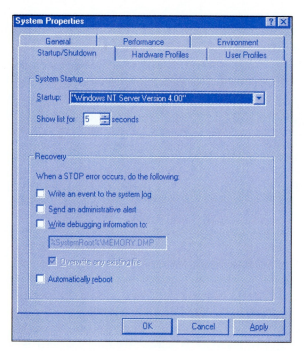

For example, during the logon process, a Windows NT Server responsible for controlling a domain will verify the user information (a user name and password) before access to the computer is allowed. The Network Neighborhood allows access to the resources available on other computers in the domain. Figure 11.12 shows the Network menu, where network components are configured.

The *network administrator* determines how the network is set up and how each of the components are configured. It is always a good idea to know whom to contact when information about a network is required. If the setting is not correct, unpredictable events may occur on the network, creating the potential for problems.

FIGURE 11.10 A list of devices

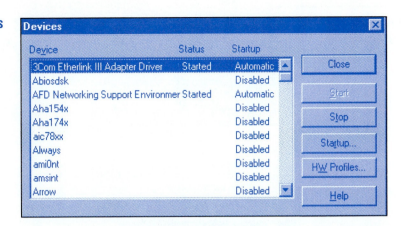

FIGURE 11.11 A list of services

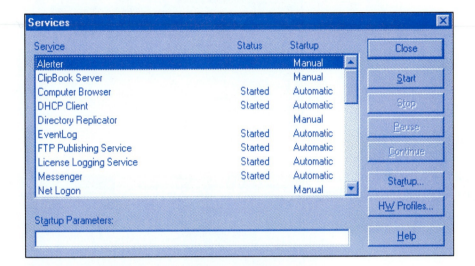

FIGURE 11.12 Currently installed Network Protocols

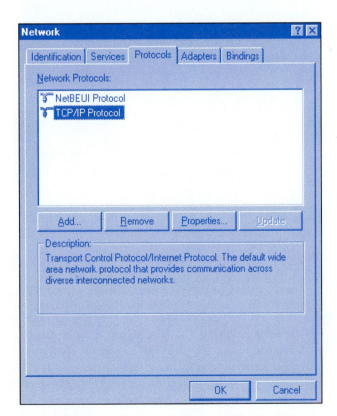

135

FIGURE 11.13 Administrative Tools menu for Windows NT

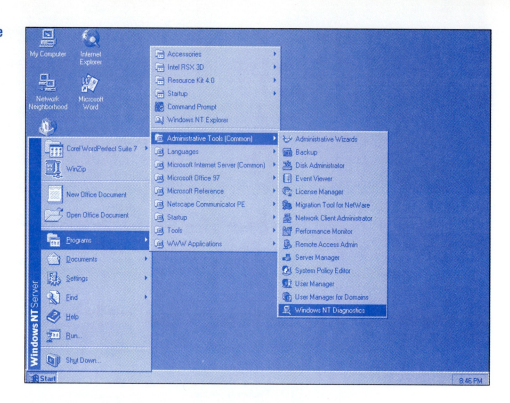

FIGURE 11.14 Windows NT Diagnostics system information

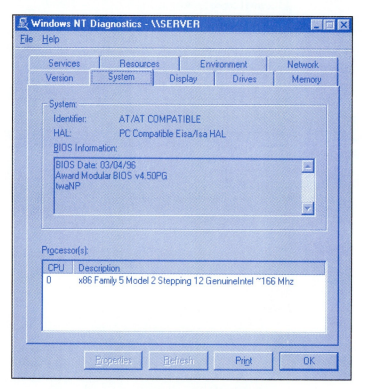

TROUBLESHOOTING TECHNIQUES

Windows NT provides many different troubleshooting aids to tackle a wide variety of common problems. The Administrative Tools menu shown in Figure 11.13 shows a list of common applications designed to properly configure a Windows NT computer. One of the most common administrative tools is the Windows NT Diagnostics application shown in Figure 11.14. The Diagnostics menu contains several different tabs, each showing a specific area of the system. When experiencing problems, it is always a good idea to examine all the information shown on the diagnostics window.

SELF-TEST

This self-test is designed to help you check your understanding of the background information presented in this exercise.

True/False

Answer *true* or *false*.

1. Logoff can be selected from the Windows NT Security menu.
2. A task that is not responding must be stopped manually.
3. The Windows NT operating system has the look and feel of a Windows 98 computer.
4. The Windows NT operating system requires an 80486 processor or better.
5. The Devices icon in the Windows NT Control Panel replaces the Add New Hardware icon in Windows 98.

Multiple Choice

Select the best answer.

6. Windows NT consists of two products:
    a. Client and server.
    b. Client and workstation.
    c. Server and workstation.
    d. Server and client workstation.
7. Windows NT uses the
    a. FAT32 file system.
    b. NTFS file system.
    c. RBFS file system.
    d. All of the above.
8. The Task Manager is used to
    a. Put the computer in a locked state.
    b. Log off from Windows NT.
    c. Create new system applications.
    d. None of the above.
9. When a computer is in a locked state,
    a. The desktop is hidden and applications are paused.
    b. The desktop is hidden and applications continue to run.
    c. The desktop is not hidden and applications are paused.
    d. The desktop is not hidden and applications continue to run.
10. The Windows NT Security menu is displayed when the
    a. Computer is booted the first time.
    b. Computer is accidentally turned off and then turned back on.
    c. User presses the Ctrl-Alt-Delete keys simultaneously.
    d. User presses Ctrl-C or Ctrl-Break repeatedly.

Completion

Fill in the blank or blanks with the best answers.

11. The _____ option will cause a Windows NT computer to perform a system reset operation and restart the operating system.
12. The Windows NT task can be stopped when a process is _____ _____.
13. If a Windows NT task must be stopped, it is best to use the _____ tab.
14. Disk _____ is an example of an administrative tool.
15. When experiencing problems, it is a good idea to examine the _____ window.

FAMILIARIZATION ACTIVITY

Using a Windows NT computer, perform the following activities:

1. Open the Control Panel.
2. Run each of the Control Panel applications to determine what type of settings may be controlled.

3. Close the Control Panel.
4. Examine the administrative tools.
5. Review online help information.

QUESTIONS/ACTIVITIES

1. Create a new Windows NT user using the User Manager administrative tool.
2. Look at the Windows NT Event Viewer application.

REVIEW QUIZ

Within 10 minutes and with 100% accuracy,

1. Identify the differences between Windows 95/98 and Windows NT.
2. Identify the key features of Windows NT.
3. Identify any common features of Windows 95/98 and Windows NT.

12 The Desktop

INTRODUCTION

Jeff Page, who was in charge of all Web development at RWA Software, ran into Joe Tekk's office. He was covered with sweat. "Joe, you've got to come and help me. My machine is trashed."

Joe followed as Jeff ran back to his office. "I've rebooted my machine five times and get the same result each time. My desktop is completely blank. There are no icons for anything."

Joe looked at the display. The taskbar was visible at the bottom of the display screen, but the rest of the screen was a large empty patch of green background. Joe tried to reassure Jeff. "Believe it or not, Jeff, I've seen something like this before. I think I might know what is wrong."

Joe sat down and grabbed the mouse. Before long, he had clicked his way into the Recycle Bin, which contained, among other things, all of the missing desktop icons. Within moments the icons were restored.

Jeff wondered how Joe had identified the problem so quickly. Joe explained, "Last week one of the network guys brought his daughter with him while he did some work. She ran over to an open laptop and deleted every icon on the desktop. I think she figured out what happens when you right-click, and learned how to delete things to make them invisible. Apparently she knows just enough about Windows to cause trouble with it."

"Do you think she's still here?" Jeff asked, looking worried. "I have to leave for a basketball game in 20 minutes and I need to leave my machine on."

Joe laughed and shook his head. "Password protect your screen saver. That will keep her out."

PERFORMANCE OBJECTIVES

Upon completion of this exercise, within 10 minutes and with 100% accuracy, you will be able to

1. Demonstrate the different features of the Start menu.
2. Identify items on the taskbar.
3. Change the appearance and properties of the desktop.
4. Create and use shortcuts.
5. Explain the function of the standard desktop icons.

BACKGROUND INFORMATION	Since the desktop is common to all three advanced Windows operating systems (95, 98, NT), we will refer to it simply as the Windows desktop. Desktop features unique to a single operating system will be discussed as necessary. The desktop is the centerpiece of the Windows operating system. The desktop typically contains a number of standard icons provided by the operating system as well as user-defined icons. As with the Windows 3.x operating system, an application is launched by double-clicking the program icon, although single-clicking may also be used in Web-style desktop settings.

The desktop also contains the taskbar, a system tray, any open folders and applications, and an optional background image called a wallpaper. Figure 12.1 shows the contents of a typical desktop. All the icons down the left-hand side of the desktop are automatically created and placed on the desktop when Windows is installed. These icons are entry points to the many built-in features of Windows. All the other icons are *shortcuts,* user-defined links to programs or other files that are used frequently.

In the upper right corner of the desktop shown in Figure 12.1 is an open *folder* called "Road Runner," which contains icons for six objects. Recall that Windows is an object-oriented operating system. Double-clicking on any of the icons in the Road Runner folder will start their associated applications. Note that the taskbar at the bottom of the desktop contains an entry for the Road Runner folder.

The lower right-hand corner of the desktop contains the *system settings area,* where several background tasks (virus protection, speaker volume, and the time-of-day clock) are represented by small icons. The icons may be hidden if necessary, to make more room on the taskbar.

Finally, the background of the desktop itself is an object that has properties that we can alter, such as what type of image to display, the overall display resolution, and the number of colors available. In the following sections we will discuss each of the many desktop components in greater detail.

THE START BUTTON

Microsoft made it easy to begin using Windows by placing the Start button at the lower left-hand corner. Everything it is possible to do in Windows is accessible through the Start button. Left-clicking on the Start button generates the menu shown in Figure 12.2. The Shut Down, Run, and Help items are selected by left-clicking on them. The other four items

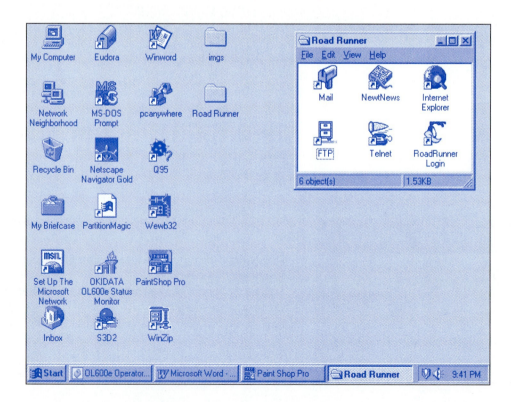

FIGURE 12.1 Sample Windows desktop

FIGURE 12.2 Start menu

FIGURE 12.3 Windows 95 Shut Down menu

produce submenus when selected by the mouse (which only needs to pass over the menu item to create the associated submenu). Let us look at each menu item and its operation.

The Shut Down item, when left-clicked, produces the display shown in Figure 12.3. The display is darkened, except for a bright Shut Down menu. You may back out of Shut Down by left-clicking the No button (or the Close box). You may also choose to restart the computer in the MS–DOS mode of operation.

The Run menu item allows a program to be executed by typing the file name in the text box or by selecting it from a graphical file menu accessed through the Browse button. Figure 12.4 shows a Run dialog box with the file name "scandisk" entered in the text box. The Run menu gives you one method to access programs that are not contained on the desktop.

The next Start menu item, Help, provides access to the help facility, which is vastly improved over that of Windows 3.x. The Help menu is shown in Figure 12.5. Do not let the short list of topics mislead you. There is so much help available that you could spend days reading the various topics provided. There is help for Dial-Up Networking, managing printers, customizing your desktop and Windows environment, and much more. You are encouraged to spend some time looking through the Help information; you may pick up some valuable tips on the way.

FIGURE 12.4 Run dialog box

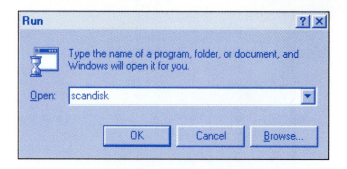

FIGURE 12.5 Windows 95 Help menu

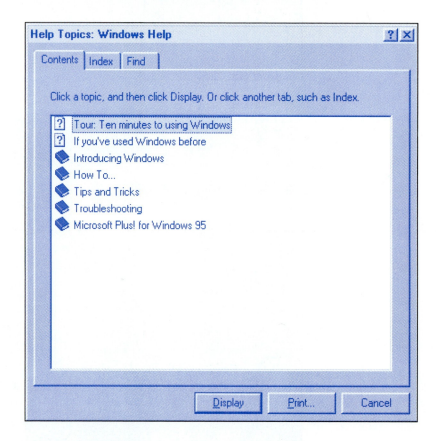

FIGURE 12.6 Windows 95 Settings submenu

    The Settings menu item opens up a submenu to three other applications, as illustrated in Figure 12.6. The Control Panel is a powerful set of utilities that are used to control how the hardware and software work under Windows. The Control Panel is the subject of Exercise 13. The Printers item brings up the Printers folder, which we will examine in Exercise 15. The third item, Taskbar, opens up the Taskbar Properties window, which we will look at in the next section. The contents of the Start menu are controlled using the Taskbar option from the Settings submenu.

    The Documents menu item produces a submenu containing links to the last 15 documents opened. As shown in Figure 12.7, files such as Microsoft Word documents appear on the Documents submenu, placed there automatically when opened by the user.

FIGURE 12.7 Documents submenu

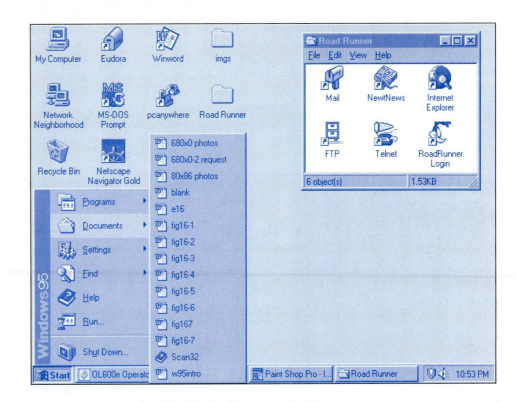

FIGURE 12.8 Programs submenu

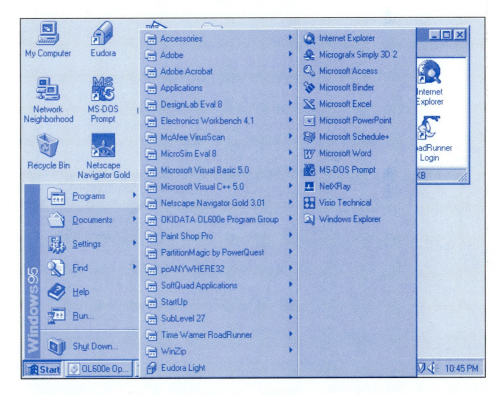

The last item on the Start menu is the Programs menu item, the starting point when starting a new application. Placing the mouse over the Programs menu item generates a submenu similar to that illustrated in Figure 12.8. The Programs submenu consists of folders and program icons. Each folder icon and associated folder name shown in the list of applications indicates that a submenu exists for that particular item. Also note the small black triangles that point to the right on many of the menu items. These indicate that

143

submenus exist for the menu items. As the mouse is dragged over this type of menu item, a new submenu will appear.

When the mouse is dragged over a program icon (NetXRay, for example), the icon becomes highlighted, and its associated application can be launched by simply left-clicking on it.

As new applications are installed, they are automatically added to the list of applications on the Programs submenu.

If you change your mind and do not want to run an application from the Programs submenu, simply left-click anywhere on the desktop outside the submenu area or choose a different menu item from the Start menu.

THE TASKBAR

The taskbar is used to switch between tasks running on the desktop. If no tasks are running, the taskbar contains only the Start button and the System Settings area. When other applications are running, or folders are open on the desktop, their icons will be displayed on the taskbar. In Figure 12.1 the taskbar contains four entries. The first three are for applications, the fourth is for the open Road Runner folder. Program and folder entries in the taskbar appear from left to right in the order they were started or opened. The taskbar can be resized, moved, and set up to automatically hide itself when not in use. Figure 12.9(a) shows a taskbar with so many applications displayed on it, it is impossible to read the names of any of them. In Figure 12.9(b) the same taskbar has been resized to allow more space for each application to be identified. To resize the taskbar, move the mouse near the top of the taskbar. It will change into an up/down arrow. Left-click and hold to drag the taskbar to a larger or smaller size, and then release.

The taskbar can also be moved to any of the four sides of the display. Figure 12.10 shows the taskbar moved to the right-hand side of the display. To move the taskbar, left-click and hold on any blank portion of the taskbar, then drag the mouse to the desired side of the display and release.

The taskbar has its own set of properties (such as its Auto hide feature) that can be adjusted. Recall that the Settings menu item of the Start menu displays a submenu containing a taskbar item. This is the entry point to the Taskbar Properties window. You can also right-click on a blank area of the taskbar and get a context-sensitive menu containing a Properties item you can click. Figure 12.11 shows the contents of the Taskbar Properties window.

You should experiment with each of the settings to see what they do and to settle on a format for the taskbar that you find pleasing. For example, if the Always on top box is unchecked, it is possible for an application to completely cover up the taskbar when it is opened up to full screen size. The only way to get the taskbar back is to minimize all applications or move applications around on the desktop to uncover the taskbar. Checking the Always on top box allows Windows to do the housecleaning for you, and keep the taskbar visible no matter how many applications are open (unless the Auto hide feature is enabled).

THE BACKGROUND

The look of the desktop background can be adjusted for a pleasing appearance. Right-clicking anywhere on an empty portion of the desktop brings up the Display Properties

FIGURE 12.9 Two views of the taskbar

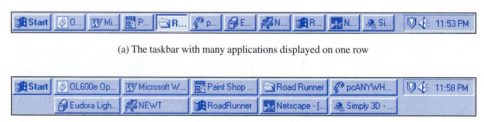

(a) The taskbar with many applications displayed on one row

(b) The same taskbar resized to two rows

FIGURE 12.10 A different location for the taskbar

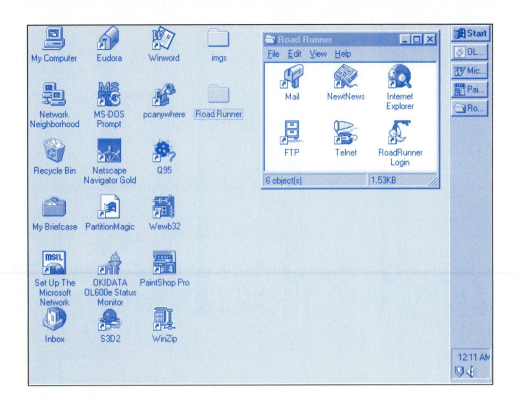

FIGURE 12.11 Taskbar Properties window

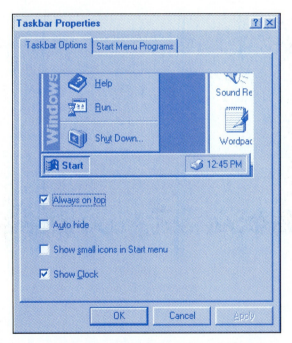

window shown in Figure 12.12. The Background controls allow you to choose a pattern or wallpaper to paint over the background. Both may be selected by double-clicking on their entry. Custom wallpapers can be selected using the Browse button, which allows selection of a *bitmap* file (.BMP extension), a standard Windows file format used for graphic images. The bitmap files displayed in the list are located in the directory where the Windows operating system is installed.

When a pattern or wallpaper type is selected with a single click, the display icon shows an example of what you will see. This is illustrated in Figure 12.13. Note that the pattern chosen will completely cover the background of the desktop. If a wallpaper is

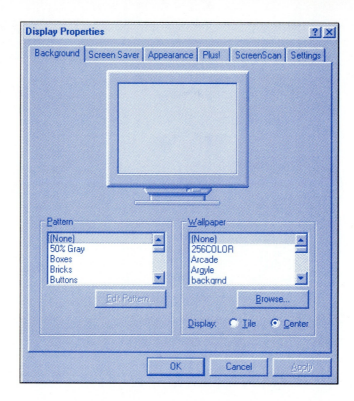

FIGURE 12.12 Display Properties window showing Background controls

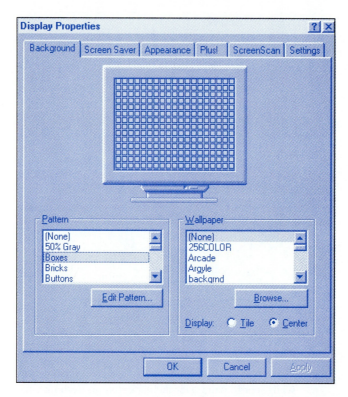

FIGURE 12.13 Choosing the Boxes pattern

selected instead, you may choose to tile the wallpaper (repeat the wallpaper pattern over and over to cover the entire desktop) or simply center the wallpaper pattern, leaving the rest of the desktop filled with a blank background or a patterned background. All of these options are shown in Figure 12.14.

In Figure 12.14(a), notice that icons are drawn on top of the background pattern in a special way so that you can still read their names. In Figure 12.14(b), icons and folders are drawn on top of the background wallpaper, which is centered. The desktop area not

FIGURE 12.14 (a) Boxes pattern, no wallpaper; (b) Forest wallpaper (centered), no pattern

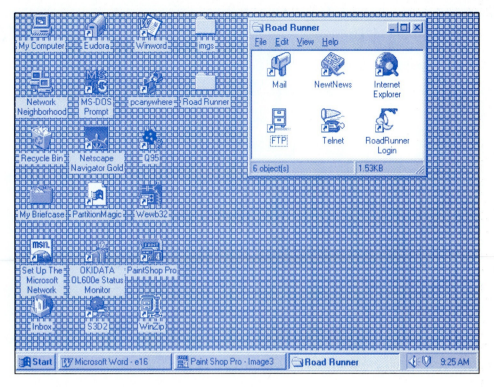

(a)

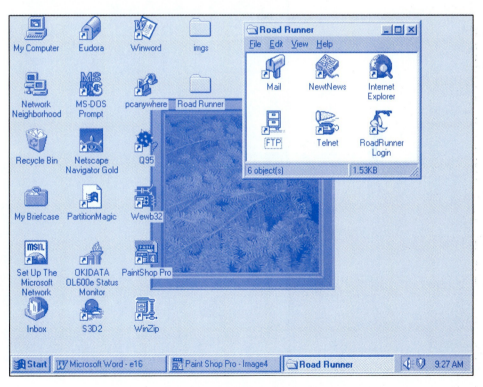

(b)

covered by a small wallpaper image can be patterned, as shown in Figure 12.14(c). If the wallpaper is tiled, the entire surface of the desktop is covered by repeating the wallpaper image horizontally and vertically as many times as necessary. This last option is shown in Figure 12.14(d).

FIGURE 12.14 *(continued)* **(c)** Bricks pattern and Forest wallpaper (centered), and **(d)** Forest wallpaper (tiled)

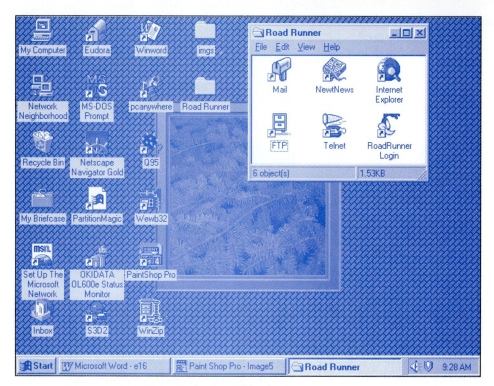

(c)

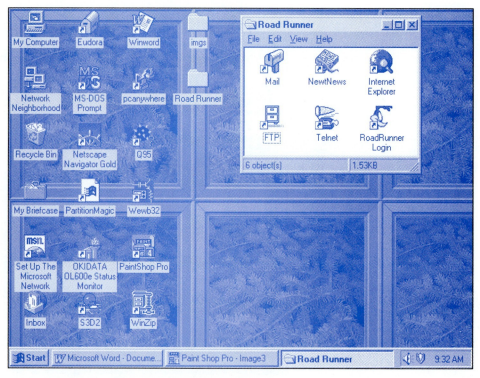

(d)

THE SYSTEM SETTINGS AREA

The system settings area of the taskbar typically contains a speaker icon (for controlling the speaker volume) and a 24-hour clock. Just placing the mouse near the time display causes a small pop-up window to appear with the full date displayed. This is illustrated in Figure 12.15.

FIGURE 12.15 Pop-up date window

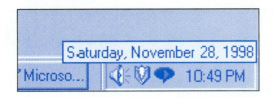

FIGURE 12.16 Date/Time Properties window

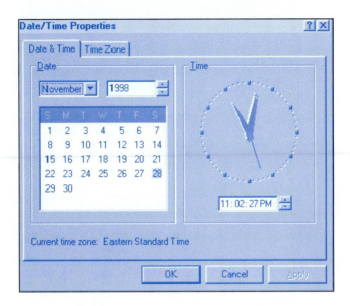

Double-clicking on the time display brings up the Date/Time Properties window illustrated in Figure 12.16. It is easy to change any part of the time or date with a few left-clicks or keystrokes. Windows will even adjust its clock automatically twice a year to properly "spring ahead" or "fall back" as required.

Some applications place their icons in the system area displayed to the left of the time. By placing the mouse over these icons, their current status is displayed. The applications are accessed by single-clicking or double-clicking them. We will examine these items in detail when we discuss accessories in Exercise 16.

CREATING AND USING SHORTCUTS

Some applications add their shortcut icons to the desktop automatically during installation. A shortcut to any application installed on a computer can be added to the desktop by simply right-clicking on an empty portion of the desktop and selecting Shortcut from the New submenu. The Create Shortcut dialog window is displayed as shown in Figure 12.17.

Simply type the name of the file to be added, or click the Browse box to locate the desired file. When the Browse box is selected, the Browse window is displayed. Figure 12.18 shows a sample Browse window for the C: drive. From this dialog box, any program located on the local hard drive or network hard drive can be located. For example, suppose we want to add a shortcut to the Calculator program. Search for the application using the Browse window. After the Calculator program is located and selected, simply click the Open button to return to the Create Shortcut dialog box. Notice that the correct path and program name is displayed in the Command line text box shown in Figure 12.19.

By selecting the Next button, Windows provides an opportunity to enter a name for the icon (which is displayed under the icon on the desktop). The name of the file is used by default but can be changed by simply typing a new name in the text box, as illustrated in Figure 12.20.

When the Finish button is selected, Windows creates the shortcut and adds the program icon to the desktop. Figure 12.21 shows the desktop with the new Calculator icon.

FIGURE 12.17 Create Shortcut dialog window

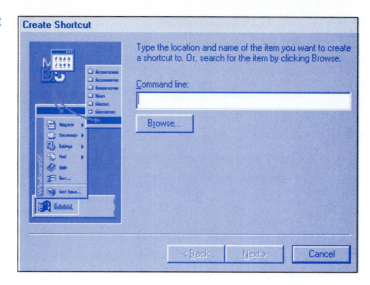

FIGURE 12.18 Browse window

FIGURE 12.19 Create Shortcut window with file selected

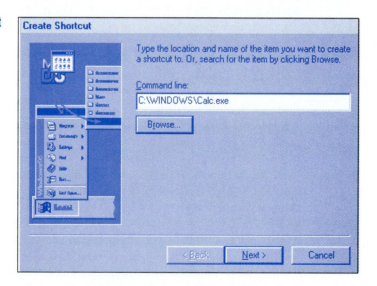

CHANGING OTHER DESKTOP PROPERTIES

Several other desktop properties you may want to experiment with are the Screen Saver, Appearance, and Settings items. Screen savers are programs that run in the background, only performing their screen saving when there has been no mouse or keyboard activity for a predetermined time period. A typical screen saver draws an interesting shape on the

FIGURE 12.20 Naming the shortcut

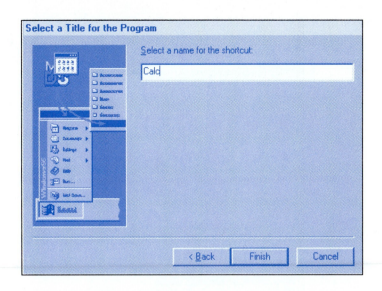

FIGURE 12.21 The desktop with the new shortcut

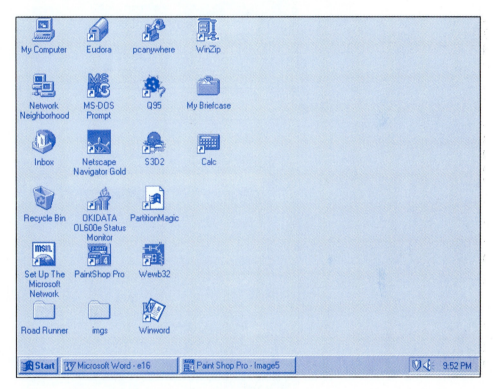

screen or performs some other graphical trick, keeping most of the screen blank. This is done to prevent an image from being burned into the phosphor coating on the display tube. When a key is pressed, or the mouse is moved, the screen saver restores the original display screen and goes back into the background.

Screen savers can also be password protected, to prevent unauthorized access to your desktop if you are away from your computer. Figure 12.22 shows the Screen Saver control window, with a small example of the selected screen saver being displayed.

The appearance of the desktop, from the color used in the title bar of a window, to the font of the desktop text, can be adjusted using the Appearance controls in the Display Properties window. Practically everything you see in the Appearance window illustrated in Figure 12.23 is clickable, including the scroll bar shown in the active window, which allows its width to be set. If the defaults for each item are not acceptable, spend some time tailoring them to suit your needs.

The Settings tab on the Display Properties window provides access to another important set of controls. As shown in Figure 12.24, the Settings controls allow you to change the size

FIGURE 12.22 Selecting a screen saver

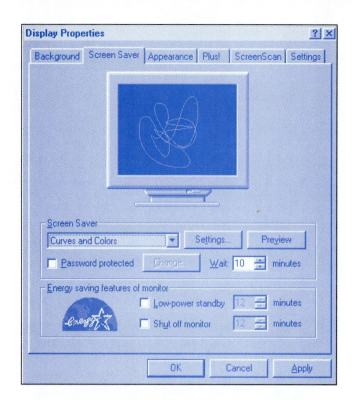

FIGURE 12.23 Desktop appearance controls

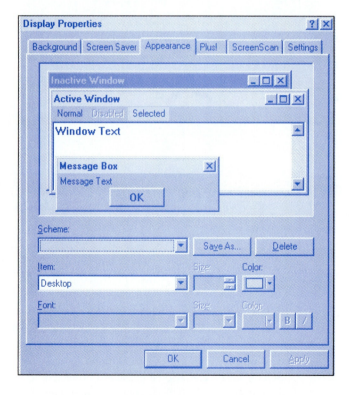

(also called *resolution*) of the desktop, the number of colors available, and the type of display connected to the computer. A small sample of the desired desktop is displayed to assist you when adjusting the settings.

MANAGING THE DESKTOP CONTENTS

The degree of control you have over the appearance of the desktop can be adjusted in many ways. For example, if you like to place new desktop icons (shortcuts or newly installed

FIGURE 12.24 Display Properties window

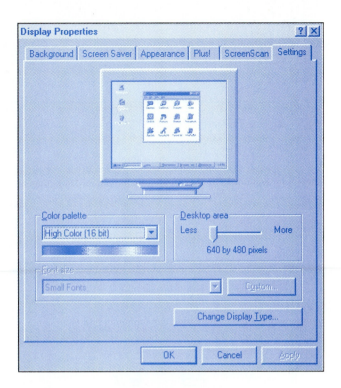

FIGURE 12.25 Arrange Icons submenu

application folders) in specific locations, you may do so by left-clicking on them and holding the mouse button down as you drag the icons to their new positions on the desktop (drop them by releasing the mouse button). If you want Windows to manage your icons for you, or if you need to prevent inexperienced users from messing up the appearance of the desktop, select the Auto Arrange option from the Arrange Icons submenu shown in Figure 12.25. Right-clicking on a blank portion of the desktop brings up the context-sensitive menu.

When Auto Arrange is enabled, icons dragged from their positions will snap back in place when released. The desktop icons will remain in an orderly arrangement.

Many applications install folders on the desktop (such as the Road Runner folder shown in many of the figures in this exercise). The desktop itself is actually a special folder, and you can view its contents using Windows Explorer. This is demonstrated in Figure 12.26. In Exercise 14 you will see exactly how to do this, and many other useful operations, such as creating folders and moving folders to different locations.

MY COMPUTER

Double-clicking the My Computer icon opens up the folder shown in Figure 12.27. The contents will vary from machine to machine, depending on the actual hardware installed on each one. Double-clicking on any of the drive icons will open a directory window for the selected drive. The properties of a disk can be displayed by right-clicking on any of the drive icons and selecting Properties from the pop-up menu. The Disk Properties window shown in Figure 12.28 shows the disk label and current usage. The used space, free space, and capacity are displayed in bytes.

FIGURE 12.26 Viewing the desktop with Windows Explorer

FIGURE 12.27 Contents of My Computer

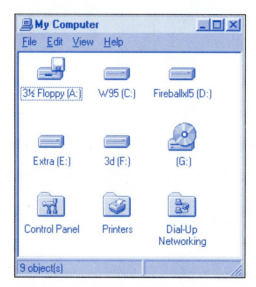

FIGURE 12.28 Disk Properties window

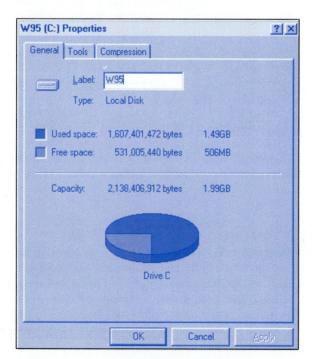

FIGURE 12.29   Disk tools

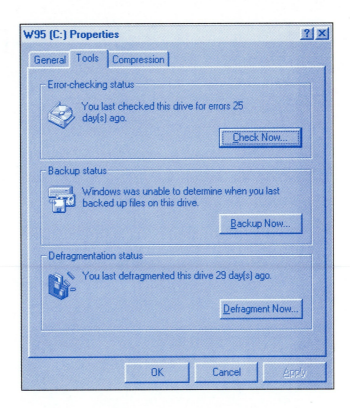

The Tools tab illustrated in Figure 12.29 shows the current status of each of the built-in disk tools provided by Windows. Each tool displays how long it has been since the utility has been run.

The Compression tab controls disk compression options. These utility programs will be examined in Exercise 16. The other system features are accessed and controlled through the Control Panel, Printers, and Dial-Up Networking folders, all of which we will cover in detail in Exercises 13, 15, and 17, respectively.

THE NETWORK NEIGHBORHOOD

If your computer is connected to a network, via a network interface card or a serial PPP (point-to-point protocol) connection, the Network Neighborhood icon, when clicked, allows you to examine other machines connected to the same network. Note that the other machines must be running a protocol called NetBEUI for them to be included in the network neighborhood. NetBEUI is part of Windows for Workgroups 3.11 and one of many protocols used with Windows.

As Figure 12.30 shows, the Network Neighborhood looks similar to a directory tree. In fact, files and printers may be shared among computers participating in the Network Neighborhood. Thus, one laser printer can serve the needs of a small laboratory or office.

THE RECYCLE BIN

Whenever anything is deleted in Windows, from executable programs to text files, images, and desktop icons, it is not yet gone for good. The exception to this rule are files deleted while running inside a DOS window. They are simply gone, unless the old UNDELETE command is still active. When you delete an item, Windows prompts you with a question, allowing you to change your mind, if necessary (as indicated in Figure 12.31).

The first destination of a deleted item is the Recycle Bin. Double-clicking the Recycle Bin icon on the desktop brings up the Recycle Bin window, which lists all files, if any, that have been deleted. Figure 12.32 shows the Recycle Bin window containing several deleted

FIGURE 12.30 Network Neighborhood

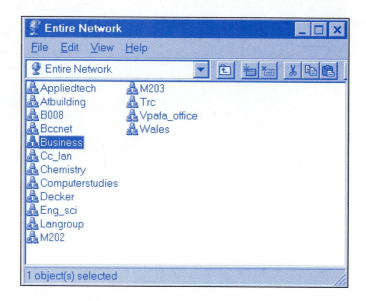

FIGURE 12.31 One of the Windows deletion safeguards

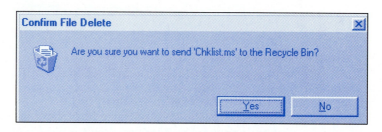

FIGURE 12.32 Contents of the Recycle Bin window

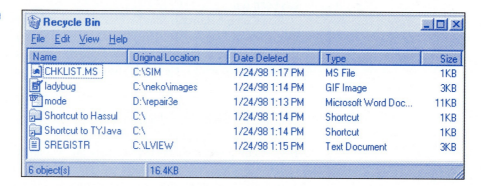

files. Under the File menu, the Empty Recycle Bin option is used to finally delete the files from the hard drive. Windows provides one final confirmation window, to make sure you really want to delete the items in the Recycle Bin. To get a file back, simply select it and choose the Undo Delete option from the Edit menu (or Restore from the File menu).

One thing to keep in mind is that the amount of free disk space does not change until the Recycle Bin is emptied. This is because a copy of the deleted file is stored in the Recycle Bin, which stores its contents on the hard drive.

MY BRIEFCASE

When you need to work on one or more files on two different computers (home and office machines), the My Briefcase utility can be used to organize and properly update the files. Files are copied into the Briefcase by dragging them onto the My Briefcase icon while in Windows Explorer. The entire Briefcase is copied onto a floppy disk by dragging its icon onto the icon for the floppy drive.

After the Briefcase files have been modified on a different computer, it is necessary to update the original files. Dragging the Briefcase icon from the floppy drive back to the desktop will cause Windows to evaluate the contents of the Briefcase and determine which files need updating.

THE INBOX

The Inbox is the central location for all e-mail activity. The first time the Inbox is selected, the Inbox Setup Wizard will begin the configuration process. The Microsoft Network, Microsoft Mail, and Microsoft Fax selections are available. Figure 12.33 shows the Inbox Setup Wizard window.

When the Inbox is selected after the services have been configured, the Microsoft Exchange application will start and begin to perform all e-mail and fax services. Figure 12.34 shows a Microsoft Exchange mail window with several e-mail messages.

E-mail is more than just letters and numbers. You can send and receive all types of files, from graphic images to executable programs. Typically, these types of files are called *attachments*. The paper clip on the fourth e-mail message in Figure 12.34 indicates an attachment exists for that message. In general, double-clicking the attachment brings up the application associated with the attachment's file type. Or, if the attachment is an executable program, it is executed when double-clicked.

All in all, e-mail provides services that are essential to modern-day computing, business, and education. Spend some time learning to use this feature of the Windows operating system.

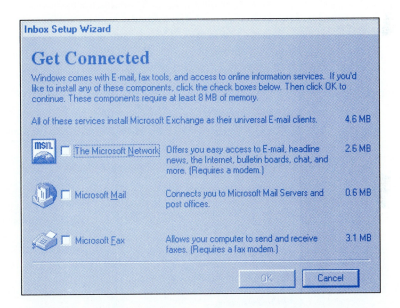

FIGURE 12.33 Inbox Setup Wizard window

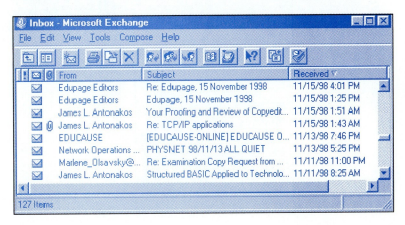

FIGURE 12.34 Microsoft Exchange mail window

GETTING BACK TO THE DESKTOP FROM DOS

If you are running in a full-screen MS-DOS window, you can exit back to the desktop by entering EXIT at the DOS prompt. If you want to keep the DOS session active, but also to be able to access the desktop, pressing ALT-ENTER will automatically resize your DOS environment to a window and bring back the desktop.

WINDOWS 98 HELP

One of the first differences between Windows 95 Help and Windows 98 Help is the appearance of the Help window. As shown in Figure 12.35, the Windows 98 Help window displays the Windows 98 logo alongside the Contents window. When a Help item is clicked in the Contents window, the associated Help information replaces the Windows 98 logo. This is a nice change from Windows 95 Help, which opened a new Help-specific window, making it awkward to return to the Contents window.

WINDOWS 98 SETTINGS MENU

Figure 12.36 shows the Settings menu found in the Windows 98 Start menu. Three new selections are available: Folder Options, Active Desktop, and Windows Update. The Folder Options selection allows you to choose how folders appear and function (single-click to open versus double-click), and handles file associations. Additional desktop properties can be examined or modified using the Active Desktop selection.

Selecting Windows Update automatically connects your computer to Microsoft's Windows update site on the Web. Once connected, you can pick and choose the updates

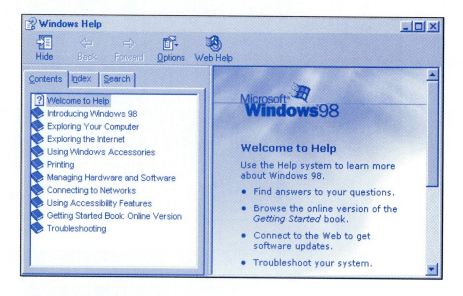

FIGURE 12.35 Windows 98 Help window

FIGURE 12.36 Windows 98 Settings menu

your Windows system requires. Software patches are downloaded and executed to make the necessary changes.

WINDOWS 98 EXPLORER

Under Windows 98, the Windows Explorer has been updated to work with the Web in addition to the desktop. Figure 12.37 illustrates how Explorer integrates the Web as just another item on the desktop. In addition to displaying lists of files and folders, Explorer is also designed to display Web content. Look along the left-hand side of the Explorer screen to see that the Voltage Regulator HTML document displayed is found in the Internet Explorer folder. The address of the HTML document is shown in the Address field.

WINDOWS 98 WEB-STYLE DESKTOP

To further integrate the Web into the Windows 98 computing environment, you have the option of configuring your desktop to act like a Web page. Folders can be opened or applications launched with a single left-click, instead of the usual double-click. When applications are launched in this way, the application is selected by simply moving the mouse over the desktop icon. As Figure 12.38 shows, the names under each icon have the familiar underline associated with clickable links on a Web page. The Web-style desktop can also be configured so that the underline is displayed when the icon is selected. With all the available choices, no two desktops look the same.

WINDOWS NT DESKTOP EVOLUTION

The Windows NT desktop has evolved along the same path as the Windows desktop. The first version, Windows NT 3.5, took on the appearance of the Windows 3.1 desktop. This is because the first version of Windows NT was an industrialized version of Windows 3.1.

The desktop of the second version of Windows NT, version 4.0, looks like the Windows 95 desktop. The next release of Windows NT, Windows 2000, is expected to look a lot like Windows 98.

FIGURE 12.37 Windows 98 Explorer window displaying a Web page

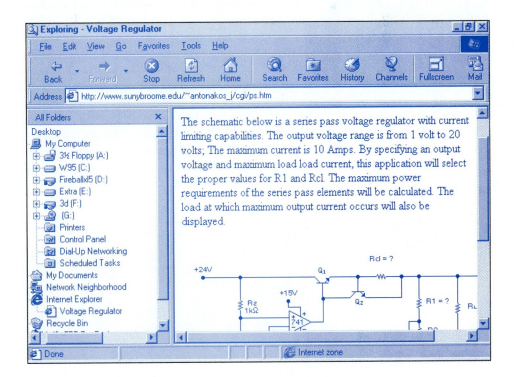

FIGURE 12.38 Windows 98 Web-style desktop

The release of Windows Explorer Version 4.0 offered the Windows 98 look and feel to the Windows 95 and Windows NT desktops. To determine the current operating system, look at the Start menu. Along the left-hand side is a graphic indicating the operating system version.

TROUBLESHOOTING TECHNIQUES

It is important to be patient with Windows. For example, suppose you have a number of applications open, and have just clicked the Close button on one of them, expecting to see the application close instantly; instead, you see nothing happen for a long time.

An impatient user may decide that Windows has died and simply turn the computer off. This is not the recommended approach (there may be important information that needs to be backed up to the hard drive). In this case, where Windows has "gone away" for a long time, attempt to switch to another application on the taskbar (or press Ctrl-Alt-Del to bring up the Close Program window). Just seeing the mouse pointer move when you use the mouse is a good sign that the operating system is still listening.

You may find that, eventually, Windows finishes closing the "dead" application, or reports an error message like the one shown in Figure 12.39. There is not really much that can be done for the application if an illegal operation is detected, but at least Windows catches these problems and does not crash because of them.

FIGURE 12.39 Illegal operation detected

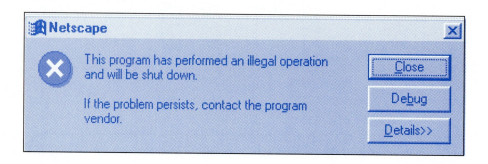

SELF-TEST

This self-test is designed to help you check your understanding of the background information presented in this exercise.

True/False

Answer *true* or *false*.

1. The taskbar contains only the Start button, running applications, and the system settings area.
2. A shortcut for any application can be added to the desktop.
3. Context-sensitive help is displayed whenever the mouse is simply moved on top of any desktop icon.
4. The Documents menu item on the Start menu contains important help information for Windows.
5. When looking for help, enter HELP on the keyboard.
6. Ctrl-Alt-Del instantly reboots the computer.
7. The background may contain a pattern and a wallpaper.
8. The size of the desktop is fixed.
9. Right-clicking a desktop icon will cause its application to start up.
10. The name and position of each desktop icon may be changed.

Multiple Choice

Select the best answer.

11. The Windows desktop is
    a. Part of the installation process.
    b. An essential component of the operating system.
    c. An optional application program.
12. When files are deleted in a DOS window, they are
    a. Gone for good.
    b. Sent to a temporary directory.
    c. Sent to the Recycle Bin.
13. When files are deleted in Windows, they are
    a. Gone for good.
    b. Sent to a temporary directory.
    c. Sent to the Recycle Bin.
14. The My Computer icon displays information about
    a. Each disk drive.
    b. Folders for the printer and control panel.
    c. Both a and b.
15. Wallpaper files are image files called
    a. Pixmaps.
    b. Bitmaps.
    c. Pic files.
16. To quickly bring up a context-sensitive display menu
    a. Left-click anywhere on the desktop.
    b. Right-click anywhere on the desktop.
    c. Press M on the keyboard.
17. Left double-clicking a disk icon in My Computer
    a. Formats the disk.
    b. Displays the disk directory.
    c. Displays the disk properties.
18. If Windows goes away for a long time
    a. Just turn the computer off.
    b. Press every key on the keyboard and move the mouse.
    c. Press Ctrl-Alt-Del and see what happens.

19. When Auto Arrange is enabled, the desktop icons
    a. Periodically move themselves to new locations.
    b. Snap back into place if moved.
    c. Converge at the center of the screen.
20. Windows help is
    a. Minimal due to lack of disk space.
    b. Only available on CD-ROM.
    c. Online and significantly better than Windows 3.x.

Completion

Fill in the blank or blanks with the best answers.

21. The Start button is located on the _____.
22. The desktop contains the _____ Neighborhood icon.
23. Wallpaper may be centered or _____.
24. The Programs submenu is located on the _____ menu.
25. The _____ is used to maintain a set of files that are used on more than one computer.
26. When a file is deleted, it goes to the _____ _____.
27. A(n) _____ takes over the display screen when there is no user activity for a length of time.
28. Electronic mail and fax services are provided by selecting the _____ icon.
29. The size of the display is set in the Display Properties _____ menu.
30. To exit from a DOS shell but keep the session running, press the _____ and _____ keys.

FAMILIARIZATION ACTIVITY

1. Create a shortcut on the desktop to the Calendar application in the Windows directory and demonstrate its use.
2. Experiment with the desktop appearance by choosing five different patterns and five different wallpapers.
3. Compare the centered and tiled options for three different wallpapers.
4. Move the taskbar to all four corners of the display and vary the size. Explain your preference for the normal position of your taskbar.
5. Create several small text files with a text editor or word processor. Put a copy of each file into the Briefcase. Drag the My Briefcase icon onto the icon for your floppy drive. Take your floppy to a different computer and drag the Briefcase from the floppy onto the desktop. Edit the files in the Briefcase. Drag the Briefcase back to the floppy. Put the floppy in the original computer and drag the Briefcase back onto the desktop. Double-click the My Briefcase icon and examine its contents. Update the files in any manner you choose.
6. Click the time display on the taskbar. Use the up and down arrows to determine the first and last year recognized by Windows. Can you grab the minute hand and move it around the clock?
7. Enable the Auto hide feature of the taskbar and experiment with the mouse to determine how and when the taskbar will reappear. Resize the taskbar. How large and small can it be?
8. Create a folder on the desktop and name it Temp. Drag some of the other desktop icons into it. Open the Temp folder and drag the icons back to the desktop one at a time. Do the icons snap into place or can you put them anywhere you choose?
9. Open the Recycle Bin. If there are already files in it, determine when they were put there and whether it is safe to delete them. Open a DOS window. Do a directory listing of the current drive and make note of the free space on the drive. Now empty the Recycle Bin. Has the free space changed?
10. Slowly move your mouse over the items in the Programs submenu as if choosing an application to start up. Do you encounter any difficulty trying to select any of the applications?

QUESTIONS/ACTIVITIES

1. If you were a Windows 3.x user, how would you explain the similarities between Program Manager and the Windows desktop?
2. Why is it good to be patient with Windows?
3. If the operating system looks dead, what might you do to check for signs of life?
4. Does the taskbar have to be at the bottom of the screen? Does it have to be on the screen at all?
5. What is the Network Neighborhood?
5. Explain how to produce the Display Properties menu.

REVIEW QUIZ

Within 10 minutes and with 100% accuracy,

1. Demonstrate the different features of the Start menu.
2. Identify items on the taskbar.
3. Change the appearance and properties of the desktop.
4. Create and use shortcuts.
5. Explain the function of the standard desktop icons.

13 The Control Panel

INTRODUCTION

Joe Tekk was sitting at his desk, eating lunch in front of his computer. His Windows 98 desktop showed the open Control Panel folder. Joe selected an icon from the bottom row, ODBC, and examined all of its various submenus and options, never actually selecting any of them.

Don, Joe's manager, saw Joe's screen and was concerned. "What are you doing in Control Panel? Is there something wrong with your system?"

"No, Don," Joe answered. "I'm just checking out some of the operations I never use. Every now and then I discover one that has some new feature I am looking for."

PERFORMANCE OBJECTIVES

Upon completion of this exercise, within 10 minutes and with 100% accuracy, you will be able to

1. Identify several of the many useful icons in the Control Panel and explain their functions.
2. Use the Control Panel to modify system properties such as time/date, sounds, the display format, and keyboard/mouse functionality.
3. Show how printers and modems are configured.

BACKGROUND INFORMATION

The Control Panel is one of the most important tools included in Windows. Just like the Control Panel included with Windows 3.x, the Windows Control Panel allows you to fine-tune and customize the hardware and software of your system. The degree of control and the number of features available is much improved over Windows 3.x. Figure 13.1 shows three views of the Control Panel. Note the similarities in all three Control Panels. Windows 95 and Windows 98 are the most similar. The Windows NT 4.0 Control Panel shares many of the same icons, but has several icons that are unique to the Windows NT 4.0 environment, such as Devices, Licensing, and Services. As indicated in Figure 13.1, nearly every aspect of the operating system can be controlled, examined, or configured through one of the many Control Panel icons. Let us examine the operation of several important icons.

FIGURE 13.1 (a) Windows 95 Control Panel, (b) Windows 98 Control Panel, and (c) Windows NT 4.0 Control Panel (Web-style format)

(a)

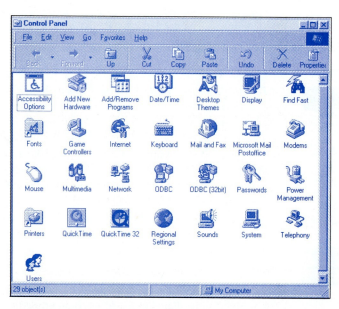

(b)

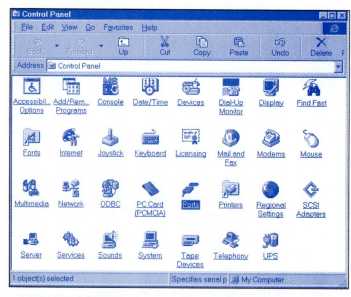

(c)

FIGURE 13.2 Accessibility Properties menu

ACCESSIBILITY

Figure 13.2 shows the Accessibility Properites menu. For users who want to customize how the keyboard, sound, display, and mouse operate, this menu provides the way. As indicated in Figure 13.2, a lot of attention has been paid to how each key on the keyboard behaves. Other accessibility options involve generation of short tones when Windows performs an operation, helpful pop-up messages, using the arrow keys to control the mouse, and changing the display colors to high-contrast mode for easier viewing.

ADD NEW HARDWARE

The process used to add new hardware to Windows 95 or 98 usually involves two steps. First, the new hardware device must be physically added to the computer. This requires removing the cover from the computer, identifying the proper location to install the new hardware, and then performing the actual installation procedure. The second step involves configuring the software to properly communicate with the new device. This is accomplished using the Add New Hardware Wizard shown in Figure 13.3. This process is discussed in detail in Exercise 19. Adding new hardware in Windows NT is different and will also be discussed in Exercise 19.

FIGURE 13.3 Add New Hardware Wizard window

FIGURE 13.4 Add/Remove Programs Properties menu

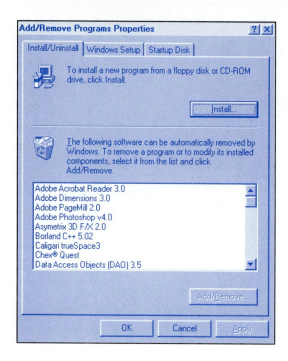

ADD/REMOVE PROGRAMS

When installing software applications in Windows, many software vendors require the installation to be performed from the Add/Remove Programs Properties menu illustrated in Figure 13.4. When the software is installed, it is registered with Windows. If at some point it is necessary to remove an application from the system, the user can simply return to the menu, select the application, and press the Add/Remove button. In addition to user application programs, the Windows operating system can also be modified by adding or removing components. This topic is discussed in Exercise 18.

DATE/TIME

This menu can be opened from inside Control Panel or by double-clicking the time display in the taskbar. The Date/Time Properties menu is shown in Figure 13.5. The user can easily change the time or date with a few left clicks. The Time Zone submenu allows you to select the time zone your computer is located in. Windows will then automatically adjust the clock for daylight saving time when required.

FIGURE 13.5 Date/Time Properties menu

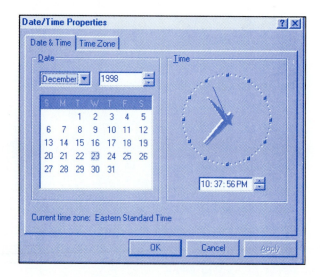

FIGURE 13.6 Display Properties menu

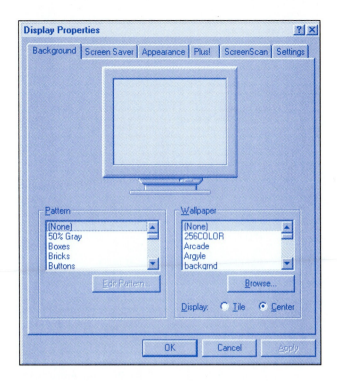

DISPLAY

The Display Properties menu is shown in Figure 13.6. Recall from Exercise 12 all of the various ways we can adjust the display. The menu in Figure 13.6 is also reachable by right-clicking on a blank area of the desktop and selecting Properties.

KEYBOARD

The Keyboard Properties menu is displayed in Figure 13.7. The Speed tab is used to set the repeat delay and repeat rate when a character is pressed and held down on the keyboard and

FIGURE 13.7 Keyboard Properties menu

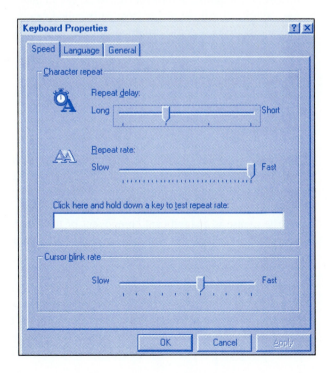

169

FIGURE 13.8 Modems Properties menu

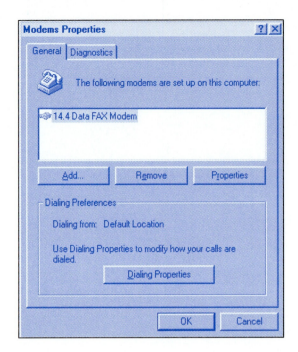

the rate at which the cursor blinks in a data entry field. Many people need to change these settings because of individual keyboarding styles. We will investigate the consequences of changing these settings during the activities at the end of this exercise.

The Language tab is used to indicate the keyboard language, and the General tab is used to identify the specific type of keyboard used, such as the Microsoft Natural keyboard.

MODEMS

Modems are very easy devices to work with in Windows. The Modems Properties menu (Figure 13.8) identifies the specific type of modem installed in your computer. The hardware installation of a modem will be covered in detail in Exercise 44.

Windows can be set up to dial from many different dialing locations. For example, many offices may require a 9 to be pressed to access an outside line; other dialing locations may not require a 9. All the places frequently called can be given a name and can be selected very easily. Even specific communication details, such as the number of data bits and the type of parity that is used, can be adjusted.

The Diagnostics submenu is useful for interrogating the modem and examining its response to commonly used modem commands.

MOUSE

This menu provides all the functional control over the mouse. The mouse can be set up for left- or right-handed operation, its double-click speed adjusted, mouse trails enabled, and appearance changed by choosing one of several scenes (3-D pointer, for example). The initial Mouse Properties menu is shown in Figure 13.9.

The mouse is a significant component of the Windows operating system. The mouse properties you choose for yourself can make your system easier to use.

MULTIMEDIA

The Multimedia Properties menu is used to control the multimedia hardware and software installed on the system. Windows comes equipped with the ability to display MPEG (Motion Pictures Experts Group) video in real time, and provides virtual device drivers for

FIGURE 13.9 Mouse Properties menu

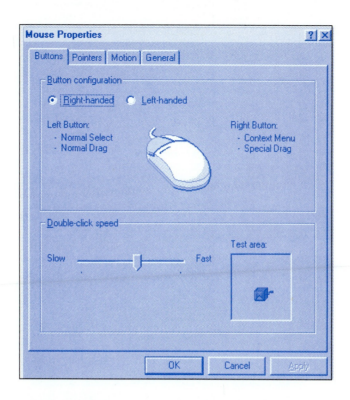

FIGURE 13.10 Multimedia Properties menu

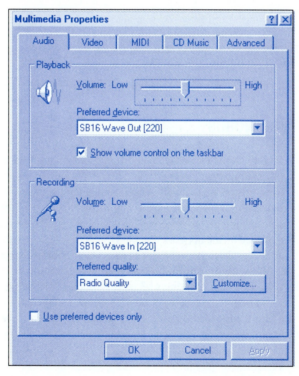

many popular sound cards and graphics accelerators. Windows can also take advantage of the new MMX technology available in the Pentium family of microprocessors. Figure 13.10 shows the initial Multimedia Properties menu.

We will examine multimedia devices in Exercise 46.

NETWORK

The Network menu allows you to add, modify, or remove various networking components, such as protocols (NetBEUI, TCP/IP), drivers for network interface cards, and Dial-Up

171

FIGURE 13.11 Network menu

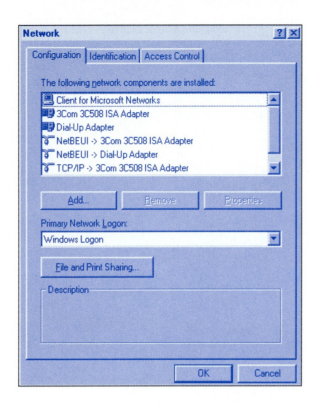

Networking utilities. You can also specify the way your machine is identified on the network, as well as various options involving file and printer sharing, and protection. Figure 13.11 shows a sample Network menu. Selecting any of the network components allows its properties to be examined.

Networking will be covered in detail in Exercise 17 as well as in Unit IV.

PASSWORDS

The Passwords Properties menu found in Windows 95 and Windows 98 is used to maintain information about passwords and security when a computer is shared among different people. Passwords are maintained in files with a .PWL extension if the computer does not participate in a Windows NT domain. The file name portion consists of the first eight letters of the user name entered at the Windows logon screen. If you are required to supply passwords for services used on the network, the passwords may be saved in a .PWL file. Unfortunately, you are allowed to delete any .PWL file you want, effectively removing all password protection for the affected user. This may be a factor in a security-conscious network.

Windows 95, Windows 98, and Windows NT can be configured to allow each computer user to maintain individual desktop settings plus other related preferences all protected by an initial password. Figure 13.12 illustrates the Passwords Properties menu. Windows NT Server can be configured as a domain controller responsible for maintaining passwords for all computers in the network.

PRINTERS

The Printers window shown in Figure 13.13 shows all the printers currently installed on a particular computer. The Add Printer icon is used to add a brand-new local or network printer to Windows. Although printers are covered in detail in Exercise 15, it is important to note the starting point for printer operations. The system tray portion of the taskbar will indicate when a printer is in use and the status of the current print job. Right-clicking on an installed printer will allow you to bring up the Properties window and change printer parameters as necessary.

FIGURE 13.12 Passwords Properties menu

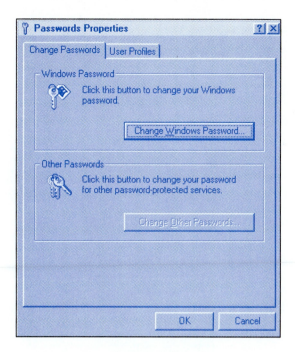

FIGURE 13.13 Printers window

REGIONAL SETTINGS

The Regional Settings menu is used to configure Windows to conform to the many different international standards. For example, the display format for numbers, currency, date, and time can all be modified. Figure 13.14 illustrates the initial Regional Settings menu. To make the job simple, simply select the specific region, and all the individual items are automatically configured. If you travel internationally, your system can automatically adjust to the new region with a few clicks of the mouse.

SOUNDS

Windows uses sounds to accompany many typical operations, such as closing an application or shutting down the computer. The sound, if any, that is played for an event is specified using the Sounds Properties menu illustrated in Figure 13.15.

Sounds are stored in .WAV files in the main Windows subdirectory. Many different sound editors are available that allow you to create or modify sounds to fully customize your Windows environment.

FIGURE 13.14 Regional Settings Properties menu

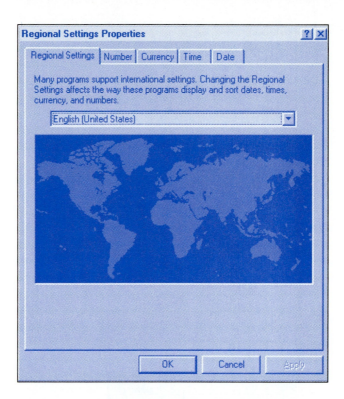

FIGURE 13.15 Sounds Properties menu

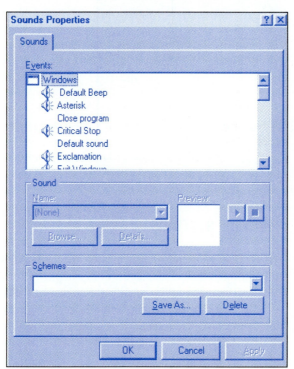

SYSTEM

The System Properties menu is the central location for all system-related information. The initial properties window for Windows 98 and Windows NT is shown in Figure 13.16. The Windows 95 menu is virtually the same as the Windows 98 menu. The operating system version, processor type, and amount of RAM are some of the system properties displayed.

The other submenus (Device Manager, Hardware Profiles, and Performance, to name a few) provide detailed access to the inner workings of the installed hardware as

FIGURE 13.16 (a) Windows 98 System Properties menu, and (b) Windows NT System Properties menu

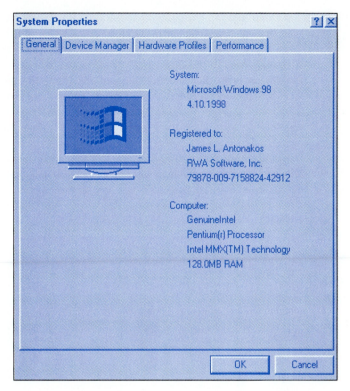

(a)

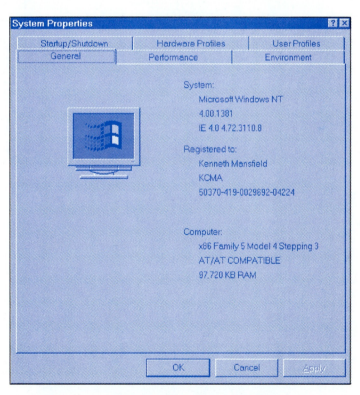

(b)

well as critical Windows variables, such as virtual memory settings and file system configuration.

There are many icons yet to be examined. You are encouraged to examine their properties on your own.

TROUBLESHOOTING TECHNIQUES

When troubleshooting on a Windows 95 or 98 computer, the System Properties menu can be used to investigate and diagnose many types of problems. The Device Manager submenu provides a single location to examine all the hardware components installed on a system. After installing a new piece of hardware, it is good practice to view the new hardware properties and make note of any configuration parameters.

On some occasions, by viewing these screens, problems may be discovered. This is indicated in Figure 13.17. When reviewing the System devices category, the exclamation point next to the Plug and Play BIOS item indicates a problem exists.

By double-clicking the Plug and Play BIOS item, the associated Properties menu is displayed, as shown in Figure 13.18. The device status portion of the menu indicates that the device drivers for the Plug and Play BIOS have not been installed correctly and then goes on to suggest that the user click on the Drivers submenu tab to change the drivers. When the problem is resolved, the error indication will be removed from the Device Manager display.

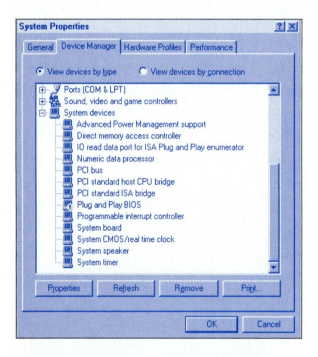

FIGURE 13.17 System Properties reports an error

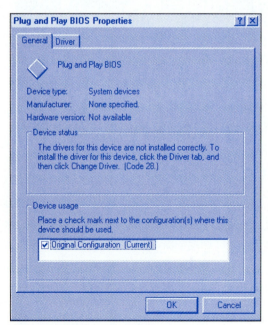

FIGURE 13.18 A course of action is suggested

It pays to become familiar with the System Properties menu when troubleshooting problems using Windows 95 or 98.

In Windows NT, hardware errors are written to a log file, which can be examined for details concerning the nature of the problem.

SELF-TEST

This self-test is designed to help you check your understanding of the background information presented in this exercise.

True/False

Answer *true* or *false*.

1. The Control Panel is used to control hardware only.
2. Only one language can be used with the keyboard.
3. Windows takes advantage of MMX technology.
4. Each user of a computer can maintain his or her own desktop.
5. The Printers window shows all active printers.

Multiple Choice

Select the best answer.

6. Passwords are stored in
    a. PWL files.
    b. PAS files.
    c. DAT files.
7. Changing the way numbers are displayed is performed in
    a. Multimedia Properties.
    b. Display Properties.
    c. Regional Settings.
8. Control Panel icons
    a. Are identical in all versions of Windows.
    b. Are only found in Windows NT.
    c. Establish control over almost all hardware and software properties.
9. The Systems Properties menu shows
    a. Information about the system printers.
    b. The version of the operating system.
    c. Information about system performance.
    d. All of the above.
10. The Accessibility options are used to
    a. Network several Windows machines together.
    b. Customize how the system and user interact.
    c. Control how the hard drive is accessed.

Completion

Fill in the blank or blanks with the best answers.

11. Windows can automatically play _____ video files.
12. To connect to a remote computer, we may use the _____ or _____.
13. The Network menu is used to change the _____ for the network interface card.
14. The _____ _____ submenu is used to display information about all the hardware components installed in a Windows 95 or Windows 98 system.
15. Sounds are stored in _____ files.

FAMILIARIZATION ACTIVITY

1. Go through each of the Control Panel operations covered in this exercise. Be sure to click on every tab to see the full extent of control you have over the operating system and the computer.

177

2. Briefly examine the remaining operations provided by Control Panel. Determine what ODBC stands for. Find out how to send a fax.
3. Make a Windows startup disk using Add/Remove Software. Windows NT calls this disk an emergency repair disk.

QUESTIONS/ACTIVITIES

1. How must Windows be configured to allow each user to maintain his or her own desktop settings?
2. Why is it necessary to have different dialing locations maintained inside of Dial-Up Networking?
3. How does Windows indicate problems with hardware settings?

REVIEW QUIZ

Within 10 minutes and with 100% accuracy,

1. Identify several of the many useful icons in the Control Panel and explain their functions.
2. Use the Control Panel to modify system properties such as time/date, sounds, the display format, and keyboard/mouse functionality.
3. Show how printers and modems are configured.

14 Windows Explorer

INTRODUCTION

Joe Tekk was busy using Windows Explorer to organize his hard drive. He moved directories, renamed them, created a folder of his most often used shortcuts, and copied several files from his hard drive to a floppy disk. Then he installed an application from a shared CD-ROM on RWA Software's Pentium II file server.

Don, Joe's manager, was watching. "Joe, do you ever close Explorer?" Don asked.

Joe laughed. "Sometimes, Don, but usually I leave it minimized on the taskbar, just in case I need it for something."

PERFORMANCE OBJECTIVES

Upon completion of this exercise, within 10 minutes and with 100% accuracy, you will be able to

1. Explain the basic features of Windows Explorer.
2. Show how to start an application.
3. Create a shortcut, move, find, and delete files and folders.
4. Map a network drive.

BACKGROUND INFORMATION

Windows Explorer can be thought of as the *command post* of Windows, performing duties similar to the Program Manager and File Manager utilities in Windows 3.x.

Figure 14.1 shows a typical Explorer display of the folders on drive C:. Although there are indeed similarities between Windows 95 Explorer and Windows 98 Explorer (shown, respectively, in Figures 14.1(a) and 14.1(b) for comparison), Windows 98 Explorer offers additional features because Internet Explorer 4.0 has been integrated into the desktop. For example, Web pages can be viewed in the rightmost panel, without having to first open a browser. We will examine the new Windows 98 Explorer features at the end of this exercise. Bear in mind that much of what follows regarding Windows 95 Explorer operation also applies to Windows 98 Explorer. Furthermore, Windows NT Explorer is practically identical to Windows 95 Explorer, although you may install Internet Explorer 4.0 and gain the enhancements for your Windows NT environment. For now, let us take a detailed look at Windows 95 Explorer.

FIGURE 14.1 (a) Windows 95 Explorer window, and (b) Windows 98 Explorer Window

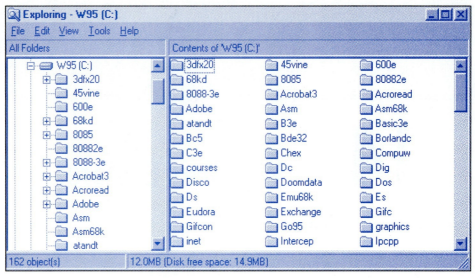

(a)

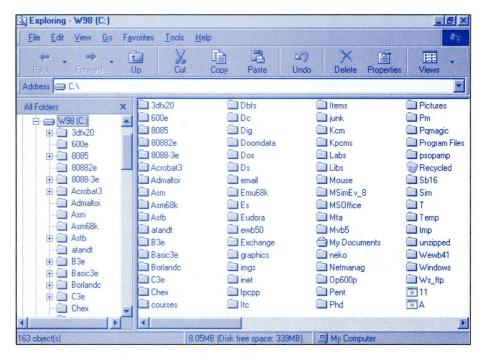

(b)

The Windows 95 Explorer window can be broken down into four areas: the pull-down menus, the folder display window, the file display window, and the status bar. The pull-down menus provide a way to access the features of Explorer, to configure it, and do many other useful things, such as map network drives. The folder display window allows the selection of any resource on the computer. This includes all the drives (floppy, hard, CD-ROM, network), special folders (Control Panel, Printers, Dial-Up Networking), and other system and user items.

The file display window shows a list of the folders and files in the currently selected location. For example, in Figure 14.1(a), the file display window shows the contents of the root directory of drive C:. Note the drive label "W95" at the top of the window. The status bar, located at the bottom of the Explorer window, shows the number of objects in the file display window as well as the amount of disk space used by the objects and the disk free space. Single-clicking on an item in the folder display window will show the contents of the item in the file display window.

FIGURE 14.2 Explorer display with graphical toolbar

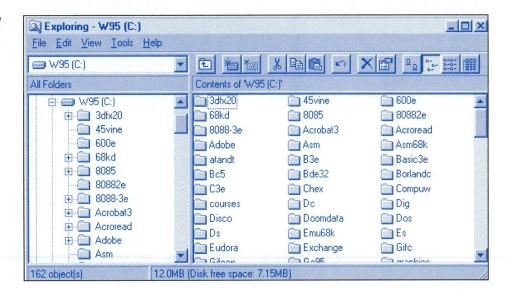

FIGURE 14.3 Explorer toolbar functions

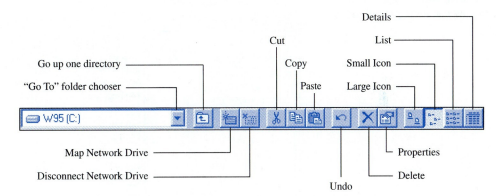

CHANGING THE VIEW

The Windows 95 Explorer interface can be customized in many ways. First, let's add a graphical toolbar to the Explorer window. This is accomplished by selecting the Toolbar option in the View menu. The new Explorer display window should look similar to Figure 14.2. Many of the pull-down menu items are represented in the graphical toolbar. Holding the mouse still over a toolbar button will produce a small pop-up description window, such as "Delete" or "Map Network Drive." The function of each button is shown in Figure 14.3.

Another way to change Explorer's view is to adjust the method used to show files and folders. These methods are as follows:

- By large icon
- By small icon
- As a list
- With details

Figure 14.4 shows the large icon format in the file display window. The files displayed in the window scroll vertically, and are listed in alphabetical order with folders first.

Figure 14.5 shows an example of small icon format. This display mode also scrolls vertically.

When files and folders are displayed using the list option, the display looks similar to that shown in Figure 14.6. Notice that the list scrolls *horizontally*, not vertically, and uses the small icon format.

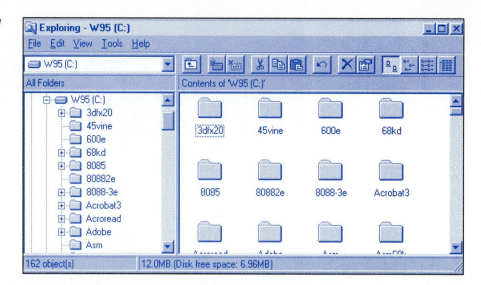

FIGURE 14.4 Explorer display window showing large icons

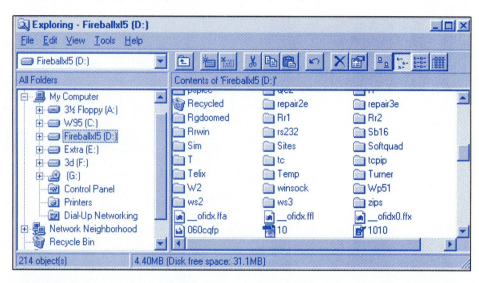

FIGURE 14.5 Explorer display window showing small icons

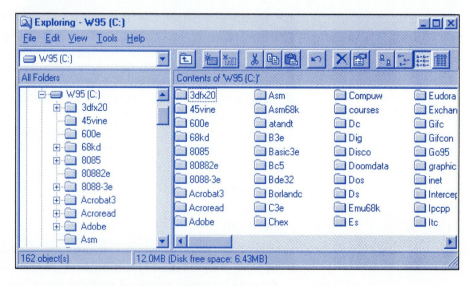

FIGURE 14.6 Explorer display window showing icon list

It may be necessary to view the details of each file. This includes the type of file or folder, its size, and last modification date. Figures 14.7 and 14.8 show examples of this display mode.

The arrangement of files and folders in the file display window is flexible (by name, date, size, and so on) and can be limited to certain types if desired. The Options selection in the View menu is used to adjust these features.

FIGURE 14.7 Explorer display window showing folder details

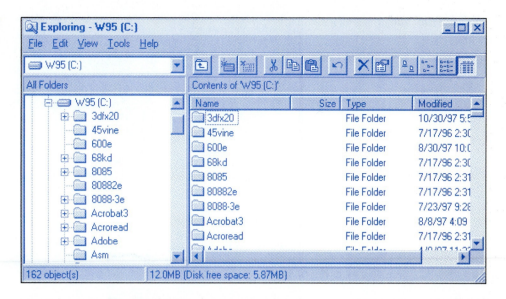

FIGURE 14.8 Explorer display window showing file details

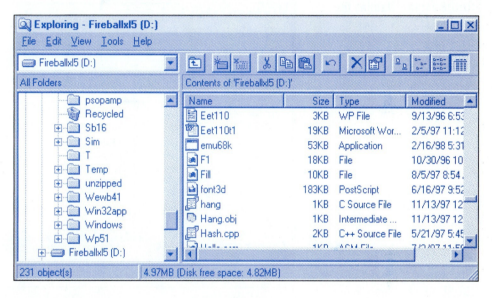

CREATING NEW FOLDERS

To help organize files, it is often necessary to create folders. By selecting "File, New, Folder," Windows 95 automatically creates the new folder in the current directory (the directory shown in the file display window). You are prompted to enter a name for the new folder.

The destination of the new folder is important. For example, if you want the folder to be in the root directory of drive E:, you must make this directory the current directory. This is easily done by left-clicking on the icon for drive E: in the folder display window. Figure 14.9 shows a new folder being created. Windows 95 automatically names the folder New Folder and gives you the opportunity to rename it by typing in a new name. Figure 14.10 shows the new folder renamed to Ken's Stuff.

To create a new folder inside an existing folder, left-click on the existing folder to select it and then create the folder as you normally would.

DELETING A FOLDER

Occasionally it becomes necessary to delete the contents of a folder. This is typically a result of the hard disk's running out of free space. By selecting the folder, and then selecting "Delete," Windows 95 will prepare to delete the contents of the folder. First, as Figure

FIGURE 14.9 Creating a new folder

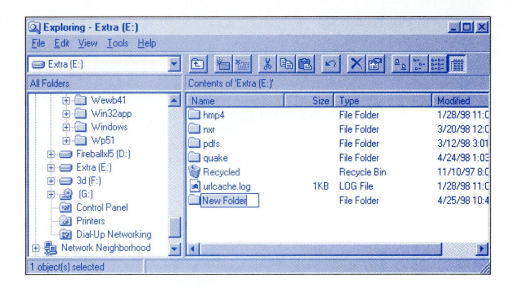

FIGURE 14.10 Naming the new folder

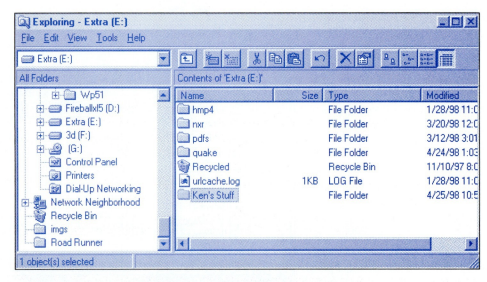

FIGURE 14.11 Confirmation window

14.11 illustrates, Windows 95 will issue a confirmation message to make sure you really want to move the folder to the Recycle Bin. This is because you might have selected the wrong menu item or pressed a wrong button by accident. Or, being human, you may simply have changed your mind.

Windows 95 makes doubly sure that a file you have chosen for deletion should really be deleted. Figure 14.12 shows the warning window that is displayed when a program is deleted. Other, selected file types, such as drivers or fonts, may require confirmation as well. In addition, Windows 95 usually knows when a file you try to delete is in use by an application (or itself), and may disallow the delete operation.

While files are being moved into the Recycle Bin, Windows 95 displays an animation of trash being thrown into a wastebasket. This is shown in Figure 14.13.

The entire folder can be recovered from the Recycle Bin at a later time, if necessary, by clicking File . . . Restore inside the Recycle Bin.

FIGURE 14.12 Confirming deletion of a program

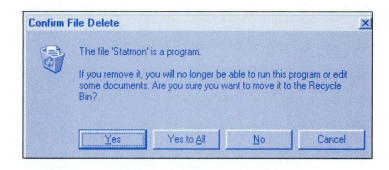

FIGURE 14.13 Deleting the files in a folder

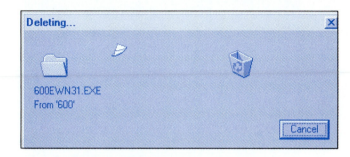

CREATING SHORTCUTS

A shortcut is an icon that you can double-click to start an application, rather than navigating to the application using Windows Explorer and double-clicking it there. For ease of use, shortcuts to often-used applications are usually placed on the desktop, so you can instantly access the applications without opening Explorer or using the Run or Programs menu.

To create a shortcut, use Explorer to navigate to the folder where the application is stored. Select the application by left-clicking it. Then choose File, Create Shortcut to create the shortcut icon, which is placed into the current folder. The name of the shortcut is automatically "Shortcut to . . . ," although you may rename it if you want. Drag and drop the shortcut onto the desktop for easy access to it.

By right-clicking the shortcut icon and selecting Properties from the menu that appears, you can examine or set its properties, such as the type of window it starts the application in (minimized, maximized), its attributes (hidden, read-only), and the working directory. Figure 14.14 shows the initial Properties display for a shortcut to an MS-DOS application called ASM. The Program tab displays the properties shown in Figure 14.15. Note that the shortcut can be started with the same command line parameters you would use while in DOS, just by entering them in the Cmd line box.

For appearance, you may want to change the icon used by a shortcut. Left-clicking the Change Icon button brings up the window shown in Figure 14.16. The new icon can be chosen from the group provided, or you can use Browse to select an icon from a different location.

The Font and Screen tabs control the appearance of the window the shortcut application runs in, and the Memory tab allows you to adjust the way memory is allocated to the application. Last, the Misc tab provides control over other important properties, such as the ability to use a screen saver with the application, and how the application may be terminated.

Note: If you right-click a non-DOS application's shortcut icon, such as Word, and then select Properties, you get a screen with only two tabs (General and Shortcut)—not six tabs as in Figure 14.14. This is further proof that menus are context sensitive.

CHECKING/SETTING PROPERTIES

All the items in the Folder and File display windows can be examined in great detail by looking at the item properties. For example, by right-clicking a disk drive and selecting Properties from the menu, the hard disk Properties window is displayed, as illustrated in Figure 14.17.

FIGURE 14.14 Initial shortcut properties screen

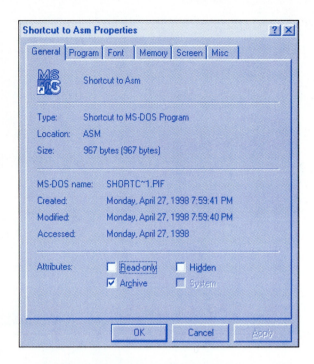

FIGURE 14.15 Program properties for shortcut

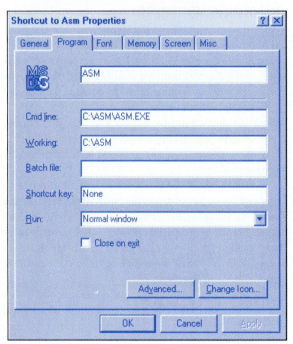

FIGURE 14.16 Choosing a shortcut icon

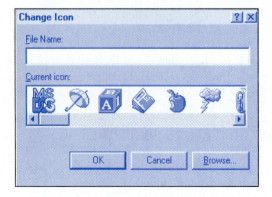

FIGURE 14.17 Disk Properties window

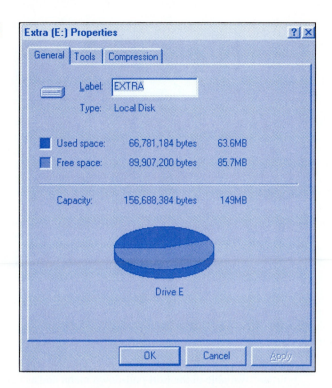

FIGURE 14.18 A Folder Properties menu

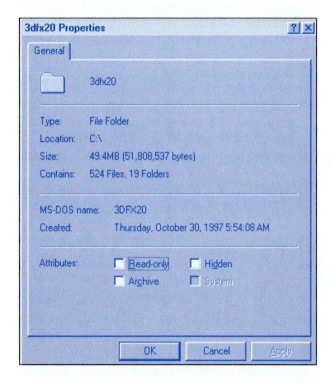

From the disk Properties menu, any of the hard disk properties can be modified or disk tools run. Similarly, if we right-click on a folder and select Properties from the menu, folder properties are displayed. Examine Figure 14.18 to view the properties of a folder. The folder properties include items such as the file name, creation date, file location, size of all the files stored inside the folder, and file attributes. The file attributes can be changed as necessary.

Many property windows have an Apply button, which causes the changes made in the Properties window to take effect. Properties are typically set or cleared by clicking check boxes or radio buttons, or by entering data into a text box.

FIGURE 14.19 Selecting multiple files

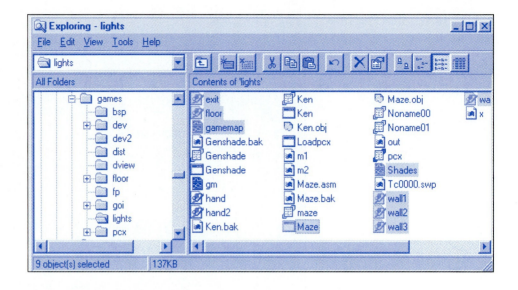

EDITING FEATURES

Explorer provides four essential editing features: Cut, Copy, Paste, and Undo. These operations are available in the Edit pull-down menu and as buttons on the toolbar. To cut a file or folder, left-click it and then select Edit, Cut, or click the Cut button. Cutting is typically used when you want to move a file or folder to another location. Unlike Delete, nothing is placed into the Recycle Bin when you cut an object. To move the object that has been cut to another location, navigate to the destination, and click the Paste button (or choose Paste from the Edit menu).

Clicking the Copy button (or selecting Copy from the Edit menu) instead of Cut leaves the original file or folder in place, storing a copy when Paste is clicked.

To select more than one file from a folder, or multiple folders, press and hold the Ctrl (Control) button on the keyboard, and left-click on each file or folder you want to select. If you change your mind, simply left-click anywhere in a blank portion of the file display window to deselect everything.

Figure 14.19 shows an example of selecting multiple files in a folder.

The Undo button is used to back up and erase the results of the last operation. For example, if you accidentally move a folder to the wrong location, Undo will move it back for you. You may need to refresh the display to see the results.

FINDING THINGS

Windows Explorer has many options to assist you in finding a file or set of files on your machine. You can search subfolders, search by the date of creation or last access, or search for files of a certain size. You can even search for files containing specific text strings.

Clicking on the Find option under the Tools menu allows you to choose "Files or Folders" or "Computer" as the search location. Choosing "Files or Folders" produces the search window shown in Figure 14.20. The file ASM.C on drive C: is the subject of the search. You may enter any legal file name in the Named box, including file names with wild card characters. Clicking on "Find Now" begins the search, which can be stopped at any time by clicking "Stop." Figure 14.21 shows the results of the search. As indicated, two copies of ASM.C were found, each in a different directory on drive C:.

Clicking the Date Modified tab brings up the search options shown in Figure 14.22. Notice that you can search for a recently created or modified file by selecting a *time frame* for the search.

FIGURE 14.20   Finding a file

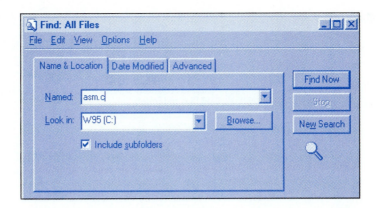

FIGURE 14.21   Search results

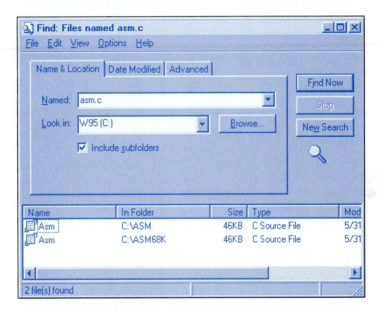

FIGURE 14.22   Using the date to limit the search

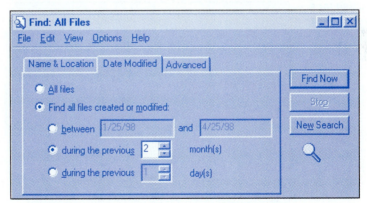

Advanced search options, illustrated in Figure 14.23, include searching files for a string of text (the word "microprocessor" in Figure 14.23), searching for files of a particular size, or searching files associated with a specific application (illustrated in Figure 14.24).

Combinations of each search option may also be used to further restrict the scope of the search.

The second option on the "Tools . . . Find . . ." menu is "Computer." Instead of searching for a file or folder, you can search the network your computer is connected to (even if you are using Dial-Up Networking) for a specific machine. In Figure 14.25 a machine called "Waveguide" is found using this search method.

Altogether there are many ways to search for an object using Windows.

FIGURE 14.23 Searching for text

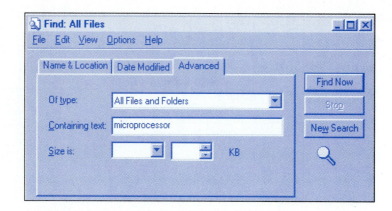

FIGURE 14.24 Selecting a file type to search for

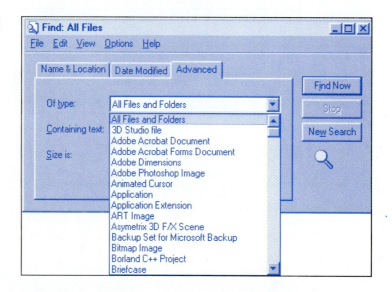

FIGURE 14.25 Searching for a computer on a network

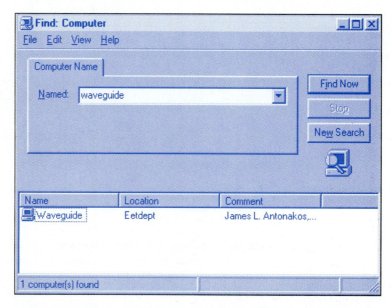

WORKING WITH NETWORK DRIVES

If you have a connection to a network (dial-up PPP or network interface card), you can use Explorer to *map* a network drive to your machine. This is done by selecting Map Network Drive on the Tools menu. Figure 14.26 shows the menu window used to map a network drive.

FIGURE 14.26 Mapping a network drive

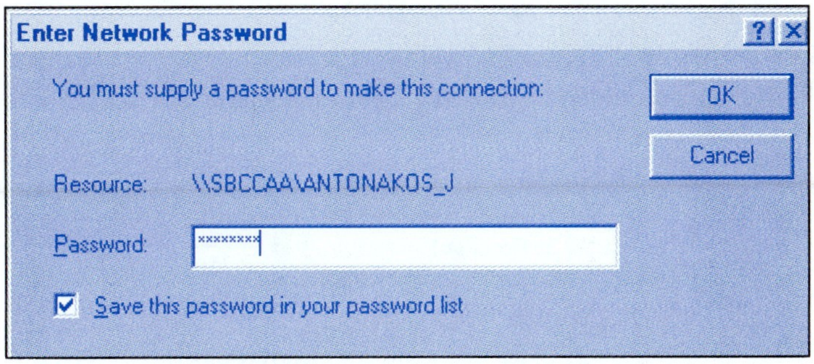

FIGURE 14.27 Supplying a network password

FIGURE 14.28 Contents of network drive H:

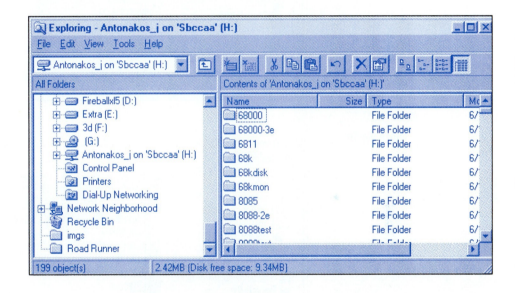

The computer automatically picks the first free drive letter (you can pick a different one) and requires a path to the network drive. In Figure 14.26 the path is \\SBCCAA\ANTONAKOS_J. The general format is \\machine-name\user-name.

Access to the network drive may require a password, as indicated in Figure 14.27. If an invalid password is entered, the drive is not mapped.

If the drive is successfully mapped, it will show up in Explorer's folder display window. Figure 14.28 shows the contents of the mapped drive. Note that drive H: has a different icon from the other hard drives.

When you have finished using the network drive, you can disconnect it (via the Tools menu). This is illustrated in Figure 14.29.

USING GO TO

When the toolbar is turned on, the Go To folder chooser (above the folder display window) is available; this provides access to all components of the computer. Figure 14.30 shows many of the typical items found in the Go To list.

191

FIGURE 14.29 Disconnecting a network drive

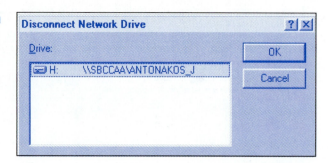

FIGURE 14.30 Using the Go To menu to select an item

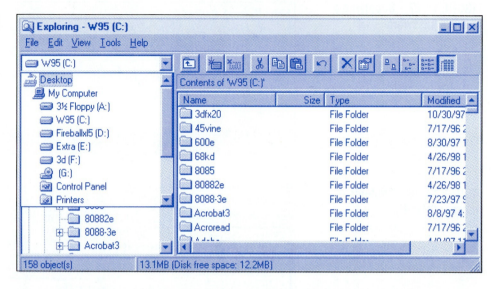

FIGURE 14.31 More Go To items

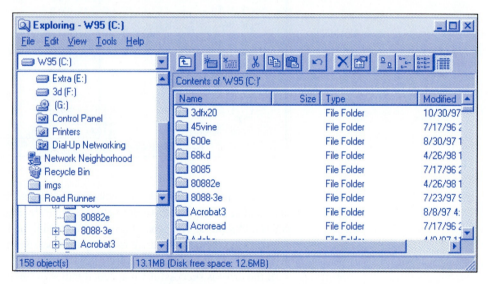

Notice that the first item in the list is the Desktop. Built-in desktop components such as My Computer, Network Neighborhood, and the Recycle Bin are in the Go To list, as well as other desktop items added by the user. This is illustrated in Figure 14.31, which shows the remaining items in the Go To list. The imgs and Road Runner folders, which reside on the desktop, are part of the Go To list as well.

Left-clicking on an item (such as My Computer) in the Go To list displays its contents in the file display window. This is illustrated in Figure 14.32.

So almost anywhere you want to go on your machine is only a click away with the Go To chooser.

FIGURE 14.32 The contents of My Computer

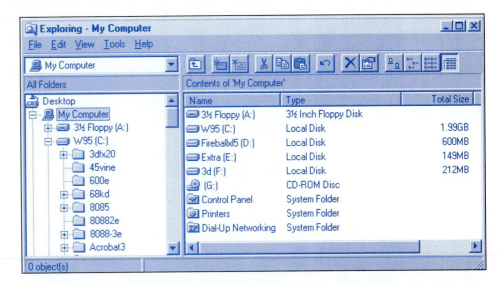

FIGURE 14.33 Initial Help screen

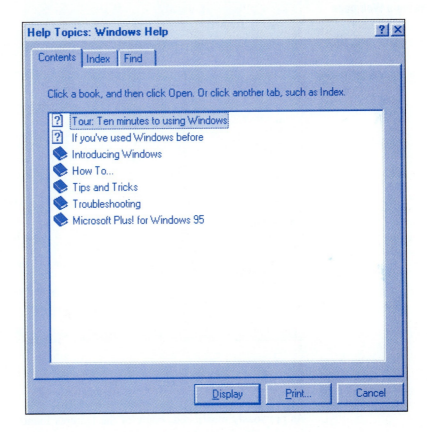

HELP

Help on almost any Windows 95 topic is provided through the Help menu (previously examined in Exercise 10). This is the same Help that is found on the Start menu. The initial Help screen is shown in Figure 14.33.

In addition, selecting the About Windows 95 item on the Help menu brings up the information window shown in Figure 14.34.

The Help menu completes a set of Explorer menus that provide a large amount of control over the Windows 95 environment. Spend some time getting familiar with all the features of Explorer.

FIGURE 14.34 The About Windows 95 Help window

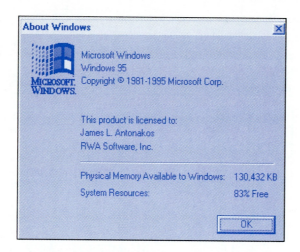

FIGURE 14.35 Windows 98 Explorer showing a Web page

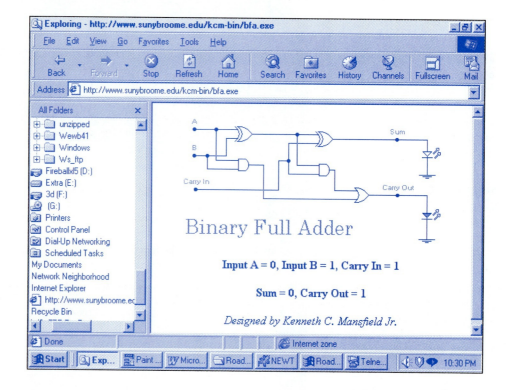

WINDOWS 98 EXPLORER

As previously mentioned, many of the features found in Windows 95 Explorer are also found in Windows 98 Explorer. You can still map drives, navigate around directories, and change views. However, many new features are provided to improve the Windows 98 experience. One enhancement is illustrated in Figure 14.35, where a Web page is being viewed in the rightmost panel. Compare this with Figure 14.1(b), which displays files and folders in the same panel.

In general, the features of Windows 98 Explorer are provided by Internet Explorer 4.0. The toolbar is context dependent with larger buttons that are more visible and convenient. The Address Bar eliminates the need for the Go To option in Windows 95 Explorer. The path to a directory folder or file, or the WWW address of a Web page, can be entered in the Address Bar (with Explorer 4.0 automatically completing the entry if it has been entered before). Furthermore, you may easily add a file, folder, or Web page to your list of favorite locations (with Web pages automatically updated in a background process as frequently as desired).

The context-sensitive aspects of Explorer 4.0 can be seen by comparing the graphical toolbars in Figure 14.35 and 14.1(b). In Figure 14.35 the buttons provide options normally available when browsing the Web, such as Stop, Refresh, and Home. The graphical toolbar in Figure 14.1(b) provides the typical editing functions necessary when working with folders and files, such as Cut, Copy, and Paste.

WINDOWS NT

The Windows NT 4.0 operating system comes standard with the same interface as the Windows 95 operating system. All the discussion about Windows 95 Explorer earlier in this exercise is applicable.

Fortunately, the newer features available in Windows 98 Explorer can also be added to Windows NT 4.0 by upgrading to Windows Explorer 4.0. The upgrade can be downloaded directly from Microsoft or when installing other Microsoft software (such as Visual C++ 6.0). In either case, it is easy to get the new software and take advantage of the new features.

TROUBLESHOOTING TECHNIQUES

While the graphical toolbar contains buttons for many useful operations, it is important to remember that some operations are missing. For example, you may get frustrated searching the toolbar for the Create Folder button, only to find that there is no button for the operation, which is available only as a pull-down menu item. It pays to periodically reexamine the meaning of each toolbar button. You may find that you've been using the Cut and Paste operations from the pull-down menu quite frequently, instead of their built-in toolbar buttons.

SELF-TEST

This self-test is designed to help you check your understanding of the background information presented in this exercise.

True/False

Answer *true* or *false*.

1. The Windows Explorer window contains three separate areas.
2. The graphical toolbar contains a limited set of the available features from the pull-down menus.
3. Using the List option scrolls the files vertically.
4. Deleting a file consists of moving the file contents into the Recycle Bin.
5. New folders are always created in the currently selected drive's root directory.

Multiple Choice

Select the best answer.

6. The file display shows
   a. The number of objects in the window and the amount of disk space used.
   b. The graphical toolbar.
   c. A list of files and folders in the currently selected location.
   d. Help topics.
7. New folders are created in
   a. The root directory on the C: drive.
   b. The Windows subdirectory on the D: drive.
   c. The currently selected drive's root directory.
   d. The currently selected location.
8. When deleting a folder, Explorer
   a. Moves the folder into the Recycle Bin immediately.
   b. Prompts two times to be sure all files are to be deleted.
   c. Prompts one time to be sure all files are to be deleted.
   d. Prompts multiple times to be sure all files are to be deleted.

9. Explorer performs four essential editing features
   a. Cut, Copy, Paste, Redo.
   b. Cut, Edit, Move, Rename.
   c. Cut, Paste, Undo, Redo.
   d. Cut, Copy, Paste, Undo.
10. When mapping a network drive
    a. The computer automatically picks the last free drive letter.
    b. The disk shows up in the Explorer display just like the local drives.
    c. Use of the Go To menu item is required to identify the specific network resources.
    d. None of the above.

Completion

Fill in the blank or blanks with the best answers.

11. The list of items on the desktop contains the built-in desktop components _____, _____, and _____.
12. _____-clicking on an item displays its contents in the file display window.
13. Many of the _____ menu items are located in the graphical toolbar.
14. _____ properties include file name, creation date, file location, and total size.
15. The _____ _____ button begins the search when finding files.

FAMILIARIZATION ACTIVITY

1. Practice using all the display modes of Explorer to determine which one you prefer.
2. Use Explorer to map a network drive. What kinds of software are available on the network drive?
3. Practice creating and renaming files and folders, moving them to different locations, copying them, and navigating through the drives on your system.
4. Create shortcuts to your frequently used applications.

QUESTIONS/ACTIVITIES

1. What is the value of a mapped network drive?
2. Make a list of ten things you normally do while using your computer (for example, rename or copy files, run applications). Explain how these chores can be performed by Windows Explorer (if possible).

REVIEW QUIZ

Within 10 minutes and with 100% accuracy,

1. Explain the basic features of Windows Explorer.
2. Show how to start an application.
3. Create a shortcut, move, find, and delete files and folders.
4. Map a network drive.

15 Managing Printers

INTRODUCTION

Joe Tekk lugged a heavy laser printer into his office and set it down on his desk. He plugged the power cable in, loaded paper into the paper tray, and connected the printer to his Windows 98 machine. The printer was an old HP LaserJet Series II that Joe had bought at a hamfest for $25.

Joe ran the Add Printer Wizard and loaded the drivers for the laser printer. However, when he printed a test page, nothing happened. After several minutes of troubleshooting, Joe found the problem: a bent pin on the printer connector. He straightened the pin with a set of needlenose pliers and tried the test page again.

His $25 laser printer worked just fine.

PERFORMANCE OBJECTIVES

Upon completion of this exercise, within 10 minutes and with 100% accuracy, you will be able to

1. Check the properties of any available printers.
2. Install a new printer.
3. Connect to a network printer.

BACKGROUND INFORMATION

Windows has many built-in features to assist you in using your printer, from installing it to sharing the printer on a network. Figure 15.1 shows the starting point for many printer operations, the Printers folder. You can open the Printers folder by left-clicking Start, Settings, and Printer. Or, from inside the Control Panel, you can double-click the Printers folder. The printers installed on the system appear in the Printers folder as separate printer icons. Figure 15.1 shows just one installed printer, the Okidata OL-400e.

PRINTER PROPERTIES

Printer properties can be examined by right-clicking the desired Printer icon. Doing so opens up a context-sensitive menu that provides some basic printer control (pause/purge print jobs, set default printer) and a properties section. The Properties window for the Okidata laser printer is shown in Figure 15.2(a).

FIGURE 15.1 Printers folder

FIGURE 15.2 (a) Initial Windows 95/98 printer Properties window, and (b) Windows NT printer Properties

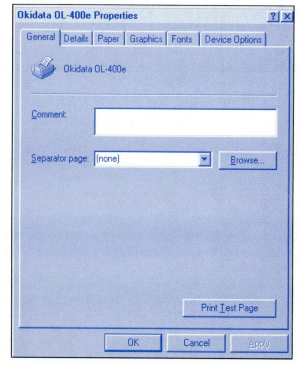

(a)

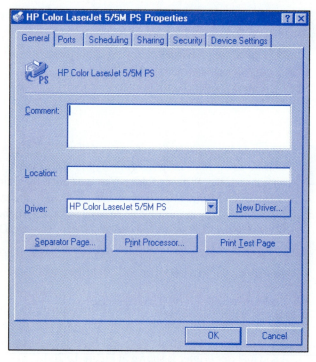

(b)

FIGURE 15.3 Built-in printer help

Figure 15.2(a) illustrates the Windows 95/98 Properties window for the Okidata printer. Note that the format of the tabs displayed along the top of the Properties window may contain a Sharing tab if printer sharing is enabled. Figure 15.2(b) shows the Windows NT properties menu. Although much of the same information is available on Windows NT, it is located on different tabs. You are encouraged to familiarize yourself with these screens.

If the printer has been installed correctly, left-clicking the Print Test Page button will cause the printer to print a test page. The test page contains a graphical Windows logo and information about the printer and its various drivers. A dialog box appears asking whether the test page printed correctly. If the answer is no, Windows starts a printer help session. Figure 15.3 shows the initial Help window.

Windows will ask several printer-related questions to help determine why the printer is not working. The causes are different for network printers, so Windows provides two different troubleshooting paths (network vs. local).

In Windows 95/98 the Details tab brings up the window illustrated in Figure 15.4(a). Many of the software and hardware settings are accessed from this window. Notice that LPT1 is set up as an ECP Printer Port. ECP stands for extended capabilities port, a parallel port standard that allows 8-bit bidirectional data flow. This allows ECP to support other peripherals over the parallel printer port, such as external CD-ROM drives or scanners.

The Ports tab shown in Figure 15.4(b) shows a comparable Windows NT menu that indicates an HP Color LaserJet printer is attached on LPT2. The Enable printer pooling

FIGURE 15.4 (a) Windows 95/98 Printer details menu

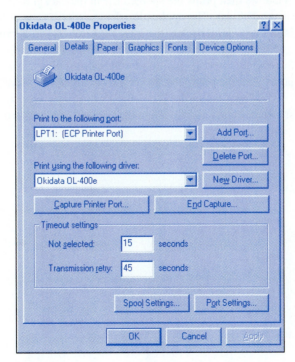

(a)

(continued on the next page)

FIGURE 15.4   (continued)
(b) Windows NT printer Ports menu

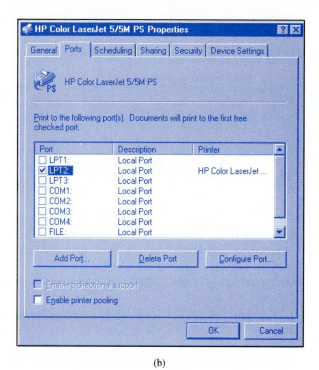

(b)

feature in Windows NT allows for several printers to work together to handle a heavy print workload. Print jobs in the queue are assigned to available printers in the pool so that each of the printers in the pool can be used to service many different print jobs simultaneously.

Left-clicking the Spool Settings button (or the Scheduling tab in Windows NT) allows you to check/set the spool settings for the printer. These settings are indicated in Figure 15.5. Print spooling is a technique that speeds up the time required by an application to send data to the printer. The hard disk is used as a temporary holding place for the print job. The application prints everything very quickly to a spooling file on the hard drive. Windows then prints the spooled file in the background, while the user continues working on other things. The application does not have to wait for each page to be printed before returning control back to the user. This process is diagrammed in Figure 15.6.

Notice that in Figure 15.5(b), other useful information is available on the Windows NT Scheduling menu in addition to the spool settings. The times that the printer is available, priority for the print jobs, and other various settings can be modified as required.

The Paper tab in the Windows 95/98 printer Properties window brings up the window shown in Figure 15.7(a). Here the size and orientation of the paper are selected, as well as

FIGURE 15.5   (a) Windows 95/98 Spool Settings window

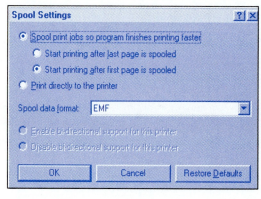

(a)

FIGURE 15.5 *(continued)* **(b) Windows NT printer scheduling menu**

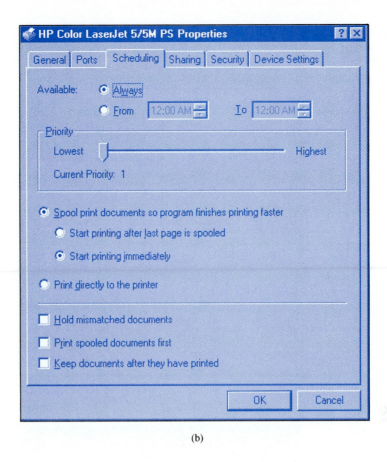

(b)

FIGURE 15.6 Operation of a print spooler

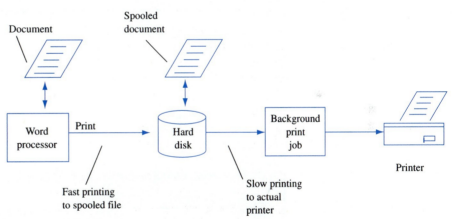

the default number of copies. Normally, pages are printed in Portrait mode. Selecting Landscape turns the printed document 90 degrees, which is useful for printing wide documents such as spreadsheets. In Figure 15.7(b) the Device Settings tab shows the paper settings as well as the Additional Postscript Memory setting for the HP Color LaserJet printer running under Windows NT.

The Graphics settings menu illustrated in Figure 15.8 allows the resolution, *dithering*, and intensity of the printer to be adjusted. Dithering is a method used to represent a particular color by using one or more colors that are similar. This is a necessary step when printing graphical documents that may contain many more colors (or shades of gray) than the printer supports.

You should experiment with the graphics settings until you find the right combination for your printer. Note that the printer properties window does not contain a Graphics tab in Windows NT.

FIGURE 15.7 (a) Windows 95/98 Paper settings menu, and (b) Windows NT printer Device Settings menu

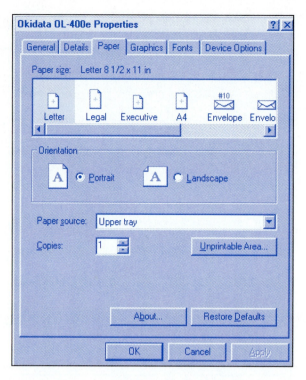

(a)

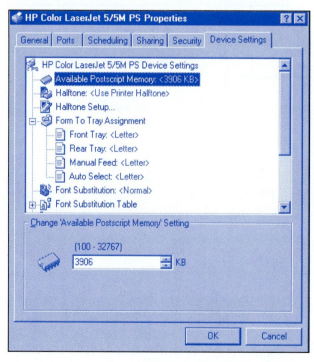

(b)

ADDING A NEW PRINTER

To add a new printer, double-click the Add Printer icon in the Printers folder shown in Figure 15.1. This will start up the Add Printer Wizard, an automated process that guides you through the installation process.

The first choice you must make is shown in Figure 15.9. A local printer is local to your machine. Only your machine can print to your printer, even if your computer is networked. A network printer can be printed to by anyone on the network who has made a connection to

FIGURE 15.8 Windows 95/98 Graphics settings menu

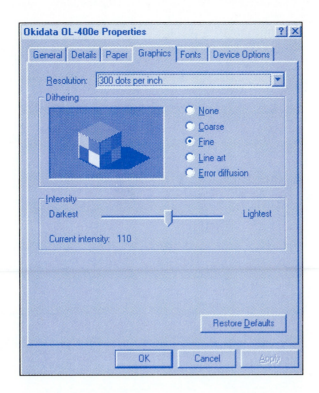

FIGURE 15.9 Choosing local/network printing

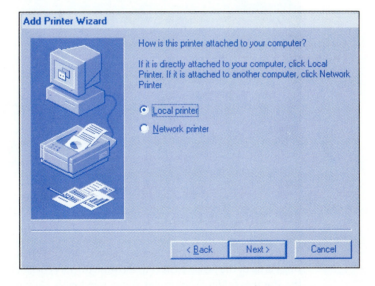

that printer. A network printer is also a local printer to the machine that hosts it. If you are installing a network printer, the next window will look like that shown in Figure 15.10. The printer being mapped is an HP LaserJet II (named "hplaserii" on the network) connected to the machine "deepspace." You can also browse the Network Neighborhood to select a network printer. DOS accessibility to the network printer is controlled from this window as well.

Next, the manufacturer and model of your printer must be chosen. Windows has a large database of printers to choose from. Figure 15.11 shows the initial set of choices. If your printer is not on the list, you must insert a disk with the appropriate drivers (usually supplied by the printer manufacturer).

Once the printer has been selected, the last step is to name it (as in the network printer "hplaserii").

If only one printer is installed, it is automatically the default printer for Windows. For two or more printers (including network printers), one must be set as the default. This can be done by right-clicking on the Printers icon and selecting Set As Default. You can also access printer properties and change the default printer from inside the printer status window, using the Printer pull-down menu.

203

FIGURE 15.10 Mapping a network printer

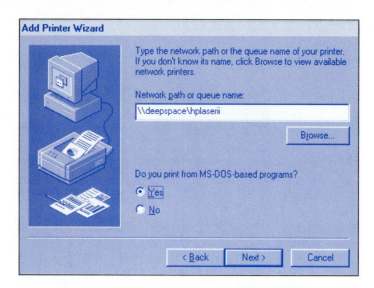

FIGURE 15.11 Choosing a printer manufacturer/model

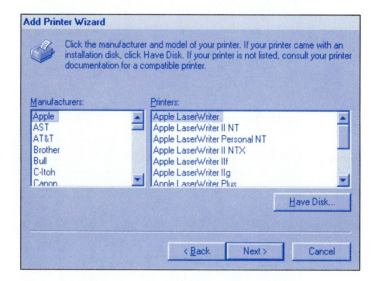

CHECKING THE PRINTER STATUS

To check the status of the current printer, double-click the Printer icon in the system settings area of the taskbar (next to the time display) or double-click the desired Printer icon in the Printers folder. The printer status window is shown in Figure 15.12. The printer status indicates that a document is being printed by the printer.

If Windows detects a problem with the printer (out of paper, offline, powered off), it will report the error without having to bring up the status window. Figure 15.13 shows a typical printer error window.

If a document cannot be printed for some reason before the system is shut down, Windows will save the print job and try to print it the next time the system is booted.

PAUSING A PRINT JOB

To pause a print job (possibly to add more paper), select Pause Printing from the Printer or Document pull-down menus in the printer status window. This will shut off the flow of new data to the printer. If several pages have already been sent, they will be printed before the pause takes effect.

FIGURE 15.12 Printer status window

FIGURE 15.13 Printer error window

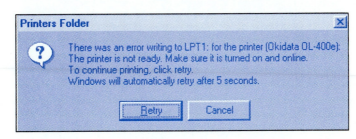

RESUMING A PRINT JOB

To resume a paused print job, select Pause Printing again (it should have a black check mark next to it when paused).

DELETING A PRINT JOB

Individual print jobs can be deleted by selecting them with a single left-click and then choosing Cancel Printing from the Document pull-down menu. *All* the jobs in the printer can be deleted at the same time by selecting Purge Print Jobs from the Printer pull-down menu.

NETWORK PRINTING

To make the printer on your machine a network printer, you need to double-click the Network icon in Control Panel and then left-click the File and Print Sharing button. This opens up the window shown in Figure 15.14. The second box must be checked to allow network access to your printer.

After a network printer connection has been established, you may use it like an ordinary local printer. Windows communicates with the network printer's host machine using NetBEUI. What this means is that jobs sent to a network printer are sent in small bursts (packets) and typically require additional time to print due to the network overhead. In a busy environment, such as an office or college laboratory, printer packets compete with all the other data flying around on the network, and thus take longer to transmit than data traveling over a simple parallel connection between the computer and the printer.

FIGURE 15.14 Giving network access to your printer

205

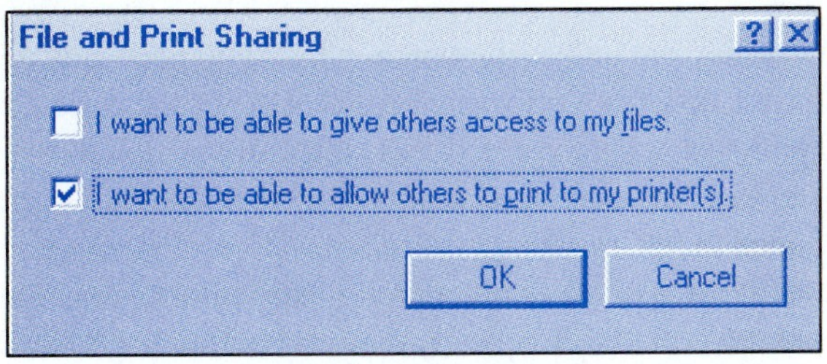

FIGURE 15.15 Windoes NT printer Security menu

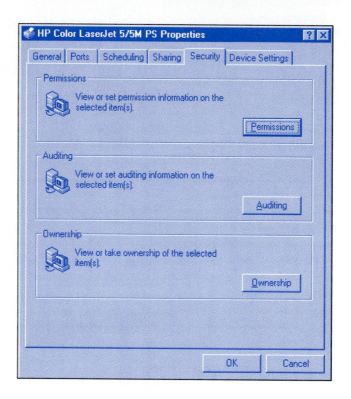

WINDOWS NT PRINTER SECURITY

The Security tab in Windows NT printer properties offers several options not available in Windows 95/98. Figure 15.15 shows the window that allows access to view or change the Permissions, Auditing, and Ownership properties of a Windows NT printer. The system user can assign the printer permissions so that individuals, groups, or everyone in the entire domain may access a particular printer. The type of access (No Access, Print, Manage Documents, or Full Control) is specified for each user or group.

Windows NT printer auditing can be configured to retain certain events. These events (Print, Full Control, Delete, Change Permission, and Take Ownership) are grouped into two categories: success and failure. In general, Auditing involves tracking certain users or examining the data to determine a trend such as paper usage, job length, or who prints the most.

The ownership of a printer can be changed by selecting the Ownership menu button. The owner of a printer can modify permission settings and grant permission to other users. The type of access to the printer determines if ownership of a printer can be taken.

By setting the Permissions, Auditing, and Ownership options appropriately, the system administrator can manage every aspect of the printing process for every computer user in the Windows NT domain. You are encouraged to examine each of these menus in detail to determine what types of settings are currently selected on your computer.

TROUBLESHOOTING TECHNIQUES

You may run into a situation in which you need new drivers for your printer, but you do not have a disk from the manufacturer. One of the easiest solutions is to search the Web for the printer manufacturer and look for a driver download page for your printer. Figure 15.16 shows a Netscape window of a portion of Okidata's driver page. The printer drivers are contained in self-extracting executables. Download the one you need and run it. Be sure to read the README file or other preliminary information. Some printer manufacturers require the old drivers to be completely removed before beginning a new installation or update.

FIGURE 15.16 Downloading a printer driver

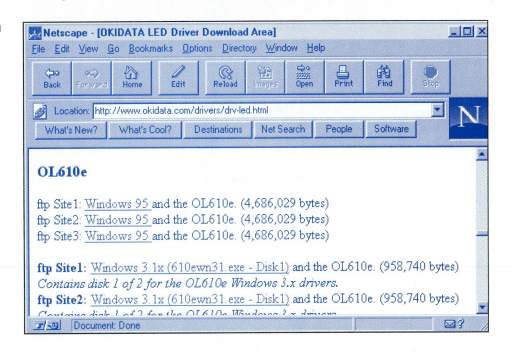

SELF-TEST

This self-test is designed to help you check your understanding of the background information presented in this exercise.

True/False

Answer *true* or *false*.

1. Only one printer may be installed at a time.
2. Print spooling refers to the coiling of the printer cable around a spool.
3. One printer driver will work for many different printers.
4. Print jobs can be paused.
5. ECP stands for easily connected printer.
6. Windows NT has no security advantages over Windows 95/98.

Multiple Choice

Select the best answer.

7. A print job is deleted by
    a. Dragging it to the Recycle Bin.
    b. Selecting it and choosing Cancel Printing.
    c. Turning the printer off.
8. When Purge Print Jobs is selected,
    a. The current job is purged.
    b. Only network printer jobs are purged.
    c. All print jobs are purged.
9. A network print job typically takes
    a. Less time than a local print job.
    b. The same time as a local print job.
    c. More time than a local print job.
10. When you pause a print job,
    a. Paper immediately stops coming out of the printer.
    b. The current page finishes printing.
    c. All complete pages sent to the printer finish printing.

207

11. In the network printer path \\waveguide\riko, the name of the printer is
    a. Waveguide.
    b. Riko.
    c. Neither.
12. The Windows NT printer Security menu shows
    a. Administrator, Access, and Permission.
    b. Owner, Permissions, and Manager.
    c. Permissions, Auditing, and Ownership.

Completion

Fill in the blank or blanks with the best answers.

13. A technique that attempts to match colors and gray levels during printing is called _____.
14. Data is sent to network printers using the _____ protocol.
15. The two types of printer connections are network and _____.
16. Saving a print job on the hard drive and printing it in the background is called _____.
17. One printer is always set as the _____ printer.
18. Windows NT printer auditing groups events into two categories: _____ and _____.

FAMILIARIZATION ACTIVITY

1. Examine the properties of all the printers connected to your computer, including any network printers.
2. Install a new local printer.
3. Install a new network printer.
4. Connect to a network printer.
5. Send a ten-page document containing text and graphics to a network printer. Keep track of how long it takes to finish printing. Repeat the print job three more times and compare the timing results.
6. Download a new printer driver from the Web and install it.

QUESTIONS/ACTIVITIES

1. Make a list of all the printers in the lab. How many of them can be found in the installation list of manufacturers and models (Figure 15.11)?
2. Go to a local computer store and check the prices of various printers. What is available for less than $200?

REVIEW QUIZ

Within 10 minutes and with 100% accuracy,

1. Check the properties of any available printers.
2. Install a new printer.
3. Connect to a network printer.

16 Accessories

INTRODUCTION

Joe Tekk stopped into his office for a few minutes on the way to a meeting. He sat down at his computer and checked his schedule with the Calendar utility, opened WordPad to write a quick note to a friend, and examined a graphical display of network traffic using System Monitor. Then his telephone rang. It was Jeff Page, wondering what a $550 motherboard would cost with a 17% discount. Joe brought up the Calculator utility, performed the calculation, and gave Jeff the answer.

As he closed the Calculator window, Joe wondered what he would do without all the utilities provided by Windows.

PERFORMANCE OBJECTIVES

Upon completion of this exercise, within 10 minutes and with 100% accuracy, you will be able to

1. Describe the basic contents of the Accessories folder.
2. Demonstrate the operation of several utilities, such as the Calculator, Notepad, and Paint.
3. Explain the importance of the System Tools and Administrative Tools folders.

BACKGROUND INFORMATION

Windows provides a large number of accessory applications designed to help perform all the small chores we might require when using a computer to simplify our lives. Figure 16.1 shows the Accessories menu, which is full of useful applications. Notice that there are several folders in the submenu as well.

Notice the similarities between the Windows 95/98 Accessories menu in Figure 16.1(a) and the Windows NT Accessories menu in Figure 16.1(b). Many of the discussions that follow apply to Accessories applications found in both operating systems.

In this exercise we will take a brief look at many of the accessory applications. You are encouraged to spend additional time learning how to use any applications that are useful to you or recommended by your instructor.

FIGURE 16.1 (a) Typical Windows 95/98 Accessories menu, and (b) Windows NT Accessories menu

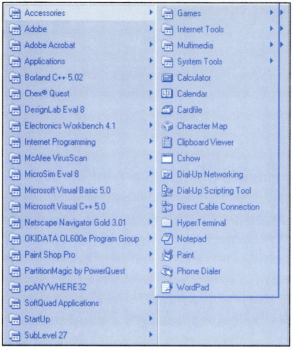

(a)

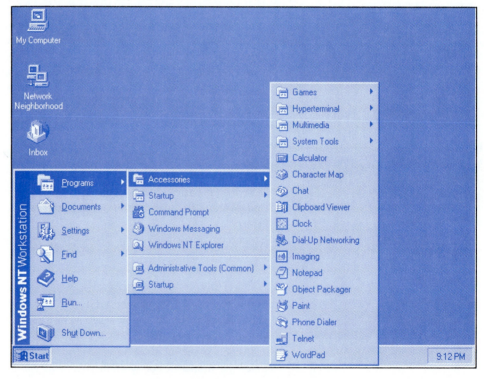

(b)

GAMES

Windows comes equipped with several games, just as Windows 3.x did. These games can be useful tools when introducing a new user to the Windows environment. The Solitaire game shown in Figure 16.2 can be used to introduce a user to the mouse. Experienced Windows users rely on the mouse regularly, so why not get used to using the mouse in a fun way?

FIGURE 16.2 Solitaire game window

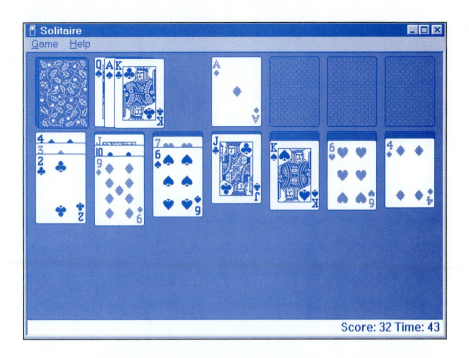

FIGURE 16.3 Internet Explorer window

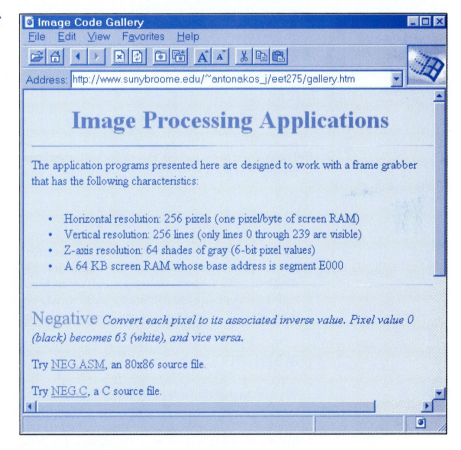

INTERNET TOOLS

The Internet (covered in depth in Exercise 26) is a worldwide collection of networks that all communicate with each other, sharing files and data. One very popular form of posting and sharing information is through a *Web page*. A Web page is written in a language called HTML (hypertext markup language) and is viewed using a *browser,* an application capable of interpreting HTML and displaying a graphical page layout, as illustrated in Figure 16.3.

FIGURE 16.4 Multimedia applications

FIGURE 16.5 System Tools menu

The browser being used in Figure 16.3 is the Internet Explorer, which can be downloaded for free over the Web, and is an automatic part of Windows.

MULTIMEDIA

There are several multimedia applications included in Windows. They provide features such as a CD player and a sound recorder, to name just a few. Figure 16.4 shows the contents of the Multimedia folder in the Accessories menu. Since there is such a large variety of both multimedia software and multimedia hardware add-ons for Windows, Exercise 46 is devoted to covering all of these types of multimedia applications and devices.

SYSTEM TOOLS

The system tools provide many important features for controlling, monitoring, and tuning your Windows system. Note that many of the Windows 95/98 tools are also found in Windows NT. Exceptions to this rule will be noted accordingly. Figure 16.5 shows the selections available in the System Tools menu. Let us look at several of these applications.

Windows 95/98 Disk Defragmenter

When a hard drive is used for a long period of time, many files are created, modified, and deleted. The nature of the file allocation mechanism eventually causes the files of the hard drive (or even a floppy) to become *fragmented*. A fragmented file is spread out over many different areas of the disk, rather than being stored in one big block. This fragmentation increases the time required to read or write to the file, and can lower performance significantly if a large portion of the disk becomes fragmented.

FIGURE 16.6 Selecting a drive to defragment

FIGURE 16.7 Defragmentation in progress

Fortunately, for Windows 95/98 users, the Disk Defragmenter application is included in the system tools to automatically defragment a disk drive. The files are read one by one and written back unfragmented (space is made available before writing the file back). Figure 16.6 shows the initial window for the Disk Defragmenter. Any or all of the disk drives may be selected. Depending on the size of the drive and the amount of fragmentation, the defragmentation process can be quite time-consuming. A simplified status window reports the current progress, as shown in Figure 16.7. Clicking Show Details brings up a multicolored graphical display of the drive being defragmented, with different colors used to identify files being moved, files that cannot be moved, and free space. Stopping the process does not destroy any data on the drive.

Windows NT does not come with a defragmenting program of its own, but you may install a third-party product for disk maintenance, such as Diskeeper, from Executive Software (www.execsoft.com). Diskeeper works with NTFS, the native file system structure used by Windows NT.

The old-fashioned method of defragmentation was to back up the hard disk to tape, format the hard disk, and restore from the backup. Cautious users may still prefer this method, the reward being a *permanent* copy of the hard drive data.

Resource Meter

Three important resources used by Windows 95/98 are system, user, and GDI. The amount of each resource available determines how well Windows 95/98 can handle a new event, such as the launch of a new application. The Resource Meter, shown in Figure 16.8, displays the free percentage of each resource in a bar-graph display. The display is updated as resources are allocated and deallocated.

ScanDisk

Occasionally, a file or directory may develop a problem that prevents Windows 95/98 from reading or writing the file, or accessing the directory. This could be the result of having shut the computer down improperly, or having an application run amok and cause some damage. Or you may simply have bumped the computer by accident, causing a head crash in the hard drive.

No matter what the cause, it is always a good idea to run the ScanDisk utility to check the integrity of a disk. Figure 16.9(a) shows the initial ScanDisk window. ScanDisk checks the organization of the FAT (file allocation table), the directory areas, and every file on the

FIGURE 16.8 Windows 95/98 Resource Meter display

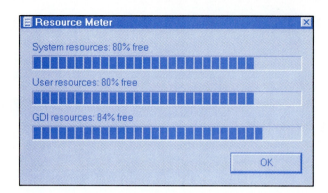

FIGURE 16.9 (a) ScanDisk control window, and (b) Sample Chkdsk execution in Windows NT

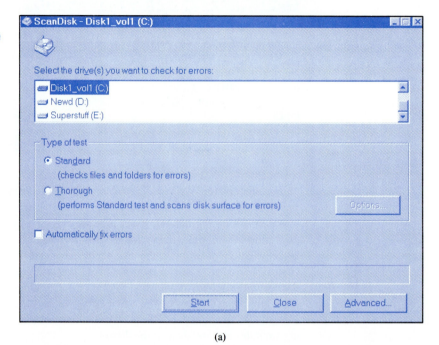

(a)

(b)

disk. The entire surface of the disk can be searched for bad sectors as well. Windows 95 version B and Windows 98 automatically run ScanDisk if they detect that the computer was not properly shut down the last time it was used.

In Windows NT Chkdsk, not ScanDisk, is used to check the structure of a disk drive. Figure 16.9(b) shows the results of running Chkdsk on a Windows NT disk.

FIGURE 16.10 System Information window

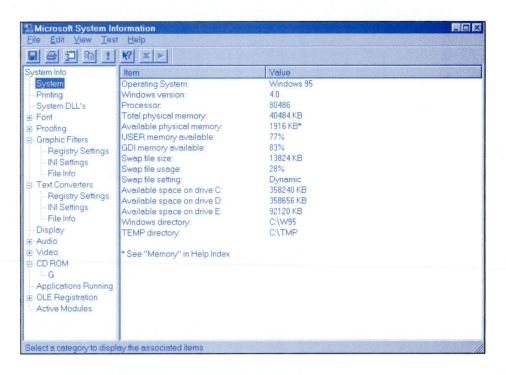

FIGURE 16.11 Monitoring system resources

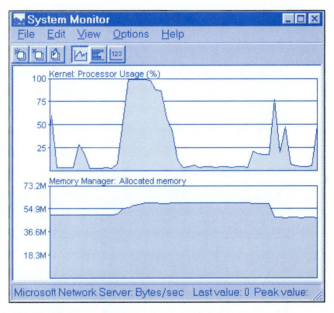

System Information

For a quick look at the information particular to your computer, use the System Information utility. Figure 16.10 shows the first screen of information, which gives a good summary of the pertinent data regarding the overall system. It would be worthwhile to spend some time looking at the different items available in the System Information window. You might be surprised at what you find, such as applications running in the background that you did not even know were there.

System Monitor

To really monitor the performance of your computer, use the System Monitor utility. With this tool, you can display the *history* of usage for one or more resources. As Figure 16.11 shows, the percentage of processor usage and the amount of allocated memory are being tracked by the System Monitor. The Edit menu allows you to add items to track, choosing from four different areas: file system, kernel, memory manager, and network.

FIGURE 16.12 Standard Calculator window

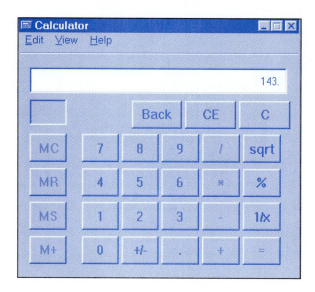

FIGURE 16.13 Scientific Calculator window

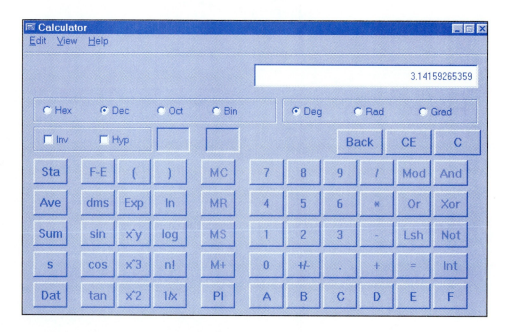

Compression Agent and DriveSpace

There are other tools in the System Tools menu that you may want to examine. One of these is Compression Agent. The Compression Agent utility is used to compress files on a *DriveSpace* compressed drive. DriveSpace is the technology used by Windows 95/98 to compress files on a hard drive or floppy disk, to increase the amount of storage capacity. There is a slight overhead required for decompression when a file is accessed, which can lower performance. Only compress your drive if you've run out of space and cannot add a new drive to your system.

CALCULATOR

If you need to do a few quick calculations, there is no need to go looking for your calculator. Windows has a built-in calculator that performs basic math functions in *standard mode* (shown in Figure 16.12), or in *scientific mode,* where the calculator has many additional features, such as the use of scientific numbers, conversion between different bases, and several transcendental functions. Figure 16.13 illustrates the scientific calculator.

FIGURE 16.14 Friday's schedule

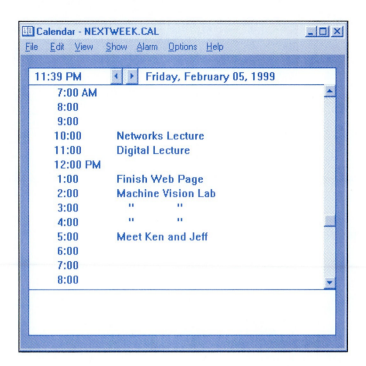

FIGURE 16.15 Sample Cardfile contents

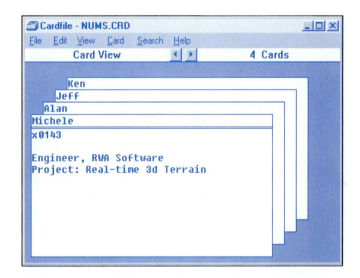

CALENDAR

The Calendar application program is used to maintain a listing of important events or appointments. Each day is displayed showing time increments along the left-hand margin and can be modified very easily by simply typing information at the appropriate time. The calendar can also be set to generate an alarm up to 10 minutes before each appointment or event. Figure 16.14 shows a typical day in the calendar application.

When many computers are networked together, electronic calendars are used to simplify the scheduling process for groups of people. This may be one reason why use of electronic calendars is becoming very widespread.

CARDFILE

The Cardfile utility is an electronic form of the popular 3-by-5-inch index card. You can label a card with a heading, put other information on the body of the card, and save groups of cards in their own file. Figure 16.15 shows a sample set of cards.

FIGURE 16.16 Character Map window

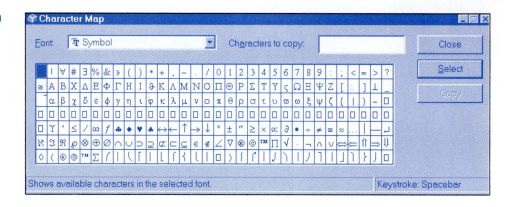

FIGURE 16.17 An image saved on the Clipboard

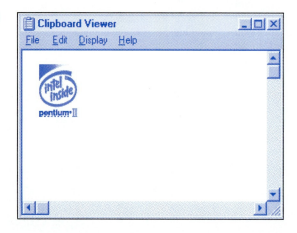

CHARACTER MAP

The Character Map tool provides a graphical display of each character available in a particular font. If you require a special character to include in a report or project, you can use Character Map to browse the installed fonts on your system and find the character. Figure 16.16 shows the contents of the Symbol font.

CLIPBOARD VIEWER

Whenever you cut or copy an object, or press the Print Screen button, data is placed on the Windows Clipboard. The Clipboard is a temporary holding place for information that you may want to exchange between applications. For example, when developing a Web page, you may use the Clipboard to cut and paste text and images from several different applications into your Web design tool. Figure 16.17 shows the Clipboard Viewer window, which contains a small graphic image placed on the Clipboard by a paint program.

MISSING SHORTCUTS

Occasionally you may come across an icon that does not have an application associated with it anymore (the application was deleted or moved to another folder or drive). In this case Windows will display a small searchlight while it attempts to locate the missing application. Figure 16.18 demonstrates this process.

A new application can be assigned by using the Browse feature. If there is no need for the shortcut any longer, it can be deleted by navigating to the Accessories folder using Windows Explorer and then right-clicking on the shortcut and choosing the Delete operation.

FIGURE 16.18 Missing Shortcut window

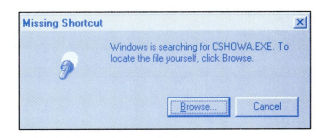

FIGURE 16.19 Dial-Up Networking window

FIGURE 16.20 Initial Direct Cable Connection window

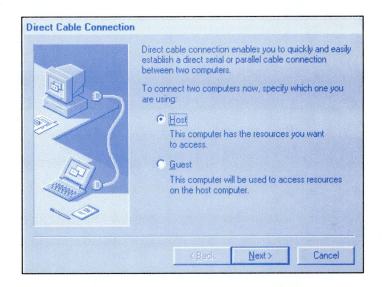

DIAL-UP NETWORKING

Dial-Up Networking (covered in detail in Exercises 17 and 27), allows your computer to use its modem to connect to a computer network (using a special protocol called PPP, for point-to-point protocol). Figure 16.19 shows the initial dial-up window. Double-clicking the My Office icon automatically dials the network mainframe computer and establishes the connection. Once connected, your computer can share files as if it were part of the actual network.

DIRECT CABLE CONNECTION

Two Windows machines can communicate with each other directly by using a direct cable connection between the serial or parallel ports of each machine. The initial window (shown in Figure 16.20) allows the user to specify Host or Guest mode. The type of connection (parallel vs. serial) and port are specified next. More information about using a direct cable connection is presented in Exercises 17 and 27.

FIGURE 16.21 Using HyperTerminal to communicate with a mainframe

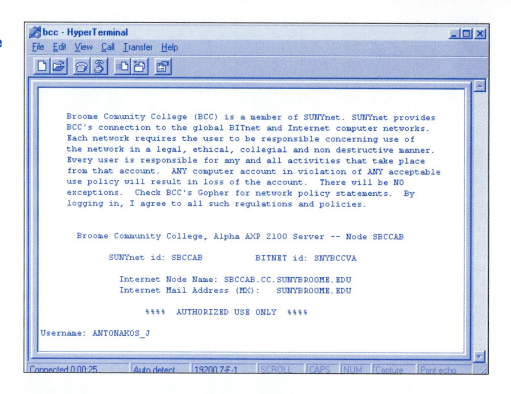

FIGURE 16.22 Notepad window

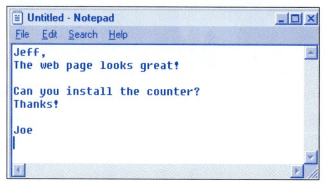

HYPERTERMINAL

If you do not own a modem communications program, Windows comes equipped with the HyperTerminal application, which emulates an ASCII data terminal and controls the modem at the same time. Figure 16.21 shows the start of a HyperTerminal session, with the user ANTONAKOS_J logging into the Broome Community College Alpha server. HyperTerminal will capture its screen to a file, transfer files between computers, and emulate several different types of terminals.

NOTEPAD

The Notepad is a screen-based text editor. You can paste information into the notepad from the Clipboard, or enter it directly from the keyboard. Figure 16.22 shows a quick note being entered. Notepad is useful for small text files (it can search for text, print, and perform simple editing chores), basically replacing the EDIT utility from DOS.

PAINT

The Paint utility allows you to create and edit graphical files (which may include text, graphics, or images). Figure 16.23 shows a simple schematic being edited. Paint saves files

FIGURE 16.23 Using Paint to create a graphics file

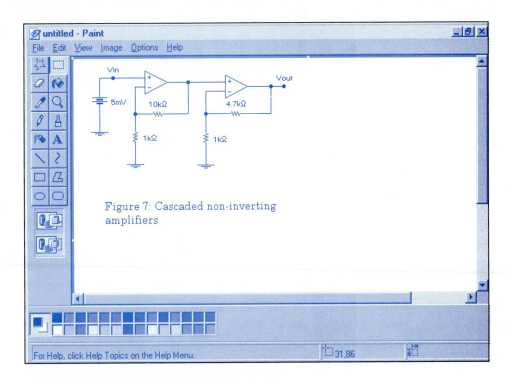

FIGURE 16.24 Phone Dialer window

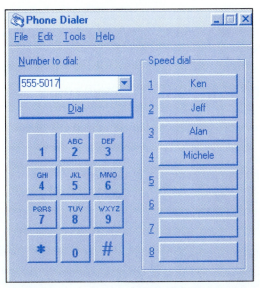

in one format: bitmap. Bitmap files are native to the Windows operating system and have a special binary format that makes them compatible with the desktop display.

PHONE DIALER

Dialing a touch-tone number is as easy as a single click when you use Phone Dialer. Numbers may be saved for speed dialing, or entered from a graphical keypad (or manually from the keyboard). The Phone Dialer window is shown in Figure 16.24. After the number has been dialed (via the internal modem), Phone Dialer instructs you to lift the handset and begin speaking.

WORDPAD

WordPad has many of the features of a full-fledged word-processing application. These include insertion of objects, expanded editing functions, and more control over the printer.

FIGURE 16.25 Editing a document using WordPad

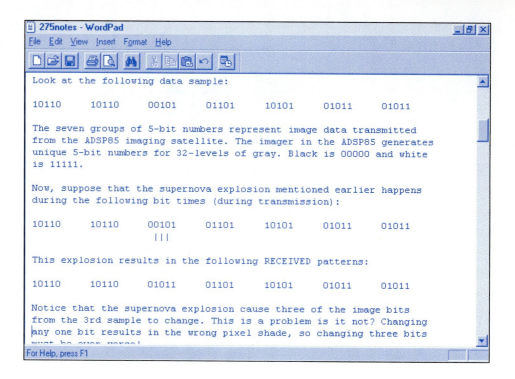

Figure 16.25 shows a sample document being edited in WordPad. Together, WordPad and Paint provide you with a significant amount of text and graphical file processing, all included automatically.

WINDOWS NT ADMINISTRATIVE TOOLS

Windows NT provides additional tools for system administration. The very names of the tools suggest a different atmosphere from the Windows 95/98 environment. In fact, Windows NT allows users to have their own accounts on the system, with varying degrees of privilege and control over the system. Figure 16.26 shows the Administrative Tools

FIGURE 16.26 Windows NT Administrative Tools menu

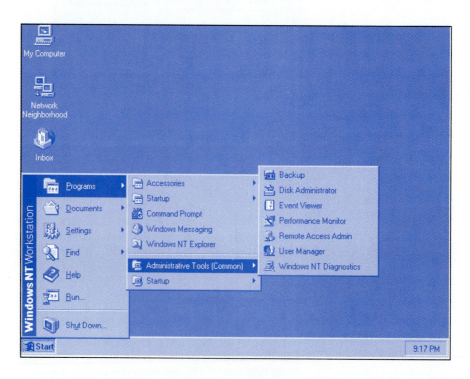

FIGURE 16.27 Event Viewer display

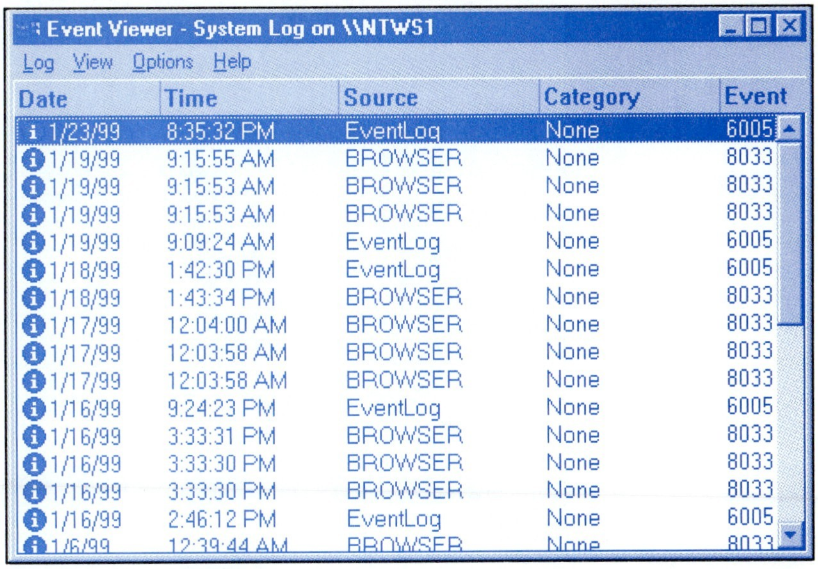

FIGURE 16.28 User Manager display

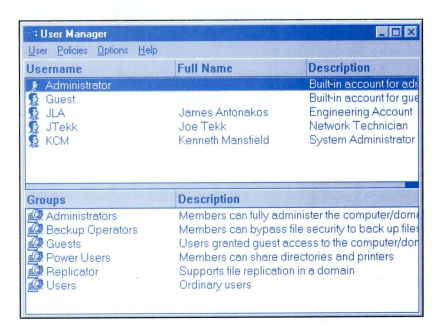

(Common) menu. The applications listed in the menu are designed to monitor and configure many different aspects of the Windows NT operating system. Let us briefly examine three of these applications. You are encouraged to spend some time exploring these tools in the Familiarization Activity section.

Event Viewer

The Event Viewer tool, shown in Figure 16.27, keeps a running log of various system events. The historical information in the log may prove helpful when troubleshooting problems.

User Manager

The User Manager tool allows user accounts to be created, modified, and deleted (if your own account has sufficient privilege). Figure 16.28 illustrates the user base of a small system. Users who are not displayed in the list cannot log on to the Windows NT system. When Windows NT is first installed, an Administrator account is created automatically that has all available privileges, including the ability to begin creating other accounts on the new system.

FIGURE 16.29 Windows NT Memory Diagnostics display

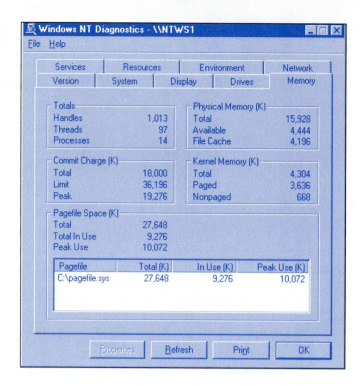

Windows NT Diagnostics

Windows NT keeps track of a significant amount of system parameters. Viewing these parameters and statistics is useful when fine-tuning the operating system for peak performance or locating a bottleneck. This activity is typically done by an experienced system dministrator. The presentation of the system data is interesting in its own right, so a tour of the individual diagnostic tabs would be worthwhile. For example, the Memory diagnostic tab display shown in Figure 16.29 uses nomenclature that is not familiar to the typical system user.

TROUBLESHOOTING TECHNIQUES

It is easy to forget that the Accessories folder contains so many useful utilities. It would be good to create shortcuts to the utilities you use the most, placing them on the desktop for quick access.

You may want to periodically review the contents of the Accessories folder. You may re-encounter a utility that you had forgotten about and now could use.

SELF-TEST

This self-test is designed to help you check your understanding of the background information presented in this exercise.

True/False

Answer *true* or *false*.

1. Games have no value for the beginning user.
2. HTML is a program that executes when browsing the Web.
3. The Character Map tool provides directions on where to download fonts on the Internet.
4. A modem is required for Dial-Up Networking.
5. A direct cable connection can be used to connect two Windows machines.
6. HyperTerminal is the only utility that can dial a number with the modem.
7. Any user can create, modify, and delete accounts in Windows NT.

Multiple Choice

Select the best answer.

8. A list of names might be stored using the
   a. Notepad.
   b. Cardfile.
   c. Calendar.
   d. Clipboard Viewer.
9. An image can be transferred between applications using the
   a. Notepad.
   b. Cardfile.
   c. Calendar.
   d. Clipboard Viewer.
10. The time/date of an important meeting can be stored using the
    a. Notepad.
    b. Cardfile.
    c. Calendar.
    d. Clipboard Viewer.
11. Windows NT systems events are viewed with the
    a. Notepad.
    b. Event Viewer.
    c. Calendar.
    d. Clipboard Viewer.

Matching

Match each utility on the right with each item on the left.

12. Used to create an image.
13. Connects with a mainframe.
14. Used to create a text document.
15. Performs numeric operations.

a. Calculator
b. Notepad
c. Paint
d. HyperTerminal

Completion

Fill in the blank or blanks with the best answers.

16. HTML stands for _____ _____ _____.
17. A disk that has each file stored in different locations, rather than one long block, is said to be _____.
18. The _____ utility checks a disk for errors (such as bad FAT entries).
19. Two editors capable of creating a text file are Notepad and _____.
20. Windows NT administrative accounts have all available _____.

FAMILIARIZATION ACTIVITY

1. Use Netscape or Internet Explorer to browse the Web. Try www.windows95.com or www.winfiles.com, two very good sites for Windows users.
2. Check the amount of fragmentation for each drive on your lab system. If you have a floppy disk, check it also.
3. If approved by your instructor, defragment each drive. Make note of how long it takes to do a single drive and the size of the drive.
4. Determine the amount of available resources (system, user, GDI).
5. Use the System Monitor to watch at least four parameters (processor usage, etc.) for 10 minutes. Perform other work on the system as you watch the display. What do you notice?
6. Use the calculator to determine how long, in minutes, a photon of light takes to travel from the Sun to Earth. The distance is 93 million miles and the speed of light is 186,000 miles/second.
7. Make a cardfile of your favorite television shows. Explain how you organized them.

8. Open the Clipboard Viewer. Is there anything on it? If so, where do you think the data came from?
9. Use Notepad to write a short letter requesting a sample catalog from a computer manufacturer. Print out the final version.
10. Open the letter from step 9 using WordPad. Use the text editing features to make some words bold, others italic, and others a different size. Center and right justify some of the text. Print out the final version.
11. Explore each of the Windows NT Administrative Tools on the Administrative Tools display menu.

QUESTIONS/ACTIVITIES

1. What other utilities might be useful in the Windows environment?
2. Search the Web for the utilities you listed in question 1. With approval, download one of your choices and install it. How well does it meet your expectations?

REVIEW QUIZ

Within 10 minutes and with 100% accuracy,

1. Describe the basic contents of the Accessories folder.
2. Demonstrate the operation of several utilities, such as the Calculator, Notepad, and Paint.
3. Explain the importance of the System Tools and Administrative Tools folders.

17 An Introduction to Networking with Windows

INTRODUCTION

"That's the computer right there."

Those were the first words Joe Tekk heard when he entered a high school laboratory maintained under contract by RWA Software. "Pardon me?" Joe asked.

The laboratory technician was a senior, ready to graduate in a few months, with little patience for computers that did not work.

"It's that one right there. It won't connect to the network." He pointed at the computer until Joe got to it. Joe walked around to the back of the computer, pulled the T-connector off the back of the network card, and looked at it closely.

"Here's your problem," he said, to the surprise of the student. "The metal pin is missing from the center of the connector."

The student looked at the connector and then back at Joe. "How did you know to look for that?"

"I always pull the connector out first. I've seen this happen before. Now, it's just a habit."

PERFORMANCE OBJECTIVES

Upon completion of this exercise, within 10 minutes and with 100% accuracy, you will be able to

1. Identify hard disk resources available on a network computer.
2. Identify printer resources available on a network computer.
3. Create a Dial-Up Networking connection.

BACKGROUND INFORMATION

Windows offers many different ways to connect your machine to one or more computers and plenty of applications to assist you with your networking needs. In this exercise we will briefly examine the basics of networking in Windows, leaving the rest for Unit IV, Computer Networks.

MICROSOFT NETWORKING

Although Windows supports many different types of common networking protocols (covered in detail in Exercise 24), the backbone of its network operations is **NetBEUI**

(NetBIOS Extended User Interface), a specialized Microsoft protocol used in Windows for Workgroups, Windows 95/98, and Windows NT. NetBEUI allows small (up to 200 nodes) networks of users to share resources (files and printers).

THE NETWORK NEIGHBORHOOD

The Network Neighborhood is a hierarchical collection of the machines capable of communicating with each other over a Windows network. Note that systems running Windows for Workgroups have the ability to connect to the network as well.

Figure 17.1 shows a typical Network Neighborhood. The three small PC icons named At213_tower, Nomad, and Waveguide all represent different machines connected to the network. Each machine is also a member of a *workgroup,* or *domain* of computers that share a common set of properties.

Double-clicking on Waveguide brings up the items being shared by Waveguide. As indicated in Figure 17.2, Waveguide is sharing two folders: pcx and pub.

The Network Neighborhood gives you a way to graphically navigate to shared resources (files, CD-ROM drives, printers).

NETWORK PRINTING

A network printer is a printer that a user has decided to share. For the user's machine it is a local printer. But other users on the network can map to the network printer and use it as if it were their own printer. Figure 17.3 shows a shared printer offered by a computer named Nomad. Nomad is offering an hp 890c.

It is necessary to install the printer on your machine before you can begin using it over the network.

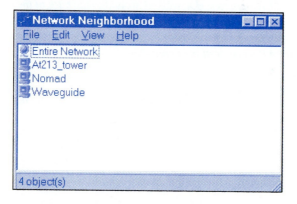

FIGURE 17.1   Network Neighborhood window

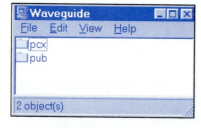

FIGURE 17.2   Items shared by Waveguide

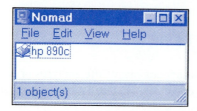

FIGURE 17.3   A printer shared by Nomad

FIGURE 17.4 Indicating a shared drive

FIGURE 17.5 Sharing Properties window for drive D:

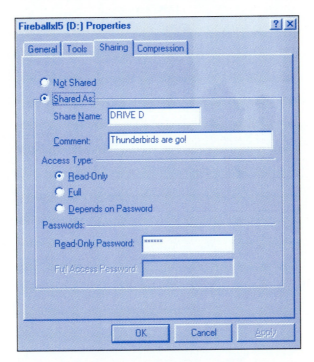

SHARING FILES OVER A NETWORK

A computer can share its disks with the network and allow remote users to map them for use as an available drive on a remote computer. The first time a disk is shared and a connection is established, it is necessary to provide a password to gain access to the data. The password is typically provided by the network administrator. This password is usually stored in the password file for subsequent access to the disk if it is reconnected after a reboot. Figure 17.4 shows the contents of My Computer. The small hand holding drive D: (Fireballxl5) indicates the drive is shared.

The user sharing the drive controls the access others will have to it over the network. Figure 17.5 shows the sharing properties for drive D: (right-click on the drive icon and select Properties). Clearly, the user has a good deal of control over how sharing takes place.

DIAL-UP NETWORKING

Dial-Up Networking is designed to provide reliable data connections using a modem and a telephone line. Figure 17.6 shows two icons in the Dial-Up Networking folder (found in

FIGURE 17.6 Dial-Up Networking icons

FIGURE 17.7 Making a New Connection window

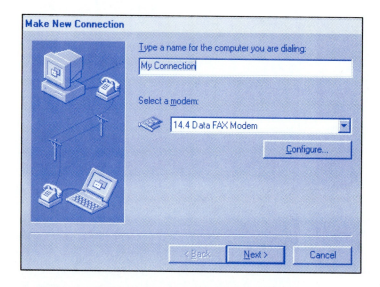

FIGURE 17.8 Information required to access host

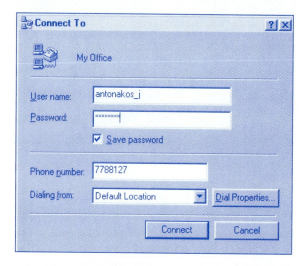

Accessories on the Start menu). Double-clicking the Make New Connection icon will start the process to make a new connection as shown in Figure 17.7. The name of the connection and the modem for the connection are specified.

It is also necessary to provide an area code and telephone number during the configuration process. This number must be for a machine capable of supporting a PPP (point-to-point protocol) connection.

Once the connection has been created, it is activated by double-clicking it. To connect to a remote host, it is necessary to supply a user name and a password. This can be done automatically by the Dial-Up Networking software. Figure 17.8 shows the connection window for the My Office icon.

FIGURE 17.9 Active Dial-Up Networking connection

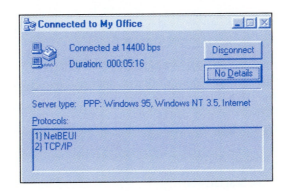

After the information has been entered, the Connect button is used to start up a connection. When the connection has been established, Windows displays a status window showing the current duration of the connection and the active protocols. Figure 17.9 shows the status for the My Office connection. Left-clicking the Disconnect button shuts the connection down and hangs up the modem.

CONNECTING TO THE INTERNET

Besides a modem or a network interface card and the associated software, one more piece is needed to complete the networking picture: the *Internet service provider* (ISP). An ISP is any facility that contains its own direct connection to the Internet. For example, many schools and businesses now have their own dedicated high-speed connection (typically a T1 line, which provides data transfers of more than 1.5 million bits/second).

Many users sign up with a company (such as AOL or MSN) and then dial in to these companies' computers, which themselves provide the Internet connection. The company is the ISP in this case.

Even the local cable company is an ISP now, offering high-speed cable modems that use unassigned television channels for Internet data. The cable modem is many times faster than the fastest telephone modems on the market.

Once you have an ISP, the rest is up to you. You may design your own Web page (many ISPs host Web pages for their customers), use e-mail, browse the Web, Telnet to your school's mainframe and work on an assignment, or download a cool game from an ftp site. Exercise 26 provides additional details about the Internet.

TROUBLESHOOTING TECHNIQUES

Troubleshooting a network connection requires familiarity with several levels of operation. At the hardware level, the physical connection (parallel cable, modem, network interface card) must be working properly. A noisy phone line, the wrong interrupt selected during setup for the network interface card, incompatible parallel ports, and many other types of hardware glitches can prevent a good network connection.

At the software level there are two areas of concern: the network operating system software and the application software. For example, if Internet Explorer will not open any Web pages, is the cause of the problem Internet Explorer or the underlying TCP/IP protocol software?

Even with all of the built-in functions Windows automatically performs, there is still a need for human intervention to get things up and running in the world of networking. Unit IV will take you on a more detailed tour of this topic, so that you have more control over the operation of your networked computer.

SELF-TEST

This self-test is designed to help you check your understanding of the background information presented in this exercise.

True/False

Answer *true* or *false*.

1. NetBEUI is a protocol only used by Windows NT.
2. Dial-up connections work with ordinary dial-up phone numbers.
3. The cable company is an example of an ISP.
4. Network printers can be used as soon as you map them.
5. Anyone who wants to can delete all the files on a shared drive.

Multiple Choice

Select the best answer.

6. NetBEUI is a protocol used
    a. Only for network printers.
    b. Only for file sharing.
    c. For sharing files and printers.
7. A workgroup is a set of users that
    a. Share common properties.
    b. Use the same printer.
    c. Work as a team on projects.
8. The Network Neighborhood shows
    a. The networked computers within 20 meters of your machine.
    b. Every computer on the entire network.
    c. Machines sharing resources.
9. What is required for Dial-Up Networking?
    a. A modem.
    b. A network interface card.
    c. A direct cable connection.
10. The Network Neighborhood shows
    a. Every computer on the network at the same time.
    b. Hierarchical groups of networked computers.
    c. All the computers on the Internet.

Completion

Fill in the blank or blanks with the best answers.

11. NetBEUI stands for NetBIOS _____ _____ _____.
12. Another term for workgroup is _____.
13. Dial-Up Networking is accessed via the _____ folder in the Start menu.
14. The Dial-Up Networking connection uses the _____ protocol.
15. ISP stands for _____ _____ _____.

FAMILIARIZATION ACTIVITY

Do one or both of the following activities:

Activity 1

1. Set up a new modem connection in Dial-Up Networking. Your instructor will supply the Dial-Up Networking number and other parameters.
2. Begin a Dial-Up Networking session. How long does it take to connect?
3. Use the Network Neighborhood to view the machines reachable over your connection.
4. Try to find three machines sharing files or printers.

Activity 2

1. Establish a network connection using the network interface card. This may be automatically done at boot time.
2. If possible, determine the number of machines on your Windows network by counting the icons in the Network Neighborhood windows.

QUESTIONS/ACTIVITIES

1. Go to a local business that advertises on the Web. Ask them to describe their network connection. Do they have dial-up service? What is the cost? Who maintains their systems?
2. Search the Web for information on how satellites are used in Internet connections.

REVIEW QUIZ

Within 10 minutes and with 100% accuracy,

1. Identify hard disk resources available on a network computer.
2. Identify printer resources available on a network computer.
3. Create a Dial-Up Networking connection.

18 Installing New Software

INTRODUCTION

Joe Tekk was very excited. He was downloading a demo version of a game over the Internet. He had considered buying the game, but he did not really know if it was worth the money.

Joe finished downloading a 12MB self-exploding .EXE file that contained all the installation files in a compressed format. He planned to install the software on his D: drive, since the C: drive was quite low in disk space.

However, as he ran the installation procedure, it failed, stating the disk was out of room. Joe noticed the files were being extracted to the TEMP directory on the C: drive during the installation process.

Joe ended up having to reclaim some space on the C: drive so he could install the program. As it turned out, the 12MB of compressed files grew to more than 40MB during the installation process. When the C: drive finally contained enough free space, the installation was successful.

Joe had never run into this problem before, but it was one he would not soon forget, since he had wasted quite a bit of time.

PERFORMANCE OBJECTIVES

Upon completion of this exercise, within 10 minutes and with 100% accuracy, you will be able to

1. Determine which Windows applications are currently installed.
2. Locate the updates for the Windows operating system on the Microsoft Web site.
3. Discuss the general steps required to perform an application installation and deinstallation.

BACKGROUND INFORMATION

The process of installing software involves moving files from a floppy disk or CD-ROM to a suitable hard disk location (local or network). Every single file located on a computer disk has been installed, one way or another. The files actually consist of executable images, data files, initialization files, dynamic link libraries, and other custom files necessary for the computer or application to run. The files are placed in a directory structure determined by an application's installation program.

What types of software do we install? The operating system itself, plus every single application. Note that many applications register themselves with the operating system during the installation process. Specific application settings are stored in the Registry.

EXISTING WINDOWS 3.X SOFTWARE

Most data files and application programs are upward compatible. This means that Windows 3.x applications can be installed on Windows 95, 98, and NT. Unfortunately, Windows 95, 98, and NT applications cannot be installed on a Windows 3.x computer. When a Windows 3.x computer is upgraded to Windows 95 or 98, all installed applications are also upgraded, by placing information from the old .INI files to the new system Registry. There is no upgrade path to Windows NT from any of the other Windows operating systems.

INSTALLING NEW SOFTWARE FROM THE WINDOWS CD-ROM

The Windows operating system can be configured in many different ways. There are likely to be many files that were not installed during the initial Windows installation. To see the current system configuration, double-click the Add/Remove Programs icon in the system Control Panel. Then select the Windows Setup tab as shown in Figure 18.1. Just by looking at the window, it is apparent what components are installed.

Check boxes along the left margin of the component window identify three situations. First, if a box is checked and the inside of the box is white, all components of the category have been installed. If a box is checked and the inside of the box is gray, only some of the components from that particular category are installed. The Details button is used to show the specific status of each component. If the Windows operating system components are changed, the computer will request the Windows CD-ROM or floppy installation media to copy the additional files.

The Windows files are stored in .CAB files on the CD-ROM drive. These files contain the Windows operating system components in compressed form for distribution. Each file is extracted from the .CAB file using a special procedure built into the system.

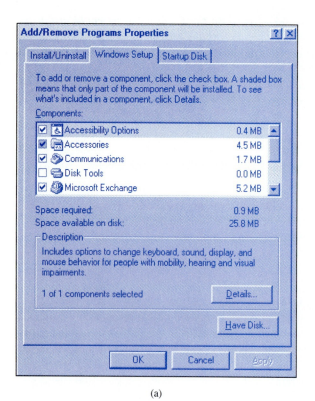
(a)

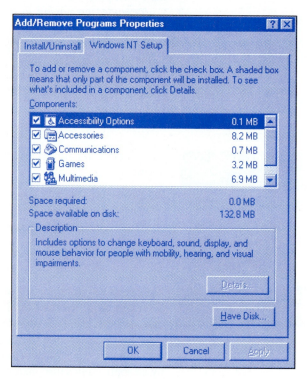
(b)

FIGURE 18.1 (a) Windows 95/98 Setup menu and (b) Windows NT Setup menu

GETTING THE LATEST UPDATES FROM MICROSOFT

As improvements are made in the Windows operating system, they are posted on the World Wide Web. Figure 18.2 shows the downloads page for support drivers, patches, and service packs from Microsoft. As you can see, each category contains specific applications such as Word, Exchange, and the different Microsoft operating systems.

INSTALLING THIRD-PARTY SOFTWARE

Application software is usually installed with a custom software installation wizard. Let us examine an installation process under Windows 95. Windows 98 and NT operate in a similar manner. The first step in performing a software installation is identifying which installation program to run. Figure 18.3 shows the Run window with the path specified to a setup file. When the OK button is selected, the Setup Wizard begins the installation process. Figure 18.4 shows the Norton Utilities For Windows 95 Setup Wizard installation screen. The installation program asks for the user name and company. This information is usually displayed by the application each time it is run. Each window has option buttons that allow the program user to move back and forward through the screens presented during the installation.

The software producers do not actually sell their code, or executables—they license them. The license allows the user to run the product within the scope of the license agreement. It is impossible to install the software without agreeing with the license terms, as shown in Figure 18.5.

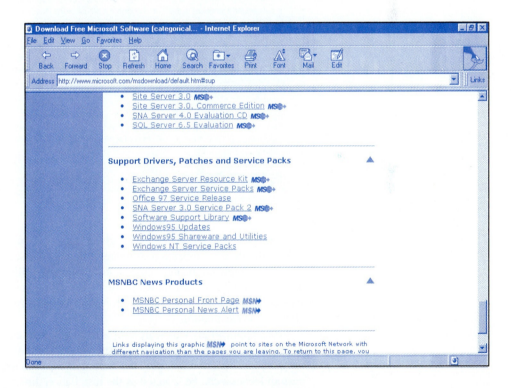

FIGURE 18.2 Operating system update from the Web

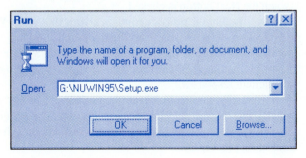

FIGURE 18.3 Location and name of the installation file

FIGURE 18.4 Installation Setup Wizard

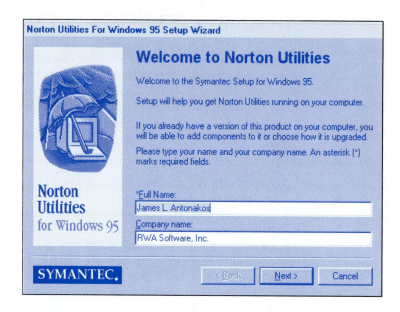

FIGURE 18.5 License agreement

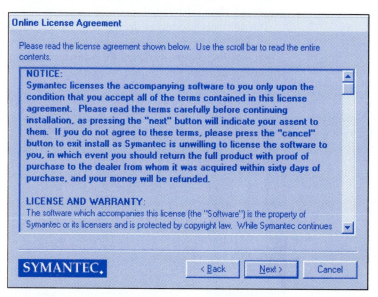

During an installation, the user is given choices about how the Setup Wizard is to install or reconfigure the application software. A complete installation usually installs all components of an application, and a custom installation may allow for one or more components of a product to be installed individually. Figure 18.6 shows a typical Setup option window indicating that a complete installation is to be performed. The installation proceeds by clicking the Next button.

Some applications will search for a previous installation of the application to identify which particular components are installed. When the user is prompted to confirm the program location, it will display the current installation locations. Figure 18.7 shows how a custom location can be selected as the installation directory. The directory will be created when the Next button is clicked.

The installation process continues with the Setup Wizard asking about how the new application software is to be configured, such as adding special features to the Recycle Bin, as shown in Figure 18.8 and whether to run the System Doctor when the system boots, as shown in Figure 18.9. These options are usually set to a default answer. When in doubt, use the default responses.

Eventually, the Setup Wizard begins the process of copying files from the installation media and the selected installation location. The progress meter shown in Figure 18.10 shows a percentage of the how close the installation is from being complete and the current file being processed.

FIGURE 18.6 Installation options

FIGURE 18.7 Selecting a custom storage location

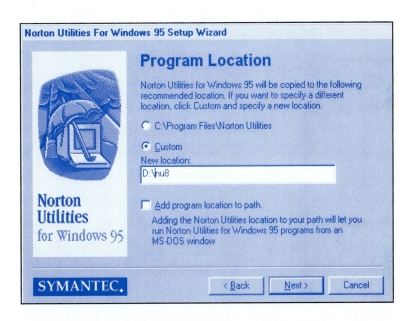

FIGURE 18.8 More installation options

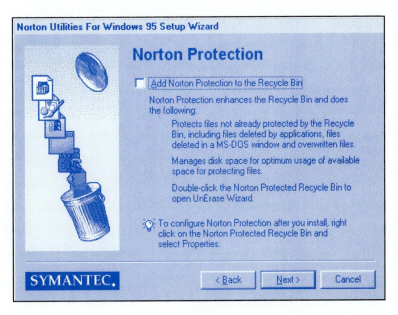

FIGURE 18.9 More installation options

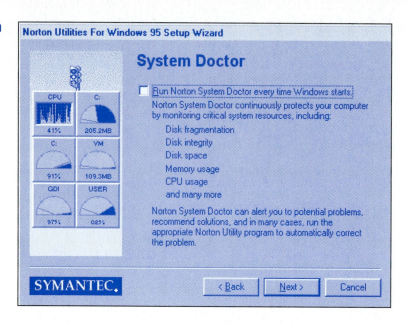

When the installation process completes, it may require the system to be rebooted. This allows for new drivers to be loaded during system initialization. Figure 18.11 shows a congratulatory message for successfully completing the installation.

REMOVING APPLICATION SOFTWARE

Many situations may require a software application to be removed from a computer. Because an installation program may install application programs in many different directories

FIGURE 18.10 Files being copied to the hard disk

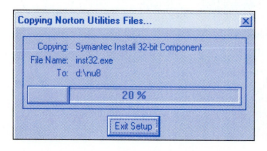

FIGURE 18.11 Installation process requires a system reboot

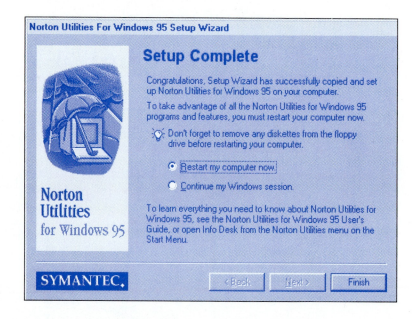

on a disk, an application cannot be deleted by simply deleting the files in the installation directory. This action deletes many files and will remove the ability to run the program but leaves behind a trail of other files, which remain on the hard disk for no purpose.

To solve this dilemma, newer Windows applications provide an uninstall feature. Using the Add/Remove Programs icon in the Control Panel, the Install/Uninstall tab shows all installed products, as shown in Figure 18.12. These are the applications that can be uninstalled. To uninstall any product, simply select it from the list and press the Add/Remove button. Confirmation prompts will double-check to make sure the choice to remove a product is correct. With positive acknowledgment, the item is removed from the computer. Figure 18.13 shows the current step being performed by the uninstall process and a progress display indicating the status of the procedure. Exercise extreme caution when uninstalling software. Unless the files have been backed up, it may be impossible to retrieve deleted data files.

FIGURE 18.12 Application selection window for Install/Uninstall feature

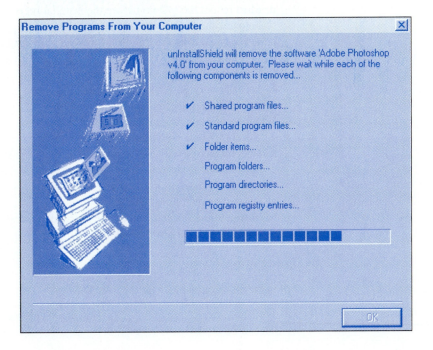

FIGURE 18.13 Progress display during removal of software

241

TROUBLESHOOTING TECHNIQUES

When a software installation fails, it can usually be attributed to a lack of available resources on the computer, such as disk space or memory. Refer to the product literature to determine the application requirements and be sure your system meets the requirements before you begin an installation.

Be cautious when trying to make room for new applications. Do not delete directories manually or run an uninstall program without making quite sure that any files that may be deleted can be replaced if necessary. Remember, if you delete the wrong files, your system can be rendered unusable.

SELF-TEST

This self-test is designed to help you check your understanding of the background information presented in this exercise.

True/False

Answer *true* or *false*.

1. All the optional Windows operating system components are installed when Windows is installed.
2. Due to software licensing issues, program updates cannot be made available over the Web.
3. The Windows operating system is stored in compressed form on the Windows installation CD-ROM.
4. Every installation wizard must do the same thing, regardless of the application being installed.
5. A checked box with a white background on the Windows Setup menu indicates only a portion of a component is installed.
6. Windows 3.x may be upgraded to Windows NT.

Multiple Choice

Select the best answer.

7. Windows updates can be retrieved from the
   a. World Wide Web.
   b. Electronic mail attachments.
   c. Newest version of the Windows CD-ROM.
8. When installing software, the Setup Wizard
   a. Must reboot the system before the installation process can proceed.
   b. Prepares the files on the CD-ROM for transfer.
   c. Copies files from the CD-ROM to a suitable installation location.
9. A .CAB file contains
   a. Compressed Windows installation files.
   b. Security information and passwords.
   c. Application program data files.
10. Windows 95 applications are
    a. Not compatible with new Windows operating systems.
    b. Not compatible with old Windows 3.x operating systems.
    c. Not a factor as far as the operating system is concerned.
11. When a program is uninstalled, the application files are
    a. Deleted by the uninstall process.
    b. Moved to the Recycle Bin by the uninstall process.
    c. Moved into the TEMP directory.

Completion

Fill in the blank or blanks with the best answers.

12. The _____ _____ on the Windows Setup menu show which components are installed.

13. Files are copied to a hard drive during a(n) _____ process.
14. _____ compatibility means Windows 3.x software can be installed on a Windows 95, 98, or NT system.
15. The system Registry contains information about each registered _____ _____.
16. Access to the Add/Remove Programs icon is found on the _____ _____.

FAMILIARIZATION ACTIVITY

1. Determine the status of each component in the Windows Setup window. Which ones are completely installed, partially installed, or not installed at all?
2. Determine what application programs have been installed.
3. Visit the Microsoft Web site to examine support information and data.

QUESTIONS/ACTIVITIES

1. What are .CAB files?
2. What actions does a Setup Wizard program perform?
3. What actions are performed by an uninstall program?

REVIEW QUIZ

Within 10 minutes and with 100% accuracy,

1. Determine which Windows applications are currently installed.
2. Locate the updates for the Windows operating system on the Microsoft Web site.
3. Discuss the general steps required to perform an application installation and deinstallation.

19 Installing New Hardware

INTRODUCTION

Joe Tekk had just returned from a troubleshooting assignment. Don, Joe's manager, asked him how it went.

"Well, I've never encountered this problem before. The customer complained that her keyboard was not working properly. I thought it might be a bad connector, a stuck key, or dirt inside the keyboard, but everything looked fine."

Don's eyebrows shot up. "So, what was it?"

"I have no idea, Don. I was poking around inside Device Manager and I decided to delete the keyboard from the system. I rebooted, reinstalled the keyboard, rebooted again, and the problem went away."

Don thought about Joe's story. "I never would have tried that."

Joe laughed. "Me neither. It was the act of a desperate man."

PERFORMANCE OBJECTIVES

Upon completion of this exercise, within 10 minutes and with 100% accuracy, you will be able to

1. Remove and install a piece of hardware.
2. Examine/adjust hardware settings.
3. Verify correct operation of the new hardware.

BACKGROUND INFORMATION

In this exercise we will watch what happens as we install a new piece of hardware. If you use your computer a great deal, eventually you will feel a need for improvement. Adding a faster processor (on a new motherboard) or an additional hard drive, upgrading your video adapter to one that supports DirectX, or installing more DRAM all lead down the same path. The installation either goes well or something goes wrong that sets you back hours when the job should have taken 20 minutes.

STARTING OFF

Unit V has plenty of exercises that describe how different types of hardware are installed and what the requirements are. The purpose of this exercise is to examine the built-in hardware installation support Windows offers. To demonstrate, we will work through the installation of a modem. Physically, the modem must be plugged into its associated slot, secured, and properly connected to the telephone lines.

When Windows 95/98 comes up for the first time after the modem has been installed, it may find a new plug-and-play device and proceed with the installation automatically, in accordance with the plug-and-play guidelines. If the modem is not a plug-and-play device, it is necessary to run the Add New Hardware utility in Control Panel. The first window should look like Figure 19.1. Notice the Back and Next buttons. You are able to move back and forth in the installation process until you are satisfied with the results. Hardware installation in Windows NT will be covered at the end of this exercise.

DETECTING NEW HARDWARE

Figure 19.2 shows the next installation screen. Be careful when you let Windows 95/98 detect the new hardware by itself. Windows 95/98 is not a perfect operating system. It may get the model or make of your network card wrong, or not detect a plug-and-play device. Windows 95/98 may hang during the detection, forcing you to simply reboot without being able to properly shut down first. The good thing is that Windows 95/98 is always on the watch for new hardware, so if it does not see the device the first time you reboot, it may the second time. Figure 19.3 is evidence that your installation may not go as smoothly as planned.

If you click the No button in the screen shown in Figure 19.2 and do not let Windows 95/98 detect the new hardware, you will have to navigate through a series of menus asking what type of hardware is being installed and other details. Figure 19.4 shows the menu you get if you click "No." Windows 95/98 has a wealth of hardware device types ready to go.

While the detection phase is going on, the display will look similar to Figure 19.5.

FIGURE 19.1 The first Add New Hardware dialog box

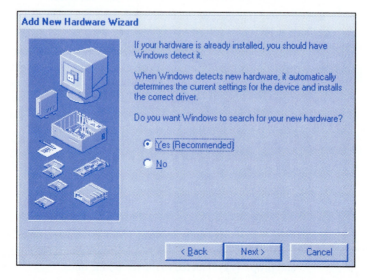

FIGURE 19.2 Add New Hardware Wizard menu

FIGURE 19.3 Add New Hardware Wizard informational message

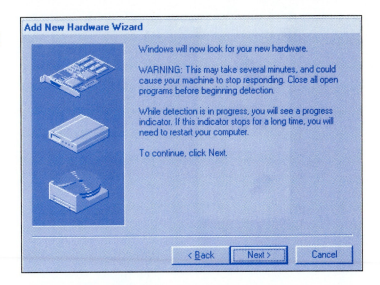

FIGURE 19.4 Choosing a hardware device to add

FIGURE 19.5 The Detection progress meter

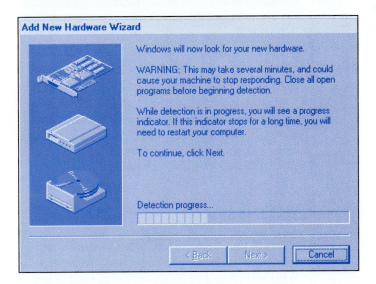

THE RESULTS

When Windows 95/98 finishes detecting the new hardware, the window shown in Figure 19.6 will appear. It is a good idea to click the Details button to see what Windows 95/98 actually found. Figure 19.7 indicates that a 14.4 Data FAX Modem was detected.

FIGURE 19.6 New hardware has been detected

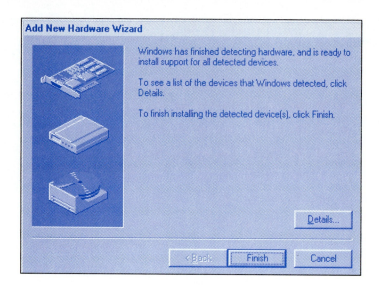

FIGURE 19.7 The new hardware details

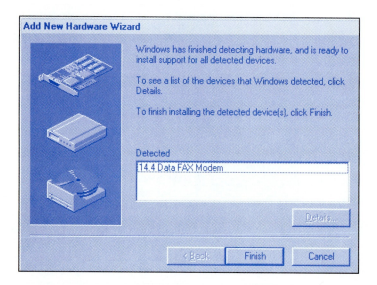

FINISHING UP

Clicking the Finish button completes the hardware installation and verifies that the modem was installed correctly. Figure 19.8 indicates that you are given one last chance to back out and not install the new hardware, or change the detected device. If you click the Change button, Windows 95/98 presents you with a menu of modems to choose from, the same menu you get if you skip the hardware detection phase by answering "No" in the first screen. Figure 19.9 shows the menu of modems.

GENERAL COMMENTS

No matter what hardware is being installed, there are two types of software drivers available for use, typically *real-mode* device drivers and *protected-mode,* or *virtual,* device drivers. Since Windows likes to do as much work as possible in protected-mode (the most powerful mode of operation in the Intel processors; see Exercises 56 and 58 for details), a large number of virtual device drivers (.VXD extension on Windows 95/98 machines) are provided for all types of hardware. If you have an older hardware device that requires a real-mode device driver, or your hardware is not supported by Windows (no .VXD file), you have no choice but to use a real-mode driver. In general, real-mode device drivers lower the performance of Windows, which

FIGURE 19.8 Verifying the new hardware information

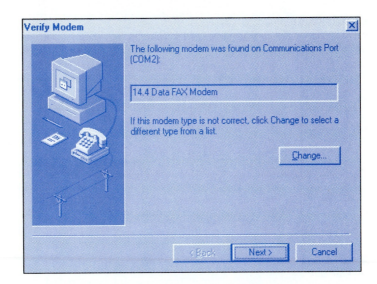

FIGURE 19.9 Choosing a modem

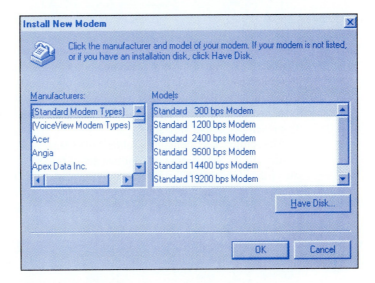

must switch from protected mode to real mode to use the device driver (and back when finished). The mode-switching overhead could be a factor if the driver is used often (as in a modem).

In addition, there is bound to come a time when an interrupt or I/O location assigned to a new device conflicts with an existing assignment (for example, interrupt number 5 or port number 330 being used by two devices). Windows does a good job of catching these types of conflicts, as indicated in Figure 19.10. The Conflict information box indicates that the interrupt 10 setting for the hardware being configured conflicts with the network interface card. It should be possible to change the interrupt and avoid the conflict.

Some new devices require less effort to install than others. Adding another 64MB of DRAM is as easy as sliding the SIMM or DIMM into its socket and pushing it into place. BIOS will find the additional memory the next time the machine boots, as will Windows.

INSTALLING HARDWARE IN WINDOWS NT

Installing new hardware on a Windows NT computer is done much differently from the Windows 95/98 environment, which provides an Add New Hardware icon in the Control Panel. Device types and drivers that can be added to Windows NT are provided on the Windows NT CD-ROM.

To begin the installation process, it is necessary to open the Windows NT CD-ROM, whose contents are shown in Figure 19.11. The Drvlib folder contains 12 folders that

FIGURE 19.10 Interrupt 10 is in conflict

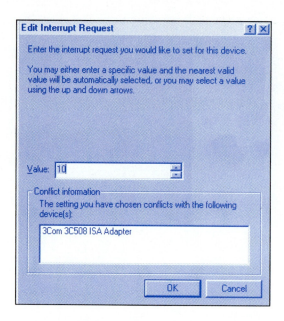

categorize the different types of hardware that can be added. These folders are shown in Figure 19.12. One of the most common types of devices to add to a Windows NT computer is a sound card, which is located in the Audio folder. The Windows NT CD-ROM provides drivers that support ESS and Sound Blaster hardware. Other Windows NT drivers are supplied by the individual manufacturers using either a floppy disk, CD-ROM, or directly from their respective Web sites.

The Audio folder contains a folder called Sbpnp (Sound Blaster Plug and Play shown in Figure 19.13), which contains another set of folders that provide the necessary drivers for all of the Windows NT support hardware platforms. These platforms include the Alpha, Mips, Power PC, and Intel (and compatible) microprocessors. The *Readme.txt* file provides the specific details to install the device on any of the supported platforms. To install a

FIGURE 19.11 Windows NT CD-ROM contents

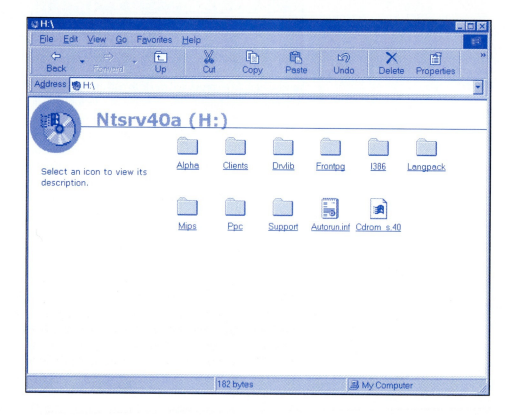

FIGURE 19.12 Drvlib folder

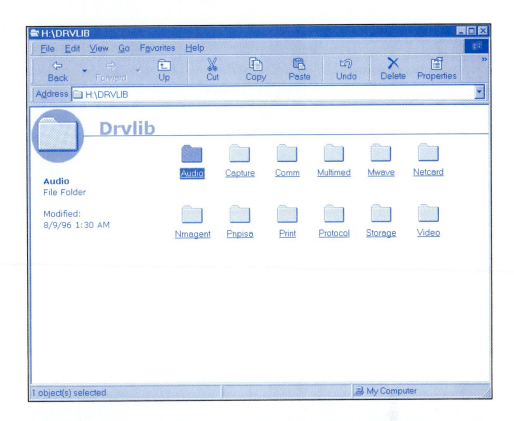

FIGURE 19.13 Sound Blaster files

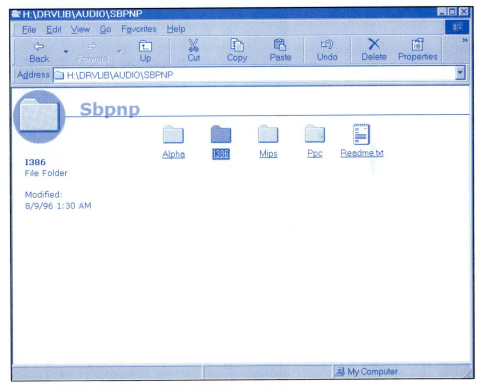

Sound Blaster audio device on an Intel microprocessor platform, it is necessary to right-click on the *Sbpnp.inf* icon located in the I386 folder and then select the install option by left-clicking on install as shown in Figure 19.14. The installation process copies all the required files to their proper locations, and prepares for the driver files to be loaded after the system is rebooted. You are encouraged to explore each of the device folders found in the Drvlib folder.

FIGURE 19.14 Installing the new hardware

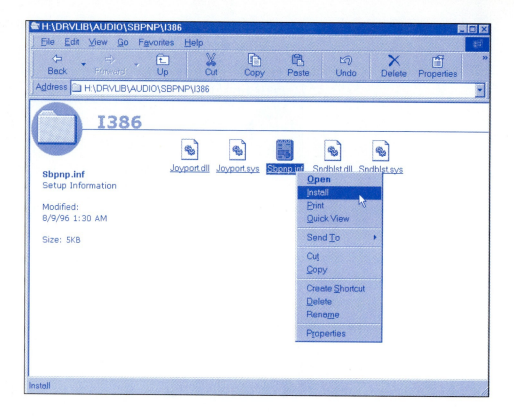

TROUBLESHOOTING TECHNIQUES

A few things to keep in mind when installing new hardware:
- Be prepared to reboot a number of times.
- Do not be shocked if a plug-and-play card is not recognized or does not work when first installed.
- Avoid DOS (real-mode) device drivers if possible, using a virtual device driver (.VXD file) supplied by Windows.
- Sometimes it helps to delete a troublesome hardware device. The next reboot will reinstall it, hopefully solving the problem.
- Check the interrupt and I/O settings for conflicts.
- Windows contains many useful built-in troubleshooters, accessed via the Help menu. Follow the recommendations of the troubleshooter when trying to resolve conflicts.

SELF-TEST

This self-test is designed to help you check your understanding of the background information presented in this exercise.

True/False

Answer *true* or *false*.

1. Plug-and-play cards always work when first installed.
2. It is sometimes necessary to reboot when Windows is detecting hardware.
3. Windows contains a large database of hardware drivers.
4. It is good to use real-mode device drivers.
5. Windows cannot help at all when hardware conflicts arise.

Multiple Choice

Select the best answer.

6. Windows detects new hardware
   a. Every time a machine boots up.
   b. Only when you use a driver disk.
   c. Only during installation of the operating system.

7. When installing new hardware, the install wizard allows you to
   a. Click a single button that does everything.
   b. Always use a virtual device driver.
   c. Go back and change your mind.
8. If "No" is clicked on the menu in Figure 19.2, Windows will
   a. Abort the install process.
   b. Install a generic driver.
   c. Allow you to choose your hardware from a list.
9. After Windows detects new hardware it
   a. Verifies its operation.
   b. Runs a performance test to optimize the device.
   c. Disables all other hardware until the install is complete.
10. Virtual device drivers operate in
    a. Real mode.
    b. Imaginary mode.
    c. Protected mode.

Completion

Fill in the blank or blanks with the best answers.

11. VXD stands for _____ _____ _____.
12. You may need to _____ when detecting new hardware.
13. Two possible sources of conflict are _____ and I/O locations.
14. Real-mode device drivers require Windows to _____ back and forth from protected mode.
15. Real-mode device drivers _____ the performance of Windows.

FAMILIARIZATION ACTIVITY

1. If your lab machine has a modem, use Device Manager to select the modem and then click the Remove button to delete it from the system (a confirmation window will pop up). Power down and reboot. If your modem is a plug-and-play modem, Windows should detect it and begin installing it. Finish the installation and demonstrate that the modem works properly.
2. Repeat step 1 for any other piece of hardware your instructor deems acceptable.

QUESTIONS/ACTIVITIES

1. If your familiarization activity used a plug-and-play card, how smooth was the installation? Were you really able to plug and play?
2. What is the problem with real-mode device drivers?

REVIEW QUIZ

Within 10 minutes and with 100% accuracy,

1. Remove and install a piece of hardware.
2. Examine/adjust hardware settings.
3. Verify correct operation of the new hardware.

20 Other Network Operating Systems

INTRODUCTION

Joe Tekk visited his friend Marlene Hall, an educational planner for a high-technology consulting firm. Marlene was busy preparing color transparencies for a presentation.

"Hi, Joe," she said, handing him a thick pile of transparencies. "Look through these and tell me what you think."

Joe examined the transparencies with interest. Marlene had put together a detailed comparison of Windows NT and NetWare. "Who are these for?" he asked.

Marlene gave Joe a quick stare and then replied, "The president of our European division."

Two weeks later, Joe received a call from Marlene. The presentation had gone so well the president chose both operating systems, and hired Marlene to oversee their integration in the European office.

PERFORMANCE OBJECTIVES

Upon completion of this exercise, within 10 minutes and with 100% accuracy, you will be able to

1. Compare the features of NetWare with Windows NT.
2. Discuss the file organization, protocols, and security available in NetWare.
3. Briefly describe the Linux operating system.

BACKGROUND INFORMATION

In this exercise we will examine two additional network operating systems: NetWare and Linux. Let us start with a look at the features of NetWare.

NETWARE

The NetWare operating system originated in the early days of DOS, allowing users to share information, print documents on network printers, manage a set of users, and so on. One important difference between NetWare and Windows is in the area of application software. NetWare does not provide the 32-bit preemptive multitasking environment found in Windows. Applications written for Windows will not run on NetWare. They may, however, communicate over the network using NetWare's proprietary IPX/SPX protocol. The following sections describe many of the main features of NetWare.

INSTALLING/UPGRADING NETWARE

Unlike Windows 95, 98, and NT, the NetWare operating system runs on top of DOS (as Windows 3.x did). So before NetWare can be installed on a system, DOS must be up and running. NetWare 4.x and above provide a DOS environment as part of the installation process. Versions of NetWare older than 3.1x must be upgraded to 3.1x before they can be further upgraded to NetWare 4.x and above.

There are two ways to perform an upgrade: through *in-place migration* and through *across-the-wire migration*. In-place migration involves shutting down the NetWare server to perform the upgrade directly on the machine. Across-the-wire migration transfers all NetWare files from the current server to a new machine attached to the network. The new machine must already be running NetWare 4.x or above. This method allows the older 3.1x server to continue running during the upgrade.

NETWARE-WINDOWS TIME LINE

Table 20.1 shows the NetWare and Windows operating system releases during the last decade. Both operating systems have matured to provide significant network support, management, and productivity features.

NDS

The Network Directory Service (NDS) is the cornerstone of newer NetWare networks. Network administrators can manage all users and resources from one location. NDS allows users to access global resources regardless of their physical location using a single login. The NDS database organizes information on each object in the network. These objects are users, groups, printers, and disk volumes. Typically these objects are organized into a hierarchical tree that matches the internal structure of an organization. Figure 20.1 shows a typical tree structure for a small two-year college. Each major area of the organization (Administration, Academic, Student Support, and Alumni) has its own unique requirements. The requirements are applied to all the users who are associated with each specific area. Over time, as the requirements of the organization change, elements in the hierarchical tree are added, modified, or removed very easily.

NetWare uses *data migration* to move data from one location to another to maintain effective use of available hard drive space. Large files are moved to a secondary storage system (such as a *jukebox*) and copied back (demigrated) to the hard drive when needed. Files that have been migrated still show up in directory listings.

Accurate timekeeping plays an important role in the operation of NDS. Multiple servers must agree on the network time so that file updates are performed in the correct sequence. NetWare uses several kinds of time servers to maintain an accurate Universal Coordinated Time (UTC). These servers are called reference, primary, secondary, and single-reference. Reference servers use a connection to an accurate time source (such as the U.S. Naval

TABLE 20.1 NetWare–Windows timeline

NetWare		Windows			
Version	Year	3.x/95/98		NT	
		Version	Year	Version	Year
3.x	1989	3.0	1990	3.1	1993
4.x	1993	3.1	1992	3.5	1994
5	1998	3.11	1993	3.51	1995
		95	1995	4.0	1996
		98	1998	5.0 beta	1997

FIGURE 20.1 Typical tree structure

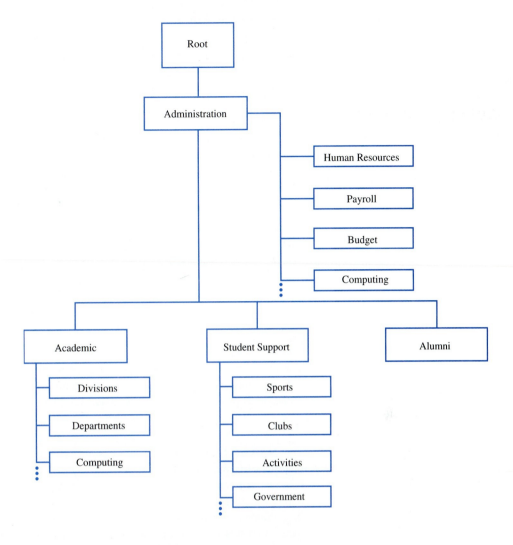

Observatory's Atomic Clock) to provide the network time. Primary and reference servers negotiate with each other to determine the network time. Secondary servers provide the time to NetWare clients. Single-reference time servers are designed for use on small networks where one machine has total control over the network time.

NDS add-ons are also available for Windows NT and Unix environments allowing those systems to fully participate in the NetWare environment.

HCSS

The High Capacity Storage System (HCSS) provided by NetWare allows for tremendously large volumes to be created that span up to 32 physical hard drives. When hard drives were quite small in comparison to the sizes available today, HCSS allowed for the creation of volumes up to 32GB in size. NetWare 5 introduced increased volume sizes up to 8 Terabytes (still much larger than disks typically available today). In conjunction with a configuration of RAID (Redundant Array of Inexpensive Disks), data is protected even if one of the disks in the volume fails.

NETWORK PROTOCOLS

Figure 20.2 shows an outline of the major protocols used in the NetWare environment. These protocols are introduced in Exercise 24. Let us take a look at them and see what their roles are in the NetWare environment.

FIGURE 20.2 NetWare protocol suite

ISO-OSI Layer	Protocols		
7 = Application	NCP	SAP	RIP
6 = Presentation	NCP	SAP	RIP
5 = Session			
4 = Transport	SPX		
3 = Network	IPX		
2 = Data Link	LSL 802.2 / MLID		
1 = Physical	802.3 Ethernet / 802.5 Token Ring		

IPX

The Internetwork Packet Exchange (IPX) protocol is similar to UDP under TCP/IP, providing connectionless, unreliable network communication. IPX packets are used to carry higher-level protocols such as SPX, RIP, NCP, and SAP.

NCP

The NetWare Core Protocol (NCP) is the workhorse of NetWare, responsible for the majority of traffic on a NetWare network. Spanning three layers of the OSI protocol stack (session, presentation, and application), NCP carries all file system traffic in addition to numerous other functions, including printing, name management, and establishing connection-oriented sessions between servers and workstations. NCP provides connection-oriented packet transmission allowing for positive acknowledgment of each packet. A more efficient implementation called NCPB (NetWare Core Protocol Burst) was added to reduce the amount of control traffic necessary when using NCP.

SPX

The Sequenced Packet Exchange (SPX) protocol is similar to TCP under TCP/IP, providing connection-oriented communication that is reliable (lost packets are retransmitted). SPX packets contain a sequence number that allow packets received out of order to be reassembled correctly. Flow control is used to synchronize both ends of the connection to achieve maximum throughput.

SAP

The Service Advertising Protocol (SAP) is a broadcast protocol that is used to maintain a database of servers and routers connected to the NetWare network.

RIP

Like SAP, the Routing Information Protocol (RIP) is also a broadcast protocol. RIP is used by routers to exchange their routing tables. Multiple routers on the same network discover each other (during a RIP broadcast), and build entries for all networks that can be reached through each other. When conditions on the network change (a link goes down or is added), the change will be propagated from router to router (using RIP packets).

MLID

The Multiple Link Interface Driver (MLID) provides the interface between the network hardware and the network software. A specification called *ODI (Open Data-link Interface)*

TABLE 20.2 NetWare Rights

Rights Name	Rights Description
Access Control	May control the rights of other users to access files and directories
Create	May create new file or subdirectory
Erase	May delete existing files or directories
File Scan	May list the contents of a directory
Modify	May rename and change file attributes
Write	May write data into an existing file

is supported by MLID. ODI allows NetWare clients to use multiple protocols over the same network interface card. The LSL (Link Support Layer) provides the ODI facilities.

Menus, Login Scripts, and Messaging

Several of the most important issues for the user involve the system menus, shared access to a common set of data, and electronic messaging capabilities. Access to the software located on each system is created using the menu generation program. Access to items in the menu is made available during the login process using a login script.

The login script contains a list of commands that are executed when each user logs in to the network. The commands are typically used to establish connections to network resources such as mapping of network drives. A login script is a property of a container, Profile, or User object. If a login script is defined for each of these objects, all associated login scripts will execute when a user logs in, allowing for a great deal of control over each user's environment.

Electronic messaging is provided for all users using information available through NDS. Each user's specific information is centrally maintained in the NDS database. The Message Handling Service (MHS) provides access to the X.400 standard implementation for e-mail. FirstMail client software is provided with Novell NetWare 4.1 and above, which is used to access the x.400 messaging services. Add-on products such as GroupWise offer more sophisticated support for electronic messaging. GroupWise also provides document management, calendaring, scheduling, task management, workflow, and imaging.

Security

The security features of the NetWare operating system offer the system and/or network administrator the ability to monitor all aspects of the system operation from a single location. There are two types of security: file system security and NDS security.

File System Security

Table 20.2 shows the various types of rights that may be assigned to a NetWare user. Note that similar rights are available under Windows NT.

Rights are inherited and/or modified via filters or masks that designate permissible operations.

NDS Security

In addition to encryption of login passwords, NDS security provides auditing features that allow one user to monitor events caused by other users (changes to the file system, resource utilization).

MANAGEMENT

Management of any network involves many different activities to ensure quality control. Some of these items include:

- Monitor network traffic to develop a baseline from which to make network-related decisions.
- Unusual activity monitoring such as successive login failures, or file creation or file write errors.
- Disk resource utilization issues.

- Software and hardware installation and upgrades.
- Backup scheduling.
- Help desk support for problem resolution.
- Short-range and long-range planning.

Many other important items can be added to this list, depending on the organization. Some of these issues will be explored in the problems located at the end of this exercise.

PRINT SERVICES

A core component of NetWare consists of the services available for managing and using printers. Print jobs are first sent to a printer queue, where they are temporarily stored until the assigned printer is available. Printers may be attached to workstations, print servers, or even directly connected to the network.

NetWare 5 expands print services to allow for notification of print job completion or status, enhanced communication between printers and clients (printer features are shared), and support for multiple operating systems. An online database of printer drives is also provided to assist with new printer installations.

NetWare Client Software

In addition to one or more NetWare servers, there will be many NetWare clients on the network taking advantage of the services provided. Windows users can install NetWare client software and have access to the power of NetWare while still being part of a Windows network environment.

Figure 20.3 shows the new items found in the Network Neighborhood window after the NetWare client software has been installed. NetWare's folders exist side by side with Windows 98's icon for the Raycast machine.

Figure 20.4 shows the context-sensitive menu that appears after right-clicking on the NetWare icon (N) in the system tray. Note all the different features accessible from the menu. Other NetWare properties can be examined and/or modified by selecting Novell NetWare Client properties in the Network window under Control Panel.

THE LINUX OPERATING SYSTEM

The Linux operating system is a free version of the popular Unix operating system used on many different workstations and servers. The Linux operating system was written by Linus Torvold, a Swedish computer science student, as a personal project that grew into a worldwide cooperative development effort. Complete with all of the bells and whistles of a proprietary Unix operating system, Linux provides a stable environment on which to run many of the popular applications.

Like Windows, Linux is a stand-alone network operating system that does not run on top of DOS. When Windows and Linux are installed on the same system, Linux allows access to the Windows drives. A significant feature of Linux is that the complete source code of the entire operating system is included on the CD-ROM. This allows for custom modifications to be made, if necessary, and modifications necessary to add new hardware devices as they become available.

Linux, like Unix, uses TCP/IP as the built-in network protocol, allowing any Linux-based computer the ability to be used as a server on the Internet. All the required elements such as Domain Name Services (DNS), Dynamic Host Configuration Protocol (DHCP), and the complete suite of server software for HTTP, FTP, Telnet, and so on, are also provided.

Because of its unique status as a fully functional, free-operating system, Linux is quickly gaining wide popularity. In addition, support for Linux is provided by the entire Linux community.

FIGURE 20.3 Network Neighborhood contents after NetWare installation

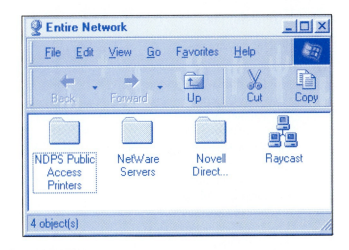

FIGURE 20.4 NetWare controls

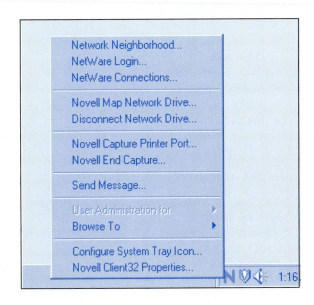

TROUBLESHOOTING TECHNIQUES

Selecting a network operating system is based on many factors, including the features provided and the complexity of the installation and management procedures. Is the operating system centralized or distributed? Will the network be strictly TCP/IP or should multiple protocols be supported? What file system properties are desired?

Whatever the answers are, the end result is an operating system that will require specific troubleshooting methods to diagnose and repair problems.

SELF-TEST

This self-test is designed to help you check your understanding of the background information presented in this exercise.

True/False

Answer *true* or *false*.

1. NetWare is a distributed operating system.
2. NDS allows users global access to resources.
3. SPX is an unreliable protocol.
4. The source code for NetWare is provided on the installation CD-ROM.
5. Linux runs on top of DOS.

Multiple Choice

Select the best answer.

6. Two ways to upgrade a NetWare server are in-place migration and
   a. Out-of-place migration.
   b. Across-the-wire migration.
   c. Parallel migration.
7. SPX packets contain
   a. Sequence numbers for proper reassembly.
   b. Routing information.
   c. Service broadcast data.
8. Login passwords are
   a. Stored as plain text.
   b. Stored as encrypted text.
   c. Not stored.
9. Using a single login, each user
   a. Gains access to container files only.
   b. Gains access to all network resources.
   c. Gains access to local files only.
10. NetWare print services support
    a. Multiple operating systems.
    b. Notification of job completion.
    c. Both a and b.

Completion

Fill in the blank or blanks with the best answers.

11. Two types of security are file system security and _____ security.
12. Linux is based on _____.
13. The main protocol used for NetWare operations is _____.
14. RAID stands for _____ _____ _____ _____.
15. NetWare uses several types of _____ servers to synchronize network events.

FAMILIARIZATION ACTIVITY

NetWare

1. Try to measure the delay associated with demigrating a large file.
2. Watch the IPX network traffic with a protocol analyzer. What types of packets do you see?

Linux

1. Install Linux on a machine that already has an operating system on it. How are both operating systems accessible?
2. Search the Web for Linux download information. What is available?

QUESTIONS/ACTIVITIES

1. Find a local business that uses NetWare. How many users are supported? Why did they choose NetWare? What version are they running?
2. Can you purchase NetWare in your local computer software store?

REVIEW QUIZ

Within 10 minutes and with 100% accuracy,

1. Compare the features of NetWare with Windows NT.
2. Discuss the file organization, protocols, and security available in NetWare.
3. Briefly describe the Linux operating system.

UNIT IV Computer Networks

EXERCISE 21 What Is a Computer Network?
EXERCISE 22 Network Topology
EXERCISE 23 Networking Hardware
EXERCISE 24 Networking Protocols
EXERCISE 25 Network Applications
EXERCISE 26 The Internet
EXERCISE 27 Windows NT Domains
EXERCISE 28 An Introduction to Telecommunications

21 What Is a Computer Network?

INTRODUCTION

Joe Tekk was visiting his friend Julie Plume, an instructor at a local community college. Julie was interested in setting up a network in her classroom.

"Joe," she began, "I need to know a number of things. How much will it all cost? Where do I buy everything? Who can set it up for me?"

Joe laughed. "Hold on, Julie, one thing at a time. The cost depends on how many computers you want to network, the type of network used, and who you buy your equipment from. I have a number of networking catalogs you can look at, and you can also browse the Web for networking products."

Joe looked around the room. There were 14 computers, two laser printers, and a color scanner. "You could probably buy a 16-port Ethernet hub that would take care of this entire room. One network interface card for each PC, some UTP cable, and that's about it. Probably a few hundred dollars will do it. I could set it up with you some afternoon."

Julie had more questions. "Will I need to buy software?"

"Most of the stuff you'll want to do, such as network printing and sharing files, is already built into Windows. You may need to purchase special network versions of some of your software."

"Just one more question, Joe," Julie said. "How does it all work?"

PERFORMANCE OBJECTIVES

Upon completion of this exercise, within 15 minutes and with 100% accuracy, you will be able to

1. Sketch and discuss the different types of network topologies and their advantages and disadvantages.
2. Sketch and explain examples of digital data encoding.
3. Discuss the OSI reference model.
4. Explain the basic operation of Ethernet and token-ring networks.

BACKGROUND INFORMATION

A computer network is a collection of computers and devices connected so that they can share information. Such networks are called local area networks or LANs (networks in office buildings or on college campuses) and wide area networks or WANs (networks for very large geographical areas). Computer networks are becoming increasingly popular. With **Internet** spanning the globe, and the **information superhighway** on the drawing boards, the exchange of information among computer users is increasing every day. In this exercise we will examine the basic operation of a computer network, how it is connected, how it transmits information, and what is required to connect a computer to a network. This exercise lays the foundation for the remaining exercises in this unit.

COMPUTER NETWORK TOPOLOGY

Topology has to do with the way things are connected. The topology of a computer network is the way the individual computers or devices (called *nodes*) are connected. Figure 21.1 shows some common topologies.

Figure 21.1(a) illustrates a *fully connected network*. This kind of network is the most expensive to build, because every node must be connected to every other node in the network. The five-node network pictured requires 10 connections. A 20-node network would require 190 connections. The advantage of the fully connected network is that data need only traverse a single link to get from any node to any other node.

Figure 21.1(b) shows the *star network*. Note that one node in the network is a centralized communications point. This makes the star connection inexpensive to build, since a minimum number of communication links are needed (always one less than the number of nodes). However, if the center node fails, the entire network shuts down. This does not happen in the fully connected network.

The *bus network* is shown in Figure 21.1(c). All nodes in the bus network are connected to the same communication link. One popular bus network is **Ethernet,** which we will be covering shortly. The communication link in an Ethernet network is often a coaxial cable connected to each node through a T-connector. The bus network is inexpensive to build, and it is easy to add a new node to the network just by tapping into the communication link. One thing to consider in the bus network is the maximum distance between two nodes, because this affects the time required to send data between the nodes at each end of the link.

The last topology is the *ring network,* shown in Figure 21.1(d). This connection scheme puts the nodes into a circular communication path. Unlike the other topologies, the communication links between nodes may be one-way links in a ring network. Thus, as in the bus network, the maximum communication time depends on how many nodes there are in the network.

FIGURE 21.1 Topologies for a five-node network

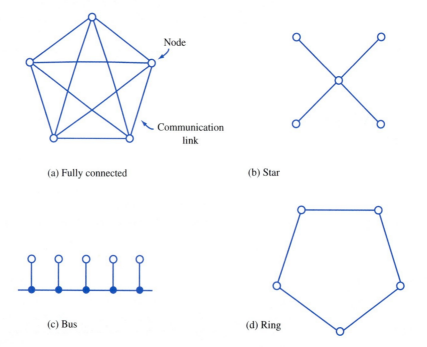

REPRESENTING DIGITAL DATA

The information exchanged between computers in a network is of necessity digital, the only form of data with which a computer can work. However, the actual way in which the digital data is represented varies. Figure 21.2 shows some of the more common methods used to represent digital data.

When an analog medium is used to transmit digital data (such as through the telephone system with a modem), the digital data may be represented by various forms of a *carrier-modulated* signal. Two forms of carrier modulation are amplitude modulation and frequency-shift keying. In amplitude modulation, the digital data controls the presence of a fixed-frequency carrier signal. In frequency-shift keying, the 0s and 1s are assigned two different frequencies, resulting in a shift in carrier frequency when the data changes from 0 to 1 or from 1 to 0. A third method is called *phase-shift keying,* where the digital data controls the phase shift of the carrier signal.

When a digital medium is used to transmit digital data (between COM1 of two PCs, for example), some form of digital waveform is used to represent the data. A digital waveform is a waveform that contains only two different voltages. Inside the computer, these two voltages are usually 0 volts and 5 volts. Outside the computer, plus and minus 12 volts are often used for digital waveforms. Refer again to Figure 21.2. The nonreturn to zero (NRZ) technique simply uses a positive voltage to represent a 0 and a negative voltage to represent a 1. The signal *never* returns to zero.

Another popular method is Manchester encoding. In this technique, phase transitions are used to represent the digital data. A one-to-zero transition is used for 0s and a zero-to-one transition is used for 1s. Thus, each bit being transmitted causes a transition in the Manchester waveform. This is not the case for the NRZ waveform, which may have long periods between transitions. The result is that Manchester encoding includes both data *and* a clock signal, which is helpful in extracting the original data in the receiver.

COMMUNICATION PROTOCOLS

Just throwing 1s and 0s onto a communication link is not enough to establish coherent communication between two nodes in a network. Both nodes must agree in advance on what the format of the information will look like. This format is called a *protocol* and is firmly defined. Figure 21.3 shows one of the accepted standards governing the use of protocols in computer networks. The Open Systems Interconnection (OSI) reference model defines seven layers required to establish reliable communication between two nodes. Different protocols are used between layers to handle such things as error recovery and information routing between nodes. A handy way to remember the names of each layer is contained in a simple statement: All Packets Should Take New Data Paths.

FIGURE 21.2 Methods of representing digital data

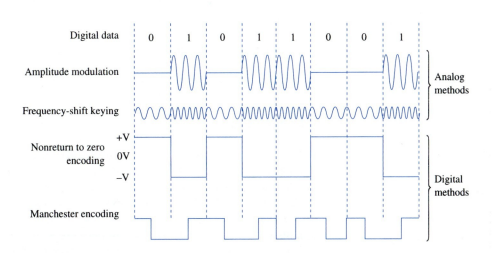

FIGURE 21.3 OSI reference model

Not all of the seven layers are always used in a computer network. For example, Ethernet uses only the first two layers. The OSI reference model is really just a guide to establishing standards for network communications.

The Physical Layer

The Physical layer (layer 1) controls how the digital information is transmitted between nodes. In this layer, the encoding technique, the type of connector used, and the data rate, all of which are *physical* properties, are established.

The Data-Link Layer

The Data-Link layer (layer 2) takes care of error detection and correction.

The Network Layer

The Network layer (layer 3) is responsible for assembling blocks of data into **packets.** A typical data packet consists of control and addressing information, followed by the actual data.

The Transport Layer

The Transport layer (layer 4) is the first layer that is not concerned with how the data actually gets from node to node. Instead, the Transport layer assumes that the physical data is error-free, and concentrates on providing correct communication between nodes from a *logical* perspective. For example, the Transport layer guarantees that a large block of data transmitted in smaller chunks is reassembled in the proper order when received.

The Session Layer

The Session layer (layer 5) handles communication between processes running on two different nodes. For example, two mail programs running on different nodes must communicate with each other.

The Presentation Layer

The Presentation layer (layer 6) deals with matters such as text compression and encryption.

The Application Layer

The Application layer (layer 7) is where the actual user program executes.

ETHERNET

One of the most popular communication networks in use is Ethernet. Ethernet was developed jointly by Digital Equipment Corporation, Intel, and Xerox in 1980. Ethernet is referred to as a *baseband system,* which means that a single digital signal is transmitted. Contrast this with a *broadband system,* which uses multiple channels of data.

Ethernet transmits data at the rate of 10 million bits per second (which translates to 1.25 million bytes per second). This corresponds to a bit time of 100 ns. Manchester encoding is used for the digital data. New 100Mbit and 1000Mbit Ethernet is already being used.

FIGURE 21.4 Eleven-user Ethernet LAN

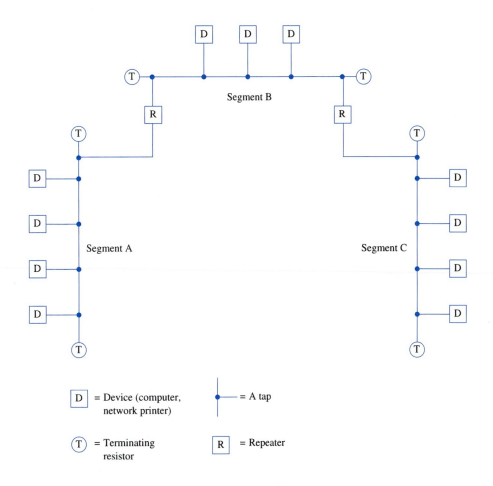

Each device connected to the Ethernet must contain a **transceiver** that provides the electronic connection between the device and the coaxial cable commonly used to connect nodes. Figure 21.4 shows a typical Ethernet installation. The 11 devices on the Ethernet are grouped into three *segments*. Each segment consists of a coaxial cable with a *tap* for each device. It is very important to correctly terminate both ends of the coaxial cable in each segment; otherwise, signal reflections will distort the information on the network and result in poor communications. Segments are connected to each other through the use of *repeaters,* which allow two-way communication between segments.

Each Ethernet device has its own unique binary address. The Ethernet card in each device waits to see its own address on the coaxial cable before actually paying attention to the data being transmitted. Thus, when one device transmits data to another, every device listens. This is called *broadcasting,* much like the operation of a radio. However, Ethernet contains special hardware that detects when two or more devices attempt to transmit data at the same time (called a *collision*). When a collision occurs, all devices that are transmitting stop and wait a random period of time before transmitting the same data again. The random waiting period is designed to help reduce multiple collisions. This procedure represents a protocol called Carrier Sense Multiple Access with Collision Detection (CSMA/CD).

The format in which Ethernet transmits data is called a *frame*. Figure 21.5 details the individual components of the Ethernet frame. Recall that the physical and data-link layers are responsible for handling data at this level. Note that the length of the data section is limited to

FIGURE 21.5 Ethernet frame format

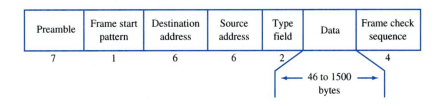

a range of 46 to 1500 bytes, which means that frame lengths are also limited in range. Because of the 10Mbit/second data rate and the format of the Ethernet frame, the lengths of the various segments making up an Ethernet LAN are limited to 500 meters. This guarantees that a collision can be detected no matter which two nodes on a segment are active.

TOKEN-RING NETWORKS

Token-ring networks are not as popular as Ethernet but have their own advantages. The high collision rate of an Ethernet system with a lot of communication taking place is eliminated in a token-ring setup.

The basic operation of a token-ring network involves the use of a special token (just another binary pattern) that circulates between nodes in the ring. When a node receives the token, it simply transmits it to the next node if there is nothing else to transmit. But if a node has its own frame of data to transmit, it holds onto the token and transmits the frame instead. Token-ring frames are similar to Ethernet frames in that both contain source and destination addresses. Each node that receives the frame checks the frame's destination address with its own address. If they match, the node captures the frame data and then retransmits the frame to the next node. If the addresses do not match, the frame is simply retransmitted.

When the node that originated the frame receives its own frame again (a complete trip through the ring), it transmits the original token again. Thus, even with no data being transmitted between nodes, the token is still being circulated.

Unfortunately, only one node's frame can circulate at any one time. Other nodes waiting to send their own frames must wait until they receive the token, which tends to reduce the amount of data that can be transmitted over a period of time. However, this is a small price to pay for the elimination of collisions.

TROUBLESHOOTING TECHNIQUES

Troubleshooting a network problem can take many forms. Before the network is even installed, decisions must be made about it that will affect the way it is troubleshot in the future. For example, Ethernet and token-ring networks use different data encoding schemes and connections, as well as different support software. Each has its own set of peculiar problems and solutions.

Troubleshooting a network may take you down a hardware path (bad crimps on the cable connectors causing intermittent errors), a software path (the machine does not have its network addresses set up correctly), or both. There may even be nothing wrong with the network, the failure coming from the application using the network. So, a good deal of trial and error may be required to determine the exact nature of the problem. In the remaining exercises, many of these troubleshooting scenarios will be discussed.

SELF-TEST

This self-test is designed to help you check your understanding of the background information presented in this exercise.

True/False

Answer *true* or *false*.

1. The Internet is a computer network.
2. Both analog and digital media can be used to transmit digital data.
3. All seven layers of the OSI reference model are always used for communication in a network.
4. Ethernet uses collision detection to handle transmission errors.

Multiple Choice

Select the best answer.

5. The term LAN stands for
   a. Logical access node.
   b. Local area network.
   c. Large access network.
6. Phase transitions for each bit are used in
   a. Amplitude modulation.
   b. Carrier modulation.
   c. Manchester encoding.
   d. NRZ encoding.
7. Ethernet transmits data in
   a. Continuous streams of 0s and 1s.
   b. Frames.
   c. Blocks of 256 bytes.

Matching

Match a description of the topology property on the right with each item on the left.

8. Fully connected   a. The whole network shuts down when the central node fails.
9. Star   b. All nodes connect to the same communication link.
10. Bus   c. Communication links are one-way.
11. Ring   d. Most expensive to build.

Completion

Fill in the blank or blanks with the best answers.

12. The _____ of a network concerns how the nodes are connected.
13. A(n) _____ _____ network provides the fastest communication between any two nodes.
14. Using two different frequencies to represent digital data is called _____ _____.
15. The layer responsible for error detection and recovery is the _____ layer.
16. Collisions are eliminated in the _____ network.

QUESTIONS/ACTIVITIES

1. Visit the computer center of your school. Find out who the network administrator is and discuss the overall structure of the school's network with him or her.
2. Visit a local computer store and find out how much it would cost to set up a 16-user LAN.

REVIEW QUIZ

Within 15 minutes and with 100% accuracy,

1. Sketch and discuss the different types of network topologies and their advantages and disadvantages.
2. Sketch and explain examples of digital data encoding.
3. Discuss the OSI reference model.
4. Explain the basic operation of Ethernet and token-ring networks.

22 Network Topology

INTRODUCTION

Joe Tekk met Don, his manager, at 6 A.M., outside the doors of a local high school.

"Are you ready, Joe?" Don asked. Joe had never accompanied Don on a site upgrade before.

"Sure, Don," Joe replied. "I'm looking forward to it."

For the next four hours, Joe crawled around on the floor, poked his head up into drop ceilings and underneath benches, and traced cables down long corridors, between floors, and down into the boiler room. When he finished, he was tired, bruised, and dirty.

"Well, Joe," Don said, "We've mapped the whole network out. Now we can begin the upgrade."

"Now?" Joe asked wearily.

"No," Don laughed, "later. We have to order the network components first. Go get some rest. You did a good job today."

PERFORMANCE OBJECTIVES

Upon completion of this exercise, within 10 minutes and with 100% accuracy, you will be able to

1. Describe the difference between physical topology and logical topology.
2. Sketch the physical topologies of bus, star, ring, and fully connected networks.
3. Explain what is meant by network hierarchy.

BACKGROUND INFORMATION

Topology concerns the structure of the connections between computers in a network. Figure 22.1 shows three computers (A, B, and C) and a network *cloud,* a graphic symbol used to describe a network without specifying the nature of the connections. The network cloud may comprise only the network found in a small laboratory, or may represent a wide area network (WAN) such as the digital telephone network.

The three computers in Figure 22.1 are connected in two different ways: physically and logically. Let us examine these two types of connections.

FIGURE 22.1 Network cloud connecting three machines

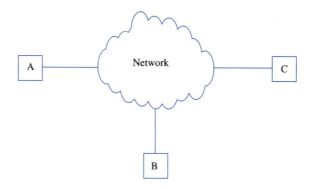

PHYSICAL TOPOLOGY VERSUS LOGICAL TOPOLOGY

Figure 22.2 shows the details of the connections inside the network cloud. Four intermediate network nodes (W, X, Y, and Z) are responsible for relaying data between each of the three machines A, B, and C. Five connections exist between the four intermediate nodes. This is the physical topology of the network. We will cover the details of each type of physical topology in the next few sections.

The logical topology has to do with the path a packet of data takes through the network. For example, from machine A to machine C there are three different paths. These paths are as follows:

1. Link 3
2. Link 1 to link 2
3. Link 1 to link 4 to link 5

Clearly, data sent on link 3 will get from machine A to machine C in the shortest time, while the third path (links 1, 4, and 5) takes the longest. Due to the nature of the network, we cannot guarantee that link 3 is the one that is always used. It may become too busy; its noise level may unexpectedly increase, making it unusable; or a tree might have fallen on the fiber carrying link 3's data. Thus, packets of data may take different routes through the network, arriving *out of order* at their destination. It is the job of the network software protocol to properly reassemble the packets into the correct sequence. Exercise 24 provides the details of the many network protocols in use.

When a large amount of data must be sent between machines on a network, it is possible to set up a *virtual circuit* between the machines. A virtual circuit is a prearranged path through the network that *all* packets will travel for a particular session between machines. For example, for reasons based on the current state of the network, a virtual circuit is established between machines B and C through links 4 and 2. All packets exchanged between B and C will take links 4 and 2.

FIGURE 22.2 Physical network topology

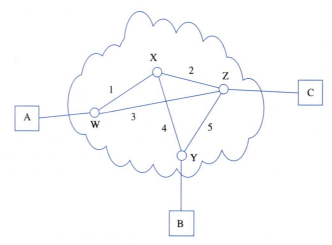

FIGURE 22.3 Typical network

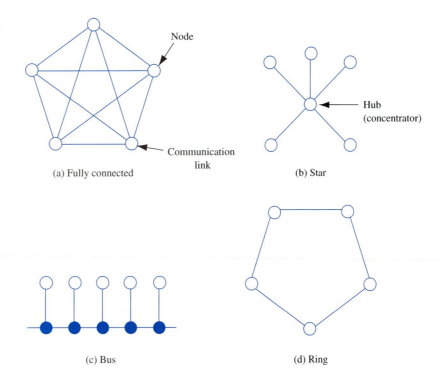

FULLY CONNECTED NETWORKS

Figure 22.3 shows four basic types of network connections. The fully connected network in Figure 22.3(a) is the most expensive to build, for each node has a link (communication channel) to every other node. Just adding one more node (for a total of six nodes) brings the number of links to 15. Seven fully connected nodes require 21 links. It is easy to see that the number of links required quickly becomes unmanageable. Even so, fully connected networks provide quick communication between nodes, for there is a one-link path between every two nodes in the network. Even if a link goes down, the worst-case path only becomes two links long. So, fully connected networks are very reliable and somewhat secure, since many links have to fail for two nodes to lose contact.

STAR NETWORKS

Figure 22.3(b) shows a star network. All nodes connect to a central **hub** (also called a *concentrator*). For small networks, only a single hub is required. Four, five, eight, even sixteen connections are available on a single hub. Large star networks require multiple hubs, which increase the hardware and cabling costs. On the other hand, if a node on the network fails, the hub will isolate it so that the other nodes are not affected. Entire groups of nodes (machines) can be isolated at a time by disconnecting their hub. This helps narrow down the source of a network problem during troubleshooting.

BUS NETWORKS

A bus network uses a single common communication link that all nodes tap into. Figure 22.3(c) shows the bus connection; 10base2 and 10base5 Ethernet use coaxial cable as the common connection. All nodes on the common bus compete with each other for possession, broadcasting their data when they detect the bus is idle. If two or more nodes transmit data at the same time, a *collision* occurs, requiring each node to stop and wait before retransmitting. This technique of sharing a common bus is known as Carrier Sense Multiple Access with Collision Detection (CSMA/CD), and is the basis of the Ethernet communication system.

Wiring a bus network is not too difficult. Suitable lengths of coaxial cable, properly terminated with BNC connectors on each end, are daisy-chained via T-connectors into one long *segment* of nodes. Each T-connector plugs into a network interface card. The problem with the daisy-chain bus connection is that bad crimps on the BNC connectors, poor connections in the T-connectors, or just an improperly terminated cable (no 50 ohm terminating resistor) can cause intermittent or excessive collisions; these problems can be difficult to find as well. A special piece of equipment called a *time domain reflectometer* (TDR) is used to send a pulse down the coaxial cable and determine where the fault is (by displaying a response curve for the cable).

In terms of convenience, the bus network is relatively easy to set up, with no significant hardware costs (no hubs are required). With 200 meters of cable possible in a segment (for 10base2 Ethernet), a large number of nodes can be wired together. Individual segments can be connected together with **repeaters** (more on this in Exercise 23).

RING NETWORKS

The last major network topology is the ring. As Figure 22.3(d) shows, each node in a ring is connected to exactly two other nodes. Data circulates in the ring, traveling through many intermediate nodes if necessary to get to its destination. Like the star connection, a ring requires only one link per node. The difference is that there is no central hub concentrating the nodes. Data sent between nodes will typically require paths of at least two links. If a link fails, the worst-case scenario requires a message to travel completely around the ring, through every link (except the one that failed). The increase in time required to relay messages around the bad link may be intolerable for some applications. The star network does not have this problem. If a link fails, only the node on that link is out of service.

Token-ring networks, although logically viewed as rings, are connected using central multistation access units (MAUs). The MAU provides a physical star connection.

HYBRID NETWORKS

A hybrid network combines the components of two or more network topologies. As Figure 22.4 indicates, two star networks are connected (with three additional nodes) via a bus. This is a common way to implement Ethernet, with coax running between classrooms or laboratories, and hubs in each room to form small subnetworks. Putting together a hybrid network takes careful planning, for there are various rules that dictate how the individual components may be connected and used. For example, when connecting Ethernet segments, a maximum of four repeaters may be used with five segments. Furthermore, if a 4Mbit/second token-ring network is interfaced with a 10Mbit/second Ethernet, there are performance issues that must be taken into consideration also (since any Ethernet traffic is slowed down to 4Mbit/second on the token ring side). In addition, the overall organization of the hybrid network, from a logical viewpoint, must be planned out as well. This will become clearer in the next section.

FIGURE 22.4 Hybrid network

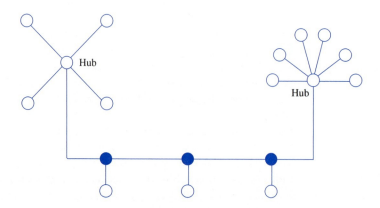

FIGURE 22.5 Hybrid network with hierarchy

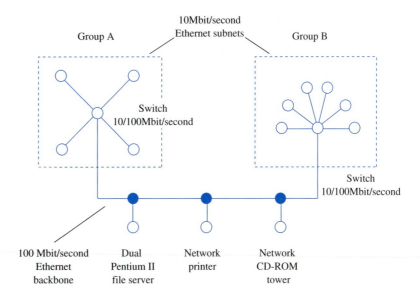

NETWORK HIERARCHY

The machines networked together in Figure 22.4 are not organized into a hierarchy. Data transmitted by any machine is broadcast through both hubs. Everyone connected to the network sees the same data and competes with everyone else for bandwidth.

The same network is illustrated in Figure 22.5, except for a few hardware modifications. Both hubs have been replaced by 10/100Mbit/second switches, which act like hubs except they only broadcast data selectively. For example, any machine in group A can send data to any other machine in group A, through the switch, without the data being broadcast on the 100Mbit/second backbone. The same is true for machines in group B. Their traffic is isolated from the important, high-speed 100Mbit/second backbone.

Now, when a machine from group A requests service from the file server, which in turn accesses the CD-ROM tower, the hierarchy of the network allows the file server to communicate with the CD-ROM tower at a high speed, while sending data back to the machine in group A only when it needs to. Machines from both groups can communicate with each other over the backbone as well, without significantly interfering with the other backbone traffic (since all machines in each group operate at 10Mbit/second). The switches enforce hierarchy by learning which data packets should be broadcast and which should not (based on their destination addresses). Since the backbone is the main communication link in the hybrid network, its 100Mbit/second speed allows each network component to communicate at its best speed.

SUBNETS

In Exercise 24 you will cover the structure and use of *IP addresses,* 32-bit numbers used to locate and identify nodes on the Internet. For example, the IP of one computer in a classroom might be 192.203.131.137. Other computers on the same subnet may begin with 192.203.131 but have different numbers at the end, as in 192.203.131.130. This type of subnet is called a *class C subnet*. We will examine subnets in more detail in Exercise 24.

TROUBLESHOOTING TECHNIQUES

It is a fact of life that we must worry about intentional harm being done to our network. In terms of security and reliability, we must concern ourselves with what is required to *partition* our network, breaking it up into at least two pieces that cannot communicate with each other. Figure 22.6 shows how bus, star, ring, and fully connected networks are partitioned. Note that the star network is completely partitioned (all nodes isolated) if the central hub fails.

FIGURE 22.6 Partitioning a network

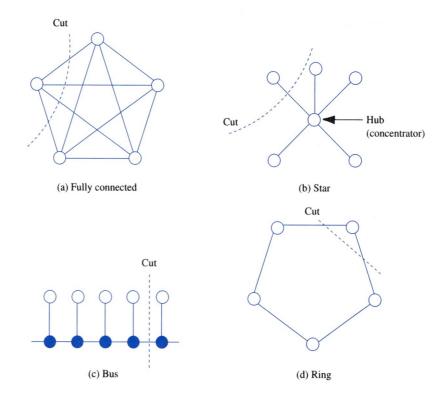

When troubleshooting a network, knowledge of its topology, both physical and logical, is essential to proper partitioning, so that testing and repairing can proceed smoothly.

SELF-TEST

This self-test is designed to help you check your understanding of the background information presented in this exercise.

True/False

Answer *true* or *false*.

1. Packets never exit once they enter a network cloud.
2. Fully connected networks require more links than star networks.
3. 10base2 Ethernet allows for 2000-meter segments.
4. A collision is required to exchange data over an Ethernet cable.
5. A hub is used to enforce network hierarchy.

Multiple Choice

Select the best answer.

6. A fully connected network of 10 nodes requires
    a. 10 links.
    b. 45 links.
    c. 90 links.
7. Star networks
    a. Require hubs.
    b. Are limited to 16 nodes.
    c. Both a and b.
8. A portion of an Ethernet bus is called a(n)
    a. CSMA module.
    b. Etherpath.
    c. Segment.

9. Assuming any single link fails, the worst-case path through the ring network of Figure 22.3(d) is
   a. Three links.
   b. Four links.
   c. Five links.
10. What is required to partition a fully connected network of six nodes?
    a. Cut one link.
    b. Cut one link at each node.
    c. Cut all links at a single node.

Matching

Match a description of the topology property on the right with each item on the left.

11. Bus
12. Star
13. Ring
14. Fully connected

a. All paths are one link long.
b. Each node has exactly two links.
c. All nodes share the same link.
d. All nodes share a central node.

Completion

Fill in the blank or blanks with the best answers

15. A prearranged connection between computers is called a(n) _____ circuit.
16. Hubs are also called _____.
17. CSMA/CD stands for _____ _____ _____ _____ with _____ _____.
18. Breaking a link in a network may _____ it.

FAMILIARIZATION ACTIVITY

1. Sketch all the different network configurations possible using six communication links.
2. Search the Web for FDDI (fiber distributed data interconnect). What type of network topology does FDDI use? What is FDDI used for?

QUESTIONS/ACTIVITIES

1. What are all the paths from machine B to machine C in the network of Figure 22.2?
2. Draw a fully connected eight-node network. How many links are there?
3. Repeat step 2 for a star network. Assume each hub has four connections available.
4. Find a laboratory or classroom that is networked. Make a diagram of the network, showing the various nodes and what they actually represent (computers, printers, etc.).

REVIEW QUIZ

Within 10 minutes and with 100% accuracy,

1. Describe the difference between physical topology and logical topology.
2. Sketch the physical topologies of bus, star, ring, and fully connected networks.
3. Explain what is meant by network hierarchy.

23 Networking Hardware

INTRODUCTION

Don, Joe Tekk's manager, had his hands full of UTP cable. He had carefully stripped off 1 inch of outer insulation and was gingerly holding all eight twisted-pair conductors in a neat row by pinching them between his thumb and forefinger. In his other hand he held a clear plastic RJ-45 crimp-on connector and was slowly pushing all eight wires into their thin grooves in the connector.

Joe walked up and slapped Don on the back in a friendly way. "What are you doing, Don?" he asked.

Don let out a long-suffering sigh as all eight wires popped out of the connector. "Starting over again, Joe," he replied.

PERFORMANCE OBJECTIVES

Upon completion of this exercise, within 10 minutes and with 100% accuracy, you will be able to

1. List and describe the basic networking hardware components.
2. Explain the differences in 10base2 Ethernet and 10baseT Ethernet.

BACKGROUND INFORMATION

In this exercise we will examine many of the different hardware components involved in networking. You are encouraged to look inside your machine to view your network interface card, around your lab to locate hubs and trace cables, and around your campus (especially the computer center) to see what other exotic hardware you can find.

ETHERNET CABLING

We begin our hardware presentation with Ethernet cabling. Ethernet cables come in three main varieties. These are:

1. RG-58 coaxial cable, used for 10base2 operation (also called *thinwire*)
2. RG-11 coaxial cable, use for 10base5 operation (also called *thickwire*)
3. Unshielded twisted pair (UTP), used for 10baseT operation

There are other, specialized cables, including fiber (10baseFL), that are used as well.

FIGURE 23.1 Coaxial cable construction

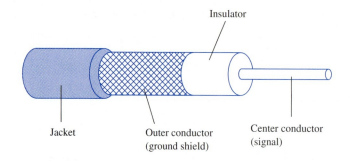

FIGURE 23.2 Assorted connectors and cables (*photograph by John T. Butchko*)

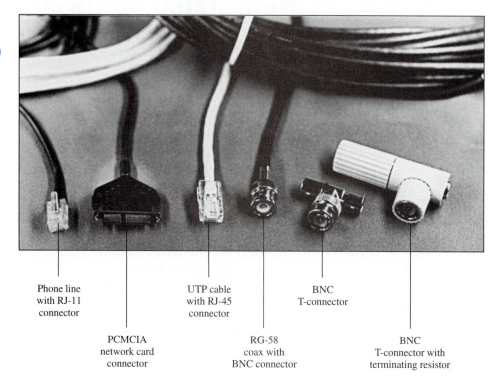

RG-58 cable is typically used for wiring laboratories and offices, or other small groups of computers. Figure 23.1 shows the construction of a coaxial cable.

The maximum length of a thinwire Ethernet segment is 185 meters (600 feet), which is due to the nature of the CSMA/CD method of operation, the cable attenuation, and the speed at which signals propagate inside the coax. The length is limited to guarantee that collisions are detected when machines that are far apart transmit at the same time. BNC connectors are used to terminate each end of the cable. Figure 23.2 shows several different cables and connectors, including BNC T-connectors (one containing a terminating resistor).

When many machines are connected to the same Ethernet segment, a daisy-chain approach is used, as shown in Figure 23.3(a). The BNC T-connector allows the network interface card to tap into the coaxial cable, and the coax to pass through the machine to the next machine. The last machines on each end of the cable (or simply the cable ends themselves) must use a terminating resistor (50 ohms) to eliminate collision-causing reflections in the cable. This connection is illustrated in Figure 23.3(b).

RG-11 coaxial cable is used as a *backbone* cable, distributing Ethernet signals throughout a building, an office complex, or other large installation. RG-11 is thicker and more sturdy than RG-58 coax. Thickwire Ethernet segments may be up to 500 meters (1640 feet) long with a maximum of five segments connected by repeaters. RG-11 cable is typically orange, with black rings around the cable every 2.5 meters to allow taps into the cable. The taps, called *vampire taps,* are used by transceivers that transfer Ethernet data to and from the cable.

FIGURE 23.3 10base2 Ethernet wiring

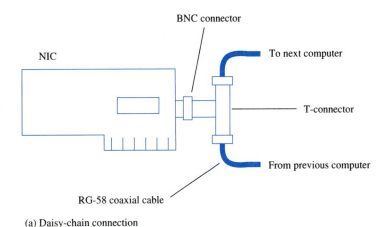

(a) Daisy-chain connection

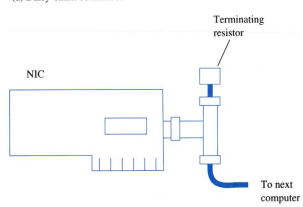

(b) Terminating connection (required at each end of the cable)

UTP cable, used with hubs and other 10baseT equipment, uses twisted pairs of wires to reduce noise and allow higher-speed data rates (100Mbit/second category 5 UTP for Fast Ethernet). The twists tend to cause the small magnetic fields generated by currents in the wires to cancel, reducing noise on the signals. UTP cable length is limited to 100 meters (328 feet) and RJ-45 connectors are used for termination.

The structure of the 8-pin RJ-45 connector is shown in Figure 23.4. Its modular format is similar to the telephone companies' 6-pin RJ-11 connector.

Table 23.1 shows the wire color combinations used in UTP cabling. Note that only two pairs are required for 10baseT operation.

UTP cables are wired as straight-through or crossover cables. Figure 23.5 shows the wiring diagrams for each type of cable. Straight-through cables typically connect the computer's network interface card to a port on the hub. Crossover cables are used for NIC-to-NIC communication, and for hub-to-hub connections when no crossover port is available. Sample wiring configurations are shown in Figure 23.6.

Table 23.2 compares each cabling system.

THE NIC

The network interface card (NIC) is the interface between the PC (or other networked device) and the physical network connection. In Ethernet systems, the NIC connects to a segment of coaxial or UTP cable (fiber NICs are available but not very common yet). As with any other type of adapter card, NICs come in ISA, PCMCIA, and PCI bus varieties. Figure 23.7 shows a typical Ethernet NIC. Since the NIC contains both BNC and RJ-45 connectors, it is called a *combo* card. The NE2000 Compatible stamp indicates that the NIC supports a widely accepted group of protocols.

FIGURE 23.4 RJ-45 (10baseT) connector

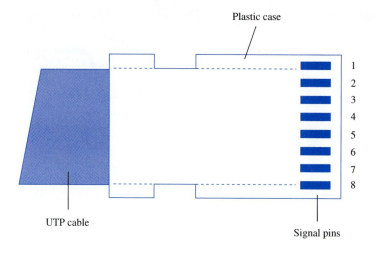

(a) Top view

(b) Side view

TABLE 23.1 RJ-45 pin assignments

Pin	Color	Function	Used for 10baseT
1	White/Orange	T2	✔
2	Orange/White	R2	✔
3	White/Green	T3	✔
4	Blue/White	R1	
5	White/Blue	T1	
6	Green/White	R3	✔
7	White/Brown	T4	
8	Brown/White	R4	

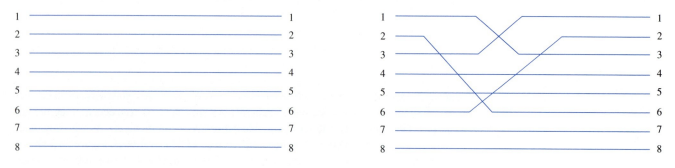

(a) Straight through (b) Crossover

FIGURE 23.5 RJ-45 cabling

FIGURE 23.6 10baseT Ethernet wiring

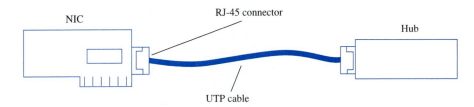

(a) Individual machine connection

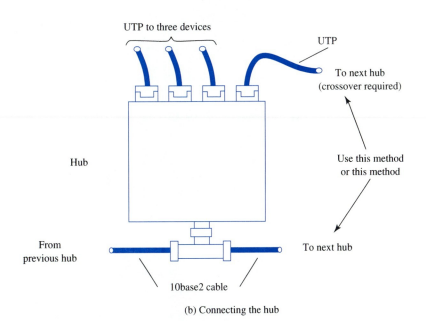

(b) Connecting the hub

TABLE 23.2 Comparing cabling systems

	10base5	10base2	10baseT
Cable Type	RG-11	RG-58	UTP
Maximum Segment Length	500 m (1640 feet)	185 m (606 feet)	100 m (328 feet)
Max Nodes	100	30	2
Max Segments	5	5 (3 with nodes)	1024

The NIC in Figure 23.7 is an ISA adapter card. PCI networking cards are available in both non-bus-mastering and bus-mastering varieties. Figure 23.8 shows a PCMCIA Ethernet NIC and cable.

The NIC is responsible for operations that take place in the physical layer of the OSI network model. It is only concerned with sending and receiving 0s and 1s, using the IEEE 802.3 Ethernet standard (or IEEE 802.5 token ring).

Windows 95 identifies the installed NIC in Network Properties. Figure 23.9 shows the 3Com NIC entry. Note that the NetBEUI and TCP/IP protocols are *bound* to the 3Com adapter. To use a protocol with a NIC you must bind the protocol to the adapter card. This is typically done automatically when the protocol is added.

Double-clicking the 3Com 3C508 entry brings up its Properties window, which is shown in Figure 23.10. The indicated driver type is NDIS, Microsoft's network driver interface specification, which allows multiple protocols to use a single NIC. An ODI (open data-link interface, developed by Novell) driver performs the same function for multiple protocol stacks used with the NetWare network operating system. Figure 23.11 shows the NDIS/ODI

FIGURE 23.7 Network interface card (Ethernet) (*photograph by John T. Butchko*)

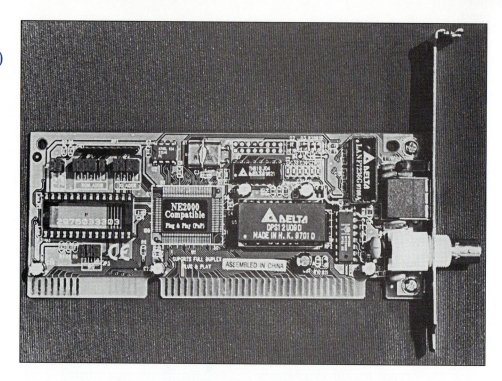

FIGURE 23.8 PCMCIA Ethernet card with cable (*photograph by John T. Butchko*)

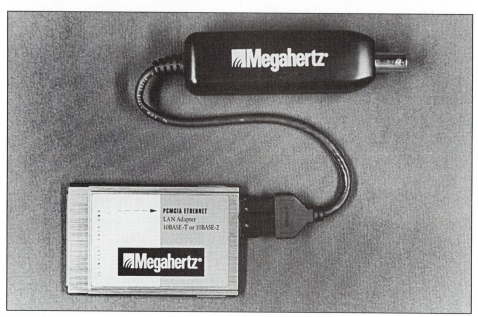

interface. Both are designed to *decouple* the protocols from the NIC. The protocols do not require any specific information about the NIC. They use the NDIS/ODI drivers to perform network operations with the drivers responsible for their specific hardware.

It is important to mention that all NICs are manufactured with a unique 48-bit MAC address (for example, 00-60-97-2B-E6-0F). You can view your NIC's MAC address using the WINIPCFG utility (from the Run menu). Figure 23.12 shows the initial WINIPCFG screen.

In the next exercise (Exercise 24) we will see how the ARP and RARP protocols work with the MAC address.

TOKEN RING

The IEEE 802.5 standard describes the token-ring networking system. IBM developed the initial 4Mbit/second standard in the mid-1980s, with 16Mbit/second token ring also available.

FIGURE 23.9 3Com 3C508 NIC entry

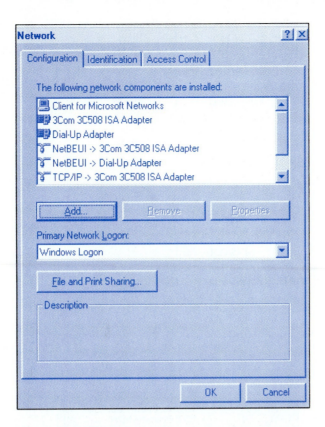

FIGURE 23.10 NIC Properties window

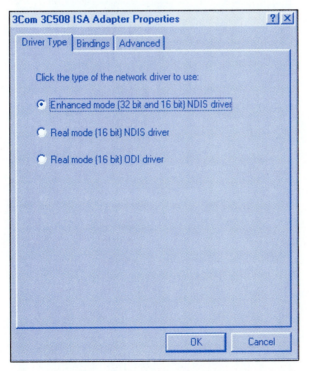

Token-ring networks use a multistation access unit (MAU), which establishes a logical ring connection even though the physical connections to the MAU resemble a star. Figure 23.13 shows the basic operation of the MAU. Computers in the ring circulate a software *token*. The machine holding the token is allowed to transmit data to the next machine on the ring (even if the data is not meant for that machine). One machine (typically the first to boot and connect) is identified as the *active monitor* and keeps track of all token-ring operations. If the active monitor detects that a machine has gone down (or been shut off), the connection to that machine is bypassed. If the active monitor itself goes down, the other

FIGURE 23.11 NDIS/ODI interface

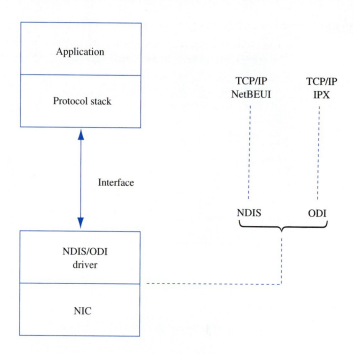

FIGURE 23.12 Viewing the NIC's MAC address

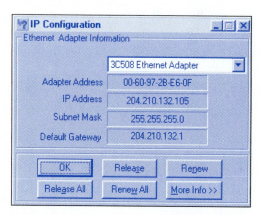

FIGURE 23.13 Token-ring network

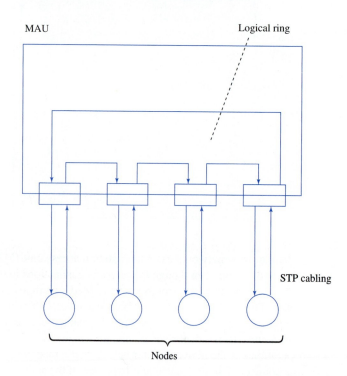

machines vote to elect a new active monitor. Thus, we see that token-ring networks are *self-healing,* unlike Ethernet, which is only capable of resolving collisions.

Token-ring connections are made using STP (shielded twisted pair) cables. STP cable contains a metal shield around the twisted pairs that provides isolation from external crosstalk and noise. In general, do not substitute STP for UTP.

REPEATERS

A repeater connects two network segments and broadcasts packets between them. Since signal loss is a factor in the maximum length of a segment, a repeater is used to amplify the signal and extend the usable length. A common Ethernet rule is that no more than four repeaters may be used to join segments together. This is a physical limitation designed to keep collision detection working properly.

TRANSCEIVERS

A transceiver converts from one media type to another. For example, a 10base2-to-fiber transceiver acts like a repeater, except it also interfaces 10base2 coaxial cable with a fiber optic cable. It is common to use more than one media type in an installation, so many different kinds of transceivers are available.

HUBS

Hubs, also called *concentrators,* expand one Ethernet connection into many. For example, a four-port hub connects up to four machines (or other network devices) via UTP cables. The hub provides a star connection for the four ports. Many hubs contain a single BNC connector as well to connect the hub to existing 10base2 network wiring. The hub can also be connected via one of its ports. One port is designed to operate in either straight-through or crossover mode, selected by a switch on the hub. Hubs that can connect in this fashion are called *stackable* hubs. Figure 23.14 shows an eight-port stackable Ethernet hub. Port 8 is switch selectable for straight through or crossover (cascade).

A hub is similar to a repeater, except it broadcasts data received by any port to all other ports on the hub. Most hubs contain a small amount of intelligence as well, examining received packets and checking them for integrity. If a bad packet arrives, or the hub determines that a port is unreliable, it will shut down the line until the error condition disappears.

FIGURE 23.14 Ethernet hub (*photographs by John T. Butchko*)

(a) Front view

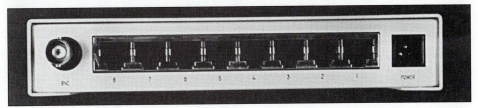

(b) Rear view

FIGURE 23.15 Connecting five segments with hubs

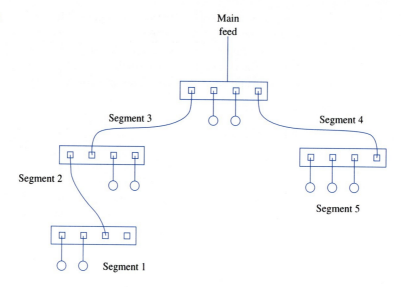

Note that a hub also acts like a repeater. Because of its slight delay when processing a packet, the number of hubs that may be connected in series is also limited. Figure 23.15 shows how several hubs are used to connect five Ethernet segments, within the accepted limits.

BRIDGES/SWITCHES

When a network grows in size, it is often necessary to partition it into smaller groups of nodes to help isolate traffic and improve performance. One way to do this is to use a *bridge,* whose operation is indicated in Figure 23.16. The bridge keeps segment A traffic on the A side and segment B traffic on the B side. Packets from segment A that are meant for a node in segment B will cross the bridge (the bridge will permit the packet to cross). The same is true for packets going from B to A. The bridge learns which packets should cross as it is used. Network performance may decline slightly when the bridge is first installed until it has built up its internal tables.

A switch is similar to a bridge, with some important enhancements. First, a switch may have multiple ports, thus directing packets to several different segments, further partitioning and isolating network traffic in a way similar to a router. Figure 23.17 shows an eight-port N-way switch, which can route packets from any input to any output. Some or all of an incoming packet is examined to make the routing decision, depending on the switching method that is used. One common method is called *store and forward,* which stores the received packet before examining it to check for errors before retransmitting. Bad packets are not forwarded.

FIGURE 23.16 Operation of a bridge

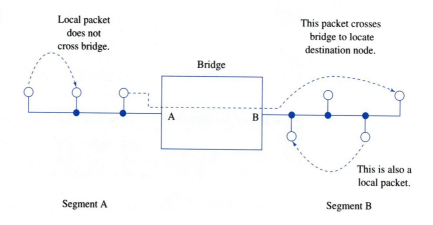

FIGURE 23.17 One configuration in an eight-port N-way switch

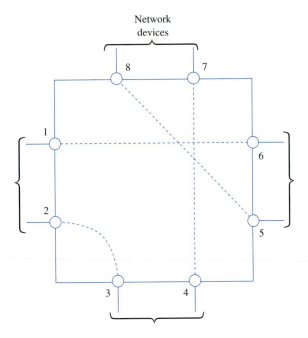

In addition, a switch typically has auto-sensing 10/100Mbit/second ports and will adjust the speed of each port accordingly. Furthermore, a *managed* switch supports SNMP for further control over network traffic.

ROUTERS

A router is the basic building block of the Internet. Each router connects two or more networks together by providing an interface for each network to which it is connected. Figure 23.18 shows a router with an interface for an Ethernet network and a token-ring network. The router examines each packet of information to determine whether the packet must be translated from one network to another, performing a function similar to a bridge. Unlike a bridge, a router can connect networks that use different technologies, addressing methods, media types, frame formats, and speeds.

A router is a special-purpose device designed to interconnect networks. For example, three different networks can be connected using two routers, as illustrated in Figure 23.19.

FIGURE 23.18 Router with two interfaces

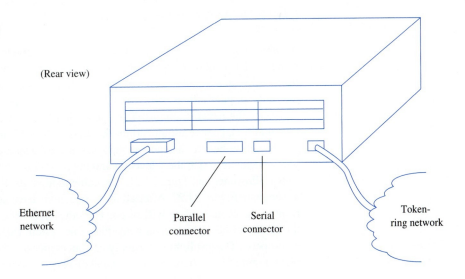

FIGURE 23.19 Two routers connecting three networks

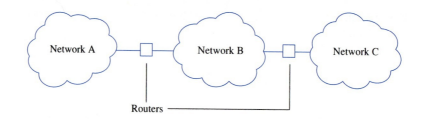

FIGURE 23.20 Packet routing

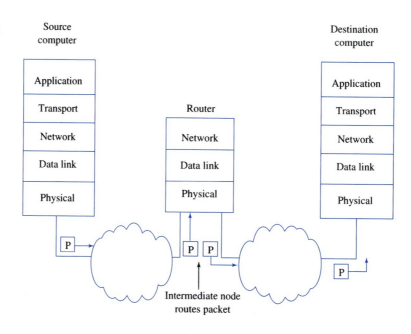

If a computer in network A needs to send a packet of information to network C, both routers pass the packets from the source network to the destination network.

Routers maintain routing tables in their memories to store information about the physical connections on the network. The router examines each packet of data, checks the routing table, and then forwards the packet if necessary. Every other router in the path (between a source and a destination network) performs a similar procedure, as illustrated in Figure 23.20. Note that a router does not maintain any state information about the packets; it simply moves them along the network.

CABLE MODEMS

A cable modem is a high-speed network device connected to a local cable television provider. The cable television company allocates a pair of channels (one for transmit, one for receive) on the cable system to transmit data. At the *head-end* of the network, located at the cable supplier offices, a traditional Internet service provider (ISP) service is established to service the network clients. The connection from the cable system ISP to the Internet uses traditional telecommunications devices, such as T1 or T3 lines.

The subscribers to the cable modem service use a splitter to create two cable wires. One wire is reconnected to the television and the second is connected to the new cable modem. This is illustrated in Figure 23.21. The cable modem itself requires just a few connections, as shown in Figure 23.22. After all the connections have been made, the power light on the front of the cable modem will be on. The other lights show the cable modem status. Both the cable and PC lights are on when the cable system ISP and the PC network card are set up properly. The test light is normally off, but comes on after a reset or when power is reapplied. Figure 23.23 shows the front panel display of a typical cable modem.

FIGURE 23.21 Cable service connections

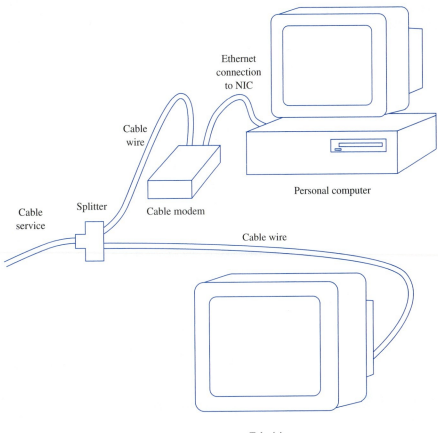

FIGURE 23.22 Cable modem connections

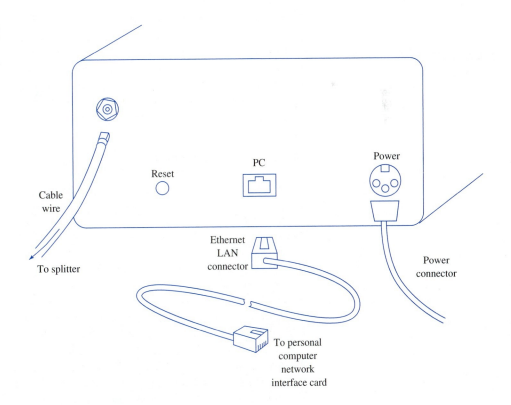

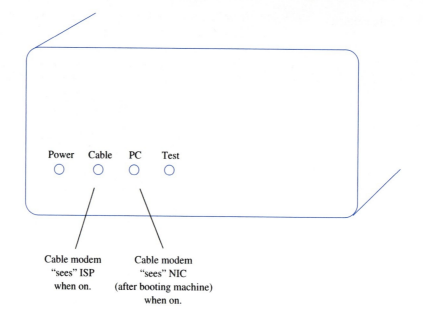

FIGURE 23.23 Cable modem indicator lights

There is no maintenance for the cable modem subscriber other than providing adequate ventilation and keeping the power applied to the cable modem at all times. The cable system ISP may update the internal software or run tests on them during off-peak hours.

SATELLITE NETWORK SYSTEM

The Hughes Corporation offers a unique solution to low-speed Internet connections. For a few hundred dollars, you can buy their DirecPC Internet satellite networking system. Figure 23.24 shows the basic operation. Internet data comes to your PC via satellite at

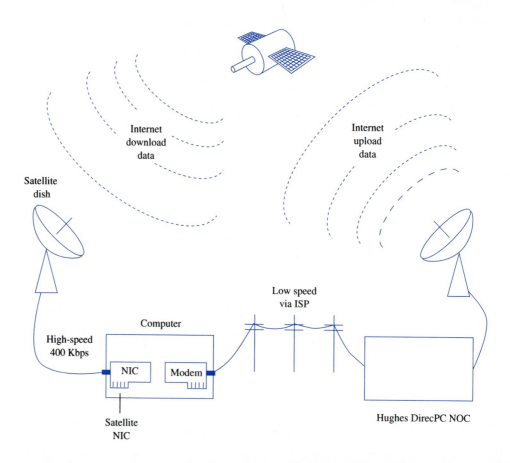

FIGURE 23.24 Satellite Internet

400 Kbps. Through your modem and ISP, data goes back to Hughes's network operations center (NOC) where it is uploaded to the Internet. This is an ideal situation for browsing, when you need to receive information fast (if a Web page contains many images) but only send information out (clicking on a new URL to load a new page) occasionally. A low-speed line for transmitted packets is acceptable, unless you are uploading large files to an FTP site or sending e-mail with large attachments. If there is no cable where you live or work (for a cable modem connection), DirecPC may be the answer for you.

TROUBLESHOOTING TECHNIQUES

One of the most important phases of a network installation is making the cables required for all the nodes. It is much less expensive to buy BNC connectors and spools of coax and make custom-length cables than it is to buy finished cables. This may not be the case for UTP cable, which is harder to terminate (four or eight wires) due to the rigid requirements of the UTP specifications.

A valuable tool to have at your disposal when preparing or checking cables is a *cable tester*. Figure 23.25 shows an electronic cable tester, capable of performing these (and many other) tests on UTP cable:

- Passive and active profiles
- Continuity
- Cable length
- NEXT (near-end crosstalk)
- Attenuation
- Noise

Other, more sophisticated network test equipment, such as the Fluke LANmeter, capture and diagnose network packets of many different protocols, gather statistics (collisions, packets sent), perform standard network operations such as ping and tracert (see Exercise 25), and can transmit packets for troubleshooting purposes. The power of this type of network analyzer is well worth the cost.

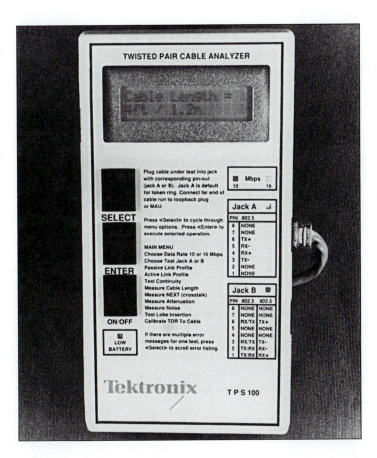

FIGURE 23.25 Electronic cable analyzer (*photograph by John T. Butchko*)

SELF-TEST

This self-test is designed to help you check your understanding of the background information presented in this exercise.

True/False

Answer *true* or *false*.

1. Only RG-58 coax is used in Ethernet systems.
2. Vampire taps are used with RG-11 cable.
3. All NICs have the same MAC address (for broadcasting).
4. Token-ring networks use an MAU.
5. Transceivers are only used with fiber and UTP.

Multiple Choice

Select the best answer.

6. Hubs are also called
   a. Repeaters.
   b. Transceivers.
   c. Concentrators.
7. Hubs act like
   a. Repeaters.
   b. Transceivers.
   c. Routers.
8. A bridge between networks C and D
   a. Broadcasts all packets between C and D.
   b. Broadcasts selected packets between C and D.
   c. Broadcasts all packets from C to D, selected packets from D to C.
9. A router
   a. Connects two different networks.
   b. Ties all hubs together.
   c. Is not used for Internet connections.
10. Cable testers check
    a. Continuity.
    b. Crosstalk.
    c. Frequency response.
    d. a and b only.

Matching

Match each cable type on the right with each network system on the left.

11. 10base2   a. Fiber
12. 10base5   b. RG-11
13. 10baseT   c. RG-58
14. 10baseFL   d. UTP
15. IEEE 802.5   e. STP

Completion

Fill in the blank or blanks with the best answers.

16. Connecting to RG-11 coaxial cable requires a(n) _____ tap.
17. A NIC that contains both types of connectors is called a(n) _____ card.
18. Protocols must be _____ to a NIC before they can be used.
19. The number of Ethernet segments that may be connected via repeaters is _____.
20. One technique used by switches is store and _____.

FAMILIARIZATION ACTIVITY

1. Go to a local computer store and ask to speak with their network technician. If the technician has a service kit that he or she takes on installations or repair assignments, ask whether you may examine the contents. What kind of spare parts are involved? What test equipment and tools are required? Does the technician keep a log book?
2. Find out what types of networks are in use at your location. What devices (hubs, switches, routers) are used to connect them?

QUESTIONS/ACTIVITIES

1. Why do you think fiber has not totally replaced coax and UTP cable?
2. What is the difference between the following:
   a. A repeater and a hub.
   b. A hub and a switch.
   c. A switch and a router.
3. What makes a hub stackable?

REVIEW QUIZ

Within 10 minutes and with 100% accuracy,

1. List and describe the basic networking hardware components.
2. Explain the differences in 10base2 Ethernet and 10baseT Ethernet.

24 Networking Protocols

INTRODUCTION

Joe Tekk walked by Don, his manager, muttering under his breath, "Arp, rarp. Arp, rarp."

Don looked at Joe with a quizzical expression. "What are you saying, Joe?" he asked.

Joe laughed. "Arp, rarp. They're abbreviations for two networking procotols. Arp stands for 'address resolution protocol.' Rarp stands for 'reverse address resolution protocol.' You know, arp, rarp. I like the sound of it."

Don just shook his head, bewildered. "You sound like my three-year-old grandson."

Joe was shocked. "Really? Does he know anything about networking?"

PERFORMANCE OBJECTIVES

Upon completion of this exercise, within 10 minutes and with 100% accuracy, you will be able to

1. Explain the purpose of each layer of the ISO-OSI model.
2. Discuss the different protocols that make up the TCP/IP suite.
3. Describe the relationship between IP addresses and MAC addresses.

BACKGROUND INFORMATION

Whenever you use a network to transfer data, an entire set of *protocols* is used to set up and maintain reliable data transfer between the two network stations. These protocols are used for many different purposes. Some report errors, some contain control information, others carry data meant for applications. One widely accepted standard that defines the layering of the network (and thus a division of protocols) is the ISO-OSI Network Model, which is described in Table 24.1.

The ISO-OSI model breaks network communication into seven layers, each with its own responsibility for one part of the communication system. As indicated, the Physical layer is responsible only for sending and receiving digital data, nothing else. Any errors that show up in the data are handled by the next higher layer, Data Link. As we will see in this exercise, each layer has its own set of protocols.

TABLE 24.1 ISO-OSI Network Model

ISO-OSI Layer	Operation (purpose)
7 = Application	Use network services via an established protocol (TCP/IP, NetBEUI).
6 = Presentation	Format data for proper display and interpretation.
5 = Session	Establish, maintain, and teardown session between both networked computers.
4 = Transport	Break application data into network-sized packets.
3 = Network	Handle network addressing.
2 = Data Link	Flow control, reliable transfer of data.
1 = Physical	All hardware required to make the connection (NIC, cabling, etc.) and transmit/receive 0s and 1s.

THE NETBEUI PROTOCOL

The Network BIOS Extended User Interface (NetBEUI) protocol is the backbone of Windows for Workgroups, Windows 95/98, and Windows NT networking. File and printer sharing between these network operating systems is accomplished through the use of NetBEUI.

NetBEUI provides the means to gather information about the Network Neighborhood. Table 24.2 shows the advantages and disadvantages of the NetBEUI protocol. One of the main disadvantages with NetBEUI is that it is a nonroutable protocol. This means that a NetBEUI message cannot be routed across two different networks. It was designed to support small networks (up to 200 nodes) and becomes inefficient in larger installations.

THE IPX/SPX PROTOCOLS

The Internetworking Packet Exchange/Sequenced Packet Exchange (IPX/SPX) protocols were first developed by Novell for use with their NetWare network operating system. Windows supports IPX/SPX and thus connects easily with a NetWare network. Figure 24.1 shows the relationship between the ISO-OSI model and the NetWare protocol suite. The different protocols in NetWare's model are defined as follows:

- NCP: NetWare Core Protocol
- SAP: Service Advertising Protocol
- RIP: Routing Information Protocol
- SPX: Sequenced Packet Exchange
- IPX: Internetwork Packet Exchange Protocol
- LSL: Link Support Layer
- MLID: Multiple Link Interface Driver

Although the suite of NetWare protocols are used in many different applications, a different protocol suite has gained much wider acceptance, largely due to its use on the World Wide Web. This is the TCP/IP protocol suite. Let us now examine its features.

TABLE 24.2 Advantages and disadvantages of NetBEUI

Advantages	Disadvantages
Easy to implement	Not routable
Good performance	Few support tools
Low memory requirements	Proprietary
Self-tuning efficiency	

FIGURE 24.1 NetWare protocol suite

ISO-OSI Layer	Protocols		
7 = Application	NCP	SAP	RIP
6 = Presentation	NCP	SAP	RIP
5 = Session			
4 = Transport	SPX		
3 = Network	IPX		
2 = Data Link	LSL 802.2 / MLID		
1 = Physical	802.3 Ethernet / 802.5 Token ring		

THE TCP/IP PROTOCOL SUITE

The TCP/IP protocol is unquestionably one of the most popular networking protocols ever developed. TCP/IP has been used since the 1960s as a method to connect large mainframe computers together to share information among the research community and the Department of Defense. Now TCP/IP is used to support the largest computer network, the Internet. Most manufacturers now incorporate TCP/IP into their operating systems, allowing all types of computers to communicate with each other. Figure 24.2 shows the relationship between the ISO-OSI seven-layer networking model and the TCP/IP networking model. Since the protocols indicated in Figure 24.2 are so common, they will be explained in the following sections.

RFCS

The Network Information Center, or NIC, located at www.internic.net contains information about many different aspects of the Internet. One of the most important items stored at the Internic are the Request for Comments documents, or RFCs. These documents describe

FIGURE 24.2 TCP/IP protocol suite

ISO-OSI Layer	TCP/IP Protocols	
7 = Application	Telnet	
6 = Presentation	FTP	SNMP
5 = Session	SMTP	DNS
4 = Transport	TCP	UDP
3 = Network	IP	
2 = Data Link	LLC 802.2 / MAC	
1 = Physical	802.3 Ethernet / 802.5 Token ring	

TABLE 24.3 Several important TCP/IP RFCs

Protocol	RFC	Name
Telenet	854	Remote Terminal Protocol
FTP	959	File Transfer Protocol
SMTP	821	Simple Mail Transfer Protocol
SNMP	1098	Simple Network Management Protocol
DNS	1034	Domain Name System
TCP	793	Transport Control Protocol
UDP	768	User Datagram Protocol
ARP	826	Address Resolution Protocol
RARP	903	Reverse Address Resolution Protocol
ICMP	792	Internet Control Message Protocol
BOOTP	951	Bootstrap Protocol
IP	791	Internet Protocol

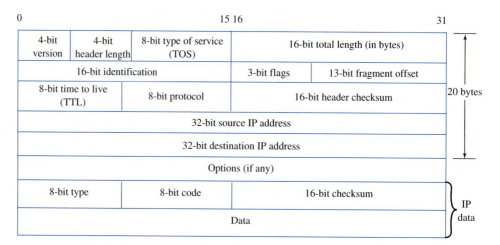

FIGURE 24.3 IP encapsulated message

how each of the protocols contained in TCP/IP are implemented. Refer to Table 24.3 for a list of RFCs associated with some of the most popular TCP/IP protocols. You are encouraged to visit the Network Information Center to become familiar with all the information and services offered, such as the RFCs.

IP

The Internet Protocol (IP) is the base layer of the TCP/IP protocol suite. All TCP/IP data is packaged in units called IP datagrams. All other TCP/IP protocols are encapsulated inside an IP *datagram* for delivery on the network. IP datagrams are eventually encapsulated inside a particular hardware frame, such as Ethernet or token ring. In general, the IP is considered to be unreliable because there is no guarantee the datagram will reach its destination. IP provides what is called *best effort* delivery. Usually, when an IP datagram runs into trouble on the network, it is simply discarded. An error contained in an ICMP message may or may not be returned to the sender. Figure 24.3 illustrates an example of how an ICMP message is encapsulated in an IP datagram. The upper layers of the IP such as TCP and UPD provide the required reliability.

IP datagrams are routed on the network using an IP address. The IP address consists of a 32-bit number. The IP address is divided into four sections, each containing 8 bits. Each

FIGURE 24.4 Class C network IP

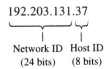

section may take on a decimal value between 0 and 255. These four 8-bit groups are called octets and are separated by periods. This is called *dotted-decimal notation*. An example of dotted-decimal notation is shown in Figure 24.4 illustrating a class C Internet address.

The Internet addresses are classified into five address classes. These addresses are shown in Figure 24.5. Each address class consists of a network ID and a host ID.

The network ID in a class A address contains a maximum number of 128 possible networks and more than 16 million hosts. A class B address allocates more bits in the IP address to the network ID and fewer bits to the host ID, creating a possible 16,384 networks and 65,536 hosts. The class C address provides more than two million networks, each with a possible 256 hosts. Class D addresses are reserved for multicast data, and class E addresses are reserved.

Some of the IP addresses are reserved for special functions. For example, the class A address 127 is reserved for loopback testing, allowing a local method to test functionality of the TCP/IP software and applications. Other addresses are used as masks to identify each network type, such as 255.255.255.0 to identify a class C address.

IP VERSION 6

One of the problems with the current version of the Internet is the lack of adequate addresses. Basically, there are no more available (through the Network Information Center). A new version of the Internet, sometimes called the next-generation Internet, has resolved many of the problems experienced by its predecessor.

For example, the address for the next-generation Internet is 128 bits as opposed to 32 bits. This additional address space is large enough to accommodate network growth for the foreseeable future. Addresses are grouped into three different categories: *unicast* (single computer), *multicast* (a set of computers with the same address), and *cluster* (a set of computers that share a common address prefix), routed to one computer closest to the sender.

Other changes include different header formats, new extension headers, and support for audio and video. Unlike IP version 4, IP version 6 does not specify all of the possible protocol features. This allows new features to be added without the need to update the protocol.

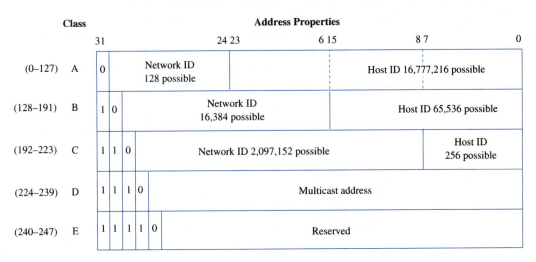

FIGURE 24.5 IP address classes

TCP

The Transmission Control Protocol (TCP) defines a standard way that two computers can communicate together over interconnected networks. Applications using TCP establish *connections* with each other through the use of predefined *ports* or *sockets*. A TCP connection is reliable, with error checking, acknowledgment for received packets, and packet sequencing provided to guarantee the data arrives properly at its destination. Telnet and FTP are examples of TCP/IP applications that use TCP.

UDP

The User Datagram Protocol (UDP) is similar to TCP except it is *connectionless*. Data is transmitted with no acknowledgment of whether it is received. UDP is thus not as reliable as TCP. But some applications do not require the additional overhead of TCP handshaking, using UDP instead. For example, a multiplayer network game might use UDP because it is simple to implement, requires less overhead to manage than TCP, and because the game may not be severely affected if a few packets are lost now and then.

DNS

Every machine on a network must be uniquely identified. On the Internet, this identification takes the form of what is called an *IP address*. The IP address consists of a 4-byte number, commonly represented in dotted-decimal notation. For example, 204.210.133.51 specifies the IP address of some machine on some network somewhere in the world. To make it easier to remember an address, a *host* name can be associated with an IP address. The Domain Name Service (DNS) provides the means to convert from a host name to an IP address, and vice versa. Instead of using the IP address 204.210.133.51, we may instead enter "raycast.rwa.com" as the host name. Typically a DNS server application running on the local network has the responsibility of converting names to IP addresses.

ARP AND RARP

Before any packet can be transmitted, the address of its destination must be known. This address is called a *MAC address*. MAC stands for media access control, and takes the form of a 48-bit binary number that uniquely identifies one machine from every other. Every network interface card manufactured responds to a preassigned MAC address.

The Address Resolution Protocol (ARP) is used in the Data-Link layer to obtain the MAC address for a given IP address. For example, an ARP request may say, "What is the MAC address for 204.210.133.51?" The ARP reply may be, "The MAC address is 00-60-97-2B-E6-0F."

The Reverse Address Resolution Protocol (RARP) performs the opposite of ARP, providing the IP address for a specific MAC address. Figure 24.6 gives an example of both protocols at work.

FIGURE 24.6 Using ARP and RARP

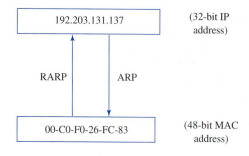

ICMP

The Internet Control Message Protocol (ICMP) uses IP packets to provide updates on network error conditions. Some ICMP messages cause other messages to be sent. Table 24.4 shows a list of the different ICMP messages.

When Netscape gives you its "host unreachable" error message, an ICMP message is responsible.

TABLE 24.4 ICMP messages

Type	Code	Description	Query	Error
0	0	Echo reply	✔	
3		Destination unreachable		
	0	network unreachable		✔
	1	host unreachable		✔
	2	protocol unreachable		✔
	3	port unreachable		✔
	4	fragmentation needed but don't-fragment bit set		✔
	5	source route failed		✔
	6	destination network unknown		✔
	7	destination host unknown		✔
	8	source host isolated (obsolete)		✔
	9	destination network administratively prohibited		✔
	10	destination host administratively prohibited		✔
	11	network unreachable for TOS		✔
	12	host unreachable for TOS		✔
	13	communication administratively prohibited by filtering		✔
	14	host precedence violation		✔
	15	precedence cutoff in effect		✔
4	0	Source quench (elementary flow control)		✔
5		Redirect		
	0	redirect for network		✔
	1	redirect for host		✔
	2	redirect for type-of-service and network		✔
	3	redirect for type-of-service and host		✔
8	0	Echo request	✔	
9	0	Router advertisement	✔	
10	0	Router solicitation	✔	
11		Time exceeded		
	0	time-to-live equals 0 during transit (Traceroute)		✔
	1	time-to-live equals 0 during reassembly		✔
12		Parameter problem		
	0	IP header bad (catchall error)		✔
	1	required option missing		✔
13	0	Timestamp request	✔	
14	0	Timestamp reply	✔	
15	0	Information request (obsolete)	✔	
16	0	Information reply (obsolete)	✔	
17	0	Address mask request	✔	
18	0	Address mask reply	✔	

FIGURE 24.7 FTP application

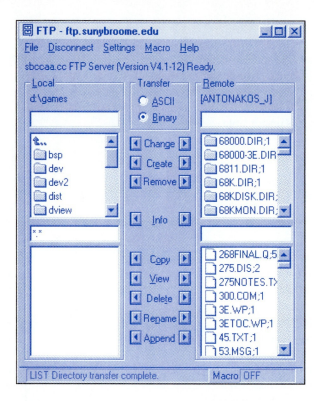

FTP

The File Transfer Protocol (FTP) allows a user to log on to a remote computer and transfer files back and forth through simple commands. Many FTP sites allow you to log on as an *anonymous* user, an open account with limited privileges but still capable of file transfers. A typical FTP application might look like the one shown in Figure 24.7. Files may be transferred in either direction.

SMTP

The Simple Mail Transport Protocol (SMTP) is responsible for routing e-mail on the Internet using TCP and IP. The process usually requires connecting to a remote computer and transferring the e-mail message, but due to problems with the network or a remote computer, messages can be temporarily undeliverable. The e-mail server will try to deliver any messages by periodically trying to contact the remote destination. When the remote computer becomes available, the message is delivered using SMTP.

TELNET

A Telnet session allows a user to establish a terminal emulation connection on a remote computer. For example, an instructor may Telnet into his or her college's mainframe to do some work. Figure 24.8 shows how the Telnet connection is set up. Once the connection has been made, Telnet begins emulating the terminal selected in the Connection Dialog window. Figure 24.9 shows this mode of operation.

SLIP AND PPP

The Serial Line Interface Protocol (SLIP) and Point-to-Point Protocol (PPP) are used to transfer IP packets over serial connections, such as those provided by modems. Windows

FIGURE 24.8 Establishing the Telnet connection

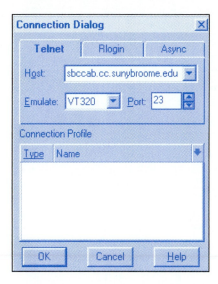

FIGURE 24.9 Sample Telnet session

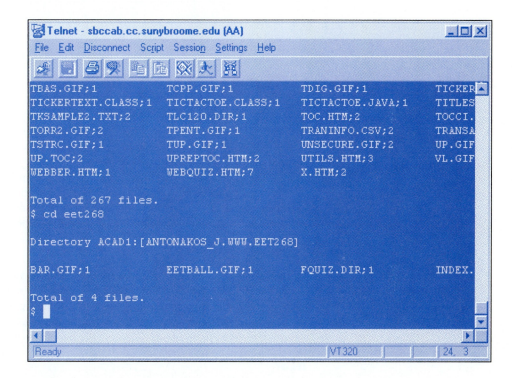

provides PPP service through its Dial-Up Networking software. SLIP, developed earlier than PPP, is limited to supporting only TCP/IP applications, whereas PPP supports TCP/IP, IPX, and other protocols at the same time. These two protocols operate in the lower two network layers, as indicated by Figure 24.10. Note that PPP operates in the Physical *and* Data-Link layers, whereas SLIP only functions inside the Physical layer. This is further evidence that PPP has additional features.

SNMP

Network managers responsible for monitoring and controlling the network hardware and software use the Simple Network Management Protocol (SNMP), which defines the format and meaning of messages exchanged by the manager and agents. The network manager (*manager*) uses SNMP to interrogate network devices (*agents*) such as routers, switches, and bridges in order to determine their status and also retrieve statistical information.

FIGURE 24.10 SLIP and PPP protocols

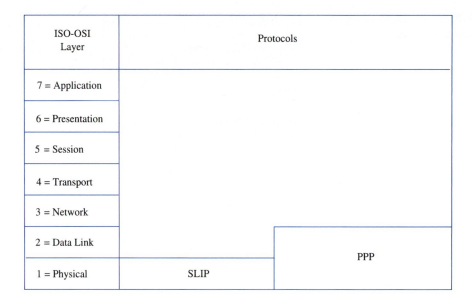

ROUTING PROTOCOLS

Routing protocols direct the flow of information within and between networks. Let us look at a small sample of the protocols in use today.

RIP

The Routing Information Protocol (RIP), used to route NetWare packets, relies on a method called *distance vector route discovery,* which determines the shortest number of hops to the destination by adding the cost of each subpath to the destination. To maintain accurate information, all routers using RIP periodically broadcast their routing tables to each other (typically once every 30 seconds). RIP is not a good routing protocol for large networks due to the large amount of traffic resulting from routing table broadcasts.

OSPF

Open Shortest Path First (OSPF) is a routing protocol used over TCP/IP that performs *link state routing.* This protocol has advantages over RIP's distance vector routing protocol, such as sharing routing tables less frequently (initially when the router powers up), performing *load balancing,* and constantly testing the state of each communication link (thus adjusting to network problems more quickly than RIP).

NLSP

Network Link Services Protocol (NLSP) is NetWare's link state routing protocol, capable of routing packets farther than RIP (127 hops versus RIP's 15 hops), load balancing, and fast response to changes in the network.

IGRP

The Interior Gateway Routing Protocol (IGRP) is similar to RIP, using distance vector route discovery and routing table broadcasting. Tables are broadcast every 90 seconds instead of every 30 seconds to reduce network traffic. This causes an increase in the time required for the network to recover from a lost route resulting from a broken connection.

IEEE 802 Standards

Several IEEE standards have been developed to support network implementations. Three of these standards are:

- IEEE 802.2 Logical Link Control
- IEEE 802.3 Ethernet (CSMA/CD)
- IEEE 802.5 Token Ring

FIGURE 24.11 LanExplorer protocol analyzer window

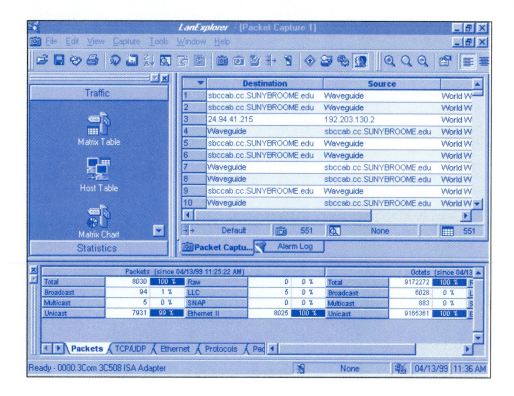

IEEE 802.2 is the initial protocol at work in the Data-Link layer. It interfaces with the Ethernet and token-ring protocols, which operate primarily in the Physical layer. Any company manufacturing network interface cards or other products must conform to the IEEE 802 standards.

The original Ethernet specification called for 10Mbit/second capability. The demands of network users has resulted in a newer 100Mbit/second Ethernet, with faster standards approaching. CSMA/CD (carrier sense multiple access with collision detection) is used to manage data transmission.

Token ring is available in speeds of 4- and 16Mbit/second. This network protocol requires that a special token be passed from one node to another. Only the node holding the token can send data over the network.

PROTOCOL ANALYZERS

Protocol analyzers (or *sniffers*) are hardware or software devices that listen to the traffic on a network and capture various packets for examination. Hardware analyzers also double as cable testers.

Figure 24.11 shows a demo version of the LanExplorer protocol analyzer at work. LanExplorer displays an ongoing update of network traffic statistics; allows packets to be captured, disassembled, and saved; and can transmit packets to facilitate testing and troubleshooting.

TROUBLESHOOTING TECHNIQUES

Windows provides a very useful utility called WINIPCFG, which you can run from the Run menu. Figure 24.12 shows the display window with all details included. Notice the various IP addresses indicated. Because the network software used by the system receives an IP address *on-the-fly,* via DHCP (Dynamic Host Configuration Protocol), the IP address of the DHCP server (204.210.159.18) must be known by the system. DHCP is not used when your system has been allocated a fixed IP address by your network administrator.

If you have difficulty with your network connection, the information displayed by WINIPCFG will be valuable to the person who is troubleshooting the connection.

FIGURE 24.12 WINIPCFG display window

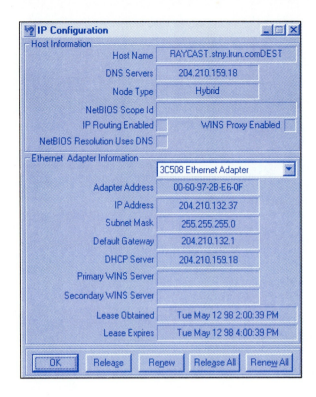

SELF-TEST

This self-test is designed to help you check your understanding of the background information presented in this exercise.

True/False

Answer *true* or *false*.

1. The IPX/SPX protocols are the backbone for TCP/IP.
2. The ISO-OSI Network Model defines the nine network layers.
3. RFCs are documents that describe how each of the TCP/IP protocols are implemented.
4. IP datagrams are guaranteed to be delivered.
5. Loopback testing allows for a remote method to test TCP/IP and application functionality.

Multiple Choice

Select the best answer.

6. TCP connections are
   a. Unreliable and must rely on the IP protocol to provide the necessary reliability.
   b. Created using predefined ports or sockets.
   c. Considered connectionless without the need to acknowledge the received data.
   d. Limited to class A addresses.
7. Dotted-decimal notation is used to
   a. Describe each of the five address classes.
   b. Identify all the networks and hosts on the Internet.
   c. Reserve addresses used for special purposes.
   d. All of the above.
   e. None of the above.
8. DNS servers provide
   a. The MAC address for every host address.
   b. The error-checking protocol for ICMP messages.
   c. A mapping from a host name to an IP address.
   d. A mechanism to transfer IP packets over serial lines.

9. PPP operates on
    a. All seven ISO-OSI networking layers.
    b. All nine ISO-OSI network layers.
    c. The Physical and Data-Link layers.
    d. Remote terminal emulation connections.
10. Routing protocols are designed to
    a. Direct the flow of information between networks.
    b. Interrogate network device agents to gather statistics.
    c. Broadcast traffic on all segments of a network.
    d. Troubleshoot TCP and UDP connection problems.

Matching

Match a description of the items on the right with each item on the left.

11. IPX/SPX
12. NetBEUI
13. TCP/IP
14. PPP
15. FTP

a. Dial-Up Networking protocol
b. Connection-Oriented File Transfer Protocol
c. NCP, SAP, and RIP protocols
d. Microsoft Windows network protocol
e. UDP, TCP, and ICMP protocols

Completion

Fill in the blank or blanks with the best answers.

16. Most FTP sites allow for _____ login.
17. SMTP is used to exchange _____.
18. A(n) _____ describes each of the TCP/IP protocols.
19. IP addresses are commonly shown in _____ _____ _____.
20. The _____ protocol is designed for use on small networks and is not routable.

FAMILIARIZATION ACTIVITY

1. Install LanExplorer from the companion CD. If possible, do the following:
   - Capture several packets and note the type of packets captured.
   - Watch the network statistics for several minutes.
   - Identify all the different protocols present on the network.
2. Visit the computer center at your school and ask the network technicians to share information about the IP classes in use, what network management protocols are used, and how many nodes are on the network.
3. Look up the term *broadcast storm* in a networking dictionary (or search the Web for it). What is a broadcast storm, what are its effects, and how can it be prevented?

QUESTIONS/ACTIVITIES

1. What is the purpose of a set of protocols?
2. Why are IP addresses used for routing instead of MAC addresses?

REVIEW QUIZ

Within 10 minutes and with 100% accuracy,

1. Explain the purpose of each layer of the ISO-OSI model.
2. Discuss the different protocols that make up the TCP/IP suite.
3. Describe the relationship between IP addresses and MAC addresses.

25 Network Applications

INTRODUCTION

Joe Tekk answered the telephone in his office. It was Jeff Page, who sounded worried.
"Joe, PING my computer for me. I'm at 206.25.117.119."
Joe entered the appropriate command. Several Request timed out *error messages appeared. "It doesn't look good, Jeff. PING doesn't see your machine."*
Jeff groaned. "Hold on, Joe."
A few minutes later Jeff said, "Try it again."
Joe ran the PING utility again. This time he got the normal display statistics.
"What did you do, Jeff?" he asked, wanting to know what the problem was.
"I don't know. I rebooted and everything works again."

PERFORMANCE OBJECTIVES

Upon completion of this exercise, within 10 minutes and with 100% accuracy, you will be able to

1. Discuss the role of PING and TRACERT.
2. Explain the numbers WINIPCFG displays.
3. Define DHCP and state its purpose.

BACKGROUND INFORMATION

In addition to the browser, e-mail, FTP, and Telnet applications previously discussed, there are a few additional network applications that deserve mention. Let us take a look at them and the network functions they perform.

PING

PING, which stands for Packet Internet Groper, is a TCP/IP application that sends datagrams once every second in the hope of an echo response from the machine being PINGed. If the machine is connected and running a TCP/IP protocol stack, it should respond to the PING datagram with a datagram of its own. PING displays the time of the return response in milliseconds or one of several error messages (such as "Request timed out" or "Destination host unreachable").

PING can be used to simply determine the IP address of a World Wide Web site if you know its host name. For example, the IP address for www.yahoo.com is 204.71.200.74, which can be found with PING by opening a DOS window, and entering

```
C> ping www.yahoo.com
```

You should see something similar to this:

```
Pinging www.yahoo.com [204.71.200.74] with 32 bytes of data:

Reply from 204.71.200.74: bytes=32 time=180ms TTL=245
Reply from 204.71.200.74: bytes=32 time=127ms TTL=245
Reply from 204.71.200.74: bytes=32 time=145ms TTL=245
Reply from 204.71.200.74: bytes=32 time=146ms TTL=245
```

So, PING performed DNS on the host name to find out the IP address, and then sent datagrams to Yahoo's host machine and displayed the responses. If you know the IP address you can enter it directly and PING will skip the DNS phase.

To get a list of PING's features, enter PING with no parameters. You should see something similar to this:

```
Usage: ping [-t] [-a] [-n count] [-l size] [-f] [-i TTL] [-v TOS]
            [-r count] [-s count] [[-j host-list] | [-k host-list]]
            [-w timeout] destination-list

Options:

    -t              Ping the specified host until interrupted.
    -a              Resolve addresses to hostnames.
    -n count        Number of echo requests to send.
    -l size         Send buffer size.
    -f              Set Don't Fragment flag in packet.
    -i TTL          Time To Live.
    -v TOS          Type Of Service.
    -r count        Record route for count hops.
    -s count        Timestamp for count hops.
    -j host-list    Loose source route along host-list.
    -k host-list    Strict source route along host-list.
    -w timeout      Timeout in milliseconds to wait for each reply.
```

You are encouraged to experiment with these parameters.

TRACERT

TRACERT (Trace Route) is a TCP/IP application that determines the path through the network to a destination entered by the user. For example, running

```
C> tracert www.yahoo.com
```

generates the following output:

```
Tracing route to www7.yahoo.com [204.71.200.72]
over a maximum of 30 hops:

  1    20 ms    19 ms    19 ms  bing100b.stny.lrun.com [204.210.132.1]
  2    10 ms    14 ms     9 ms  m2.stny.lrun.com [204.210.159.17]
  3    12 ms    24 ms    10 ms  ext_router.stny.lrun.com [204.210.155.18]
  4    46 ms    40 ms    44 ms  border3-serial4-0-6.Greensboro.mci.net [204.70.83.85]
  5    42 ms    53 ms    46 ms  core1-fddi-0.Greensboro.mci.net [204.70.80.17]
  6   109 ms   160 ms   122 ms  bordercore2.Bloomington.mci.net [166.48.176.1]
  7   123 ms   126 ms   113 ms  hssi1-0.br2.NUQ.globalcenter.net [166.48.177.254]
```

```
 8    125 ms   117 ms   115 ms   fe5-1.cr1.NUQ.globalcenter.net [206.251.1.33]
 9    114 ms   125 ms   113 ms   pos0-0.wr1.NUQ.globalcenter.net [206.251.0.122]
10    125 ms   124 ms   121 ms   pos1-0-OC12.wr1.SNV.globalcenter.net [206.251.0.74]
11    122 ms   139 ms   115 ms   pos5-0.cr1.SNV.globalcenter.net [206.251.0.105]
12    128 ms   129 ms   138 ms   www7.yahoo.com [204.71.200.72]

Trace complete.
```

The trace indicates that it took 12 *hops* to get to Yahoo. Every hop is a connection between two routers. Each router guides the test datagram from TRACERT one step closer to the destination. TRACERT specifically manipulates the TTL (time to live) parameter of the datagram, adding 1 to it each time it rebroadcasts the test datagram. Initially the TTL count is 1. This causes the very first router in the path to send back an ICMP time-exceeded message, which TRACERT uses to identify the router and display path information. When the TTL is increased to 2, the second router sends back the ICMP message, and so on, until the destination is reached (if it ever is).

It is fascinating to examine TRACERT's output. Notice that hop 5 contains a reference to *FDDI,* which means that the datagram spent some time traveling around a fiber distributed data interface network.

WINIPCFG

WINIPCFG (Windows IP Configuration) is a handy tool for examining all the numbers associated with your networking components. Figure 25.1 shows the initial WINIPCFG window, which displays some of the more important networking information. Figure 25.2 shows the results of clicking the "More Info >>" button. Note that there is an entry for DHCP Server. DHCP refers to dynamic host configuration protocol, a TCP/IP protocol that allows a DHCP client to request an IP address at boot time from a pool of free IP addresses (maintained by the DHCP server). The IP address is *leased* to the DHCP client for a limited time period (which can be renewed or extended). Every time the DHCP client boots, it may receive a different IP address. This in no way affects the way information is exchanged.

ECHO SERVERS AND CHAT SERVERS

Individuals teaching networking courses typically have on hand two applications that demonstrate how to send and receive messages over a network. An *echo server* simply sends any echo client messages back to the client unchanged. A *chat server* allows two or more users to send messages that are broadcast to all members of the group (with private messaging possible via software). These two applications are simple enough that they are used as beginning programming exercises in a course on network programming with

FIGURE 25.1 Initial WINIPCFG window

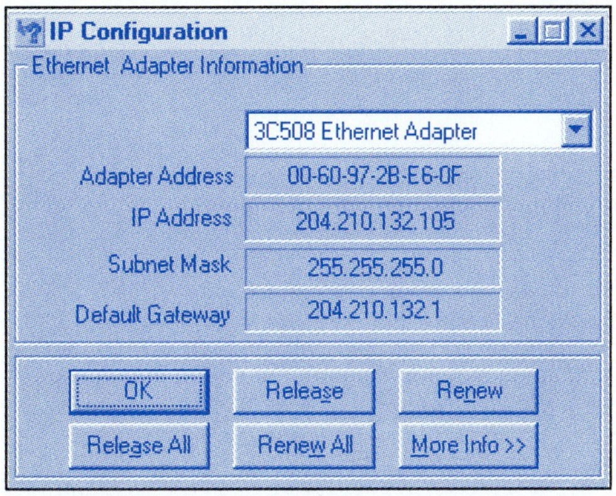

FIGURE 25.2 More WINIPCFG details

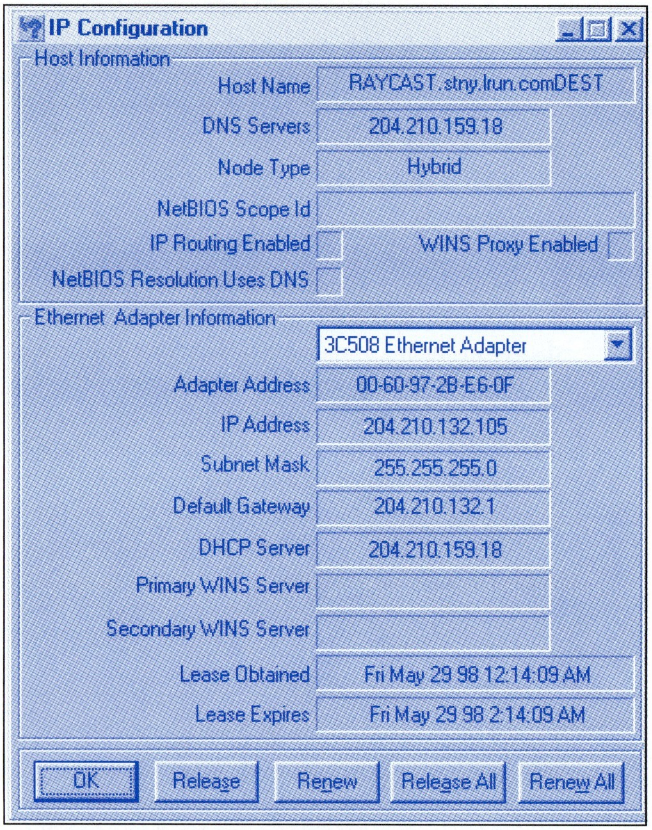

sockets. A socket is a name for a TCP/IP connection, an application that "plugs" into the appropriate "socket."

Windows contains an installable chat application called WINCHAT that allows one user to dial another by entering the machine name of the machine to chat with. Figure 25.3 shows a WINCHAT session in progress. The user James is communicating with Ken on the machine named REFLECTOR (NetBEUI must be operational for WINCHAT to work). Any machine on the Network Neighborhood that has a WINCHAT utility can be dialed by entering its name.

FIGURE 25.3 WINCHAT network utility

FIGURE 25.4 pcANYWHERE control panel

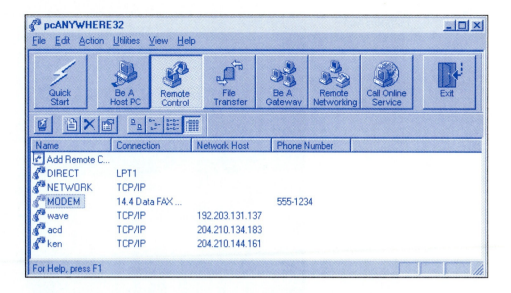

FIGURE 25.5 Connecting to the host machine

PCANYWHERE

pcANYWHERE is an excellent utility for connecting two machines over a network. One machine is set up as the host. The second machine remotely controls the host. The remote machine's pcANYWHERE window looks exactly like the desktop on the host. If you click on a program icon, the program will launch on the host. Essentially, you can do just about everything from the remote machine you could do on the host. pcANYWHERE is a TCP/IP application, so it can be used over large, routable networks (although a modem may also be used).

A screen shot of pcANYWHERE's control panel is shown in Figure 25.4. With the Remote Control button pushed, double-clicking on wave, acd, or ken will begin a network connection sequence (using TCP/IP) to the machine at the IP address indicated. Clicking wave, for example, brings up the initialization window shown in Figure 25.5.

If the connection is successful, the host desktop appears in the remote machine's pcANYWHERE window. Figure 25.6 illustrates this feature. Now you can do whatever you like with the host machine. Bear in mind that even over a high-speed network screen, updates might be sluggish due to the amount of data that must be transferred.

pcANYWHERE contains many powerful features, including built-in chat and file-transfer functions. Figure 25.7 shows the file-transfer screen, which is easy to use and navigate. The arrow at the top of the window indicates the direction of the transfer. Clicking the arrow changes its direction. Whatever files are highlighted in the file menus are transferred when the Send button is pressed. Entire directories can be sent with a single click.

Overall, pcANYWHERE adds a new dimension to remote computing.

FIGURE 25.6 Host desktop

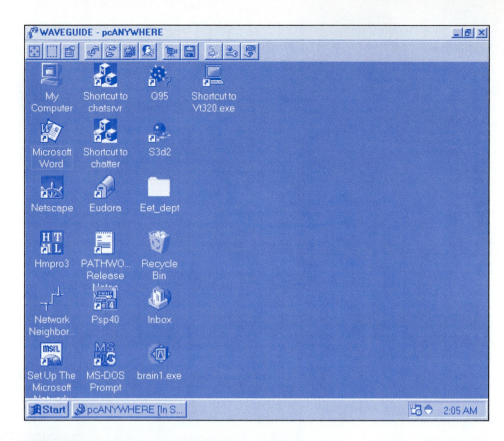

FIGURE 25.7 pcANYWHERE file-transfer screen

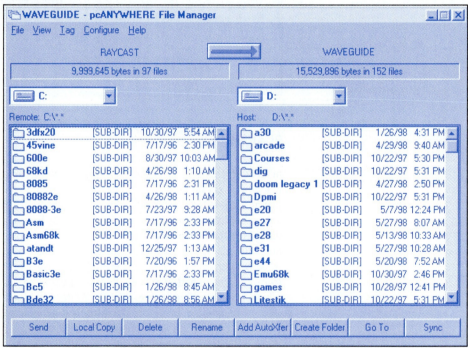

TROUBLESHOOTING TECHNIQUES

Although rebooting a machine is the last resort, in many cases troubleshooting a network problem may require several reboots. If you install a new protocol, change the name of your machine, or add a network printer, Windows recommends that you reboot so that the changes may take effect. You might as well follow Windows's advice and reboot. While the machine is doing so you have a few moments to yourself to review what is happening. Is the problem with the computer (protocol properties not set right), the network connection (external cabling, NIC interrupt or I/O settings, protocol binding problems), the machine on the other end, or the application software?

By the time the machine has rebooted you are probably waiting to try something else, a different theory, a new approach. Try it. Reboot if necessary after making adjustments, and remain patient. There is a reason the network is not there.

SELF-TEST

This self-test is designed to help you check your understanding of the background information presented in this exercise.

True/False

Answer *true* or *false*.

1. PING sends a sonar pulse onto the Internet.
2. Running TRACERT with an invalid URL will cause it to hang forever.
3. An echo server retransmits all received messages.
4. WINIPCFG only works for fixed IP addresses.
5. pcANYWHERE has built-in file-transfer capability.

Multiple Choice

Select the best answer.

6. PING uses
    a. NetBEUI packets.
    b. TCP/IP packets.
    c. PING packets.
7. How many different class C networks are shown in the TRACERT output for Yahoo?
    a. 8.
    b. 9.
    c. 10.
8. To find out the adapter address (MAC address), use
    a. PING.
    b. TRACERT.
    c. WINIPCFG.
9. Which utility is used to send text messages to another user?
    a. PING.
    b. WINCHAT.
    c. WINIPCFG.
10. DHCP is used to
    a. Obtain a fixed IP address.
    b. Obtain a leased IP address.
    c. Carry PING requests.

Completion

Fill in the blank or blanks with the best answers.

11. WINCHAT uses the _____ protocol.
12. DHCP stands for _____ _____ _____ _____.
13. pcANYWHERE uses a remote computer to connect to a(n) _____ computer.
14. To determine an unknown IP address, use _____.
15. To determine the path to an IP address, use _____.

FAMILIARIZATION ACTIVITY

1. PING the following addresses:
    - www.yahoo.com
    - www.intel.com
    - www.whitehouse.gov

- www.nasa.gov
- 192.203.131.137

2. Run TRACERT on the addresses in step 1. Comment on any interesting router names that show up.

QUESTIONS/ACTIVITIES

1. How long is the leased IP address good for in Figure 25.2?
2. Why use WINCHAT when you can easily make a call on the telephone?
3. Search the Web for products similar to pcANYWHERE.

REVIEW QUIZ

Within 10 minutes and with 100% accuracy,

1. Discuss the role of PING and TRACERT.
2. Explain the numbers WINIPCFG displays.
3. Define DHCP and state its purpose.

26 The Internet

INTRODUCTION

It was 2:45 A.M. Joe Tekk was awake, sitting in his darkened living room. The only light in the room was coming from the monitor of his computer. Joe was exhausted, but didn't want to stop browsing the Web. He had stumbled onto a Web page containing links to computer graphics, game design, and protected-mode programming. For three hours, Joe had been going back and forth from one page to another, adding some links to his bookmarks and ignoring others. When he finally decided to quit, it was not due to lack of interest, but simply time to go to sleep.

"From now on, I'm only browsing for 30 minutes," Joe vowed to himself. But he knew he would have another late-night browsing session that would last much longer. It was too much fun having so much information available instantly.

PERFORMANCE OBJECTIVES

Upon completion of this exercise, within 10 minutes and with 100% accuracy, you will be able to

1. Describe the basic organization of the Internet.
2. Explain the purpose of a browser and its relationship to HTML.
3. Discuss the usefulness of CGI applications.

BACKGROUND INFORMATION

The Internet started as a small network of computers connecting a few large mainframe computers. It has grown to become the largest computer network in the world, connecting virtually all types of computers. The Internet offers a method to achieve *universal service,* or a connection to virtually any computer, anywhere in the world, at any time. This concept is similar to the use of a telephone, which provides a voice connection anywhere at any time. The Internet provides a way to connect all types of computers together regardless of their manufacturer, size, and resources. The *one* requirement is a connection to the network. Figure 26.1 shows how several networks are connected together.

The type of connection to the Internet can take many different forms, such as a simple modem connection, a cable modem connection, a T1 line, a T3 line, or a frame relay connection.

FIGURE 26.1 Concept of Internet connections

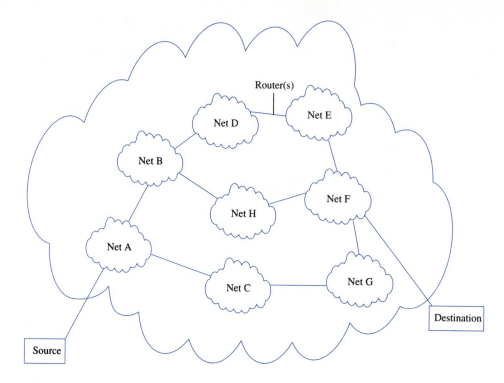

TABLE 26.1 Organization of the Internet

Domain Type	Organization Type
edu	Educational institution
com	Commercial organization
gov	Government
mil	Military
net	Network providers and support
org	Other organizations not listed above
country code	A country code, for example, .us for United States, .ca for Canada

THE ORGANIZATION OF THE INTERNET

The current version of the Internet (V4) is organized into several categories, as shown in Table 26.1. The name of an Internet host shows the category to which it is assigned. For example, the rwa.com domain is the name of a company, and the bcc.edu domain is an educational institution. Each domain is registered on the appropriate root server. For example, the domain rwa.com is known by the com root server. Then, within each domain, a locally administered Domain Name Server allows for each host to be configured.

WORLD WIDE WEB

The World Wide Web, or WWW as it is commonly referred to, is actually the Hypertext Transport Protocol (HTTP) in use on the Internet. The HTTP protocol allows for hypermedia information to be exchanged, such as text, video, audio, animation, Java applets, images, and more. The hypertext markup language, or HTML, is used to determine how the hypermedia information is to be displayed on a WWW browser screen.

The WWW browser is used to navigate the Internet by selecting *links* on any WWW page or by specifying a Universal Resource Locator, or URL, to point to a specific *page* of information. There are many different WWW browsers. The two most popular are Microsoft Internet Explorer and Netscape Navigator, shown in Figures 26.2 and 26.3,

FIGURE 26.2 Sample home page displayed using Internet Explorer

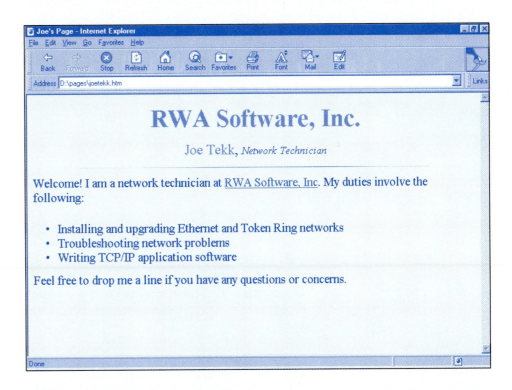

FIGURE 26.3 Sample home page displayed using Netscape Navigator

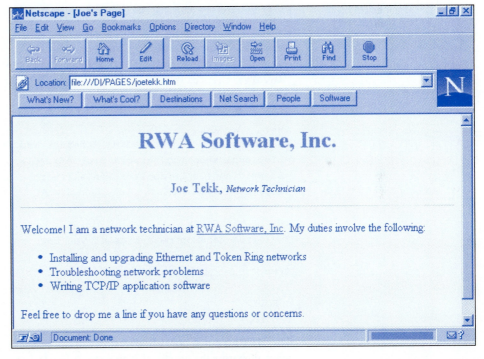

respectively. Both of these browsers are available free over the Internet, and contain familiar pull-down menus and graphical toolbars to access the most commonly used functions such as forward, backward, stop, print, and reload.

Note the differences in page layout between Figures 26.2 and 26.3. Individuals who design Web pages must take into account the different requirements of each browser so that the page looks acceptable in both browsers.

HTML

HTML stands for hypertext markup language. HTML is the core component of the information that composes a Web page. The HTML *source* code for a Web page has an overall

323

syntax and structure that contains formatting commands (called *tags*) understood by a Web browser. Here is a sample HTML source. The actual Web page for this HTML code was shown in Figure 26.3.

```
<HTML>

<HEAD>
<TITLE>Joe's Page</TITLE>
</HEAD>

<BODY BGCOLOR="#FFFF80">

<P ALIGN="CENTER">
<B><FONT SIZE="+3" COLOR="#FF0000">RWA Software, Inc.</FONT></B>
</P>

<P ALIGN="CENTER">
<FONT SIZE="+1"><FONT COLOR="#008000">Joe Tekk</FONT>,
<FONT SIZE="-1"><I>Network Technician</I></FONT></FONT>
</P>

<P ALIGN="CENTER">
<IMG SRC="bar.gif" ALT="Color Bar">
</P>

<P ALIGN="LEFT">
Welcome! I am a network technician at
<A HREF="http://www.rwasoftware.com">RWA Software, Inc</A>. My duties
 involve the following:
</P>

<UL>
<LI>Installing and upgrading Ethernet and Token Ring networks</LI>
<LI>Troubleshooting network problems</LI>
<LI>Writing TCP/IP application software </LI>
</UL>

<P ALIGN="LEFT">
Feel free to drop me a line if you have any questions or
concerns.
</P>

</BODY>
</HTML>
```

The HTML source consists of many different tags that instruct the browser what to do when preparing the graphical Web page. Table 26.2 shows some of the more common tags. The main portion of the Web page is contained between the BODY tags. Note that BGCOLOR= "#FFFF80" sets the background color of the Web page. The six-digit hexadecimal number contains three pairs of values for the red, green, and blue color levels desired.

Pay attention to the tags used in the HTML source and what actually appears on the Web page in the browser (Figure 26.3). The browser ignores whitespace (multiple blanks between words or lines of text) when it processes the HTML source. For example, the anchor for the RWA Software link begins on its own line in the source, but the actual link for the anchor is displayed on the same line as the text that comes before and after it.

Many people use HTML editors, such as HoTMetaL or Front Page, to create and maintain their Web pages. Options to display the page in HTML format, or in WYSIWYG (what you see is what you get), are usually available, along with sample pages, image editing, and conversion tools that convert many different file types (such as a Word document) into

TABLE 26.2 Assorted HTML tags

Tag	Meaning
<P>	Begin paragraph
</P>	End paragraph
<B>	Bold
<I>	Italics
<IMG SRC...>	Image source
<UL>	Unordered list
<LI>	List item
<TABLE>	Table
<TR>	Table row
<TD>	Table data
<A>	Anchor

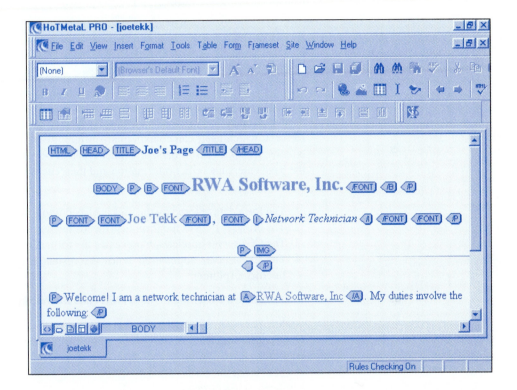

FIGURE 26.4 HoTMetaL Pro 4.0 with sample page

HTML. Demo versions of these HTML editors, and others, can be downloaded from the Web. Figure 26.4 shows HoTMetaL's graphical page editor with Joe Tekk's page loaded.

WWW pages are classified into three categories: static, dynamic, and active. The easiest to make are static and involve only HTML code. The page content is determined by what is contained in the HTML code. Dynamic WWW pages contain a combination of HTML code and a "call" to a server using a Common Gateway Interface application, or a CGI application. In this scenario, information supplied by the user into an HTML form is transferred back to a host computer for processing. The host computer then returns a dynamic customized WWW page. Active pages contain a combination of HTML code and applets. Therefore, the WWW page is not completely specified during the HTML coding process. Instead, using a Java applet, it is specified while being displayed by the WWW browser.

CGI

The Common Gateway Interface (CGI) is a software interface that allows a small amount of interactive processing to take place with information provided on a Web page. For example, consider the Web page shown in Figure 26.5. The Web page contains a FORM

FIGURE 26.5 Web page with FORM element

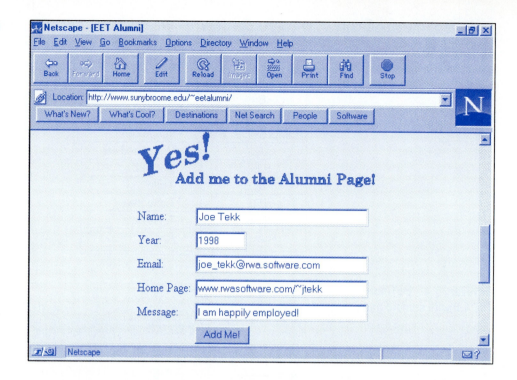

element, which itself can contain many different types of inputs, such as text boxes, radio buttons, lists with scroll bars, and other types of buttons and elements. The user browsing the page enters his or her information and then clicks the Add Me! button. This begins the following chain of events:

1. The form data entered by the user is placed into a message.
2. The browser POSTs the form data (sends message to CGI server).
3. The CGI server application processes the form data.
4. The CGI server application sends the results back to the CGI client (Netscape or Internet Explorer).

Let us take a closer look. The HTML code for the alumni page form looks like this:

```
<FORM ACTION="/htbin/cgi-mailto/eetalumni" METHOD="POST">
<P ALIGN="CENTER"><IMG SRC="yes.gif"></P>
<CENTER>
<TABLE WIDTH="50%" ALIGN="CENTER">
<TR><TD>Name:</TD>
<TD><INPUT TYPE="TEXT" NAME="name" SIZE="32"></TD></TR>
<TR><TD>Year:</TD>
<TD><INPUT TYPE="TEXT" NAME="year" SIZE="8"></TD></TR>
<TR><TD>Email:</TD>
<TD><INPUT TYPE="TEXT" NAME="email" SIZE="32"></TD></TR>
<TR><TD>Home Page:</TD>
<TD><INPUT TYPE="TEXT" NAME="home" SIZE="32"></TD></TR>
<TR><TD>Message:</TD>
<TD><INPUT TYPE="TEXT" NAME="msg" SIZE="32"></TD></TR>
<TR><TD></TD>
<TD><INPUT TYPE="SUBMIT" VALUE="Add Me!"></TD></TR>
</TABLE>
</CENTER>
</FORM>
```

The first line of the form element specifies POST as the method used to send the form data out for processing. The CGI application that will receive the POSTed form data is the cgi-mailto program in the *htbin* directory. Specifically, cgi-mailto processes the form data and sends an e-mail message to the *eetalumni* account. The e-mail message looks like this:

```
From:   SBCCVA::WWWSERVER
To:     eetalumni
CC:
Subj:

REMOTE_ADDRESS: 204.210.159.19
name: Joe Tekk
year: 1998
email: joe_tekk@rwa.software.com
home: www.rwasoftware.com/~jtekk
msg: I am happily employed!
```

Note that the identifiers (name, year, email, home, and msg) match the names used to identify the text input elements in the form.

Instead of e-mailing the form data, another CGI application might create a Web page on-the-fly containing custom information based on the form data submitted. CGI applications are written in C/C++, Visual BASIC, Java, Perl, and many other languages. The Web is full of sample forms and CGI applications available for download and inclusion in your own Web pages.

JAVA

The Java programming language is the method used to create active WWW pages using Java applets. An active WWW page is specified by the Java applet when the WWW page is displayed rather than during the HTML coding process. A Java applet is actually a program transferred from an Internet host to the WWW browser. The WWW browser executes the Java applet code on a Java virtual machine (which is built into the WWW browser). The Java language can be characterized by the following nonexhaustive list:

- General purpose
- High level
- Object oriented
- Dynamic
- Concurrent

Java consists of a programming language, a run-time environment, and a class library. The Java programming language resembles C++ and can be used to create conventional computer applications or applets. Only an applet is used to create an active WWW page. The run-time environment provides the facilities to execute an application or applet. The class library contains prewritten code that can simply be included in the application or applet. Table 26.3 shows the Java class library functional areas.

TABLE 26.3 Java class library categories

Class	Description
Graphics	Abstract window tool kit (AWT).
Network I/O	Socket level connnections.
File I/O	Local and remote file access.
Event capture	User actions (mouse, keyboard, etc.).
Run-time system calls	Access to built-in functions.
Exception handling	Method to handle any type of error condition.
Server interaction	Built-in code to interact with a server.

The following Java program is used to switch from one image to a second image (and back) whenever the mouse moves over the Java applet window. Furthermore, a mouse click while the mouse is over the applet window causes a new page to load.

```java
import java.awt.Graphics;
import java.awt.Image;
import java.awt.Color;
import java.awt.Event;
import java.net.URL;
import java.net.MalformedURLException;

public class myswitch extends java.applet.Applet implements Runnable
{
    Image swoffpic;
    Image swonpic;
    Image currentimg;
    Thread runner;

public void start()
{
    if (runner == null)
    {
       runner = new Thread(this);
       runner.start();
    }
}

public void stop() {
    if (runner != null)
    {
        runner.stop();
        runner = null;
    }
}

public void run()
{
    swoffpic = getImage(getCodeBase(), "swoff.gif");
    swonpic = getImage(getCodeBase(), "swon.gif");
    currentimg = swoffpic;
    setBackground(Color.red);
    repaint();
}

public void paint(Graphics g)
{
    g.drawImage(currentimg, 8, 8, this);
}

public boolean mouseEnter(Event evt, int x, int y)
{
    currentimg = swonpic;
    repaint();
    return(true);
}

public boolean mouseExit(Event evt, int x, int y)
{
    currentimg = swoffpic;
```

```
        repaint();
        return(true);
}

public boolean mouseDown(Event evt, int x, int y)
{
        URL destURL = null;
        String url = "http://www.sunybroome.edu/~eet_dept";

        try
        {
            destURL = new URL(url);
        }
        catch(MalformedURLException e)
        {
            System.out.println("Bad destination URL: " + destURL);
        }
        if (destURL != null)
            getAppletContext().showDocument(destURL);
        return(true);
}

}
```

Programming in Java, like any other language, requires practice and skill. With its popularity still increasing, now would be a good time to experiment with Java yourself by downloading the free Java compiler and writing some applets.

RELATED SITES

Here are a number of service, reference, and technology-based sites that may be of interest:

- www.prenhall.com — Engineering and technology textbooks
- www.yahoo.com — Search engine
- www.internic.net — Internet authority
- www.intel.com — Intel Corporation
- www.microsoft.com — Microsoft Corporation
- www.sunybroome.edu/~mansfield_k — Author's home page
- www.sunybroome.edu/~antonakos_j — Author's home page
- www.netscape.com — Netscape corporation

The Internet is full of information about every aspect of the Web page development process. Many people put a tremendous amount of information on their own Web pages. You are encouraged to learn more about Web pages and Web programming.

TROUBLESHOOTING TECHNIQUES

The Internet and the World Wide Web are not the same thing. The Internet is a physical collection of networked computers. The World Wide Web is a logical collection of information contained on many of the computers comprising the Internet. To download a file from a Web page, the two computers (client machine running a browser and server machine hosting the Web page) must exchange the file data, along with other control information. If the download speed is slow, what could be the cause? A short list identifies many suspects:

- Noise in the communication channel forces retransmission of many packets.
- The path through the Internet introduces delay.
- The server is sending data at a limited rate.
- The Internet service provider has limited bandwidth.

So, before buying a new modem or upgrading your network, determine where the bottleneck is. The Internet gets more popular every day. New home pages are added, additional files are placed on FTP sites for downloading, news and entertainment services are coming online and broadcasting digitally, and more and more machines are being connected. The 10- and 100Mbit Ethernet technology is already hard-pressed to keep up with the Internet traffic. Gigabit networking is coming, but will only provide a short respite from the ever-increasing demands of global information exchange.

SELF-TEST

This self-test is designed to help you check your understanding of the background information presented in this exercise.

True/False

Answer *true* or *false*.

1. The hypertext markup language is used to encode GIF images.
2. CGI stands for Common Gateway Interchange.
3. HTML contains formatting commands called tags.
4. The main portion of a Web page is contained between the HEADER tags.
5. Java is a Web browser produced by Microsoft Corporation.

Multiple Choice

Select the best answer.

6. CGI applications can be written using
    a. Perl, Java, C, C++, and Visual BASIC.
    b. Only the Javascript language.
    c. An HTML editor.
7. The three different categories of Web pages are
    a. Large, medium, and small.
    b. Active, passive, and neutral.
    c. Static, dynamic, and active.
8. When the network is slow,
    a. Turn off all power to the computer and perform a reset.
    b. Try to determine where the bottleneck is located.
    c. Immediately upgrade to the newest, most expensive hardware available.
9. CGI applications use FORMs to
    a. Receive the input required for processing.
    b. Post the data to an e-mail application.
    c. Send information to the browser display.
10. The concept of universal service and the Internet involves
    a. Being able to connect to a universal router on the Internet.
    b. Being able to exchange information between computers at any time or place.
    c. Allowing all users to access the universal Internet database.

Completion

Fill in the blank or blanks with the best answers.

11. The same _____ code is displayed differently using different Internet browsers.
12. A CGI application provides the ability to create _____ _____ on-the-fly.
13. Java is used to create _____ Web pages.
14. _____ information includes text, video, audio, Java applets, and images.
15. The _____ protocol is used to exchange hypermedia information.

FAMILIARIZATION ACTIVITY

WWW

1. Examine each of the pull-down menu items in Netscape Navigator and Internet Explorer.
2. Read the online help to learn about browser features.

3. Identify similarities between the two browsers discussed in this exercise.
4. Identify differences between the two browsers discussed in this exercise.

CGI

1. Search the Web to locate information about Perl.
2. Locate a source for Perl, available free of charge.
3. Download Perl.
4. Install Perl.
5. Run some of the sample Perl scripts.

Java

1. Search the Web to locate information about the Java language.
2. Locate the source of a Java compiler available free of charge.
3. Download the Java compiler.
4. Install the Java compiler.
5. Compile some of the sample Java applets.
6. Execute a sample Java applet.

QUESTIONS/ACTIVITIES

1. Determine how to clear the browser's cache memory.
2. Determine the current allocation settings for the browser cache.
3. Search the Web to locate some useful resources related to active Web page development.

REVIEW QUIZ

Within 10 minutes and with 100% accuracy,

1. Describe the basic organization of the Internet.
2. Explain the purpose of a browser and its relationship to HTML.
3. Discuss the usefulness of CGI applications.

27 Windows NT Domains

INTRODUCTION

Joe Tekk was very excited. He was finally given the opportunity to set up the new Windows NT domain for RWA Software, Inc. Joe thought that a Windows NT network was necessary because it was becoming harder and harder to maintain the workgroup that was originally installed several years ago.

Joe told Don, his manager, "RWA has grown so much since I started working here. The NT Server operating system is going make it so much easier to administer the network."

Don replied, "If you say so, Joe. I'll leave the network administration up to you." He continued, "Joe, please let me know what you are planning before we make any big changes. We don't want to make any avoidable mistakes."

Joe responded, "I'm glad you mentioned that, Don. I laid out the plan on paper so everyone can understand how the new network will operate. I have also set up a timetable to get everyone up and running."

Don smiled and said, "Great job, Joe. Keep it up!"

Joe spent most of his spare time reading about Windows NT. When he finally received his copy of the Windows NT Server CD-ROM, he could not wait to get started.

PERFORMANCE OBJECTIVES

Upon completion of this exercise, within 10 minutes and with 100% accuracy, you will be able to

1. Describe the benefits of creating a Windows NT domain.
2. Explain some different types of Windows NT domains.
3. Discuss the different types of clients able to join a Windows NT domain.

BACKGROUND INFORMATION

Any group of personal computers can be joined together to form either a workgroup or a domain. In a workgroup, each computer is managed independently, but may share some of its resources with the other members of the network, such as printers, disks, or a scanner. Unfortunately, as the number of computers in the workgroup grows, it becomes more and more difficult to manage the network. This is exactly the situation where a Windows NT domain can be used. A domain offers a centralized mechanism to relieve much of the administrative burden commonly experienced in a workgroup. A domain requires at least one computer running the Windows NT Server operating system. Table 27.1 illustrates the characteristics of a workgroup and a domain.

TABLE 27.1 Comparing a workgroup and a domain

Workgroup	Domain
Small networks	Large networks
Peer-to-peer	Client–server
No central server	Central server
Low cost	Higher cost
Decentralized	Centralized

DOMAINS

Each Windows NT domain can be configured independently or as a group where all computers are members of the same domain. Figure 27.1 shows two independent domains. Each domain consists of at least one Windows NT primary domain controller (PDC) and any number of backup domain controllers (BDC). One shared directory database is used to store user account information and security settings for the entire domain.

A BDC can be promoted to a PDC in the event the current PDC on the network becomes unavailable for any reason. A promotion can be initiated manually, causing the current PDC to be demoted to a backup. Figure 27.2 shows a domain containing two Windows NT server computers. One computer is the PDC, and the other computer is the BDC.

Windows NT can administer the following types of domains:

- Windows NT Server domains
- Windows NT Server and LAN Manager 2.x domains
- LAN Manager 2.x domains

A LAN Manager 2.x domain is a previous version of Microsoft networking software used by older MS-DOS and Windows computers.

The different types of activities that can be performed on a domain include the following:

- Create a new domain
- Modify an existing domain
- Join a domain
- Add a computer to a domain
- Remove a computer from a domain
- Synchronize files in a domain
- Promote a BDC to a PDC
- Establish trust relationships

When a system is set up as a PDC, the new domain name is required in order to proceed through the Windows NT installation process. This domain name is required by all other computer users who want to join the domain. Note that each domain can contain only *one* primary domain controller. All other Windows NT Server computers can be designated as backups or ones that do not participate in the domain control process at all.

A computer can be configured to join a domain during the Windows NT installation process, using the Network icon in the system Control Panel, or by using the Server

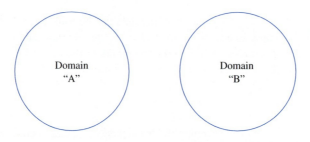

FIGURE 27.1 Independent Windows NT domains

FIGURE 27.2 Domain "A" configuration

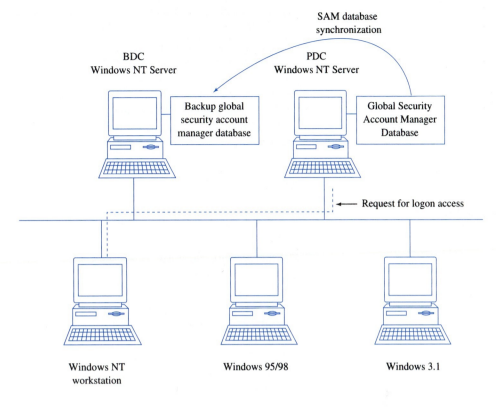

FIGURE 27.3 Domain trust relationships

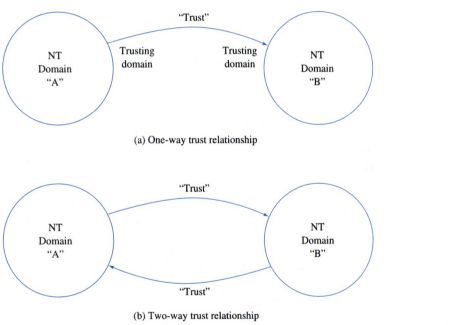

Manager tool. A computer can be removed using the Network icon in the system Control Panel or the Server Manager tool.

Synchronizing a domain involves exchanging information between a primary domain controller and any secondary or backup domain controllers as previously shown in Figure 27.2. The synchronization interval for a Windows NT computer is five minutes. This means account information entered on the primary domain controller takes only five minutes to be exchanged with all secondary computers. This synchronization is performed automatically by Windows NT.

Domains can also be set up to offer *trust relationships*. A trust relationship involves either providing or receiving services from an external domain, as shown in Figure 27.3. A trust relationship can permit users in one domain to use the resources of another domain. A trust relationship can be a one-way trust or a two-way trust, offering the ability to handle many types of requirements.

A one-way trust relationship as shown in Figure 27.3(a) identifies domain "B" as a trusted source for domain "A." A two-way trust, shown in Figure 27.3(b), involves two separate domains sharing their resources with each other. Each domain considers the other to be a trusted source. Extreme caution must be exercised when setting up trust relationships. If the trusted domain is really untrustworthy, valuable information can be lost using the "trusted" accounts.

DOMAIN CLIENTS

A Windows NT domain can support many different types of clients, such as:

- Windows NT servers
- Windows NT workstations
- Windows 95/98 clients
- Windows 3.11 clients
- Windows 3.1 clients
- MS-DOS clients
- OS/2 workstations

LOGGING ONTO A NETWORK

When a computer is configured to run in a network, each user must be authorized before access to the computer can be granted. Figure 27.4 shows a typical Windows 95/98 logon screen. Each user must supply a valid user name and a valid password in order to gain access to the computer and any network resources. In a *workgroup* setting, all password information is stored locally on each computer in PWL files. The PWL files are named using the following format: the first eight letters of the user name entered in the logon screen followed by the .PWL file extension. The PWL files contain account and password information stored in encrypted form. These files are typically stored in the Windows directory. Figure 27.5 shows the concept of a workgroup where each computer is administered independently.

In a *domain* setting, a centralized computer running Windows NT is contacted to verify the user name and password. If the information provided to the server is valid, access is granted to the local machine. If either the user name or password is invalid, access to the local computer is denied. As you might think, this method offers tremendously more flexibility as far as the administration is concerned. This concept is illustrated in Figure 27.6.

RUNNING A NETWORK SERVER

Running a network server involves installing the Windows NT operating system and then configuring it to run as a primary or secondary domain controller during the installation

FIGURE 27.4 Windows 95/98 logon screen

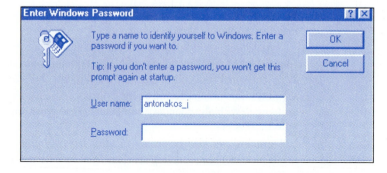

FIGURE 27.5 Workgroup concept where each computer is administered independently

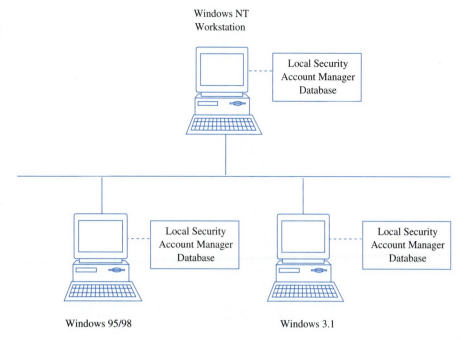

FIGURE 27.6 Domain concept

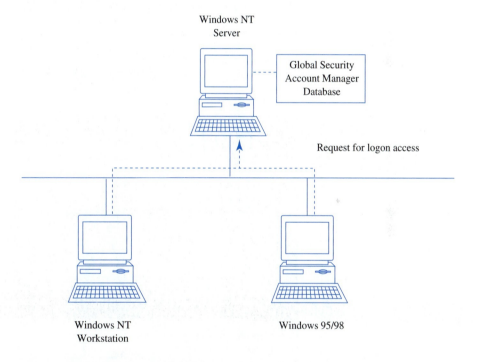

process. After a PDC is created during the installation, the domain exists on the network. Windows NT computers can then join the domain by changing the Member of domain as shown in Figure 27.7. Windows computers join the domain by changing individual settings on each computer. Figure 27.8 shows the Primary Network Logon selecting the Client for Microsoft Networks option. Then, by selecting the properties for Client for Microsoft Networks, the specific domain can be identified as illustrated in Figure 27.9. After making these changes, a system reset is necessary to make the changes active.

Network server computers are also assigned the task of running more applications to manage both the server and network. For example, a Windows NT Server may be used to add fault tolerance to disks using a Redundant Array of Inexpensive Disks (RAID) technology. A server may also run the WWW server application, Windows Internet Naming System (WINS), Dynamic Host Configuration Protocol (DHCP), and Remote Access Server (RAS). These services are usually required 24 hours a day, seven days per week.

FIGURE 27.7 Configuring a Windows NT Server

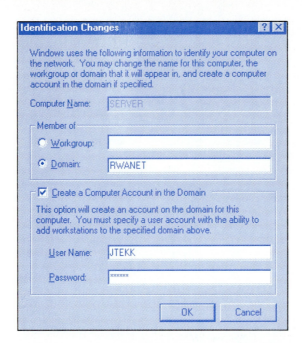

FIGURE 27.8 Windows 95/98 Network settings

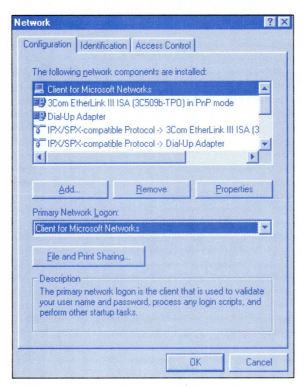

Windows NT Server computers are designed to handle the computing workload for entire organizations, corporations, or any other type of enterprise. In these cases, many servers (including a PDC and several BDCs) are made available to guarantee the availability of any required services.

USER PROFILES

In a domain, the primary domain controller maintains all user profiles. This allows for centralized control of the Security Accounts Manager (SAM) database. Two programs are provided

FIGURE 27.9 Configuring Windows 95/98 to log on to a domain

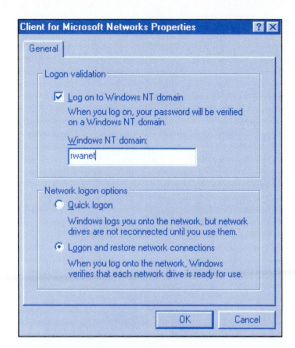

FIGURE 27.10 New User dialog box

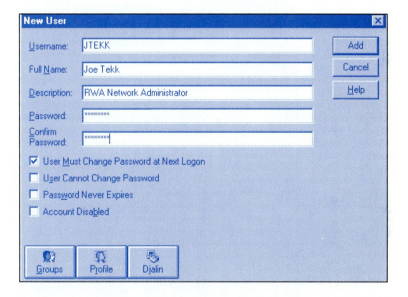

to update the SAM database. One of the programs is used in a stand-alone (no domain) environment and the other is for use where a domain is specified. Otherwise the programs operate in the same way. Let us examine what is involved when setting up a user account as illustrated in Figure 27.10. Information must be specified about each user account including user name, full name, a description of the account, and the password setting. The check boxes are used to further modify the account, such as requiring a password change during the first logon, restricting changing the password, extending the life of a password, and, lastly, disabling the account.

The three buttons at the bottom of the New User window (Figure 27.10) allow for each new account to be added to different *groups* as shown in Figure 27.11. It is a good idea to grant access to groups on an individual basis as certain privileges are granted by simply belonging to the group, such as administrator.

The User Environment Profile screen specifies the path to an individual profile and any required logon script name. Additionally, the home directory may be specified as shown in Figure 27.12.

FIGURE 27.11 Group Memberships selection screen

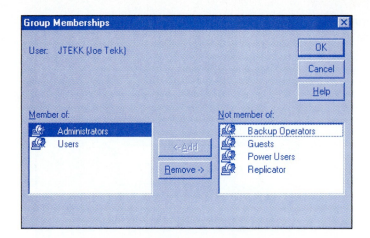

FIGURE 27.12 User Environment configuration screen

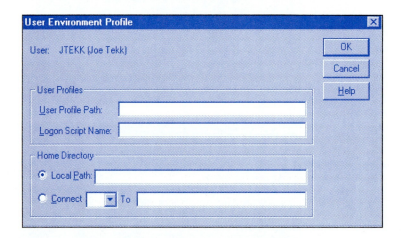

FIGURE 27.13 Dialin Information settings

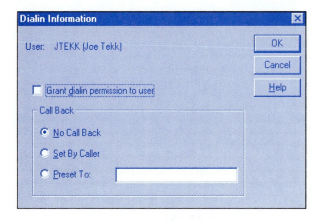

Lastly, the Dialin Information window determines if a Windows NT account has access to Dial-Up Networking. The Call Back option may also be configured to require the computer to call the user back. This is an additional security feature that may be implemented if necessary. Figure 27.13 shows these settings.

SECURITY

Windows NT is a C2 compliant operating system, when it is configured properly as defined by the National Computer Security Center (NCSC). C2 compliance involves properly configuring Windows NT to use the built-in safeguards. An application tool supplied with the operating system (C2CONFIG.EXE) examines the operating system setting against a recommended setting. Any exceptions are noted.

FIGURE 27.14 System events display

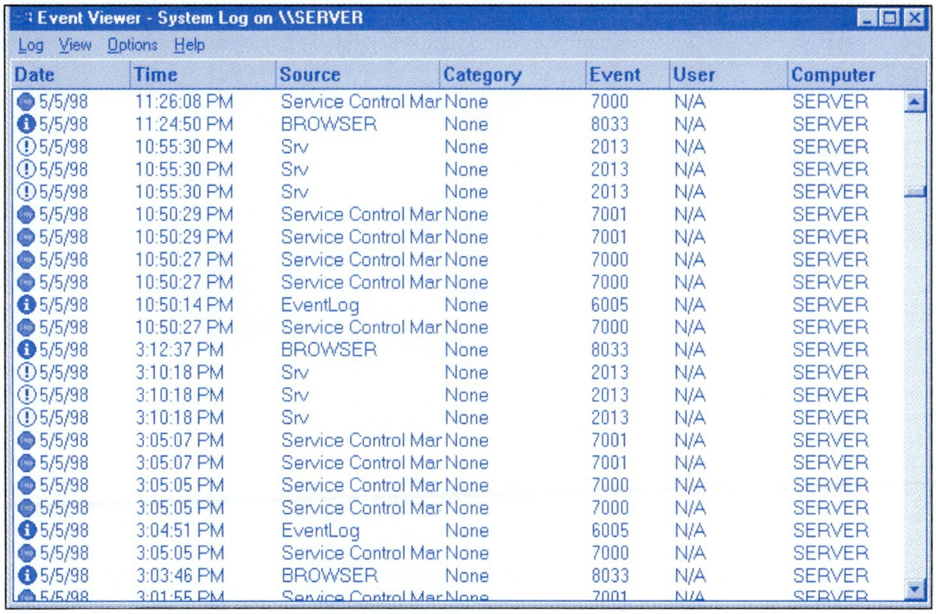

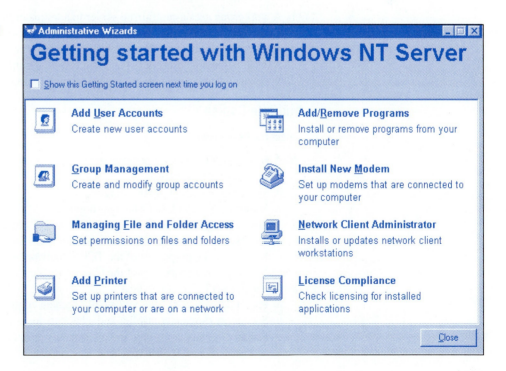

Windows NT provides security-logging features designed to track all types of system activities, such as logon attempts, file transfers, Telnet sessions, and many more. Typically the System Administrator will determine which types of events are logged by the system. Figure 27.14 shows the system log. The icons along the left margin are color coded to draw attention to more serious events. Event logs should be reviewed daily.

TROUBLESHOOTING TECHNIQUES

A networked computer environment (especially when using Windows NT) can become somewhat complex, requiring the system or network administrator to have many technical skills. Fortunately, Windows NT also provides many resources designed to tackle most networking tasks. For example, the Administrative Tools menu contains the Administrative Wizards option shown in Figure 27.15. Most of these wizards perform the activities that are necessary to get a network up and running.

FIGURE 27.15 Administrative Wizards display

341

FIGURE 27.16 Windows NT Help display

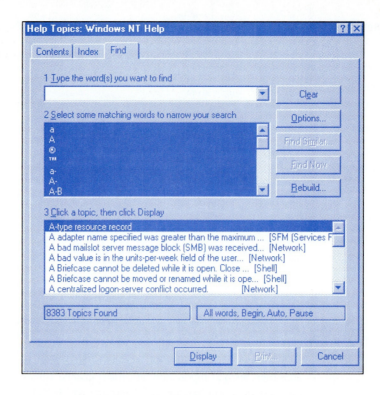

It is also a good idea to examine the online help system to get additional information, which may simplify any task. Figure 27.16 shows a Help screen that contains a total of 8383 topics, many of which contain information about networking.

SELF-TEST

This self-test is designed to help you check your understanding of the background information presented in this exercise.

True/False

Answer *true* or *false*.

1. A workgroup uses a centralized server to administer the network.
2. Each Windows NT domain can be configured independently.
3. A primary domain controller can be demoted to a backup domain controller.
4. A backup domain controller is updated every 10 minutes.
5. Windows 95 computers function only marginally in a Windows NT domain.

Multiple Choice

Select the best answer.

6. A Windows NT server can administer
   a. Windows NT domains.
   b. Windows NT and TCP/IP domains.
   c. Windows NT and LAN Manager domains.
7. Windows computers are added to a Windows NT domain by
   a. Double-clicking on the Windows NT computer in the Network Neighborhood.
   b. Modifying the properties of the TCP/IP network settings.
   c. Modifying the properties of the Client for Microsoft Network settings.
8. A Windows NT Server can be a
   a. Parent domain controller and a child domain controller.
   b. Secondary domain controller and a backup domain controller.
   c. Primary domain controller and a secondary domain controller.

9. A trusted domain
   a. Contains only one primary domain controller and no secondary controller.
   b. Is granted special access to all Windows computers in the trusted domain.
   c. Permits users in one domain to use the resources of another domain.
10. Running a network server involves
    a. Installing and configuring a Windows NT Workstation computer.
    b. Installing and configuring a Windows NT Server computer.
    c. Connecting Windows 95 computers to Windows NT workstation computers.

Matching

Match a description of the networking topic on the right with each item on the left.

11. Windows NT client
12. Windows NT server
13. Domain types
14. Trust relationships
15. Domain activity

a. Windows NT and LAN Manager
b. One-way and two-way
c. PDC, BDC, none
d. DOS, OS/2, Windows 95/98
e. Establish trust relationships

Completion

Fill in the blank or blanks with the best answers.

16. A backup domain controller is _____ to a primary domain controller.
17. A large number of computers cannot be managed effectively in a(n) _____ setting.
18. Computers administered centrally are part of a(n) _____.
19. Each domain must contain _____ primary domain controller.
20. A Windows NT Server may be either a(n) _____, _____, or not involved in the domain controller process.

FAMILIARIZATION ACTIVITY

1. Configure a Windows NT computer to function as a primary domain controller.
2. Configure a Windows NT computer to function as a backup domain controller.
3. Change the name of the domain.
4. Add some client computers to the Windows NT domain.
5. Examine networking topics using the online help system.
6. Run the Administrative Wizard applications to become familiar with account and network information.

QUESTIONS/ACTIVITIES

1. Under what circumstances can a Windows NT workstation computer become a primary domain controller? A Windows NT Server computer?
2. When should a Windows NT domain be used instead of a workgroup?
3. What is necessary for an operating system to become a network client?
4. What is C2 security?
5. Where can additional C2 information be found?

REVIEW QUIZ

Within 10 minutes and with 100% accuracy,

1. Describe the benefits of creating a Windows NT domain.
2. Explain some different types of Windows NT domains.
3. Discuss the different types of clients able to join a Windows NT domain.

28 An Introduction to Telecommunications

INTRODUCTION

Joe Tekk stood in the equipment room of a local Internet service provider. He watched silently while the ISP manager, Dave Guza, described all the hardware and software.

"We service 800 local individuals and businesses from this room," Dave explained. "The three servers on the floor are for e-mail, Web pages, and news. We have two T1 connections that constitute our main Internet connection, with a dedicated 56K baud backup for emergencies."

While Dave was speaking he moved to the back of several tall racks of equipment. "These are our modem banks. They service our 750 dial-up access lines."

Joe watched as the modem lights blinked on and off. All the modems looked busy. "Is it this busy all the time?" he asked.

Dave smiled at Joe. "No, it is usually busier."

PERFORMANCE OBJECTIVES

Upon completion of this exercise, within 10 minutes and with 100% accuracy, you will be able to

1. Describe the different telecommunication technologies.
2. Discuss reasons for choosing one technology over another.

BACKGROUND INFORMATION

The world of telecommunications is getting both larger and smaller at the same time. From a hardware standpoint, more equipment is being installed every day, connecting more and more people, businesses, and organizations.

At the same time, the pervasiveness of the World Wide Web has made it easy to communicate with someone practically anywhere on the planet. The world does not seem as large as it once did.

In this exercise we will examine the many different telecommunication technologies available, and see how they take part in our everyday communication.

FIGURE 28.1 Time-division multiplexing

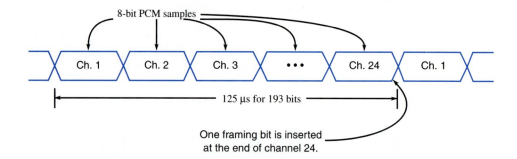

TABLE 28.1 T-carrier services

Level	Number of voice channels	Data Rate (Mbps)
1	24	1.544
2	96	6.312
3	672	44.736
4	4032	274.176

TDM

Time-division multiplexing, or TDM, is a technique used by the telephone company to combine multiple digitized voice channels over a single wire. Telephone conversations are digitized into 8-bit PCM (pulse code modulation) samples and sampled 8000 times per second. This gives 64,000 bps for a single conversation. Now, using a multiplexer, if we rapidly switch from one channel to another, it is possible to transmit the 8-bit samples for 24 different conversations over a single wire. All that is required is a fast bit rate on the single wire. Figure 28.1 shows a timing diagram for the TDM scheme on a basic carrier called a *T1 carrier*. The T1 provides 1.544 Mbps multiplexed data for twenty-four 64,000 bps channels. The 8 bits for each channel (192 bits total) plus a framing bit (a total of 193 bits) are transmitted 8000 times per second.

Table 28.1 shows the different levels of T-carrier service available.

CIRCUIT SWITCHING

In the early days of the telephone system, large rotary switches were used to switch communication lines and make the necessary connections to allow end-to-end communication. The switches completed a circuit, hence the name circuit switching.

Eventually these slow, mechanical switches were replaced with fast, electronic switches. Also called an *interconnection network,* a switch is used to direct a signal to a specific output (such as the telephone you are calling).

Figure 28.2 shows one way to switch a set of eight signals. This type of switch is called a *crossbar switch*. Connections between input and output signals are made by closing switches at specific intersections within the 8-by-8 grid of switches. Only one switch is turned on in any row or column (unless we are broadcasting). Since each intersection contains a switch that may be open or closed, one control bit is required to represent the position of each switch. The pattern for the first row of switches in Figure 28.2 is 01000000. The pattern for the second row is 00000100. A total of 64 control bits are required.

A nice feature of the crossbar switch is that any mapping between input and output is possible.

If the cost of 64 switches is too much for your communication budget, a different type of switch can be used to switch eight signals, but with less than half the number of switches.

FIGURE 28.2 Eight-signal crossbar switch

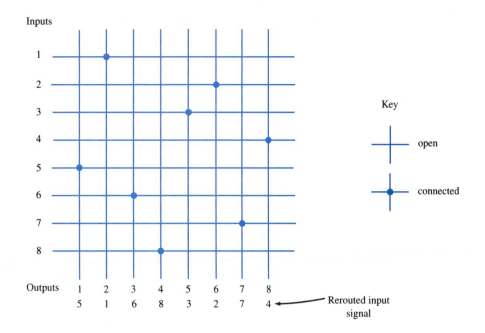

FIGURE 28.3 Eight-signal multistage switch

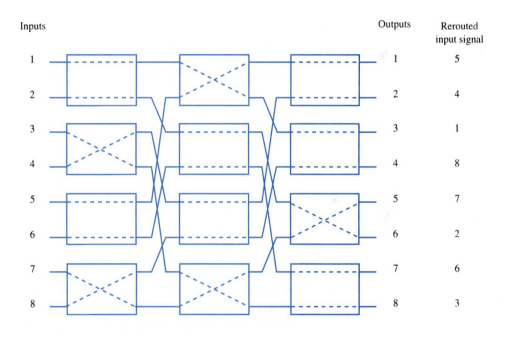

Called a *multistage switch,* it relies on several stages of smaller switches connected in complex ways. Figure 28.3 shows a sample three-stage switch capable of switching eight signals. Each smaller switch can be configured as a straight-through or crossover switch, with a single control bit specifying the mode. Now, with only 12 smaller switches, the control information has shrunk from 64 bits in the crossbar switch to only 12 bits. The number of switches is 24 (one switch for straight-through, one switch for crossover, times 12), which is less than half of the 64 required in the crossbar switch.

The price we pay for the simplified hardware in the multistage switch is a smaller number of switching possibilities. For example, in Figure 28.3 is it possible to set up the 12 smaller switches so that the output maps to 87654321? The answer is no, indicating that the multistage switch may block some signals from getting to the correct output. This problem is usually temporary, since memory buffers are typically used to store data that cannot be transmitted right away.

FIGURE 28.4 Sample WAN

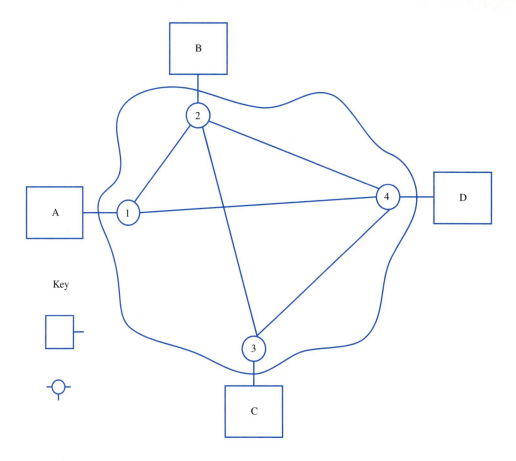

PACKET SWITCHING

Figure 28.4 shows a simple WAN connecting four networks (A, B, C, and D). Suppose that a machine on network A wants to send a large chunk of data to a machine on network D. Using packet switching, the large chunk of data is broken down into smaller blocks and transmitted as a series of packets.

Due to the nature of traffic on shared networks, some packets may go directly from router 1 to router 4 (one hop), while others may go from router 1 to router 2, then to router 4 (two hops). A three-hop route is also possible. Thus, it is likely that packets arrive at network D out of order. This is a characteristic of packet switching. Packets can be reassembled in the correct order by including a sequence number within the packet. Even so, this characteristic makes packet switching unsuitable for digitized phone conversations, which, unlike an e-mail message, cannot wait for gaps to be filled in at some unknown later time. These features provide a means for choosing between circuit switching and packet switching.

FRAME RELAY

Packet switching was designed during a time when digital communication channels were not very reliable. To compensate for errors in a channel, a handshaking arrangement of send-and-acknowledge packets was used to guarantee reliable data transfers. This error protocol added time-consuming overhead to the packet switching network, with transmitting stations constantly waiting for acknowledgments before continuing.

Frame relay takes advantage of the improvement in communication technology (fiber links, for example, have a very low error rate compared with copper links), and relies on fewer acknowledgments during a transfer. Only the receiving station need send an acknowledgment.

With fewer acknowledgments and a lower error rate, frame relay provides a significant improvement in communications technology.

FIGURE 28.5 An ATM cell

FIGURE 28.6 ATM header fields

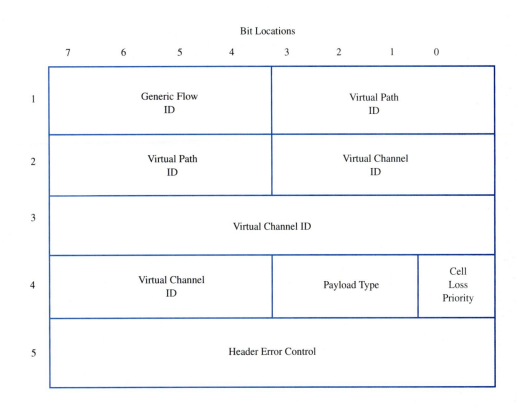

ATM

Asynchronous transfer mode (ATM), also called *cell relay,* uses fixed-size cells of data, and supports voice, data, or video at either 155.52 Mbps or 622.08 Mbps. Cells are 53 bytes each, with 5 bytes reserved for a header and the remaining 48 for data, as indicated in Figure 28.5. The reason for using fixed-size cells is to simplify routing decisions at intermediate nodes in the ATM system.

ATM uses communication connections called *virtual channel* connections. A virtual channel is set up between the end-to-end stations on the network and fixed-size cells are sent back and forth. Decisions concerning routing are resolved using information supplied in the ATM header, which is shown in Figure 28.6. Notice the entries for virtual path and virtual channel identifiers.

ISDN

The simplest Integrated Services Data Network (ISDN) connection is called a *basic rate interface,* and consists of two 64 Kbps B channels (for data) and one 16 Kbps D channel (for signaling). The design of ISDN supports circuit-switching, packet-switching, and frame operation. When ISDN is carried over a T1 line (1.544 Mbps), twenty-three 64 Kbps B channels and one 64 Kbps D channel are possible.

ISDN has yet to catch on, but it is positioned to be a major service provider in the years to come.

TABLE 28.2 SONET signal hierarchy

SONET level	Data Rate (Mbps)
STS*-1	51.84
STS-3	155.52
STS-9	466.56
STS-12	622.08
STS-18	933.12
STS-24	1244.16
STS-36	1866.24
STS-48	2488.32

*STS (Synchronous Transport Signal)

FIGURE 28.7 SONET STS-1 frame format

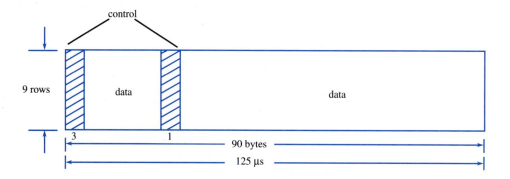

SONET

The Synchronous Optical Network (SONET) technology was designed to take advantage of the high speed of a fiber connection between networks. As Table 28.2 indicates, the lowest-speed SONET signal level (STS-1) runs at 51.85 Mbps. That is equivalent to more than 900, 57 Kbps modems running simultaneously (minus a few for overhead). STS-48 has 48 times the bandwidth of STS-1, so you can imagine how many telephone calls can be carried over a single fiber link.

There are additional benefits to using fiber: it is not susceptible to electrical noise, it can be run farther distances than copper wire before requiring a repeater to extend the signal, and it is easier to repair.

Figure 28.7 shows the format of a SONET STS-1 frame. A total of 810 bytes are transmitted in a 125-microsecond time slot. Several bytes from each row of the frame are used for control/status information, such as several 64 Kbps user channels, 192 Kbps and 576 Kbps control, maintenance, and status channels, and several additional signaling items.

FDDI

The Fiber Distributed Data Network (FDDI) was developed to provide 100 Mbps connections between LANs over a wide geographical area. Dual fiber rings are used, with the second ring serving as a backup for the first ring, called the *primary* ring. A token-passing scheme similar to token-ring technology is used to allow access to the ring.

The use of fiber allows longer distances between stations (or LANs). The FDDI physical layer allows for up to 100 fiber repeaters in the ring, with a spacing of 2 km between repeaters. Thus, the size of the FDDI ring covers a perimeter of 200 km (more than 124 miles). This is an attractive technology for long-distance communication.

TABLE 28.3 Wireless technologies applications

Wireless Technology	Application
Digital Cellular	Voice, Data
Wireless LAN	Voice, Data, Video
Personal Communication System	Voice, Data, Video, Fax, Global Positioning

MOBILE COMMUNICATION

Almost by definition, mobile communication implies the use of wireless technologies. The traditional cellular technologies are quickly migrating from analog to digital signals that offer additional features and significantly enhanced security benefits. Older geosynchronous satellite communication systems are being replaced by low earth orbit satellite communication systems that can provide wireless coverage for the entire planet.

Wireless technology is based on the concept of having transmitters and receivers. The transmission of wireless signals falls into two categories: omnidirectional and directional. Omnidirectional signals propagate from the transmitter in all directions similar to the transmitter used for an AM or FM radio station. A directional signal is focused at the receiver. Using a combination of these two types of signals, many different applications of the technology are possible. Some of these applications are shown in Table 28.3.

To accompany the new wireless technologies, the IEEE 802.11 specifications provide a software framework on which to build. A new protocol, DFWMAC (Distributed Foundation Wireless MAC), was created to work in the MAC layer of the OSI network model. A modified version of Ethernet called CSMA/CA (Collision Sense Multiple Access/Collision Avoidance) is used to transmit data in the network.

TROUBLESHOOTING TECHNIQUES

The sophistication of the wide variety of telecommunication equipment requires expertise that is typically beyond that obtained in an ordinary electronics or engineering technology program. Fully developed telecommunication degree programs are now available that train the student in all aspects of the field, using state-of-the-art equipment, such as Optical Time Domain Reflectometers, network analyzers, and digital sampling oscilloscopes. Becoming a telecommunication technician or engineer would be a challenging and rewarding pursuit.

SELF-TEST

This self-test is designed to help you check your understanding of the background information presented in this exercise.

True/False

Answer *true* or *false*.

1. Using TDM, multiple data channels are sent over a single wire.
2. A multistage switch may block a signal from getting to its destination.
3. Frame relay uses fewer acknowledgment packets than packet switching.
4. FDDI is used for networks in small geographical areas.
5. Wireless communication requires transmitters and receivers.

Multiple Choice

Select the best answer.

6. Voice conversations are sampled using
   a. AM.
   b. FM.
   c. PCM.
7. A crossbar switch allows
   a. Limited switching between input and output.
   b. Full switching between input and output.
   c. Fixed-paths only between input and output.
8. In packet switching, packets may arrive
   a. Out of order.
   b. Not at all.
   c. Both a and b.
9. ATM cells are
   a. Fixed in length at 53 bytes.
   b. Fixed in length at 64 bytes.
   c. Variable in length.
10. How large can an FDDI ring be?
    a. 2 km.
    b. 100 km.
    c. 200 km.

COMPLETION

Fill in the blank or blanks with the best answers.

11. Twenty-four 64,000 bps channels are available in a(n) _____ carrier.
12. The bit pattern needed to connect input signal 3 to output signal 7 in the crossbar switch is _____.
13. When all switches are configured as straight-through in the multistage switch, the output signal sequence is _____.
14. When all switches are configured as crossover in the multistage switch, the output signal sequence is _____.
15. ATM is also called _____ relay.

FAMILIARIZATION ACTIVITY

1. Visit a local Internet service provider. Ask for a tour of their facility. What kind of telecommunication equipment do they have? How many individuals do they serve? What kind of service do they provide?
2. Search the Web for telecommunication equipment. Compare prices and features among similar equipment.

QUESTIONS/ACTIVITIES

1. How much does your local telephone provider charge for a leased 56K baud line? How much for a T1 connection?
2. What types of telecommunication technologies are in use at your educational site?

REVIEW QUIZ

Within 10 minutes and with 100% accuracy,

1. Describe the different telecommunication technologies.
2. Discuss reasons for choosing one technology over another.

UNIT V Microcomputer Systems

EXERCISE 29 Computer Environments
EXERCISE 30 System Teardown and Assembly
EXERCISE 31 Power Supplies
EXERCISE 32 Floppy Disk Drives
EXERCISE 33 The Motherboard Microprocessor and Coprocessor
EXERCISE 34 The Motherboard Memory
EXERCISE 35 Motherboard Expansion Slots
EXERCISE 36 Power-On Self-Test (POST)
EXERCISE 37 Motherboard Replacement and Setup
EXERCISE 38 Hard Disk Fundamentals
EXERCISE 39 Hard Drive Backup
EXERCISE 40 Hard Disk Replacement and File Recovery
EXERCISE 41 Video Monitors and Video Adapters
EXERCISE 42 The Computer Printer
EXERCISE 43 Keyboards and Mice
EXERCISE 44 Telephone Modems
EXERCISE 45 CD-ROM and Sound Card Operation
EXERCISE 46 Multimedia Devices

29 Computer Environments

INTRODUCTION

Joe Tekk walked off the elevator on the 11th floor of a high-rise office building. He was on a service call to one of RWA Software's clients, a small business with a network of 20 computers.

The office workers reported frequent erratic behavior by their machines. As soon as Joe walked into the office, he was concerned, because he saw a deep shag rug carpeting the floor. Joe got a shock when he grabbed the doorknob when entering the office and another shock when he touched a table as he sat down to check a computer. He checked the fan on several mini-towers; they were clogged with dust. One keyboard looked as if a drink had been spilled on it.

Joe spoke with Debbie Grant, the office manager. "Even though there are many conditions in your office environment that need to be corrected, I still do not see the cause of the problems you are experiencing. I'm going to go check some of the other offices on your floor."

Forty-five minutes later Joe returned. "I think I have good news, Debbie. After talking to your neighbors, who are having similar problems, I spoke with the building superintendent. He told me that the elevator was repaired recently, just over a week ago. Isn't that when your problems started?"

Debbie agreed. "Yes, they were here last Tuesday. They worked on the elevator for hours, because it got stuck between floors."

Joe continued. "Apparently, they rewired the elevator's control panel, and tapped into the circuit that feeds your office and the other three on this side of the building. I think the noise from the elevator is causing your problems. Let's try adding surge protectors to your equipment to see if that helps."

PERFORMANCE OBJECTIVES

Upon completion of this exercise, you will be able to

1. Design an environmental checklist for an assigned computer workstation.
2. Perform a computer environmental check of an assigned computer workstation.

> **BACKGROUND INFORMATION**

Before going further with the hardware aspects of personal computers, this is a good time to present some of the important requirements for a computer operating environment.

The major causes of the environmental problems that can occur with personal computers are

- High temperatures
- Dust
- Corrosion
- Magnetic fields
- Electrical noise
- Electrical power variations

These causes can be classified into two major environment areas: the **physical environment** and the **electrical environment.**

PHYSICAL ENVIRONMENT

Heat

Figure 29.1 shows the relationship between the **ambient temperature** and the **internal temperature.** The ambient temperature is the temperature of the surrounding air.

When a computer system has been turned off for some time, its internal temperature is the same as its ambient temperature. However, once a computer has been on for 15 minutes or more, its internal temperature is much higher than its ambient temperature. Depending on the type of computer, the temperature difference between ambient and internal can be as much as 40°F.

The recommended ambient temperature for a PC is

Turned off: 60 to 90°F
Turned on: 50 to 110°F

Excessive internal heat in a computer can cause many serious problems to occur. Some of these problems are illustrated in Figure 29.2.

In many cases, excessive internal heat is one of the major causes of computer failure.

You can do much to prevent the problems caused by excessive heat. For example, you can recommend that air conditioning be used when the computer is to be operated in a small construction office that does not normally have air conditioning. Some of the sources of excessive heat are listed in Table 29.1 and illustrated in Figure 29.3.

Power Cycling—Thermal Shock

When a computer or any other electrical system is turned on, it undergoes a rapid increase in temperature. This increase in temperature comes about because of the normal power loss produced by electrical circuits. This power loss results in the production of heat.

FIGURE 29.1 Ambient and internal temperature

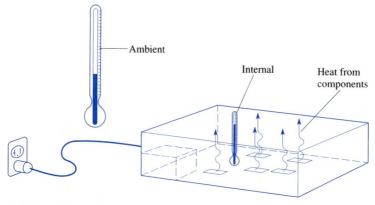

Note: Internal temperature is higher than ambient temperature.

FIGURE 29.2 Typical problems caused by excessive internal heat

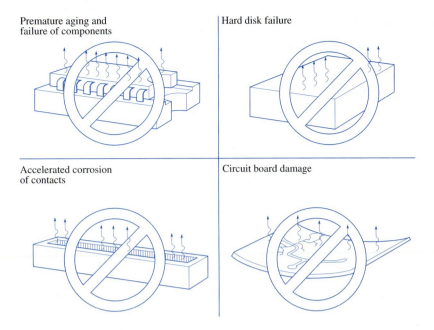

TABLE 29.1 Sources of excessive heat

Problem	Solution
High ambient temperature	Use air conditioning in the computer room or area.
Direct sunlight	Remove the computer from contact with the rays of the sun. This may be accomplished by moving the computer or by using drapes or other sun-blocking material.
Fan outlet and equipment vents blocked	Do not place the rear of the computer directly against a wall or other object. Do not cover vent holes with any material that would block the flow of air.

FIGURE 29.3 Sources of undesirable heat

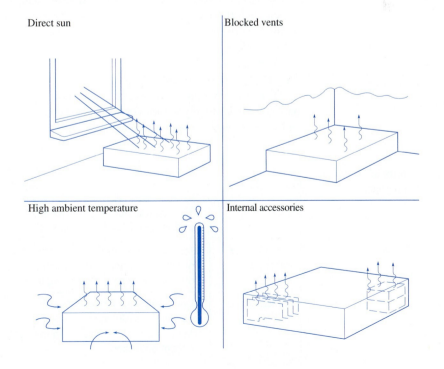

FIGURE 29.4 Concept of thermal shock

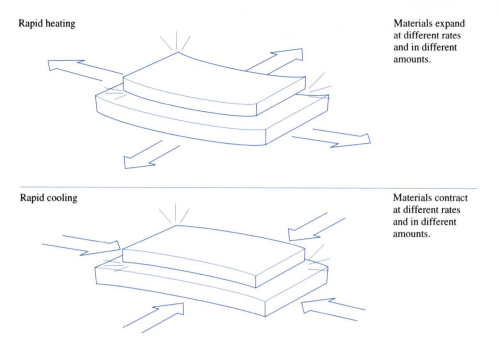

Thermal shock occurs when a rapid increase or decrease in temperature causes an undue mechanical stress on the electrical components that make up the computer circuits. Figure 29.4 illustrates the concept of thermal shock.

As shown in Figure 29.4, thermal shock is brought about by the mechanical expansion and contraction caused by changes in temperature. The greater the change in temperature, the greater the thermal shock. The amount and speed of contraction or expansion are also determined by the type of material. Every time you turn the computer on, it undergoes some degree of thermal shock. Consider, for example, coming into a cold office on a Monday morning and turning on your computer. In this case, the thermal shock may be quite large, because the computer undergoes a large increase in internal temperature. Table 29.2 lists some of the undesirable effects that can be brought about by thermal shock.

Rapid changes in system temperature can also cause problems with hard disk drives. As an example, if a hard drive is shipped during the winter and is then placed in a computer system (before having a chance to warm up), moisture can condense on the hard disks, rendering the drive useless. Some of the undesirable effects of thermal shock are illustrated in Figure 29.5.

There are several ways of preventing the harmful effects of rapid changes in temperature. These are listed in Table 29.3.

There is another point to consider about rebooting the computer by turning it off and then on again. Every time you turn your computer on, the power supply, hard disk drive motor, and other electrical components undergo surges of electrical current. Although a power

TABLE 29.2 Some undesirable effects of thermal shock

Action	Results
Chip creep	Rapid expansion or contraction of IC leads can cause them to gradually lift out of the chip sockets.
Circuit board foil separation	Because the material of the circuit board is different from the copper foil used to connect the circuits, the foil may separate and actually form hairline cracks, resulting in intermittent system operation.
Broken solder joints	Again, because of the differences in solder material and the material containing the solder, a broken solder joint can result from rapid temperature change in the system. This can lead to unreliable system operation.

FIGURE 29.5 How thermal shock affects computer components

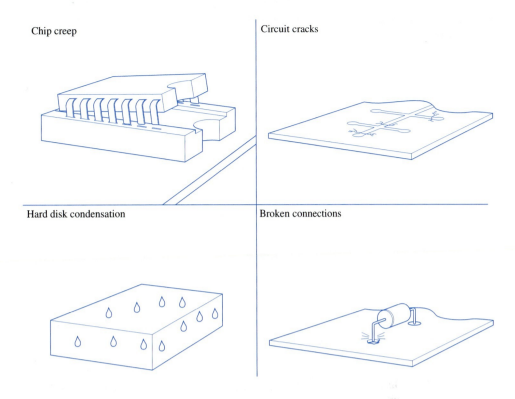

TABLE 29.3 Methods of preventing rapid changes in internal temperature

Method	Comment
Never turn the computer off.	Some computer users never turn off a computer. Doing this eliminates thermal shock. The belief here is that any long-term wear from constantly keeping the system on will be less than problems caused by thermal shock.
Warm the room and wait for the computer to reach room temperature.	Some users do not leave their computers on over the weekend or overnight. At the same time, during the winter, automated thermostats reduce the room temperature to around 50°F. When starting up in the morning, heat up the room first, allowing the computer to heat up slowly before turning it on.
Leave the computer on during the workday.	The computer should not be constantly turned on and off during the workday, even if there are long periods in the day when it is not being used.
Do not reboot by using the ON/OFF switch.	It is considered poor practice to reboot using the ON/OFF switch. If for any reason you need to reboot the system, use the Ctrl-Alt-Delete key combination (or shutdown) or a RESET button if your computer has one.

surge lasts for only a brief period of time, it can wipe out a component or power supply. Recall that when a lightbulb burns out, it usually does so just after you turn it on.

Effects of Dust

Dust can be a silent killer of your computer and its disk drives. This means that any computer system should be housed in as clean an environment as possible. There are some steps that can be taken to reduce the long-term effects of dust. One is not to burn materials (such as tobacco products) around a computer, and the other is to cover the computer with a dust cover when it is not in use.

Note: Never operate a computer with a dust cover on it because this will cause the computer to overheat.

TABLE 29.4 Methods of reducing the effects of dust

Method	Effect
Use dust covers.	Slows down the accumulation of dust inside the computer system.
Close windows.	Reduces some of the sources of dust.
Prohibit indoor fires.	Not allowing the burning of tobacco or other products indoors greatly reduces the amount of dust in the computer.
Keep foods away from work area.	Certain foods, especially the "snack" variety, can produce small particles that find their way into the computer system.
Use a cleaning schedule.	Removes dust that has already accumulated on and in the computer system.

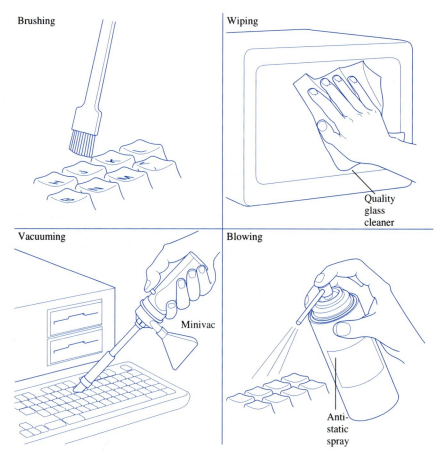

FIGURE 29.6 Methods used for removing computer dust

One of the adverse effects of dust is that, with time, it builds up a thick coating over the internal chips and other components inside your computer. This thick layer of dust can act as a heat insulator, resulting in a higher-than-normal internal temperature in your components. Another problem with dust is that even the smallest dust particle can permanently damage a disk. Table 29.4 lists some of the actions that can be taken to reduce the bad effects of dust, and Figure 29.6 shows some of the methods used to remove dust from a computer.

Effects of Corrosion

Corrosion inside a computer is the process of metal pin connectors, wires, interface cards, and chip pins undergoing chemical changes. These chemical changes can gradually eat away critical parts of the circuit, causing improper operation of the system. The main sources of corrosion are chemicals in the atmosphere and contact between computer parts and human hands.

If your computer is in an area that is subject to atmospheric pollution such as smog, you may need to use high-quality air filters on air-intake ducts leading into the area where the computers are kept. If filters are used, it is important to change them regularly in order to keep them working at their peak performance.

The sulfates found in tobacco smoke have a devastating corrosive effect on computer circuits. The burning of tobacco products should never be allowed in a closed area with a computer. Circulating air does not help here, because the increased circulation could actually cause more of the corrosive by-products of tobacco smoke to get into the computer.

Never touch metal computer parts directly with your bare hands. Doing so will leave a thin coat of your natural skin oil on the parts. This oil residue could then promote **galvanic corrosion.** Galvanic corrosion creates the effect of a small battery, causing a tiny current to flow between the metals. This process gradually eats away the metal, causing computer failure.

ELECTRICAL ENVIRONMENT

Magnetic Fields

When you were introduced to floppy disks, you were presented with some of the bad effects of magnetic fields on floppy disks. The best way to reduce the effects of magnetic fields is to be aware of where they are found and then remove the sources. Figure 29.7 shows some of the most common sources of magnetic fields that may be found around a computer station.

Electrical Noise

Electrical noise can be generally classified as any undesirable and sometimes unpredictable random changes caused by sources of electricity. There are several types of electrical noise, as characterized in Table 29.5.

Figure 29.8 illustrates various kinds of electrical noise and their relationship to the computer. All the types of electrical noise listed in Table 29.5 can have adverse effects on your computer or on other equipment such as television and radio receivers. In terms of your

FIGURE 29.7 Common sources of magnetic fields

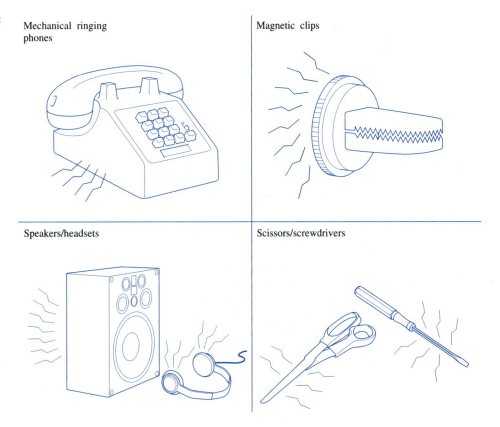

TABLE 29.5 Types of electrical noise

Type	Definition
EMR	Electromagnetic radiation is a general category of electrical noise that affects computers and other electronic equipment. It can be radiated or otherwise transmitted through space or conducted along wires or other conductors.
EMI	Electromagnetic interference is a more specific category of electrical noise that occurs in the frequency range of 1 Hz to about 10 KHz.
Transient EMI	Transient electromagnetic interference is a short-term undesirable electrical response that may appear when equipment is first turned on or off. This includes power line transients and electrical discharges from lightning.
Internal EMI	Internal electromagnetic interference is caused by the internal circuitry of the computer, such as faulty chips or improperly installed wires on the motherboard or other system components.
RFI	Radio-frequency interference is a more specific category of electrical noise that occurs at a frequency of 10 KHz and above.
Conducted RFI	Radio-frequency interference is sometimes conducted along the power line from the PC out into the power system.
Radiated RFI	Radiated radio-frequency interference is sometimes transmitted from your computer into the surrounding space.
ESD	Electrostatic discharge is like the static electricity that builds up in your body when you walk across a carpet on a dry day and is released when you touch a grounded metal object.

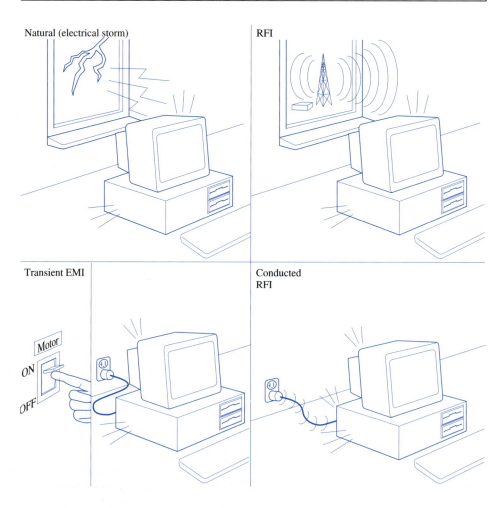

FIGURE 29.8 Various kinds of electrical noise

FIGURE 29.9 Causes of electrical interference

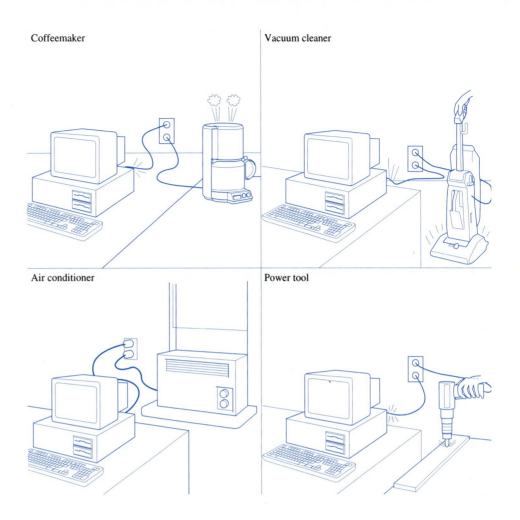

computer, electrical noise can cause the system to lock up, produce inconsistent results, damage information on disks, cause "garbage" to appear on the screen, or produce some other "mysterious" type of improper operation.

The effects of electrical interference are not always consistent and not always easy to detect. The best rule when working with them is to reduce the possibility of any of them happening. Figure 29.9 illustrates some of the causes of electrical interference.

Reducing Electrical Noise

Because of the electromagnetic radiation (EMR) that can be caused by computers, the FCC (Federal Communications Commission) has established specifications for the legal amount of radiated noise allowed to be emitted from a personal computer. The FCC has created two categories for personal computers: Class A and Class B. The Class A category is for industrial computing devices in commercial and business use and not usually sold to the general public. Class B consists of consumer computing devices. The PC is considered a Class B computing device. All PCs and compatibles must meet FCC EMR standards. Therefore, all PCs that are legally sold in the United States for personal or industrial use must be approved by the FCC. Each computer is required to have a statement affixed to it stating that it meets or exceeds the minimum requirements of the FCC.

Some of the methods of reducing the effects of electrical noise are listed in Table 29.6. If you check out a computer system using this table as a guide, you will have reduced or eliminated the majority of the causes of electrical noise.

Power Source Problems

Probably one of the most important environmental concerns for your computer is a reliable and "clean" source of electrical power. Computers are more sensitive to power line variations than most other electrical equipment. Office or room lights can tolerate wide variations in the power source with very little adverse effect on operation. This is not the

TABLE 29.6 Methods of reducing electrical noise effects

Method	Effect
Keep computer parts inside the metal enclosure.	The switching power supply is a big source of EMI. The computer's metal case helps shield against EMR and is required in order to meet minimum FCC standards.
Cover openings with metal inserts.	Do not leave openings on your computer case. This includes the openings in the rear of the system where the peripheral cards are to be inserted. Covering these openings helps reduce the effects of EMR.
Use metal honeycomb screens over cooling vents.	This practice also helps reduce the effects of EMR.
Use shielded cables.	Using shielded cables greatly reduces the effects of EMR. The shield on the connecting cable helps prevent radiation from reaching the cable.
Ground cables.	Make sure that interconnnecting cables, such as those between your computer and other devices such as the printer, have their ground wires attached as required by the manufacturer. Doing this helps reduce or eliminate a major source of electrical interference.
Use fiber optics where and when available.	One of the best methods of reducing electrical interference is to use properly installed fiber optics, which do not radiate any undesirable EMR.
Turn off cordless telephones.	Cordless telephones are actually tiny transmitters of EMR. They should be removed from the vicinity of any PC.
Avoid any high-speed digital circuits or equipment.	Digital equipment is also capable of transmitting EMR that may interfere with the operation of your computer. This includes digital thermostats, certain types of burglar alarms, and industrial controllers.
Remove any type of radio transmitters.	Radio or televison transmitting equipment is, by its nature, a major source of EMR. These systems should be removed from the vicinity of any PC.
Turn off certain types of communication receivers.	Many communication receivers use what is called the "superheterodyne" principle in their operation. These include standard AM and FM radios. These receivers, which contain a circuit called a local oscillator, can emit EMR if not properly shielded. These devices should be removed from the vicinity of a computer.
Avoid electrical machinery.	Electrical machinery, such as electrical motors, air conditioners, compressors, and heaters, is a potential source of electrical interference. Again, the computer should not be in the immediate vicinity of this type of equipment.

case with a computer. The major types of power line problems and their usual causes are listed in Table 29.7 and illustrated in Figure 29.10.

Reducing Power Problems

If you determine that power line problems are common in the area where your computer will be kept, there are some methods you can use to prevent such problems. One is conditioning the power line used by the computer, and the other is to have a power supply backup system. Both methods are illustrated in Figure 29.11. Various kinds of power-protection schemes are listed in Table 29.8.

An uninterruptible power source (UPS) can also provide protection against voltage spikes and electrical noise on the power line. It will maintain electrical power to your computer for

TABLE 29.7 Power line problems and their causes

Type of Problem	Cause(s)
Brownouts	Lowered output voltage from the wall outlet caused by an over-demand for electrical power. This overdemand can be caused by anything from a whole region using more power for air conditioning on a hot day to someone plugging in a coffeemaker on the same outlet as your computer.
Blackouts	A total loss of electrical power from the wall outlet. Blackouts are caused by a variety of conditions, such as power lines knocked down by natural or human-made disasters.
Power transients	Large and potentially dangerous voltage spikes appearing on the power line. These can be caused by lightning strikes or by the turning on or off of industrial machinery in the same building as your computer.

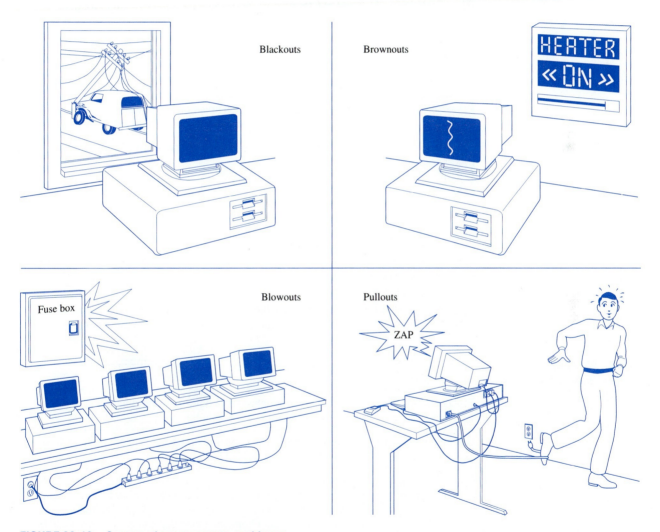

FIGURE 29.10 Causes of power source problems

a limited time when it loses its own source of power. The length of time the UPS should maintain power depends on the conditions of the area where the computer is kept. If power failures are frequent, a UPS that will supply emergency power for several hours should be considered. If power failures are less frequent, a less expensive UPS that supplies power for a few minutes may be all that is necessary. The few minutes could be used to shut down or to store data to a disk when the power has failed.

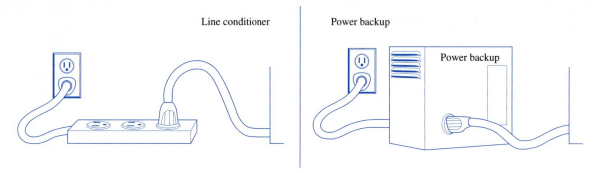

FIGURE 29.11 Methods of reducing power problems

TABLE 29.8 Types of power-protection schemes

Method	Description
Power isolators	Good at filtering out high-frequency voltage spikes. Not good at reducing the effects of a slowly changing condition, such as a brownout.
Power regulators	Good at helping maintain a constant line voltage for line voltage variations of not much more than 10%. Not effective against power spikes and brownouts.
Power filters	Good at removing electrical noise on the power line such as EMI and RFI. Does not stop voltage spikes or help with brownouts.
Uninterruptible power source (UPS)	A power source that keeps your computer on when the power is out or there is a brownout.

Electrostatic Discharge

Surprisingly, people and office furniture can accumulate large charges of static electricity. These charges can amount to several thousand volts. If this kind of voltage is discharged through your computer, it can severely damage some of the sensitive integrated circuits. Such discharge is referred to as **electrostatic discharge** (ESD). Several methods of preventing or reducing the effects of electrostatic discharge are listed in Table 29.9.

TABLE 29.9 Methods of reducing or eliminating the effects of ESD

Method	Discussion
Touch a conductive table top.	Touch the conductive table before touching anything else when first sitting down at the computer.
Use antistatic spray.	Antistatic sprays for rugs and computer equipment help reduce and control the effects of ESD.
Consider using a room humidifier.	Be cautious here. Many humidifiers help reduce the effects of static electricity but may produce undesirable corrosive side effects from the residue they leave on computers. You may have to run such a humidifier with distilled water. This could be an expensive solution.
Use antistatic floor mats.	Placing antistatic floor mats under chairs is a popular practice in offices.
Use antistatic cleaning solutions on floors.	Floors are usually cleaned with antistatic solutions in computer-manufacturing areas. One application usually lasts for several months.

TROUBLESHOOTING TECHNIQUES

Determining the cause of an environmental problem requires you to behave like a detective. You need to gather all the facts. When did the equipment stop working? Was there a major change in procedure recently?

Examine the equipment carefully. Are the connectors clean? Are they snug in their sockets? Is there adequate ventilation around the equipment? Have any antistatic precautions been taken?

Observe the individuals operating the equipment. Are they doing anything that could contribute to the problems being experienced?

Look for obvious signs that equipment has been mistreated. For example, an employee may have drilled a hole in the plaster to hang a new picture on the wall beside his computer. A coating of plaster dust on his computer would show he did not cover it when drilling the hole. There is no telling what type of physical or other damage can result from careless activities.

SELF-TEST

This self-test is designed to help you check your understanding of the background information presented in this exercise.

True/False

Answer *true* or *false*.

1. The temperature inside a computer that has been powered up is always higher than the ambient temperature.
2. The speed at which the operating temperature of a computer changes does not affect the operation of the computer.
3. Having the computer in contact with direct sunlight is not a problem as long as the room is kept cool.

Multiple Choice

Select the best answer.

4. An operating computer should not be placed directly against a backing wall because
    a. Of static electricity.
    b. The exhaust fan may become blocked.
    c. The wall may catch on fire.
    d. None of the above.
5. A mechanical stress on the computer components brought about by a rapid change in temperature is called
    a. Temperature runaway.
    b. Thermal runaway.
    c. Thermal shock.
    d. Thermal stress.
6. Undesirable electrical disturbances that cause improper computer operations may be caused by
    a. Electrical machinery.
    b. Cordless telephones.
    c. Digital thermostats.
    d. All of the above.

Matching

Match a description of electrical noise on the right with each stated type on the left.

7. RFI
8. EMI
9. EMR

a. Occurs in the frequency range of 1 Hz to 10 KHz.
b. General category of electrical noise.
c. Occurs in the frequency range of 10 KHz and above.
d. None of the above.

Completion

Fill in the blanks with the best answers.

10. A power _____ is good at filtering out high-frequency voltage spikes.
11. A(n) _____ can maintain a constant power source to your computer during an electrical outage.
12. The effect of corrosion is to cause a(n) _____ change of the material inside the computer.

FAMILIARIZATION ACTIVITY

This section contains an **environmental checklist.** This checklist is a guide to help you check out possible physical and electrical environmental hazards at a computer station. Your instructor may assign you and your lab group to an actual computer area located outside of your normal lab. This activity may be done during your normal lab time or as an outside assignment.

Keep in mind that doing an environmental check of a computer station is not always an exact activity with absolute yes or no answers. Much of what you will be doing will depend on your judgment. What is important is that you and those using the computer system are aware of what environmental factors may cause the computer to malfunction.

To use the environmental checklist, look at the "Item to Check." Then, in the "Problem" column, indicate if this item is a potential problem: Y means yes, there is a potential problem; N means no, there is not a potential problem; and ? means you cannot tell from what you are observing. The "Comments/Action" column is provided so you can record any necessary corrective action(s) or to make comments on further studies that need to be made before you can decide if there is a particular problem.

You may not have room on the checklist to write all your comments or recommended actions, so use a separate sheet of paper as needed.

After completing this checklist, turn it in to your instructor in order to complete performance objective 2 for this exercise.

Environmental Checklist

Item to Check	Problem			Comments/Action
	Y	N	?	
Ambient temperature				
Direct sun				
Blocking of fan and vents				
Power cycling				
Dust covers needed				
Windows closed				
Indoor smoke				
Food/drinks				
Potential magnetic fields				
EMI				
RFI				
ESD				
Power source				
Corrosion				

COMPUTER TECHNOLOGY
Past and Present

ENIAC computer
(Courtesy of Unisys Corporation)

Cray supercomputer
(Courtesy of Carlye Calvin)

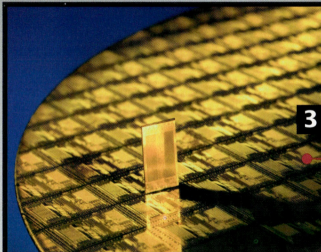

A computer chip with a silicon ingot
(Courtesy of PhotoDisc, Inc.)

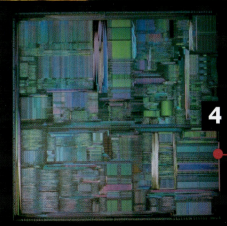

Pentium Pro processor chip
(Courtesy of Intel Museum Archives & Collection)

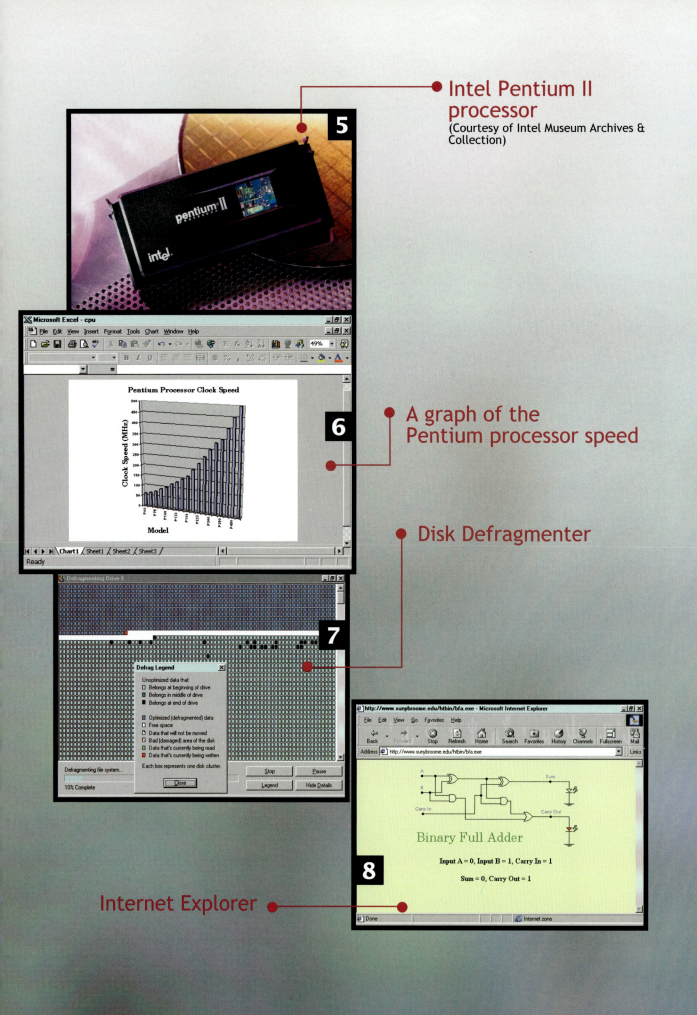

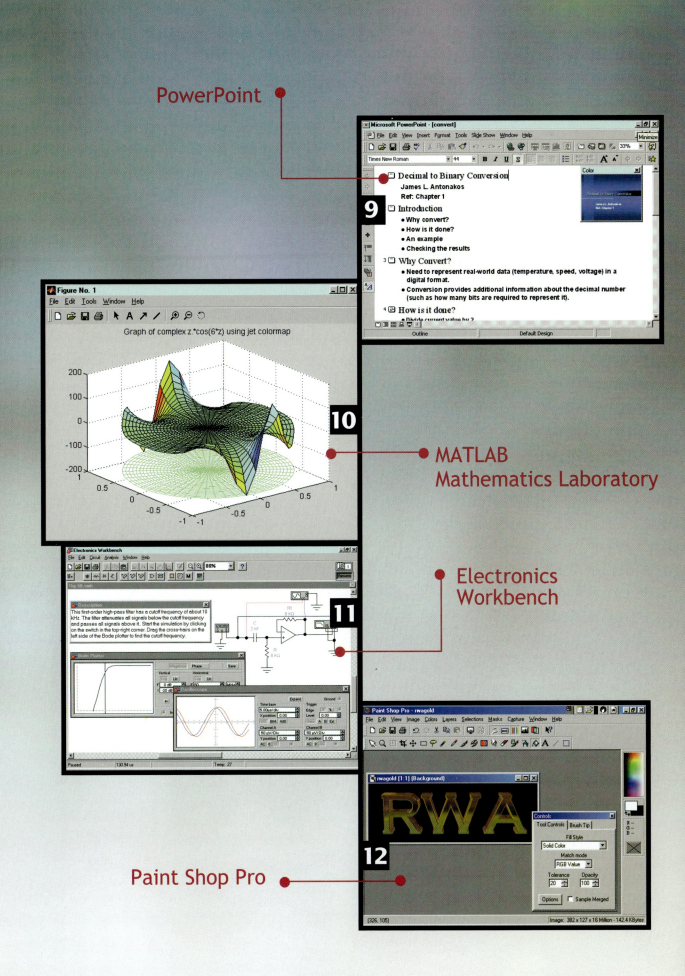

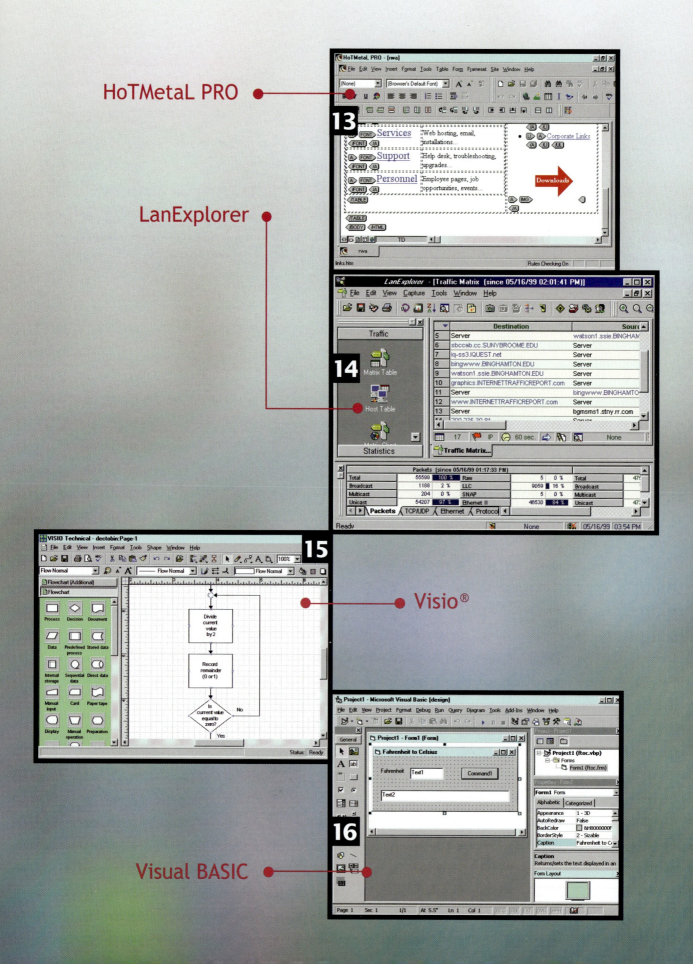

QUESTIONS/ACTIVITIES

Answer the following questions regarding the results of your environmental check in the familiarization activity for this exercise.

1. What major environmental hazards, if any, did you discover? What did you recommend as a solution?
2. What areas were you not sure of? Why?
3. Did you recommend that the computer(s) always be left on? If this was not done, what was the reason?
4. Did you discover any sources of magnetic fields? Explain.
5. When was the last time the computers were cleaned? Is there a regular cleaning program?
6. State the preventive steps that you found were used to reduce or eliminate the effects of ESD.

REVIEW QUIZ

1. Design an environmental checklist for an assigned computer workstation.
2. Perform a computer environmental check of an assigned computer workstation.

30 System Teardown and Assembly

INTRODUCTION

Joe Tekk was pushing a cart of computer components down the hallway at a local community college. He was guest lecturing in a computer hardware laboratory. The purpose of the lab was to completely build a computer from individual parts, install Windows NT, and connect to the school's network. The cart contained all the components of a multimedia PC, from power supply to CD-ROM to adapter cards.

Joe was told the lab was a three-hour lab. He looked at all the components on the cart. He hoped there was enough time.

It took the students just over two hours to do everything. They spent the third hour surfing the Web.

PERFORMANCE OBJECTIVES

Upon completion of this exercise, within 20 minutes and with 100% accuracy, you will be able to remove the cover of a personal computer and identify the following major sections:

1. Power supply
2. Floppy disk drive(s)
3. Hard disk drive(s)
4. CD-ROM drive
5. Motherboard
6. Expansion cards

BACKGROUND INFORMATION

Up to this point in the exercises, we have concentrated on specific topics related to various skills needed for computer operation. In this exercise, you are given an overview of the complete computer system. Here you discover how to take the computer system apart, identify the major sections, and explain the purpose of each of these sections.

SYSTEM OVERVIEW

Figure 30.1 shows the major sections of a personal computer, and Table 30.1 explains the purpose of each of these major sections.

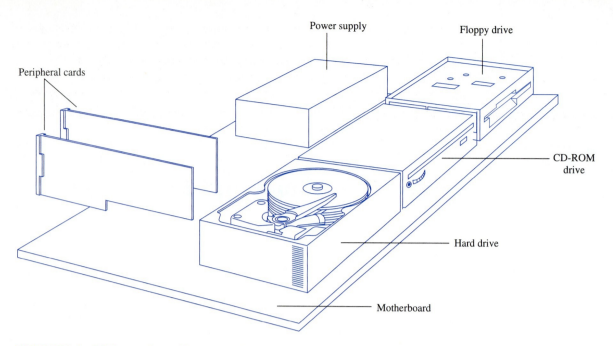

FIGURE 30.1 Major sections of a personal computer

TABLE 30.1 Purpose of each major computer section

Section	Purpose
Power supply	Converts the 120-V AC electricity from the line cord to DC voltages that are needed by the computer system.
Floppy drive(s)	Allows information to be stored and read from removable floppy disks.
Hard drive	Allows information to be stored and read from nonremovable hard disks.
CD-ROM drive	For reading CDs.
Motherboard	Holds the microprocessor and memory circuitry and provides expansion connectors.
Peripheral cards	Allow accessory features and interconnect the computer to input and output devices such as drives, printers, monitors, and other external devices.

DISASSEMBLY PROCEDURES

The process of physically disassembling and reassembling the personal computer is not difficult. It is, however, a process you should experience before going further into other specific sections of the computer. You will find that the physical arrangement of the major components (such as the power supply) is similar in almost all computer systems, even those from different manufacturers.

When taking a computer apart, it pays to be organized. Keep these points in mind as you proceed with your disassembly:

- For safety, boot the machine and enter the BIOS setup program. Record all important BIOS settings in case they are lost during the disassembly procedure.
- Turn power off and unplug the computer (both ends of the power cord).

- Draw a diagram of the motherboard. Be accurate in regard to the position and number of expansion connectors, which adapter cards are plugged in, and what cables are plugged into the motherboard. Draw the diagram large so that you can make notes on it.
- Keep all screws and loose parts in a container. If different types of screws are removed, place them in separate containers and make notes to yourself about where they came from.
- Examine each off-board connector before and after it is removed. Look for pin-1 orientation marks (red or blue stripe on floppy drive, hard drive, and CD-ROM cables) or other labeling that helps distinguish or identify the connector. Many of the small two-conductor twisted pairs that go to the panel buttons and indicators have writing on them that identifies their function (Turbo LED, Reset switch, Speaker) as well as color-coded wiring.
- If it is necessary to remove the hard drive cage to get the motherboard out, be sure to disconnect the power and signal cables to the drive before removing the cage. For multiple-drive systems, note the location of each drive so that they are returned to their original locations.
- When disconnecting the floppy drive cable, note the relationship between the red or blue stripe on the ribbon cable and the drive connector. It is important to know where pin 1 is. In a two-drive system, watch for the twist in the cable to identify drive A:.
- When pulling ribbon cable connectors (such as the 40-pin IDE connector) out of their sockets, do not yank them out by pulling on the ribbon cable itself. Instead, grab the sides of the connector and gently rock it back and forth as you pull it out. If the connector socket has a label (primary IDE, secondary IDE, COM1, and so on), read it and verify that the cable in fact goes to the indicated hardware. If the connector is not labeled, write your own label on your motherboard diagram.
- Note the shape of the keyed side of each power connector as it is removed. The important thing to remember is that the two sets of black conductors should always be in the middle of the ten power conductors.
- When removing the motherboard, look for several screws or plastic standoffs that are used to hold it in place. Do not set the motherboard down or handle it without enclosing it in an antistatic bag.

ASSEMBLY PROCEDURES

Paying attention to details will save time and effort when assembling your computer. Here are some tips you may find useful:

- Connect the power cable to each component (hard drive, CD-ROM drive, floppy drive, motherboard). Nothing is more embarrassing than forgetting to connect the power cable.
- Do not force any connections. If a cable is not seating properly, examine the cable and your diagram to help determine the problem.
- Put the hard drives back in their cages, but do not tighten the screws until the connectors have been seated.
- Check the primary and secondary IDE connectors to be sure that all 40 pins have been seated properly. It is very easy to make an improper connection and leave pins unconnected.
- Use the proper connector on the floppy drive cable to reconnect the floppy drive(s). This is especially important in a single-drive system.
- Secure each expansion card with the retaining screw.
- If the fan was removed from the processor during teardown, make sure it is reattached to the processor. Forgetting to do so may result in erratic behavior that is difficult to diagnose unless you happen to notice that the processor is really hot or that the fan is missing.
- Save the off-board front panel connections (such as the Turbo LED) for last as these are not essential to proper operation. If the machine has been reassembled correctly it will boot normally without them—you just will not see the hard drive light flash or be able to reset the machine or hear the speaker. The advantage of saving them until last is if the machine fails to boot, it cannot be attributed to these connections.
- If necessary, restore the BIOS settings recorded during teardown.

OTHER CONSIDERATIONS

If a computer is being assembled from scratch, and not from a previously torn down machine, there are probably a million reasons why it may not work when you turn it on. All of your skills as a troubleshooter are required to determine the cause of the problem. Windows may provide some assistance by telling you which interrupts or I/O devices are assigned improperly or that you may need to reinstall drivers for a particular device.

If the motherboard comes without memory, consult the motherboard manual for the type of memory used and how to install it. Typically, when memory is added or removed from the motherboard, the BIOS power-on self-test sees the change and may require action. To install a SIMM, place it into the SIMM connector at an angle. Push the SIMM slowly forward from each end until it is straight up in the socket. Small keying holds on each end of the SIMM should snap into place if the SIMM is seated properly.

If the motherboard comes without a processor, locate the appropriate Socket 5, Socket 7, or Slot 1 processor and install it. Jumpers on the motherboard are used to select the clock speed of the processor. As before, attach a cooling fan to the processor. *Note:* Pentium II and Pentium III processors, which come in the Slot 1 style, contain their own onboard cooling fan.

TROUBLESHOOTING TECHNIQUES

Whenever components from different manufacturers are mixed together during construction of a new computer, there will inevitably be times when two pieces do not fit well with each other. It may be as simple as a missing hole for a second screw on the hard-drive cage. It may be a serious problem, such as the power supply covering the RAM sockets or the motherboard mounting holes not aligning with the insulated supports on the chassis.

No matter what the situation, try to find an acceptable solution. You may have to be creative, such as making a custom mounting bracket. You may need to exchange one part for another. In the end, your efforts will be rewarded with a working system and the experience of having done it yourself.

SELF-TEST

This self-test is designed to help you check your understanding of the background information presented in this exercise.

True/False

Answer *true* or *false*.

1. All computers have three major sections: the main unit, the monitor, and the printer.
2. The purpose of the power supply is to supply power to the computer when there is a blackout.
3. A peripheral card is used to connect the printer to the main part of the computer.

Multiple Choice

Select the best answer.

4. The section of the computer that holds and electrically interconnects the major sections of the computer system is called the
   a. Power supply.
   b. Big board.
   c. Motherboard.
   d. Peripheral card.

5. One difference between a hard drive and a floppy drive is that in a hard drive, you
   a. Cannot remove the disks.
   b. Can only read information, not write it.
   c. Can only write information, not read it.
   d. None of the above.
6. When unplugging a ribbon cable
   a. Pull as hard as you can on the cable until it comes out.
   b. Pry it off with a screwdriver.
   c. Gently rock the connector back and forth as you pull up from each side.
   d. None of the above.

Matching

Match one wiring cable on the right with each item on the left.

7. Floppy drive cable
8. Hard-drive cable
9. Speaker wires

a. 40-pin connector, no twist in cable.
b. 34-pin connector, cable has a twist.
c. Two-conductor twisted pair.
d. None of the above.

Completion

Fill in the blanks with the best answers.

10. The _____ conductors on the motherboard power supply connectors should be in the middle of the ten conductors.
11. Erratic behavior in the PC may be the result of a missing _____ on the processor.
12. Before disassembling a computer, you should always unplug the _____ _____ to ensure that no electrical power can get to the computer.

FAMILIARIZATION ACTIVITY

In this activity you will actually remove the cover of your assigned computer and identify the following:

- Power supply
- Motherboard
- Peripheral cards
- Floppy disk drive(s)
- Hard disk drive(s)
- CD-ROM drive

Make sure you refer to the section of the background material that has the closest relationship to the computer with which you are working. You should have a large, flat, smooth surface upon which to work. A container such as a small, clean cup should be available for storing small pieces of hardware (such as screws). You may want to place a clean towel or other soft material over your workspace. Always remember to take your time and work carefully with the computer. Be sure to follow all the proper safety procedures.

1. Using the proper procedure for disassembling your assigned computer, remove its cover and identify the major sections for your lab partner(s).
2. Using the proper procedure for assembling your assigned computer, replace its cover.

QUESTIONS/ACTIVITIES

1. If your computer system was different from any of those discussed in this section, state how it was different.
2. Why do you think it is recommended that you unplug both ends of the power cord before disassembling the computer?
3. Name the major sections of the computer and state the purpose of each.

REVIEW QUIZ

Within 20 minutes and with 100% accuracy, remove the cover of a personal computer and identify the following major sections:

1. Power supply
2. Floppy disk drive(s)
3. Hard disk drive(s)
4. CD-ROM drive
5. Motherboard
6. Expansion cards

31 Power Supplies

INTRODUCTION

Joe Tekk was examining a computer that had suddenly stopped working. Nothing at all happened when the power was applied. Even the fan on the power supply stayed off. This bothered Joe. He took the cover off and gave the entire computer a careful visual inspection. When he looked at the display adapter card, he saw that the card was pulled slightly out of its socket, with its connector pins at a slight angle to the motherboard connector. The mounting-bracket screw that should have held the card in place was missing.

Joe reseated the card and put a screw in the card's mounting bracket. He plugged the power cord into its socket and turned the computer on. It booted normally.

Later, Joe discovered that the video monitor for the machine had been changed recently, and there had been difficulty getting the VGA connector disconnected from the display adapter card. Joe reasoned that, with no screw to hold it in place, the card had been yanked out of position during the struggle with the VGA connector.

PERFORMANCE OBJECTIVES

Upon completion of this exercise, within 20 minutes and with 100% accuracy, you will be able to

1. Remove and replace the computer power supply, using proper safety procedures.
2. Explain why different computers require different power supplies.

BACKGROUND INFORMATION

All personal computers get their electrical energy from 120-V AC wall outlets. Alternating current is used by the power company because it is more economical to transmit electricity this way. However, the circuits used by your computer must have a steady low voltage called a DC voltage.

It is the job of the computer **power supply** to convert the 120-V AC to a low and steady DC voltage that can be used by the tiny circuits inside the computer. Low-voltage DC is used for computers because it is easier to control. The function of the power supply is illustrated in Figure 31.1.

FIGURE 31.1 Function of the power supply

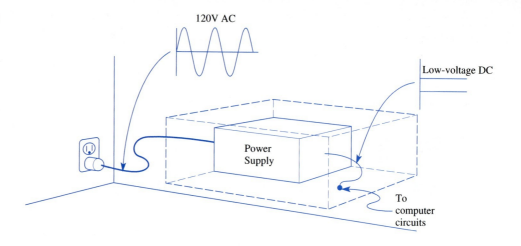

POWER SUPPLY CHARACTERISTICS

The type of power supply used in personal computers is called a **switching power supply**. A switching power supply converts the 60-Hz power line frequency into a much higher 20,000-Hz frequency. This higher frequency allows the power supply to use much smaller and more economical filtering circuits and transformers. The higher frequency used in the power supply is switched ON or OFF at a very rapid rate, according to the requirements of the system. This switching is used because it is a very efficient system for regulating the electrical energy required by the computer system, resulting in less heat loss from the power supply.

Table 31.1 lists the three electrical characteristics of power supplies.

The power used by your computer has several different output voltages available to service the requirements of various internal sections. Figure 31.2 shows a typical power supply and some of the standard output voltages along with their typical maximum current ratings.

TABLE 31.1 Electrical characteristics of power supplies

Electrical Unit	Meaning
Current	The amount of movement of electrical charge. Current is measured in amps (A). The letter symbol for current is I.
Voltage	The amount of electrical potential that can cause a current flow. Voltage is measured in volts (V). The letter symbol for voltage is E.
Power	The amount of electrical energy. Power is measured in watts (W) and is calculated by $P = IE$, where P is the power in watts, I is the current in amps, and E is the voltage in volts.

FIGURE 31.2 Typical output voltages

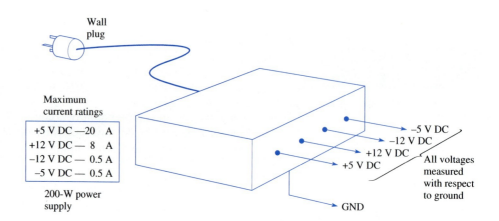

FIGURE 31.3 Circuits using the power supply

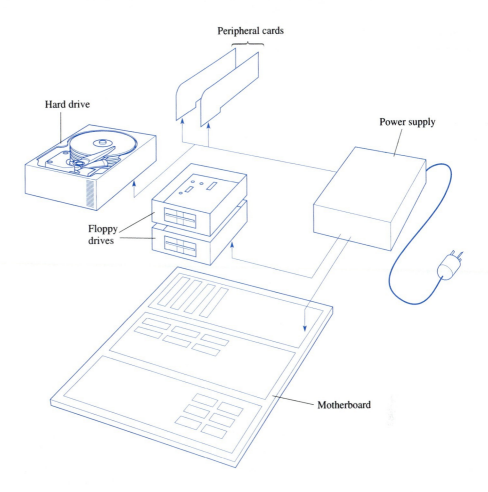

The power supply inside your computer must supply power to the following sections:

- Motherboard
- Floppy disk drives
- Hard disk drive
- Peripheral cards

These requirements are illustrated in Figure 31.3.

Different circuits inside your computer use different voltages. For example, peripheral cards use +5 V. It's important that these voltages remain steady. The amount of current delivered by each voltage from the power supply will depend on the number of circuits and disk drives the power supply must service. The more of these circuits and drives, the more current the power unit must supply and the greater will be the power demand, as illustrated graphically in Figure 31.4. This is the reason why the same power supply is not used in all computers.

As you can see from Figure 31.4, the more features that are added to the inside of a computer (including external disk drives), the greater will be the power requirements of the power supply. If the power requirements of the computer exceed those of the power supply, the power supply will shut down and the system will not operate. The solution to this is to replace the existing power supply with a power supply of a higher power (wattage) rating. Always use the following rule when replacing power supplies:

Use the same voltage values and a power rating that is the same or larger.

Figure 31.5 shows the various connectors used by some of the most common computers. Since these are *switching* power supplies, never operate the power supply without it being plugged into the motherboard and with at least one disk drive connected. This means that you should never remove a computer power supply, place it on your test bench, plug it in,

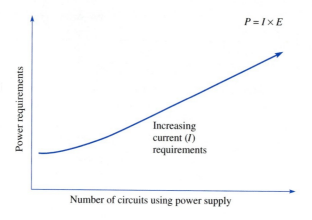

FIGURE 31.4 Relationship of system expansion and power requirements

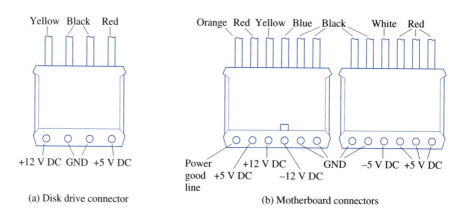

FIGURE 31.5 Various voltage connectors from the power supply

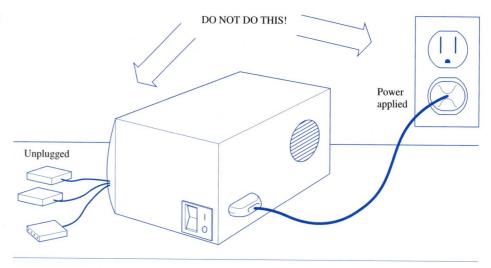

FIGURE 31.6 Possible damage to a computer power supply

and expect to measure its output voltages. Doing this will not only give you incorrect readings, it could also damage the power supply. This is illustrated in Figure 31.6.

Table 31.2 lists the different power supply capabilities for some of the most common types of computers.

Table 31.3 lists the voltage outputs and compares maximum current outputs of a 63.5-W power supply with that of a 200-W power supply.

As Table 31.3 shows, the power rating of a power supply indicates the maximum amount of current each of the output voltages can deliver. The power rating is also the sum of the individual maximum power outputs of each of the voltages. Note that the 200-W power supply is actually a 204.5-W unit.

TABLE 31.2 Common power supply power ratings

Power Rating	Computer Type
63.5 W	PC
130 W	XT
192 W	AT
94 W	PS/2 model 50
225 W	PS/2 models 60 and 80
105 W	PS/2 model 70

TABLE 31.3 Comparison of maximum current outputs

	63.5-W Unit				200-W Unit	
Voltage	Maximum Current	Power		Voltage	Maximum Current	Power
+5 V	7 A	35 W		+5 V	20 A	100 W
+12 V	2 A	24 W		+12 V	8 A	96 W
−5 V	0.3 A	1.5 W		−5 V	0.5 A	2.5 W
−12 V	0.25 A	3 W		−12 V	0.5 A	6 W
Total Wattage		63.5 W		Total Wattage		204.5 W

TROUBLESHOOTING TECHNIQUES

Keep the following tips in mind when working with power supplies:

- A power supply that suddenly quits working has done so for a specific reason. Try to recall everything that happened to the computer since the last time the power supply worked. For example, it is very easy to forget that you bumped the machine by accident when it was off, or that you just installed additional motherboard cache and managed to plug the ICs in backward.
- Erratic behavior of your computer could be a sign of a failing or overloaded power supply. There may be a significant amount of ripple voltage on the DC outputs, much larger than that tolerated by the digital circuitry. You may want to examine the outputs (with power on) with an oscilloscope, or a DMM, to determine whether the voltage level is constant and within acceptable limits.
- Troubleshooting a failed power supply down to the bad component or components requires great skill, knowledge of power supply theory (rectification, filtering, current limiting, and so on), a suitable set of test equipment, and plenty of time. It may be a lot easier to simply purchase a new power supply with a higher power rating.
- If a bad system component (on the motherboard or in any of the adapter cards) caused the power supply to fail, a new supply may fail also. Always check the system components when replacing a power supply. A good visual may help identify the problem (a burned trace or pad, an out-of-alignment connector). Adapter cards may need to be removed and reinstalled one by one to find the card that is at fault.

SELF-TEST

This self-test is designed to help you check your understanding of the background information presented in this exercise.

True/False

Answer *true* or *false*.

1. The circuits inside a computer use steady, low-voltage DC.

2. It is the job of the computer power supply to convert the 120-V AC from the wall plug to the low DC voltages used by the computer.
3. Power supplies used by computers are called *switching* power supplies and as such cannot safety be operated if they are disconnected from the circuit.

Multiple Choice

Select the best answer.

4. The use of a switching power supply results in
   a. Less heat loss.
   b. More economical components.
   c. Greater power dissipation.
   d. Both a and b.
5. Electrical power is equal to
   a. Current times voltage.
   b. Current plus voltage.
   c. Current divided by voltage.
   d. Voltage divided by current.
6. The more circuits the power supply must power,
   a. The more voltage will be required.
   b. The more current will be required.
   c. The more power will be required.
   d. Both b and c.

Matching

Match a definition on the right with each electrical unit on the left.

7. Power
8. Voltage
9. Current

a. Electrical potential.
b. Electrical energy.
c. Movement of electrical charge.
d. None of the above.

Completion

Fill in the blank or blanks with the best answers.

10. Five volts with a maximum current of 20 A will produce a maximum power of _____ W.
11. The power rating of a power supply indicates the maximum amount of _____ each of its output voltages can deliver.
12. Before removing the power supply, make sure you have completely disconnected the _____ _____.

FAMILIARIZATION ACTIVITY

Be sure to use the proper safety precautions when removing your power supply. You will need to remove the cover of the computer unit assigned to your lab group. You need to remove the cover only once for this exercise; then each lab partner can remove and insert the power supply.

1. Using the proper safety procedures, remove the power supply from the computer assigned to your lab group.
2. Reinstall the power supply in your computer.

QUESTIONS/ACTIVITIES

1. What was the power rating of the power supply you used in this exercise? How did you determine this?
2. Explain why the power requirements of a power supply increase when an extra disk drive is added to the computer system.

3. What should you look for when replacing the power supply of a computer?
4. Is it acceptable to completely remove a computer power supply from the computer, set it on the workbench, and plug it in so that you can measure its output voltages? Explain.
5. State, using diagrams where necessary, the purpose of a computer power supply.

REVIEW QUIZ

Within 20 minutes and with 100% accuracy,

1. Remove and replace the computer power supply, using proper safety procedures.
2. Explain why different computers require different power supplies.

32

Floppy Disk Drives

INTRODUCTION

Joe Tekk unlocked one of several filing cabinets in the storage room at RWA Software. It was filled with $5^{1}/_{4}"$ disks, old software archived years ago in case of fire. Joe laughed to himself. It had been a long time since he had seen a $5^{1}/_{4}"$ disk.

The $3^{1}/_{2}"$ disk is so pervasive, he thought. Just today he had seen the $3^{1}/_{2}"$ floppy used three different ways. There was a miniature floppy drive on his laptop. His friend just bought a digital CCD camera that saves images as .JPG files on a $3^{1}/_{2}"$ floppy (formatting it if necessary).

Joe thought about the third use of the $3^{1}/_{2}"$ floppy. One manager at RWA Software carries one in his shirt pocket, to every meeting. He uses the floppy to bring PowerPoint presentations to the meetings where he will be speaking.

Joe realized that the $5^{1}/_{4}"$ floppy never had a chance. It was simply too big.

PERFORMANCE OBJECTIVES

Upon completion of this exercise, within 15 minutes and with 100% accuracy, you will be able to

1. Remove and replace a $3^{1}/_{2}"$ floppy disk drive and boot the system.
2. Check the hardware settings of a floppy drive.

BACKGROUND INFORMATION

OVERVIEW

A floppy disk drive is a device that enables a computer to read and write information on a floppy disk. Details of the floppy disk were covered in Exercise 4, Using Floppy Disks.

Floppy disk drives (FDDs) are located at the front panel of the computer. The most common is the $3^{1}/_{2}"$ drive (the old $5^{1}/_{4}"$ drive is almost obsolete). These two disk drives are illustrated in Figure 32.1.

HOW A FLOPPY DISK DRIVE WORKS

Figure 32.2 is a simplified drawing of an FDD with its major components. Table 32.1 summarizes the purpose of each major component of the FDD.

FIGURE 32.1 Typical floppy disk drives

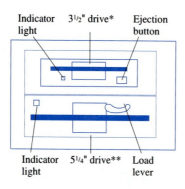

*Most frequently used now.

**Rarely used now.

FIGURE 32.2 Major components of a floppy disk drive

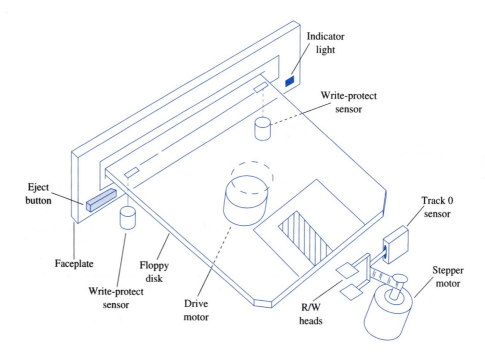

TABLE 32.1 Main parts of an FDD

Part	Purpose
Eject button	Used to eject a disk from the drive.
Write-protect sensor	Checks the condition of the floppy disk's write-protect system.
Read/write heads	Read and write information magnetically on the floppy disk. The heads move together, each working from its own side of the disk.
Track 0 sensor	Indicates when the read/write head is over track 0 of the floppy disk.
Drive motor	Spins the floppy disk inside the FDD.
Stepper motor	Moves the read/write head to different positions on the floppy disk.
Indicator light	Indicates if the disk drive is active.

OPERATING SEQUENCE

The operating sequence of a typical 3½" drive is as follows. Pushing the floppy disk into the drive causes the disk to be properly seated. The initial start-up for the drive consists of determining where track 0 is located. This is usually accomplished by a mechanical device, which is activated once the drive head is over track 0. When information is read, the stepping motor moves the read/write heads to their proper location. When information is written, the disk's write-protect status is checked and then new information is added to the disk.

DISK DRIVE SUPPORT SYSTEM

For proper operation, each part of the FDD support system must function properly.

1. *OS.* The operating system must be compatible with the media on the floppy disks.
2. *Floppy disk.* The disk itself must contain accurately recorded information in the proper format.
3. *Disk drive controller.* The drive controller conditions the signals between the motherboard and the FDD. Originally, a controller card that plugged into the motherboard was used to control the floppy drive. Most motherboards now have the floppy controllers built in.
4. *Disk drive electronic assembly.* This assembly consists of circuit boards that control the logical operations of the FDD. They act as an electrical interface between the disk drive controller and the electromechanical parts of the FDD.
5. *Disk drive mechanical assembly.* This assembly ensures proper alignment of the disk and read/write heads for reading and writing information.
6. *System power supply.* The power supply provides electrical power for all parts of the FDD, including the motors.
7. *Interconnecting cable.* Ribbon cable is used to transfer signals between the disk drive electrical assembly and the disk drive controller card.
8. *Power cable.* The DC power cable supplies electrical power to all parts of the FDD.

Let us take a closer look at many of these important components.

The OS

The operating system plays an important role in the operation of the floppy drive. Beginning with the system BIOS, all drive parameters must be known by the operating system so that data can be properly exchanged. Windows contains many applications designed specifically for disk drive operations. For example, right-clicking the drive A: icon in the My Computer window produces a Properties window similar to that shown in Figure 32.3. A pie chart is used to graphically illustrate used/free space on the drive. The volume label can be changed by entering a new one in the text box.

FIGURE 32.3 Drive A: Properties window

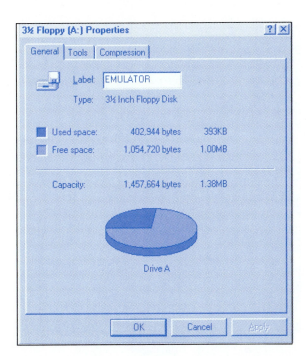

FIGURE 32.4 Floppy disk tools

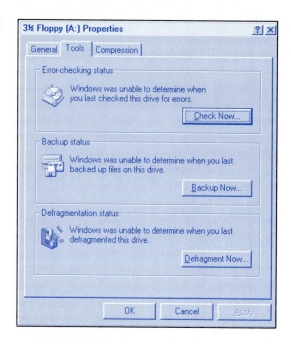

The Floppy Disk

There are several built-in Windows tools available for working with floppy disks. These tools are contained in the Tools submenu, as indicated in Figure 32.4.

The disk can be scanned for errors (using ScanDisk), backed up, or *defragmented*. A disk that has had many files created and deleted on it eventually becomes fragmented, with the files broken up into groups of sectors and scattered all over the disk (but still logically connected through the use of the FAT). This fragmentation increases the amount of time required to read or write entire files to the disk. By defragmenting the disk, all the files are reorganized, stored in consecutive groups of sectors at the beginning of the storage space on the disk.

A fourth tool is included that allows you to *compress* the data on your floppy, increasing the amount of free space available. The Compression submenu shown in Figure 32.5 indicates what will be gained by compressing the current disk.

FIGURE 32.5 Compression submenu

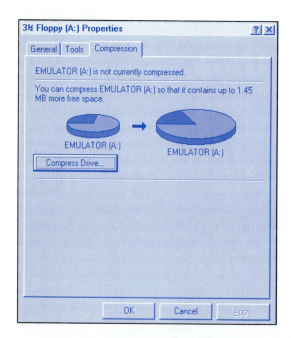

FIGURE 32.6 (a) Hardware settings for floppy drive, and (b) Windows NT Floppy Properties

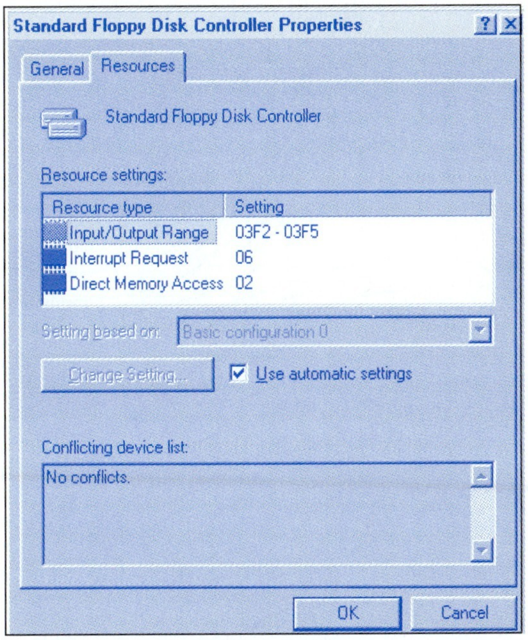

(a)

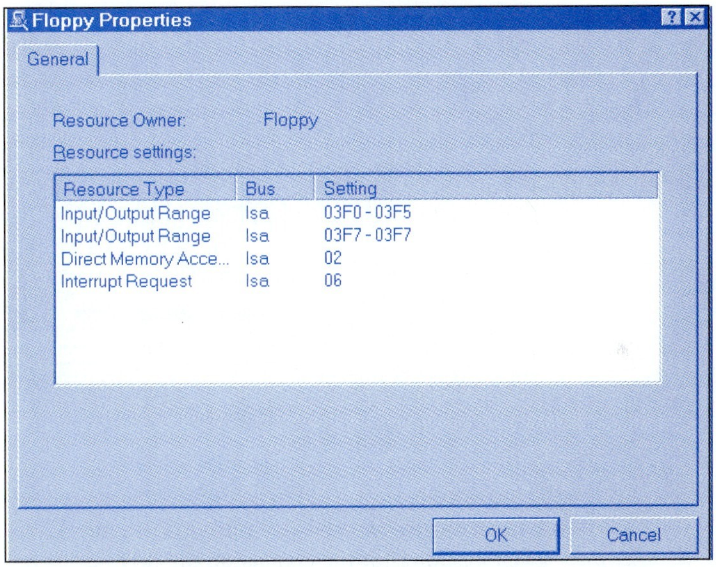

(b)

The Disk Drive Controller

The disk drive controller used to be an individual chip on a controller card. Now, the controller is just one part of a multifunction peripheral IC designed to operate the floppy and hard drives, the printer, and the serial ports. As always, I/O ports and interrupts are used to control the floppy disk drive. The settings used by Windows 95/98 can be examined/changed by using the Device Manager submenu of System Properties. The settings used by Windows NT can be examined from the Windows NT Diagnostics menu. Figure 32.6 shows the hardware configuration of a typical A: drive. These settings can be changed if necessary.

The Drive Cable

A 34-conductor ribbon cable is used to connect one or two disk drives to the controller. A twist in the cable between the two drive connectors reverses the signals on pins 9 through

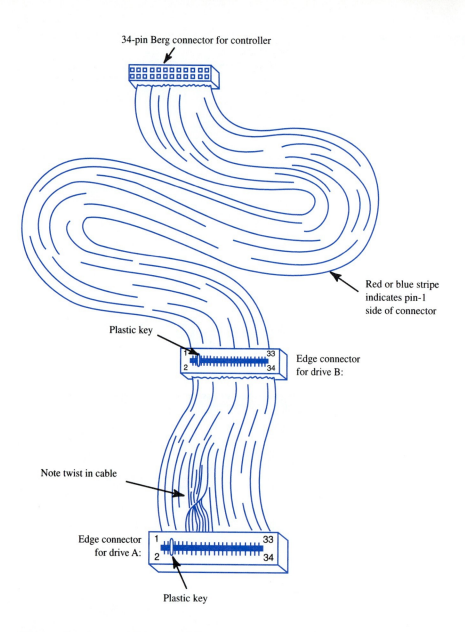

FIGURE 32.7 Floppy drive cable

16 at each connector. This twist differentiates the two drive connectors, forcing them to be used specifically for drive A: or drive B:. Figure 32.7 shows the cable details.

The meanings of the signals on the drive cable are shown in Table 32.2. Note that the signals affected by the twist in the cable are the select and enable signals for each drive.

The Power Cable

Like any other peripheral, the disk drive needs power to operate the drive and stepper motors, the read/write amplifiers and logic, and the other drive electronics. Figure 32.8 shows the pinout of the standard power connector used on the floppy drive. The connector is keyed so that it only plugs in one way.

Zip Drives

A device similar to the floppy drive is the Zip drive, manufactured by Iomega. Zip drives connect to the printer port, have removable 100MB cartridges, and boast a data transfer rate of 60MB/minute (using an SCSI connection). The 100MB disks spin at 2941 RPM, have an average access time of 29 milliseconds, and are relatively inexpensive. Newer 250 cartridges are also available.

The software driver for the Zip drive uses the signal assignments shown in Figure 32.9 to control the Zip drive through the printer port. Using the printer port to control the Zip drive allows you to easily exchange data between two computers.

TABLE 32.2 Floppy drive cable signals

Conductor (Pin)	Signal
1–33 odd	Ground
2	Unused
4	Unused
6	Unused
8	Index
10	Motor Enable A
12	Drive Select B
14	Drive Select A
16	Motor Enable B
18	Stepper Motor Direction
20	Step Pulse
22	Write Data
24	Write Enable
26	Track 0
28	Write Protect
30	Read Data
32	Select Head 1
34	Ground

FIGURE 32.8 Floppy drive power connector

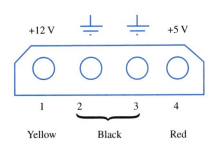

FIGURE 32.9 Zip drive parallel interface

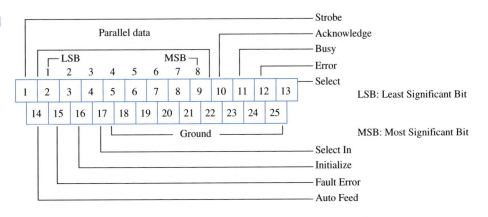

Jaz Drives

Similar to the Zip drive, the Jaz drive uses a 1GB removable cartridge that spins at 5400 RPM, has an average seek time around 10 milliseconds, and has a sustained data transfer rate of more than 6MB/second. A Jaz drive operates similarly to a hard drive, except the drive media is removable.

The 100MB SuperDisk

The SuperDisk is a new type of floppy drive with a 100MB capacity. SuperDisk drives can read/write both 100 MB SuperDisk floppies and 1.44/2.88 MB 3$^1/_2$" disks.

Working with Floppies

A few tips to keep in mind when working with floppy drives:

- Sometimes disks that are formatted on one system cannot be read by another. This may be due to differences in the read/write head alignments between both drives.
- If a floppy gives unexpected read errors, try ejecting the floppy and reinserting it.
- Run ScanDisk or some other suitable disk tool (such as Norton Utilities) to check a troublesome floppy.
- Always beware of disks given to you by someone else. Before you begin using them, scan them for viruses. This is especially important in an educational setting, where students and instructors often exchange disks as a normal part of class or lab.

TROUBLESHOOTING TECHNIQUES

SYSTEM OVERVIEW

Probable causes of what appears to be an FDD failure may be in one of the areas shown in Figure 32.10. This figure illustrates the relationship of the FDD to the entire computer system. At one end is the software on the disk; at the other extreme is the power cord connection to the electrical outlet. Every part of this system must be functioning properly for the disk drive to do its part. What is important here is to ensure that what appears to be a disk drive problem is not actually a problem caused by one of these other areas.

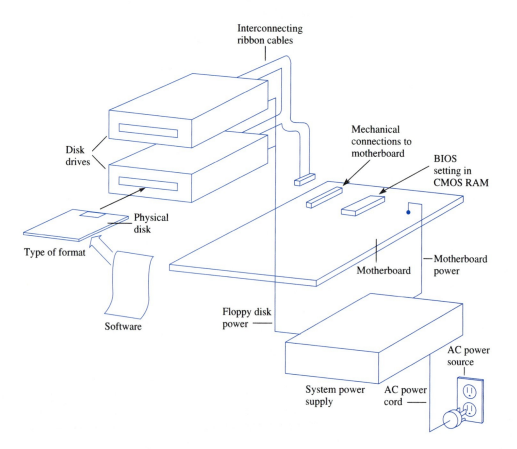

FIGURE 32.10 System relationship to floppy disk drive

TROUBLESHOOTING LOGIC

The first step in troubleshooting the disk drive is to classify the problem as occurring in one of the areas shown in Figure 32.10. Once the area at fault is determined, corrective action may be taken. Figure 32.11 is a troubleshooting diagram for determining which of these areas may be at fault.

FIGURE 32.11 FDD troubleshooting chart

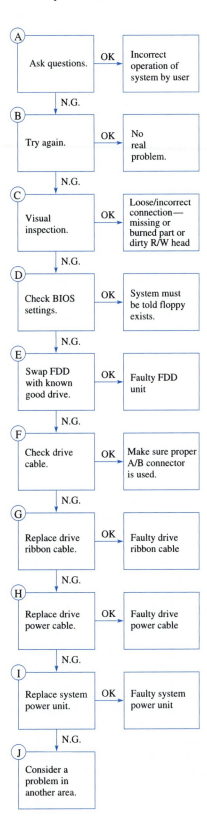

TROUBLESHOOTING STEPS

1. *Ask questions.* With all computer servicing, inquire about the history of the system. Was it recently modified? Was there an attempt to make any changes or repairs? Did the user buy the system used? For example, if the user recently installed the disk drive, it may be that certain settings need to be set differently. In this case, you will need to refer to the documentation for the system as well as the newly installed FDD. Asking questions may help you quickly spot the problem, saving you time and money.
2. *Try again.* Sometimes reseating the floppy in the drive fixes any read errors encountered. You could also try using the floppy in a different drive.
3. *Visual inspection.* A good visual inspection may reveal a burned component, an improperly seated cable, a dirty read/write head, or other mechanical evidence of the problem. Remember, a good visual inspection is an important part of any troubleshooting process.
4. *Check BIOS settings.* The system must be aware that a floppy drive exists. This is accomplished by running the BIOS setup program at boot time.
5. *Swap FDD with known good drive.* There are some precautions you need to take when doing this. It may be that the A: drive has a terminating resistor. You need to refer to documentation for the disk drive in question.
6. *Check drive cable.* In a one-drive system, make sure the connector with the twist is connected to drive A:. In a two-drive system, make sure both connectors are in the correct drives.
7. *Replace drive ribbon cable.* Here, another good visual inspection may be needed. Make sure you refer to the servicing manual to ensure that the cable was correctly installed in the first place. Replacing the cable with a known good one will help determine if the original cable or its connectors are at fault.
8. *Replace drive power cable.* This may be difficult on some systems because the power cable may be permanently attached to the power supply. If this is the case, skip this step and go on to the next one.
9. *Replace system power unit.* Doing this will help eliminate the problem of a power unit supplying the correct voltages when the power demands on it are small, but failing, as a result of the increased power demand, when a disk drive motor is activated (such as when the FDD is attempting to read a disk). Make a note of the power rating of the power supply; if it is below 100 W, substitute another known good power unit with a higher power rating.
10. *Consider a problem in another area.* If all the preceding steps fail to locate the problem, the problem is in another area of the unit. The most likely area in this case is the motherboard.

SELF-TEST

This self-test is designed to help you check your understanding of the background information presented in this exercise.

True/False

Answer *true* or *false*.

1. The most common type of floppy disk drive is the $3^{1}/_{2}$" size.
2. The purpose of a floppy disk drive is to read and write information from and to a floppy disk.
3. On a properly operating floppy disk drive, the indicator light is on only when the disk is in the writing mode.
4. The write-protect system protects information on the floppy disk from being written over.
5. The purpose of the eject button is to make sure the floppy disk is properly seated in its jacket.

Multiple Choice

Select the best answer.

6. On a floppy disk drive, information is placed on the disk by
   a. A single read/write head.
   b. Two heads, one for reading and the other for writing.
   c. Two heads, one on each side of the disk, each of which reads or writes on its side.
7. The floppy disk is turned inside its jacket by the operation of the
   a. Drive motor.
   b. Stepper motor.
   c. Spindle motor.
   d. None of the above.
8. The purpose of the twist in the ribbon cable is to
   a. Swap data in/out signals.
   b. Swap drive A/B select signals.
   c. Invert the drive data.
9. The purpose of the stepper motor is to
   a. Step the floppy disk correctly in its jacket.
   b. Turn the floppy disk in its jacket.
   c. Ensure that the floppy disk is correctly aligned.
   d. Position the read/write heads along the floppy disk.
10. The position of the floppy disk can be determined by the
    a. Track 0 sensor.
    b. Indicator light sensor.
    c. Drive assembly.
    d. Index sensor.

Matching

Match one or more conditions on the right with each item on the left.

11. Writing information
12. Reading information
13. Inserting the disk
14. Drive light on
15. Initial start-up

a. Disk is properly seated.
b. Write-protect notch is checked.
c. Stepper motor is active.
d. Heads are moved to track 0.
e. None of the above.

Completion

Fill in the blank or blanks with the best answers.

16. The floppy disk drive consists of two major sections, the _____ assembly and the electrical assembly.
17. Signals are conditioned between the motherboard and the disk drive by the _____ _____ controller.
18. A(n) _____ _____ is used to transfer information between the system and the disk drive.
19. The power cable connects the system _____ _____ and the floppy disk drive.
20. The dual-floppy drive cable has a(n) _____ between the two drive connectors.

Open-Ended

21. Describe the types of disk drives used by your system.
22. How many disk drives does your system have?
23. Who manufactures your disk drives? How did you find this out?
24. Determine if the price of one Zip disk is a bargain compared to the price of 71 floppies.
25. Is it necessary to have any software in your system for your floppy disk drives to work properly? Explain.

FAMILIARIZATION ACTIVITY

1. Survey the machines in your laboratory. How many have one floppy? How many have two? Do any of the machines have 5¼" drives?
2. Remove the floppy disk drive from your system, using the appropriate tools and safety precautions.
3. In the following space, describe the steps you used to remove the drive.

4. In the following space, make a sketch of the drive you removed, showing the following: all retaining hardware, all cable connections, any areas in which you experienced difficulty in removing the drive, and any other notes you may want to use for future reference.

5. Replace the floppy disk drive.
6. In the following space, describe the steps you used to replace the drive.

7. Have your instructor check your work before proceeding to the next step.

    Instructor OK: _____

8. Using the proper assembly procedure, reassemble the rest of your unit; if your unit was a functional one, demonstrate to your instructor that it is now working properly.

    Instructor OK: _____

9. Check the hardware settings for your floppy disk drive using BIOS setup and Device Manager.

QUESTIONS/ACTIVITIES

1. Describe, in your own words and using illustrative diagrams, the method of removing and installing the floppy disk drive on your system. Do this as if the description you are writing were to be used by another service technician for the purpose of removing and installing a floppy disk drive on the same kind of system.
2. Using a recent computer parts catalog, find the list cost of a floppy disk drive that would be an exact replacement for the one you worked with in this exercise.
3. Contact at least one local computer repair shop and find out what they would charge to replace a floppy disk drive in a computer of the type you used in this exercise.
4. If you could make a major improvement in floppy disk drives, what would it be?
5. Your instructor may require that you troubleshoot a floppy disk drive type problem in the lab. If this is a requirement, state what the problem was and how you found it. If you didn't find the problem, state what you could do differently in the future to find and correct a similar problem.

REVIEW QUIZ

Within 15 minutes and with 100% accuracy,

1. Remove and replace a 3$\frac{1}{2}$" floppy disk drive and boot the system.
2. Check the hardware settings of a floppy drive.

33 The Motherboard Microprocessor and Coprocessor

INTRODUCTION

The phone rang on Joe Tekk's desk. It was Joe's friend, a 13-year-old boy Joe met at a computer show. "Joe, it's Stephen. Are you busy?"

"No," Joe said, smiling. "What's up?" He was always happy to hear what Stephen was up to.

"Can you check the upgrade path for my 100-MHz Pentium for me? I need MMX technology for some course work at school."

Joe grabbed a computer catalog off a tall stack of catalogs piled near his desk. He leafed through it and found the microprocessor section. There was a table showing the various ways you could upgrade your microprocessor. "You can get a 166-MHz processor with MMX technology for around $100. How's that?"

Stephen was not satisfied. "That's not fast enough."

Joe reexamined the table in the catalog. "Do you have Socket 7 on your motherboard?"

Stephen said, "Hold on," and was silent for only an instant before saying, "Yes! Socket 7!" happily into the phone.

Joe was glad. "Great! You can get a 200-MHz version now for $125."

Stephen barely finished saying, "Order one, I'll pay you later," before the line went dead.

PERFORMANCE OBJECTIVES

Upon completion of this exercise, within 10 minutes and with 100% accuracy, you will be able to

1. Use a standard personal computer motherboard to explain the location, purpose, bus size, and speed of the
   - Microprocessor
   - Math coprocessor

BACKGROUND INFORMATION

DEFINITION OF THE MOTHERBOARD

The main system board of the computer is commonly referred to as the **motherboard.** A typical motherboard is shown in Figure 33.1. Sometimes the motherboard is referred to as the **system board,** or the **planar.**

FIGURE 33.1 Typical PC motherboard (*photo by John T. Butchko*)

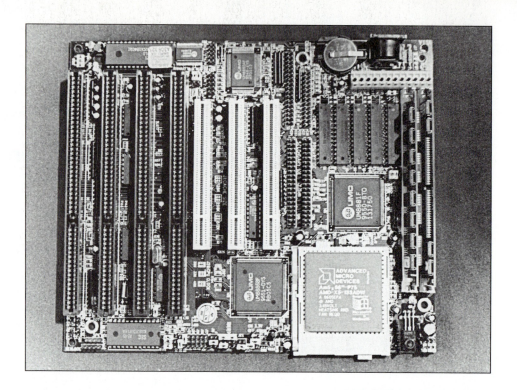

CONTENTS OF THE MOTHERBOARD

The motherboard holds and electrically interconnects all the major components of a PC. The motherboard contains the following:

- The microprocessor
- The math coprocessor (only on older 386 motherboards)
- BIOS ROM
- RAM (Dynamic RAM, or DRAM, as well as level-2 cache)
- The expansion slots
- Connectors for IDE drives, floppies, and COM ports

Table 33.1 lists these major parts and gives a brief overview of the purpose of each part.

Figure 33.2 shows a typical motherboard layout and the locations of the major motherboard parts.

In this exercise, you will have the opportunity to learn more details about the microprocessor and the coprocessor. In the following exercises, you will learn about the other areas of the motherboard.

THE MICROPROCESSOR

You can think of the **microprocessor** in a computer as the central processing unit (CPU), or the "brain," so to speak, of the computer. The microprocessor sets the stage for everything else in the computer system. Several major features distinguish one microprocessor from another. These features are listed in Table 33.2.

You can think of a **bus** as nothing more than a group of wires all dedicated to a specific task. For example, all microprocessors have the following buses:

Data bus	Group of wires for handling data. This determines the data path size.
Address bus	Group of wires for getting and placing data in different locations. This helps determine the maximum memory that can be used by the microprocessor.
Control bus	Group of wires for exercising different controls over the microprocessor.
Power bus	Group of wires for supplying electrical power to the microprocessor.

TABLE 33.1 Purposes of major motherboard parts

Part	Purpose
Microprocessor	Interprets the instructions for the computer and performs the required process for each of these instructions.
Math coprocessor	Used to take over arithmetic functions from the microprocessor.
BIOS ROM	Read-only memory. Memory programmed at the factory that cannot be changed or altered by the user.
RAM	Read/write memory. Memory used to store computer programs and interact with them.
Expansion slots	Connectors used for the purpose of interconnecting adapter cards to the motherboard.
Connectors	Integrated controller on motherboard provides signals for IDE and floppy drives, the printer, and the COM ports.

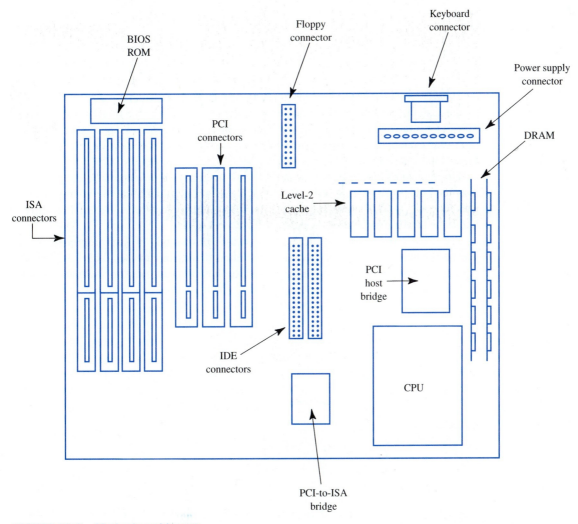

FIGURE 33.2 Motherboard layout

TABLE 33.2 Microprocessor features

Feature	Description
Bus structure	The number of connectors used for specific tasks.
Word size	The largest number that can be used by the microprocessor in one operation.
Data path size	The largest number that can be copied to or from the microprocessor in one operation.
Maximum memory	The largest amount of memory that can be used by the microprocessor.
Speed	The number of operations that can be performed per unit time.
Code efficiency	The number of steps required for the microprocessor to perform its processes.

FIGURE 33.3 Typical microprocessor bus structure

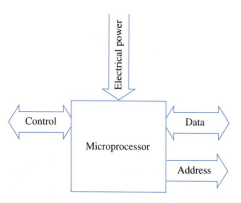

Figure 33.3 shows the bus structure of a typical microprocessor.

Since all the data that goes in and out of a microprocessor is in the form of 1s and 0s, the more wires used in the data bus, the more information the microprocessor can handle at one time. For example, some microprocessors have eight lines (wires or pins) in their data buses, others have 16, and some have 32 or 64.

The number of lines used for the address bus determines how many different places the microprocessor can use for getting and placing data. The *places* that the microprocessor uses for getting and placing data are referred to as **memory locations.** The relationship between the data and the address is shown in Figure 33.4.

The greater the number of lines used in the address bus of a microprocessor, the greater the number of memory locations the microprocessor can use. Table 33.3 lists the common microprocessors used in the PC. All of these microprocessors are manufactured by Intel Inc.

FIGURE 33.4 Relationship between data and address

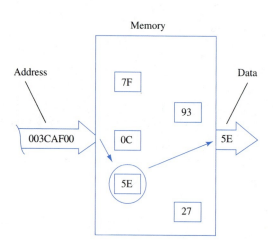

TABLE 33.3 Types of microprocessors used in the PC

Microprocessor	Lines		Maximum Clock Speed	Addressable Memory
	Data	Address		
8088	8	20	8 MHz	1MB
8086	16	20	8 MHz	1MB
80286	16	24	20 MHz	16MB
80386SX	16	24	20 MHz	16MB
80386	32	32	33 MHz	4GB
80486	32	32	66 MHz	4GB
Pentium	64	32	233+ MHz	4GB
Pentium Pro	64	36	200+ MHz	64GB
Pentium II/III	64	36	400+ MHz	64GB

Note from Table 33.3 that the greater the number of address lines, the more memory the microprocessor is capable of addressing. In the table, 1MB = 1,048,576 memory locations, and 4GB = 4,294,967,296 memory locations.

Figure 33.5 shows some of the packaging used for microprocessors.

Compatible CPUs

A number of companies manufacture processors that compete with Intel for use in PC motherboards and other applications. Two of these companies are AMD and Cyrix. Table 33.4 shows recent sets of compatible CPUs.

Having more than one processor to choose from allows you to examine pricing, chip features, and other factors of importance when making a decision.

FIGURE 33.5 Microprocessor packaging (*courtesy Intel Corporation*)

TABLE 33.4 Comparing CPUs

Intel	AMD	Cyrix
Pentium	K5	5x86*
Pentium II	K6	6x86MX

*Pentium performance, pin compatible with the 80486.

About the 80x86 Architecture

The advanced nature of the Pentium microprocessor requires us to think differently about the nature of computing. The Pentium architecture contains exotic techniques such as branch prediction, pipelining, and superscalar processing to pave the way for improved performance. Let us take a quick look at some other improvements from Intel:

- Intel has added MMX technology to its line of Pentium processors (Pentium, Pentium Pro, and Pentium II/III). A total of 57 new instructions enhance the processors' ability to manipulate audio, graphic, and video data. Intel accomplished this major architectural addition by *reusing* the 80-bit floating-point registers in the FPU. Using a method called **SIMD** (single instruction multiple data), one MMX instruction is capable of operating on 64 bits of data stored in an FPU register.
- The Pentium Pro processor (and also Pentium II/III) use a technique called *speculative execution.* In this technique, multiple instructions are fetched and executed, possibly out of order, in order to keep the pipeline busy. The results of each instruction are speculative until the processor determines that they are needed (based on the result of branch instructions and other program variables). Overall, a high level of parallelism is maintained.
- First used in the Pentium Pro, a bus technology called Dual Independent Bus architecture uses two data buses to transfer data between the processor and main memory (including the level-2 cache). One bus is for main memory, the second is for the level-2 cache. The buses may be used independently or in parallel, significantly improving the bus performance over that of a single-bus machine.
- The five-stage Pentium pipeline was redesigned for the Pentium Pro into a *superpipelined* 14-stage pipeline. By adding more stages, less logic can be used in each stage, which allows the pipeline to be clocked at a higher speed. Although there are drawbacks to superpipelining, such as bigger branch penalties during an incorrect prediction, its benefits are well worth the price.

Spend some time on the Web reading material about these changes, and others. It will be time well invested.

THE COPROCESSOR

Each Intel microprocessor released before the 80486 has a companion to help it do arithmetic calculations. This companion is called a **coprocessor.** For most software, the coprocessor is optional. However, some programs (such as CAD, computer-aided design, programs) have so many math calculations to perform that they need the assistance of the math coprocessor; the main microprocessor simply cannot keep up with the math demand.

These **math chips,** as they are sometimes called, are capable of performing mathematical calculations 10 to 100 times faster than their companion microprocessors and with a higher degree of accuracy. This doesn't mean that if your system is without a coprocessor it can't do math; it simply means that your microprocessor will be handling all the math along with everything else, such as displaying graphics and reading the keyboard.

Table 33.5 lists the math coprocessors that go with various microprocessors. Note that the 80486 and higher processors have built-in coprocessors.

For a math coprocessor chip to be used by software, the software must be specifically designed to look for the chip and use it if it is there. Some spreadsheet programs look for the

TABLE 33.5 Matching math coprocessors

Microprocessor	Math Coprocessor
8086	8087
8088	8087
80286	80287
80386	80387
80386SX	80387SX
80486DX	Built-in coprocessor enabled
80486SX	Coprocessor disabled
Pentium, Pentium Pro, and Pentium II/III	Built-in coprocessor enabled always

presence of this chip and use the microprocessor for math if the coprocessor is not present. If the coprocessor is present, the software uses it instead. Some programs, such as word-processing programs, have no use for the math functions of the coprocessor and do not use the coprocessor at all. Therefore, the fact that a system has a coprocessor doesn't necessarily mean that the coprocessor will improve the overall system performance. Improvement will take place only if the software is specifically designed to use the coprocessor and there are many complex math functions involved in the program.

TROUBLESHOOTING TECHNIQUES

Windows identifies the processor it is running on. Use System Properties in Control Panel to check the processor type, as shown in Figure 33.6. Notice that Windows 95 has identified a Pentium as the CPU. It is interesting to note that other Pentium-compatible CPUs, such as a Cyrix 5x86, are not recognized by Windows as Pentiums, and are reported in System Properties as 80486 CPUs. This kind of information is not easily found in existing documentation. It is good to talk directly to the manufacturer of the motherboard to determine how your performance might be affected.

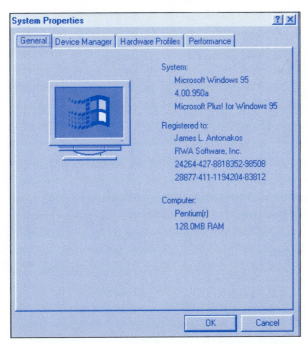

FIGURE 33.6 Processor identification, (a) Windows 95 System Properties

(a)

(continued on next page)

FIGURE 33.6 *(continued)* **(b) Windows NT System Properties**

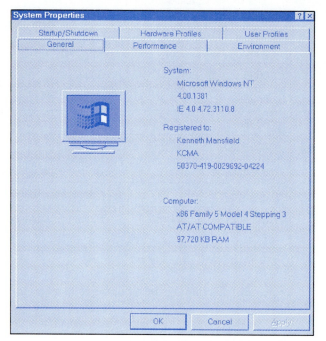

(b)

SELF-TEST

This self-test is designed to help you check your understanding of the background information presented in this exercise.

True/False

Answer *true* or *false*.

1. The motherboard is sometimes referred to as the system board.
2. All the major components in the computer are interconnected through the motherboard.
3. ROM is computer memory that is programmed at the factory and cannot be changed directly by the user.

Multiple Choice

Select the best answer.

4. You can think of the microprocessor in the computer as the
   a. CPU.
   b. "Brains of the computer."
   c. Main controlling part of the computer.
   d. All of the above.
5. The largest number that can be used by the computer in one operation is determined by the
   a. Amount of available memory.
   b. Speed of the computer.
   c. Word size.
   d. None of the above.
6. In a computer, a group of wires dedicated to a specific task is called the
   a. Bus.
   b. Track.
   c. Data path.
   d. Through path.

Matching

Match a characteristic on the right with each microprocessor on the left.

7. 80386
8. 80486
9. Pentium
10. Pentium II

a. 33-MHz clock speed.
b. 4GB of addressable memory.
c. 64GB of addressable memory.
d. None of the above.

Completion

Fill in the blank or blanks with the best answers.

11. A(n) _____ is a chip used to help the microprocessor perform mathematical computations.
12. The video technology added to the Pentium architecture uses the abbreviation _____.
13. Whether or not a math coprocessor chip is actually used is determined by the system _____.
14. The _____ and _____ processors have built-in coprocessors.

FAMILIARIZATION ACTIVITY

Use the proper procedure to open the case of your microcomputer. With your lab partner(s), locate the microprocessor and the coprocessor (if there is no coprocessor, locate the slot where the coprocessor would be located).

From the markings on the microprocessor, use the information presented in the background information section to determine the:

1. Speed of the microprocessor.
2. Number of data lines.
3. Maximum amount of addressable memory.
4. Type of numeric coprocessor to use.

QUESTIONS/ACTIVITIES

Questions

1. What are the specifications of the microprocessor used in your computer?
2. In your own words, state the purpose of the microprocessor.
3. Explain the difference in use of the microprocessor data bus and the address bus.
4. In your own words, state the purpose of a math coprocessor.
5. Explain when a math coprocessor is needed. What determines whether a math coprocessor will actually be used once it is installed?

Activities

1. Make a sketch of the motherboard used in your computer. Indicate the locations of the microprocessor and coprocessor.
2. Make a sketch of the microprocessor used in your system. Be sure to indicate all the information printed on the microprocessor chip.

REVIEW QUIZ

Within 10 minutes and with 100% accuracy,

1. Use a standard personal computer motherboard to explain the location, purpose, bus size, and speed of the
   - Microprocessor
   - Math coprocessor

34 The Motherboard Memory

INTRODUCTION

The phone rang on Joe Tekk's desk. It was Joe's friend Ken Koder. Ken was a software engineer for a local aerospace firm.

"Joe, I'm at the computer show. You know, the one that comes to the arena every two months."

"Sure, that's a great show," Joe said. Joe liked going to the shows. They always had good prices for brand-new equipment and components.

Ken continued. "They are selling 32MB EDO DRAM for $90. Do you need any?"

Joe thought a moment. He knew there were two open memory slots on his motherboard in his machine at home. He could pull the two 16MB RAMs from the other two slots and get four new 32MB RAMs, giving him 128MB at home. His office machine had two open slots as well. "Sure, Ken. Get me six if you can."

"Just six? I'm getting eight! That will give me 256MB of RAM for simulations. Anyway, I'll bring your memory over later today."

Joe thought that Ken could probably easily use 512MB for simulations. He himself was happy with 128MB just for Windows 98.

PERFORMANCE OBJECTIVES

Upon completion of this exercise, you will be able to

1. Demonstrate, within 15 minutes,
   a. Methods of increasing the memory usage of a personal computer.
   b. How to determine the organization of the computer's memory.
2. From a motherboard selected by your instructor, determine, within 5 minutes,
   a. The locations of the RAM and ROM.
   b. The size of the DRAM.
   c. The speed of the DRAM.

BACKGROUND INFORMATION

COMPUTER MEMORY

Computer memory consists of any device capable of copying a pattern of 1s and 0s that represent some meaningful information to the computer. Computer memory can be contained in *hardware*, such as in chips, or in **magneticware**, such as floppy and hard disks (or other magnetic material such as magnetic tape). Computer memory is not limited to just these two major areas. For example, a laser disk uses light to read large amounts of information into the computer; this too is a form of computer memory. For the purpose of discussion here, computer memory will be divided into two major areas: hardware memory and magneticware memory.

The hardware memory of a computer is referred to as **primary storage.** The magneticware of a computer is referred to as **secondary storage,** or **mass storage.** Here are some facts about each.

Primary Storage

- Immediately accessible to the computer.
- Any part of the memory may be immediately accessed.
- Short-term storage.
- Limited capacity.

Secondary Storage

- Holds very large amounts of information.
- Not immediately accessible.
- May be sequentially accessed.
- To be used, must be transferred to primary storage.
- Long-term storage.

In this exercise, you will see how primary and secondary computer memories are used (see Figure 34.1) and how they can work with each other to produce an almost unlimited amount of computer memory. First, let us learn about primary storage.

PRIMARY STORAGE—RAM AND ROM

There are two basic kinds of primary storage: one kind that the computer can quickly store information in and retrieve information from and another kind that the computer can only receive information from. Figure 34.2 shows the two basic kinds of primary storage memory.

The kind of memory that the computer can get information (read) from but cannot store information (write) to is called **read-only memory** (ROM). The advantage of having ROM is that it can contain programs that the computer needs when it is first turned on; these programs (called the **Basic Input/Output System,** or BIOS) are needed by the computer so it knows what to do each time it turns on (such as reading the disk and starting the booting process). Obviously, these programs should not be able to be changed by the computer user, because doing so could jeopardize the operation of the system. Therefore, ROM consists of chips that are programmed at the factory. The programs in these chips are permanent and stay that way even when the computer is turned off; they are there when the computer is turned on again.

The kind of memory that the computer can write to as well as read from is called **read/write memory.** The acronym for read/write memory is RWM, which is hard to say. Because of this, read/write memory is called RAM, which stands for **random access memory.** Both ROM and read/write memory are randomly accessible, meaning that the

FIGURE 34.1   Two major areas of computer memory

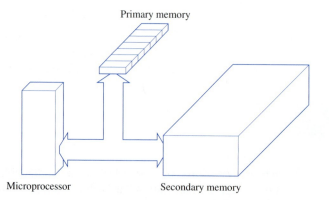

FIGURE 34.2 Two basic kinds of primary storage memory

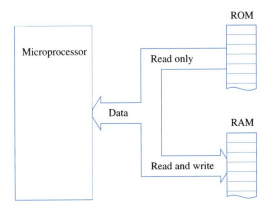

FIGURE 34.3 ROM and RAM

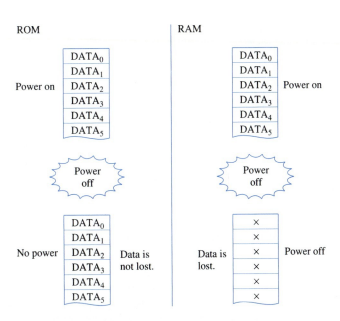

computer can get information from any location without first going through other memory locations. However, read/write memory is traditionally referred to as RAM.

Unlike ROM, RAM loses anything that is stored in it when the power is turned off. Because the information in RAM is not permanent, it is referred to as **volatile memory.** Figure 34.3 shows this difference.

The system ROM chip for the PC contains two main programs, the **Power-On Self-Test** (POST) and the **Basic Input/Output System** (BIOS). The programs in the ROM chip set the personality of the computer. As a matter of fact, how compatible a computer is can be determined primarily by the programs in these ROM chips. The ROM chips have changed over time as systems have been improved and upgraded. There have been, for example, more than 20 changes in the ROM BIOS programs by IBM for its different PCs.

You may need to update your old BIOS to use new hardware in your system (large IDE hard drives, for example). To upgrade your BIOS ROM, you may (1) replace the ROM with a new one or (2) run a special upgrade program (typically available for download off the Web) that makes changes to a *flash EPROM,* an EPROM that can be electrically reprogrammed.

BITS, BYTES, AND WORDS

Recall that a bit is a single binary digit. It has only two possible conditions: ON and OFF. Everything in your computer is stored and computed with ONs and OFFs. The bits

FIGURE 34.4 Arrangement of computer data

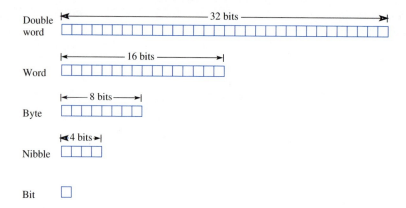

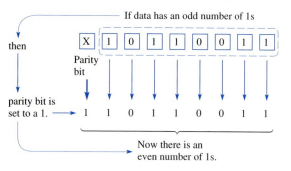
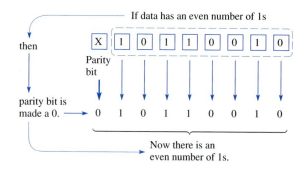

FIGURE 34.5 Even parity checking

inside your computer are arranged in such a way as to work in units. The most basic unit, or group, of bits is called a **byte.** A byte consists of 8 bits. Mathematically, 8 bits have 256 unique ON and OFF combinations. You can figure this out with your pocket calculator—just calculate 2^8, which is 2 multiplied by itself eight times. A **word** is 16 bits, or 2 bytes. When 4 bytes are taken together, such as in 32-bit microprocessors, they are called a **double word.** These different arrangements are shown in Figure 34.4.

In PCs a method called **parity checking** is used to help detect errors. There are times when, in the process of working with computer bits, a bit within a byte may accidentally change from ON to OFF or from OFF to ON. To check for such an error, parity checking uses an extra bit called the **parity bit.** IBM and most compatibles use what is called **even parity** to check their bits. Even parity means that there will always be an even number of ONs for each byte, including the parity bit. Even parity checking is illustrated in Figure 34.5.

SIMM

The **Single-In-Line Memory Module,** or SIMM, is another way of physically organizing memory. It is a small "boardlet" with several memory chips soldered to it. This boardlet is inserted into a system slot. Figure 34.6 shows a SIMM.

SIMMs came about in an attempt to solve two problems. The first problem was "chip creep." Recall that chip creep occurs when a chip works its way out of a socket as a result of thermal expansion and contraction. The old solution to this problem—soldering memory chips into the board—wasn't a good solution, because it made them harder to replace. So the SIMM was created. The only problem with the SIMM is that if only 1 bit in any of its chips goes bad, the whole SIMM must be replaced. This is more expensive than replacing only one chip. SIMMs come in 256KB, 1MB, 4MB, and 16MB sizes. A similar type of memory module, called a SIPP, contains metal pins that allow the SIPP to be soldered directly onto the motherboard.

Regarding parity bits in a SIMM, a 32-bit SIMM is nonparity, and a 36-pin SIMM stores one parity bit for each byte of data. Pentium processors incorporate parity in their address and data buses.

FIGURE 34.6 Single-in-line memory module (SIMM)

DIMM

The Dual In-Line Memory Module (DIMM) was created to fill the need of Pentium-class processors containing 64-bit data buses. A DIMM is like having two SIMMs side by side, and come in 168-pin packages (more than twice that of a 72-pin SIMM). Ordinarily, SIMMs must be added in pairs on a Pentium motherboard to get the 64-bit bus width required by the Pentium.

SDRAM

Synchronous DRAM (SDRAM) is very fast (up to 100-MHz operation) and is designed to synchronize with the system clock to provide high-speed data transfers.

EDO DRAM

Extended Data Out DRAM (EDO DRAM) is used with bus speeds at or below 66 MHz and is capable of starting a new access while the previous one is being completed. This ties in nicely with the bus architecture of the Pentium, which is capable of back-to-back pipelined bus cycles. Burst EDO (BEDO RAM) contains pipelining hardware to support pipelined burst transfers.

VRAM

Video RAM (VRAM) is a special *dual-ported* RAM that allows two accesses at the same time. In a display adapter, the video electronics needs access to the VRAM (to display the Windows desktop, for example) and so does the processor (to open a new window on the desktop). This type of memory is typically local to the display adapter card.

LEVEL-2 CACHE

Cache is a special high-speed memory capable of providing data within one clock cycle and is typically ten times faster than regular DRAM. Although the processor itself contains a small amount of internal cache (8KB for instructions and 8KB for data in the original Pentium), you can add additional level-2 cache on the motherboard, between the CPU and main memory, as indicated in Figure 34.7. Level-2 cache adds 64KB to 2MB of external cache to complement the small internal cache of the processor. The basic operation of the cache is to speed up the average access time by storing copies of frequently accessed data.

CHIP SPEED

When replacing a bad memory module, you must pay attention to the *speed* requirements of that module. If you do not, the replacement will not work. You can use memory chips that have a higher speed or the same speed as the replacement, but not a lower speed. The faster the memory, the more it costs. Adding faster memory to your system may not improve overall performance at all because the speed of your computer is determined by the system clock, among other things.

FIGURE 34.7 Using cache in a memory system

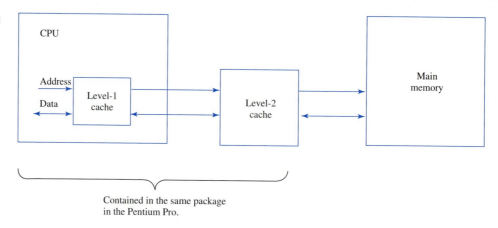

Memory chip speed is measured in **nanoseconds.** A nanosecond is 0.000000001 second. To check the speed rating of a RAM chip, look at the coding on the top of the chip. Typical DRAM speeds are 60 ns and 70 ns.

HOW MEMORY IS ORGANIZED

It is important that you have an understanding of the organization of memory in the computer. The 8088 and 8086 microprocessors are able to address up to 1MB of memory. Since some of the first PCs used the 8088 and 8086 microprocessors, when the 80286, 80386, 80486, and Pentium microprocessors were introduced, they were made **downward compatible** with their predecessors. This meant that software that worked on an older PC would still work on the newer systems.

In order to keep downward compatibility, the newer microprocessors (80286, 80386, 80486, and Pentiums) come with two modes of operation (there is one other mode that will be presented later). One mode is called the **real mode,** in which the microprocessors behave like their earlier models (and are limited to 1MB of addressable memory). The other mode, called the **protected mode,** allows the microprocessors to use the newer power designed into them (such as addressing up to 16MB for the 80286 and 4GB for the 80386, 80486, and Pentiums).

All PCs have what is called a **base memory,** which is the 1MB of memory that is addressable by the 8088, 8086, and newer microprocessors running in real mode. A common way of viewing the organization of memory is through the use of a **memory map.** A memory map is simply a way of graphically showing what is located at different addresses in memory. Figure 34.8 shows the memory map of the PC in real mode.

As you can see from the memory map, several areas of memory are designated for particular functions; not all the 1MB of memory is available for your programs. As a matter of fact, only 640KB can be used by DOS-operated systems. Table 34.1 lists the definitions of the various memory sections.

The memory above the conventional 640KB of memory is referred to as **upper memory.**

HOW MEMORY IS USED

There are three ways memory can be allocated: as **conventional memory, extended memory,** or **expanded memory.** Figure 34.9 shows the relationships among the three types of memories.

Table 34.2 explains the uses of the three different memory allocation methods. It is important to note that DOS and systems that use DOS are limited to 1MB of addressable space. The

FIGURE 34.8 Memory map of PC in real mode

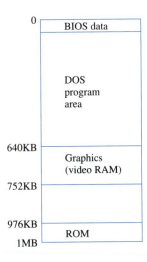

TABLE 34.1 Purpose of allocated memory

Assignment	Definition
Base memory	This refers to the amount of memory actually installed in the conventional memory area.
Conventional or user memory	This is the 640KB of memory that is usable by DOS-based programs.
Video or graphics memory	This area of memory (128KB) is reserved for storing text and graphics material for display on the monitor. As you will see in Exercise 41, the amount of this space actually used by the system depends on the requirements of the video monitor.
Motherboard ROM	This is space reserved for the use of the ROM chips on the motherboard.

FIGURE 34.9 Relationships among conventional, extended, and expanded memory

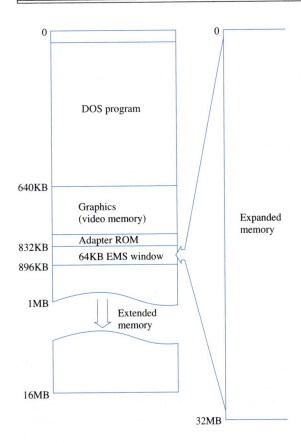

415

TABLE 34.2 Memory allocation methods

Memory Type	Comments
Conventional	1. Memory between 0KB and 1MB, with 640KB usable and 34KB reserved. 2. Completely usable by DOS-based systems. 3. Uses real mode of 8086, 8088, 80286, 80386, 80486, and Pentium.
Extended	1. Uses protected mode of 80286 (up to 16MB), 80386, 80486, and Pentium (up to 4GB). 2. Cannot be used by DOS-based systems (which are limited to 1MB of memory). 3. Can all be accessed by the IBM OS/2 operating system. 4. Can be used by a virtual disk in DOS systems. 5. Is the type of memory to use with the 80386, 80486, and Pentium microprocessors.
Expanded	1. Uses "bank-switching" techniques. 2. Requires special hardware and software. 3. Is not a continuous memory but consists of chunks of memory that can be switched in and out of conventional memory. 4. Sometimes referred to as EMS (expanded memory specification) memory.

reason for this is that DOS is made for microprocessors running in the real mode. This memory limit exists because computer and program designers thought that 1MB of memory would be all that anyone would ever need on a PC for years to come. Since 340KB of the 1MB of addressable DOS memory is reserved by the system, DOS really has only 640KB left for user programs. Thus, DOS is said to have a 640KB limit. However, as you will see, there are ways of allowing DOS to store data in an addressable memory location that is beyond this DOS limit.

BREAKING THE DOS BARRIER

The 80386, 80486, and Pentium microprocessors have a third mode of operation, called the **virtual 8086 mode.** This mode allows a single 80386, 80486, or Pentium microprocessor to divide its memory up into many "virtual" computers. The beauty of this is that each of these virtual computers can run its own program in total isolation from other programs running in the other virtual computers. This means that more than one DOS program can be run in the same computer at the same time.

You can use the DOS VDISK command to create a virtual disk in extended memory. The command is put into the CONFIG.SYS file:

    For IBM DOS: DEVICE = VDISK.SYS[size][sectors][files]/E
    For MS-DOS: DEVICE = RAMDRIVE.SYS[size][sectors][files]/E

Both of these commands will create a virtual disk in extended memory: this is the purpose of the /E modifier.

In DOS 5.0, you can force DOS to use as much of high memory as possible. This command is also put into the CONFIG.SYS file:

    DEVICE = C:\DOS\HIMEM.SYS
    DEVICE = HIGH
    DEVICE = C:\DOS\EMM386.EXE

The first line tells DOS to install its high-memory manager. The second line actually loads part of DOS into this region of memory. The third line tells DOS to load the EMM386 Expanded Memory Manager to make the expanded memory available. DOS 5.0 has a feature that releases even more of the 640KB of user memory—another command placed in the CONFIG.SYS file:

$$DOS = UMB$$

or

$$DOS = HIGH, UMB$$

Once the UMB (upper-memory support) feature is enabled, you can use the same CONFIG.SYS file to load a device driver (such as a mouse) into high memory:

$$DEVICEHIGH = [\text{path name and name of the device}]$$

If you have a TSR (terminate and stay-resident program) that you want to have loaded into high memory (such as Borland's SideKick), DOS 5.0 gives you another option. Include the following in your AUTOEXEC.BAT file:

$$LOADHIGH\ [\text{program name}]$$

Another command available in DOS 4.01 and later is

$$MEM\ /PROGRAM\ \text{or}\ MEM\ /P$$

This command will output the memory allocation for the entire system. Figure 34.10 shows a typical printout. The memory locations are given in hexadecimal values. You can use your pocket calculator to convert the hex values to decimal numbers.

FIGURE 34.10 Printout of MEM /P

```
Address    Name        Size      Type

000000                 000400    Interrupt Vector
000400                 000100    ROM Communication Area
000500                 000200    DOS Communication Area

000700     IO          0021C0    System Program

0028C0     MSDOS       008E20    System Program

00B6E0     IO          00AEA0    System Data
           ANSI        001180      DEVICE=
           CCDRIVER    002AB0      DEVICE=
           MOUSE       003540      DEVICE=
                       000380      FILES=
                       000100      FCBS=
                       0029A0      BUFFERS=
                       0001C0      LASTDRIVE=
                       000CD0      STACKS=
016590     COMMAND     000070    Data
016610     MSDOS       000040    -- Free --
016660     FASTOPEN    002780    Program
018DF0     COMMAND     001640    Program
01A440     COMMAND     000100    Environment
01A550     APPEND      001E20    Program
01C380     GRAPHICS    0014A0    Program
01D830     COMMAND     000040    Data
01D880     SHELLB      000E80    Program
01E710     COMMAND     001640    Program
01FD60     COMMAND     0000A0    Environment
01FE10     MEM         000070    Environment
01FE90     MEM         012F00    Program
032DA0     MSDOS       06D250    -- Free --


  655360 bytes total memory
  655360 bytes available
  524640 largest executable program size

  393216 bytes total extended memory
  393216 bytes available extended memory
```

FIGURE 34.11 Virtual memory

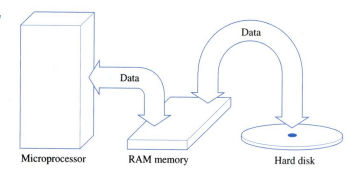

VIRTUAL MEMORY

Another method of extending memory is through the use of **virtual memory.** Virtual memory is memory that is not made up of real, physical memory chips. Virtual memory is memory made up of mass storage devices such as disks. In the use of virtual memory, the computer senses when its usable real memory is used up, stores what it deems necessary onto a disk (usually the hard disk), and then uses what it needs of the freed-up real memory. If it again needs the data it stored on the disk, it simply frees up some more real memory (by placing its contents on the disk) and then reads what it needs from the disk back into the freed-up memory. The concept of virtual memory is illustrated in Figure 34.11. Windows 95/98 uses demand-paging virtual memory to manage memory, a technique supported by features of protected mode.

MEMORY USAGE IN WINDOWS

It is not difficult to determine why an operating system performs better if it has 32MB of RAM available, rather than only 8MB. With 32MB of RAM, the operating system will be able to support more simultaneous processes without having to use the hard drive for virtual memory backup. Additional memory will also be available for the graphical user interface (multiple overlapped windows open at the same time).

It is no secret that the Windows 3.x architecture did not use memory efficiently, typically requiring at least 8MB or 16MB to get a reasonable amount of performance on a 386 or 486 CPU. Windows 95/98 also performs much better when given a large amount of RAM to work with. A minimum of 16MB or 32MB is recommended.

What does Windows 95/98 use memory for? Conventional memory, the first 640KB of RAM, still plays an important role supporting real-mode device drivers and DOS applications. For instance, if you want DOSKEY installed as part of your DOS environment under Windows 95, place its command line in AUTOEXEC.BAT as you normally would.

Upper memory, the next 360KB of the first 1MB of RAM, can be used to place DOS and other memory-resident applications above the 640KB limit, freeing more RAM for DOS applications.

Extended memory, everything above the first 1MB of RAM (4096MB total), is where Windows 95/98 runs most applications, using an addressing scheme called *flat addressing*. The flat addressing model uses 32-bit addresses to access any location in physical memory, without the need to worry about the segmented memory scheme normally used.

Windows 95 provides the Resource Meter (in the System Tools folder under Accessories) to monitor system resources. As indicated in Figure 34.12, the amount of resources available is shown graphically. The display is updated as resources are used and freed up.

Another useful tool is the System Monitor in Windows 95/98 and the Performance Monitor in Windows NT (shown in Figure 34.13), which display a running tally of resource usage over a period of time. The display format is selectable (bar, line, or numeric charts), as are the colors and type of information displayed.

FIGURE 34.12 Resource Meter display

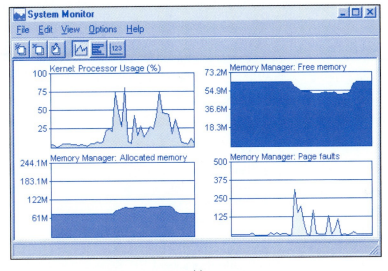

FIGURE 34.13 (a) Windows 95/98 System Monitor display and (b) Windows NT Performance Monitor display

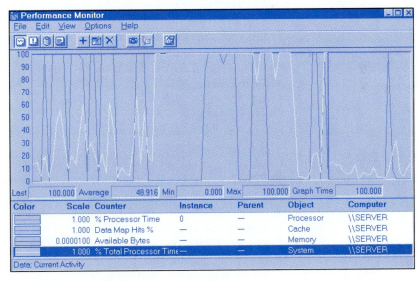

(b)

TROUBLESHOOTING TECHNIQUES

One of the simplest ways to determine whether your Windows system has enough RAM to handle its workload is to watch the hard drive light. No or little activity, except when opening or closing an application, is a good sign.

If the hard drive activates sporadically, doing a little work every now and then, the system is borderline. If the activity increases when additional applications are opened, there is a definite lack of RAM.

Frustrated with hard drive activity when only a few applications were open, one user increased the amount of RAM in his system from 32MB to 128MB (taking advantage of a drop in memory prices at that time). Now, even with a taskbar full of applications, the hard drive remains inactive.

SELF-TEST

This self-test is designed to help you check your understanding of the background information presented in this exercise.

True/False

Answer *true* or *false*.

1. All computer memory simply keeps a copy of a pattern of 1s and 0s.
2. The only useful computer memory is memory contained in computer chips.
3. When power is turned off, all computer memory is lost.
4. EDO DRAM works the same way as cache.

Multiple Choice

Select the best answer.

5. RAM is
   a. Random access memory.
   b. Read/write memory.
   c. Volatile memory.
   d. All of the above.
6. A byte is
   a. Larger than a bit.
   b. Smaller than a word.
   c. Equal to 8 bits.
   d. All of the above.
7. Base memory is
   a. All the memory that can be addressed by the microprocessor.
   b. The first 1MB of memory.
   c. The actual memory installed in the system.
   d. None of the above.
8. SDRAM stands for
   a. Static DRAM.
   b. Synchronous DRAM.
   c. Sideways DRAM.

Matching

Match a term on the right with each definition on the left.

9. Memory between 0KB and 1MB
10. Memory that uses the real mode for the 80386
11. Chunks of extra memory that can be switched in and out of conventional memory

a. Extended memory.
b. Conventional memory.
c. Expanded memory.
d. None of the above.

Completion

Fill in the blanks with the best answers.

12. _____ checking is a method used by the computer to check for errors.
13. When active memory interacts with the _____ disk, this is referred to as using _____ memory.
14. In practical applications, only _____ of memory is available for DOS systems.
15. External cache is also called _____ cache.
16. A memory wide enough for the Pentium's 64-bit data bus is the _____.

FAMILIARIZATION ACTIVITY

The first part of this familiarization activity gives you the opportunity to have your computer system use as much of its memory as possible. The second half of this activity has you examine the location and type of RAM used by the system.

1. Make sure you have DOS 5.0 or higher in your system.
2. Use the following DOS command to check how memory is used in your computer. If you have a printer available, redirect the output to the printer:

   A> MEM /PROGRAM

3. From step 2, determine whether your system has *extended* or *expanded* memory and how much. If you have problems with this, check with your instructor.
4. In this step, you force DOS to use as much higher memory as possible. Using a text editor, add the following DOS commands to your CONFIG.SYS file.

   > DEVICE = [drive][path]HIMEM.SYS
   > DEVICE = HIGH
   > DEVICE = [drive][path]EMM386.EXE

5. Now reboot your system. Again run the DOS memory program:

   A> MEM /PROGRAM

   What differences do you now see in the memory allocation of your system?
6. If your system has a device driver, such as a mouse, you can use DOS to try to force the driver into high memory. Modify your CONFIG.SYS file to add the following:

   > DOS = HIGH, UMB
   > DEVICEHIGH = [path and name of device]

7. Again, reboot your system. Using the DOS memory program

   A> MEM /PROGRAM

   determine what difference there is now in the allocation of your system memory.
8. If your system has a TSR program (such as Borland's SideKick), attempt to force as much of it as possible into high memory. Add the following to your AUTOEXEC.BAT file:

   > LOADHIGH [program name]

9. Again, reboot your system. Using the DOS memory program

   A> MEM /PROGRAM

   determine what difference there is now in the allocation of your system memory.
10. In order to use as much extended memory as possible, install a virtual disk in high memory. To do so, make sure the program VDISK.SYS (for IBM DOS) or RAMDRIVE.SYS (for MS-DOS) is on your DOS disk. Then add the following command, as appropriate to CONFIG.SYS:

    > (IBM DOS) DEVICE = VDISK.SYS/E
    > (MS-DOS)   DEVICE = RAMDRIVE.SYS/E

11. Again, reboot the system and verify that the virtual disk has been installed. Use the DOS memory program

    A> MEM /PROGRAM

    Determine what difference there is now in the allocation of your system memory.
12. Using the proper procedures, remove the case of your computer and determine the following:
    a. The locations of the RAM and ROM.
    b. The type of RAM (how much memory in each RAM chip).
    c. The speed of the RAM.
13. Check the memory statistics of your machine using the System Monitor. Watch it for 10 minutes while you open and close applications. What do you find?

QUESTIONS/ACTIVITIES

1. Does your system have any extended or expanded memory? How did you determine this?
2. Which step in the familiarization activity freed up the most user memory?
3. State how you could determine whether the virtual disk you installed was placed in high memory.
4. What is the size of the DRAM used in your system? How did you determine this?
5. What is the speed of the DRAM used in your system? How did you determine this?

REVIEW QUIZ

Within 15 minutes and with 100% accuracy,

1. Demonstrate
   a. Methods of increasing the memory usage of a personal computer.
   b. How to determine the organization of the computer's memory.
2. From a motherboard selected by your instructor, determine, within 5 minutes,
   a. The locations of the RAM and ROM.
   b. The size of the DRAM.
   c. The speed of the DRAM.

35 Motherboard Expansion Slots

INTRODUCTION

Joe Tekk was thumbing through a PC hardware catalog. One ad for a motherboard contained a description that read "Expansion: 4 PCI, 3 ISA, AGP."

Joe wondered what AGP stood for; it was a new term to him. He brought up Netscape and used Yahoo to search for "AGP." There were several hits. Joe found out that AGP stands for Accelerated Graphics Port, a high-speed hardware interface that allows 3-D graphics cards to use PC memory efficiently, for better multimedia performance.

Joe found other ads for AGP-equipped motherboards, all of which required the Pentium II/III processor. Joe laughed to himself. "Just what I needed, one more reason to replace my old motherboard."

PERFORMANCE OBJECTIVES

Upon completion of this exercise, within 5 minutes and with 100% accuracy, you will be able to

1. Identify the type of system, bus structure, and expansion slots for each of at least three computer motherboards and/or expansion cards.
2. Briefly describe the differences among the various bus architectures.

BACKGROUND INFORMATION

OVERVIEW

Expansion slots serve a very important function in personal computers. They allow you to plug in electronic cards to expand and enhance the operation of your computer. The concept of expansion slots is simple; however, in practical terms, there are many things to consider. Figure 35.1 is a simple illustration of the function of expansion slots.

An expansion slot must be able to communicate with the computer. This communication usually includes access to the microprocessor. In achieving this access, the expansion bus must not interfere with the normal operation of the microprocessor. This means that the

FIGURE 35.1 Expansion slots

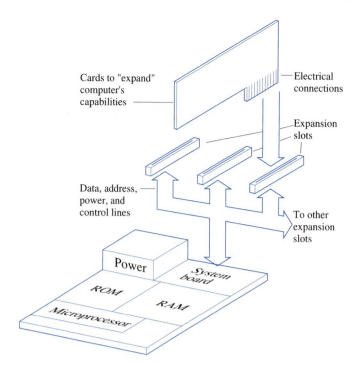

expansion bus must have access not only to the address and data lines used by the microprocessor but also to special control signals.

You need to be familiar with the functions of the various types of expansion buses that provide the means to add new features to your computer. One of the many things you will be doing for your customers is to help them make decisions about what kinds of added features they want to have for their computers. Most of these added features, such as extra memory, additional types of monitor displays, extra or different disk drives, telephone communications, and other enhancements, are added in part by electrical cards that fit into expansion slots on the computer (peripheral cards). However, as you will soon see, not all expansion slots are the same. It is important that you know their differences.

MAKEUP OF AN EXPANSION SLOT

Expansion slots have more similarities than differences. Table 35.1 lists the purposes of the different lines that are connected to the expansion slots. It should be noted that not all expansion slots use every one of these lines. The terminology used in this table does, however, apply to all expansion slots that use any of these lines.

ISA Expansion Slots

Figure 35.2 shows the ISA (Industry Standard Architecture) expansion slot. True PC compatibles also use the same kind of expansion slot. Pin assignments are shown in Figure 35.3.

The major features of the ISA expansion slots are listed in Table 35.2. The features are described in terms of a *bus*. Recall that a bus is nothing more than a group of conductors treated as a unit; as a *data bus*, it is a group of conductors used to carry data. In terms of an expansion slot, a bus can be thought of as a group of connectors that is connected to the bus on the motherboard.

Figure 35.4 shows the design of an expansion card used in a PC. Note that there are two major types of PC expansion cards: one type goes straight back from the connector and the other has a skirt that dips back down to the board level. This distinction becomes important in the design of expansion slots used in other types of computers to accommodate PC expansion cards.

TABLE 35.1 Expansion slot terminology

Connections	Purpose
Power lines	Power lines supply the voltages that may be needed by the various expansion cards. The power lines are +5 V DC, –5 V DC, +12 V DC, –12 V DC, and ground.
Data lines	Data lines are used to transfer programming information between the expansion card and the computer. One of the major differences among expansion slots in different computers is the number of data lines available.
Address lines	Address lines are used to select different memory locations. Another major difference among expansion slots is the number of address lines available.
Interrupt request lines	Interrupt request lines are used for hardware signals. These signals come from various devices, including the expansion card itself. Interrupt request signals are used to get the attention of the microprocessor. This is done so that the expansion card can temporarily use the services of the microprocessor.
DMA lines	DMA stands for *direct memory access*. DMA lines are control lines that provide direct access to memory (without having to go through the microprocessor, which tends to slow things down). DMA lines are also used to indicate when memory access is temporarily unavailable because it is being used by some other part of the system. DMA lines are used to indicate that direct memory access is being requested (called a DMA request line) and to acknowledge that request (called a DMA acknowledge).
NMI line	NMI stands for *nonmaskable interrupt*. This line is so called because it cannot be "masked," or switched off, by software. It is primarily used when a parity check error occurs in the system.
Memory-read, memory-write lines	The memory-read and memory-write lines are used to indicate that memory is either being written to or read from.
I/O read, I/O write lines	The I/O read and write lines are used to indicate that an input or output device (such as a disk drive) is to be written to or read from.
Special lines	Another one of the major differences among expansion slots in different types of computers is the number (and the types) of specialized lines used. For example, some of the OS/2 systems offer an *audio channel line* for the purpose of carrying a sound signal.

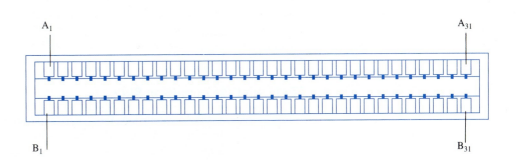

FIGURE 35.2 ISA expansion slot and pin numbering

FIGURE 35.3 ISA expansion slot pin assignments

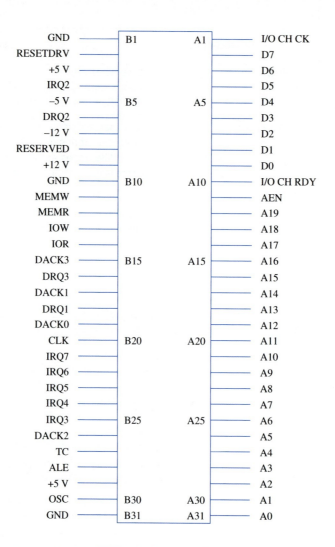

TABLE 35.2 Major features of ISA expansion slots

Type of Bus	Comments
Total pins	62 separate connectors
Data bus	8 data lines
Address bus	20 address lines (1MB addressable memory)

FIGURE 35.4 Typical ISA expansion card

AT (16-Bit ISA) Expansion Slots

Figure 35.5 shows the AT expansion slots. These expansion slots are designed to accommodate the older 8-bit ISA expansion cards as well as the newer AT expansion cards.

The reason for the two different types of slots is to accommodate the PC expansion card containing the skirt, which comes down to the board level. Note from Figure 35.5 that the expansion slots are divided into two sections. The first section has the 62 pins that are

FIGURE 35.5 Typical AT expansion slots

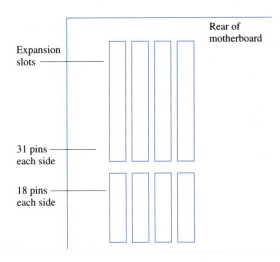

TABLE 35.3 Major features of AT expansion slots

Type of Bus	Comments
Total pins	98 separate connectors divided into two sections: one 62-pin section identical to the 8-bit ISA and one 36-pin section.
Data bus	Total of 16 data lines: first 8 located in 62-pin section, last 8 located in 36-pin section.
Address bus	Total of 24 address lines (16MB of addressable memory): first 20 located in 62-pin section, last 4 located in 36-pin section.

FIGURE 35.6 Typical AT expansion card

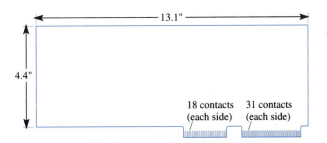

electrically identical to the 62 pins of the 8-bit ISA expansion slots. The second section contains an additional 36 pins. This gives a total of 98 electrical connections. The additional 36 pins are used to handle the additional requirements of the 80286 microprocessor used by the AT systems. Table 35.3 gives the major features of AT expansion slots.

Figure 35.6 shows a typical AT expansion card. Note that the AT expansion card has two separate rows of connectors. The second, smaller row is designed to accommodate the new requirements of the 80286 microprocessor (as well as the 16-bit 80386SX).

EXPANSION SLOT DESIGNS

Because of the need to meet the requirements of newer microprocessors such as the 80386 and beyond, an expansion slot that could accommodate a 32-bit bus was needed. There were two basic approaches to answer this need. One approach was developed by IBM in its new **micro-channel architecture** (MCA). Another approach, called the **EISA** (extended industry standard architecture), was taken by other computer manufacturers. The EISA

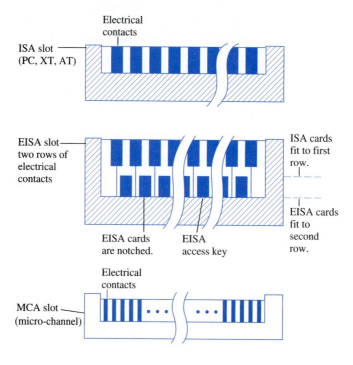

FIGURE 35.7 Three expansion slot designs

standard is compatible with the ISA bus. The main difference between the micro-channel and the EISA is that the micro-channel expansion slots are not compatible with the older ISA expansion slots. The EISA expansion slots *are* compatible with the older ISA slots. Figure 35.7 graphically illustrates the differences.

As you can see from Figure 35.7, the micro-channel expansion slots are physically smaller than the older ISA slots; their pin contacts are much closer together. The EISA slots have two rows of connectors. If an older ISA card is used, it will go down only as far as the first row of connectors; the lack of a notch on the card prevents it from going in any farther. On the other hand, if the newer EISA card is used, it will be notched and will go all the way down to the second row of connectors.

There are several different types of micro-channel expansion slots. Some are used to accommodate a 16-bit microprocessor (such as the 80386SX), whereas the others can accommodate the 32-bit microprocessors. Figure 35.8 shows the various kinds of micro-channel expansion slots.

THE MICRO-CHANNEL EXPANSION SLOT

Table 35.4 lists the key features of the 8-bit section of the micro-channel expansion slot.

Table 35.5 lists the key features of the 16-bit extension of the micro-channel expansion slot.

Table 35.6 lists the key features of the 32-bit extension of the micro-channel expansion slot.

Table 35.7 lists the key features of the video extension of the micro-channel expansion slot. This video extension is used for a **video coprocessor.** A video coprocessor is a microprocessor that is dedicated to the display, thus relieving the main system microprocessor of this chore. The result is a more detailed and quicker-responding video display for graphics and animation.

Table 35.8 lists the key features of the **matched memory extension** section of the micro-channel expansion slot. This extension is used when a higher memory transfer rate can be used by the expansion card. When an internal peripheral is capable of operating at this higher speed, the matched memory provision of the micro-channel can be used. Doing this allows data to be transferred at a 25% increase in speed.

FIGURE 35.8 Various kinds of micro-channel expansion slots

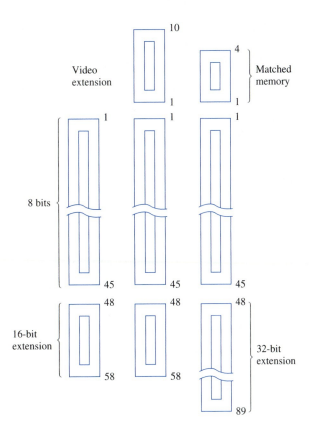

TABLE 35.4 Eight-bit section of the micro-channel expansion slot

Signals	Comments
Total pins	92 electrical connections
Address lines	24 lines; 16MB of addressable memory
Data lines	8 lines
Audio channel	Single analog audio channel for synthesized voice or music
Power lines	Several +5 V DC, +12 V DC, and −12 V DC lines with ground return (helps reduce noise interference)

TABLE 35.5 Sixteen-bit section of the micro-channel expansion slot

Signals	Comments
Total pins	24 electrical connections
Data lines	8 more data lines, increasing the total number of data lines to 16

THE LOCAL BUS

As we saw previously, the EISA connector supports 80386, 80486, and Pentium microprocessors by providing a full 32-bit data bus. Three special bus-controlling chips are used to manage data transfers through the EISA connectors. Thus, data that gets transferred between an expansion card and the CPU must go through the bus controller chip set. This effectively reduces the rate at which data can be transferred.

429

TABLE 35.6 Thirty-two-bit section of the micro-channel expansion slot

Signals	Comments
Total pins	62 electrical connections
Data lines	16 more data lines, increasing the total number of data lines to 32
Address lines	16 more address lines, increasing the total number of address lines to 32; 4GB of addressable memory

TABLE 35.7 Video extension section of the micro-channel expansion slot

Signals	Comments
Total pins	22 electrical connections
Video-control lines	Various video-control lines, such as horizontal and vertical sync, blanking, and video data lines

TABLE 35.8 Matched memory extension section of the micro-channel expansion slot

Signals	Comments
Total pins	8 electrical connections
Matched memory lines	Matched memory cycle command, matched memory cycle, matched memory request

To get around this problem, a new bus architecture was introduced, called the **local bus.** A local bus connector provides the fastest communication possible between a plug-in card and the machine by bypassing the EISA chip set and connecting directly to the CPU. Local bus video cards and hard drive controllers are popular because of their high-speed data transfer capability.

One initial attempt to define the new local bus was the VESA local bus. VESA stands for Video Electronics Standards Association, an organization dedicated to improving video display and bus technology. VESA local bus cards typically run at 33-MHz speeds, and were originally designed to interface with 80486 signals. VESA connectors are simply add-ons to existing connectors; no special VESA local bus connector exists.

THE PCI BUS

PCI stands for Peripheral Component Interconnect, and it is Intel's offering in the world of standardized buses. The PCI bus uses a *bridge* IC to control data transfers between the processor and the system bus, as indicated in Figure 35.9.

In essence, the PCI bus is not strictly a local bus, since connections to the PCI bus are not connections to the processor, but rather a special PCI-to-host controller chip. Other chips, such as PCI-to-ISA bridges, interface the older ISA bus with the PCI bus, allowing both types of buses on one motherboard, with a single chip controlling them all. The PCI bus is designed to be processor independent, plug-and-play compatible, and capable of 64-bit transfers at 33 MHz and above.

PCI connectors are physically different from all other connectors. Refer to Figure 33.2, which shows four ISA connectors and three PCI connectors. Figure 35.10 shows the pinout for a 32-bit PCI connector.

THE PCMCIA BUS

The PCMCIA (Personal Computer Memory Card International Association) bus, now referred to as *PC card bus,* evolved from the need to expand the memory available on early

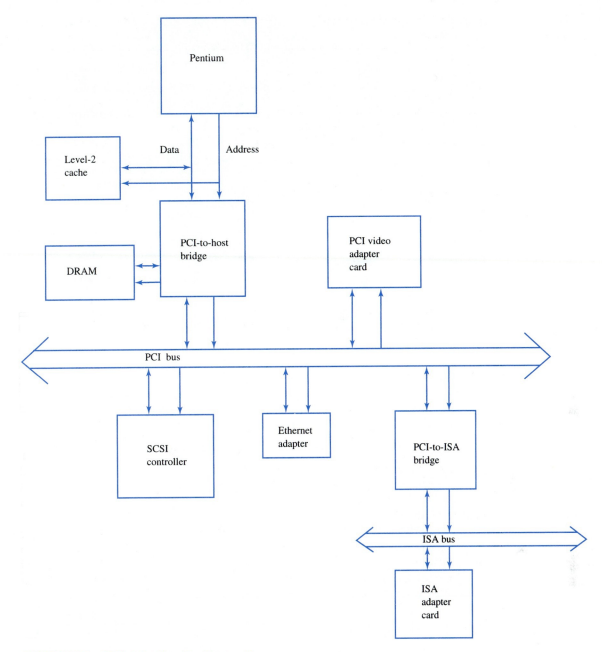

FIGURE 35.9 PCI bridge in a Pentium system

FIGURE 35.10 32-bit PCI connector

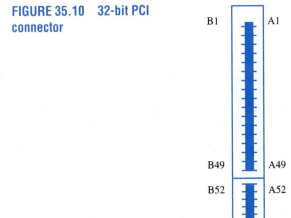

FIGURE 35.11 PCMCIA Ethernet card (*photograph by John T. Butchko*)

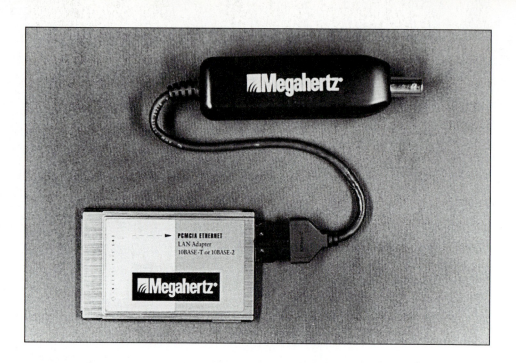

TABLE 35.9 PCMCIA slot styles

Slot Type	Meaning
I	Original standard. Supports 3.3-mm cards. Memory cards only
II	Supports 3.3-mm and 5-mm cards.
III	Supports 10.5-mm cards, as well as types I and II.
IV	Greater than 10.5 mm supported.

FIGURE 35.12 PCMCIA connector

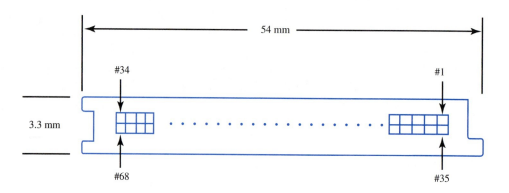

laptop computers. The standard has since expanded to include almost any kind of peripheral you can imagine, from hard drives, to LAN adapters and modem/fax cards. Figure 35.11 shows a typical PCMCIA Ethernet card.

The PCMCIA bus supports four styles of cards, as shown in Table 35.9.

All PCMCIA cards allow *hot swapping,* removing and inserting the card with power on.

A type I connector is shown in Figure 35.12. The signal assignments are illustrated in Tables 35.10 and 35.11. The popularity of laptop and notebook computers suggests the continued use of this bus.

AGP

The Accelerated Graphics Port (AGP) is a new technology that improves multimedia performance on Pentium II computers. Figure 35.13 shows where the AGP technology fits into the other bus hardware.

TABLE 35.10 PCMCIA pin assignments (available at card insertion)

Pin	Signal	Pin	Signal	Pin	Signal
1	GND	24	A_5	47	A_{18}
2	D_3	25	A_4	48	A_{19}
3	D_4	26	A_3	49	A_{20}
4	D_5	27	A_2	50	A_{21}
5	D_6	28	A_1	51	Vcc
6	D_7	29	A_0	52	Vpp2
7	CE1	30	D_0	53	A_{22}
8	A_{10}	31	D_1	54	A_{23}
9	OE	32	D_2	55	A_{24}
10	A_{11}	33	WP	56	A_{25}
11	A_9	34	GND	57	RFU
12	A_8	35	GND	58	RESET
13	A_{13}	36	CD1	59	WAIT
14	A_{14}	37	D_{11}	60	RFU
15	WE/PGM	38	D_{12}	61	REG
16	RDY/BSY	39	D_{13}	62	BVD2
17	Vcc	40	D_{14}	63	BVD1
18	Vpp1	41	D_{15}	64	D_8
19	A_{16}	42	CE2	65	D_9
20	A_{15}	43	RFSH	66	D_{10}
21	A_{12}	44	RFU	67	CD2
22	A_7	45	RFU	68	GND
23	A_6	46	A_{17}		

TABLE 35.11 PCMCIA signal differences

Pin	Memory Card	I/O Card
16	RDY/BSY	IREQ
33	WP	IOIS16
44	RFU	IORD
45	RFU	IOWR
60	RFU	INPACK
62	BVD2	SPKR
63	BVD1	STSCHG

FIGURE 35.13 AGP interface

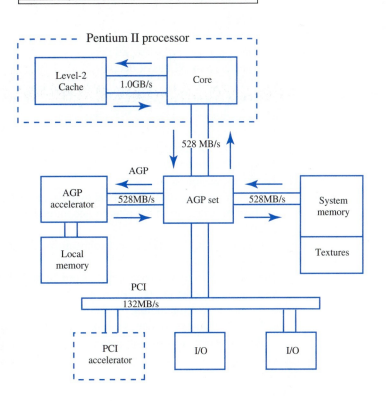

The heart of the AGP is the 440LX AGPset hardware, a *quad-ported* data switch that controls transfers between the processor, main memory, graphics memory, and the PCI bus. AGP technology uses a connector similar to a PCI connector.

With growing emphasis on multimedia applications, AGP technology sets the stage for improved performance.

TROUBLESHOOTING TECHNIQUES

Good connections between the expansion connector and the adapter card are essential to high-speed digital data communications. Get into the habit of opening your computer chassis from time to time in order to clear the motherboard, fans, and other areas where dust and other contaminants collect. There are many products available for doing cleaning of this kind, from nonconductive spray to mini-vacuum cleaners. Even the individual connector pads on an edge connector can be cleaned by rubbing them gently with a soft eraser.

Performing routine maintenance on your computer will typically save you time, effort, and money in the future.

SELF-TEST

This self-test is designed to help you check your understanding of the background information presented in this exercise.

True/False

Answer *true* or *false*.

1. Expansion slots are generally used to enhance the operation of the computer.
2. The first expansion slot design was the EISA bus.
3. The ISA connector supports 16-bit transfers.
4. All 32-bit system expansion slots are downward compatible with the 16-bit systems.
5. The IBM AT system expansion slots are compatible with the PCI expansion slots.

Multiple Choice

Select the best answer.

6. The PCI bus
   a. Uses a PCI-host bridge.
   b. Is not plug-and-play compatible.
   c. Transfers only 32 bits of data.
   d. All of the above.
7. The MCA connector
   a. Has smaller pins than the ISA connector.
   b. Supports 8-, 16-, and 32-bit buses.
   c. Has only one layer of pins.
   d. All of the above.
8. The AT expansion slots
   a. Contain 62 pins that are identical to those of the PC expansion slots.
   b. Have an extra set of 36 pins to accommodate the additional requirements of the 80286.
   c. Contain a total of 98 electrical connections.
   d. All of the above.
9. The reason that the AT expansion slots are divided into two separate sections is
   a. To accommodate PC expansion cards.
   b. To make room for cooling the cards.
   c. Because it allows for better mechanical seating of the cards.
   d. AT expansion slots are not divided into two sections.

10. All computers
    a. Use micro-channel architecture.
    b. Have EISA.
    c. Use ISA.
    d. None of the above.

Matching

Match one or more definitions on the right with each term on the left.

11. DMA
12. NMI
13. MCA
14. EISA
15. ISA

a. IBM's micro-channel.
b. PC bus system.
c. Accommodates 32 bits.
d. Accesses memory directly.
e. Causes a short interruption.
f. None of the above.

Completion

Fill in the blank or blanks with the best answers.

16. The AGP chip set requires a(n) _____ microprocessor.
17. The PCMCIA bus is now called the _____ bus.
18. The first connector to support over 1MB of memory addressing was the _____.
19. PCMCIA cards allow _____ _____.
20. PCI stands for _____ _____ _____.

Open-Ended

21. Explain the purpose of interrupt request lines.
22. Which of the following standards are compatible: ISA, EISA, MCA?
23. State how the EISA expansion slot is made compatible with the ISA expansion card.
24. Explain how micro-channel architecture can accommodate both the 16-bit 80386SX and the 32-bit 80386 microprocessors.
25. What is the distinguishing feature of micro-channel architecture?
26. Why is the local bus faster than all the other bus types?

FAMILIARIZATION ACTIVITY

Your instructor will have several different types of computers available for you to observe. These computers may be accessible to you in the lab or outside the lab. Some of these computers may be located in a retail outlet, where their observation may be part of a homework assignment. In any case, make sure you understand the background information section of this exercise well enough so that you can successfully identify the types of computers presented to you in the performance objective for this exercise.

QUESTIONS/ACTIVITIES

1. What kind of computer systems were available for you to observe as part of this lab exercise?
2. Was an AT-type computer available? If so, how many expansion slots did it have? How many of those slots were made to accommodate a PC expansion card with a "skirt" that came down to the board?
3. Was a computer with an EISA expansion slot available as part of this exercise? If so, what make of computer was it? What kind of microprocessor did it have?
4. As a part of this exercise, was an IBM computer with micro-channel architecture available? If so, what model was it? What kind of microprocessor did it have?
5. In your own words, explain the differences among ISA, EISA, and MCA.
6. If your computer contained a local bus slot, what type of expansion card was plugged into it, if any?

REVIEW QUIZ

Within 5 minutes and with 100% accuracy,

1. Identify the type of system, bus structure, and expansion slots for each of at least three computer motherboards and/or expansion cards.
2. Briefly describe the differences among the various bus architectures.

36 Power-On Self-Test (POST)

INTRODUCTION

Joe Tekk pressed the power switch on an old computer for the fifth time. Each time he tried to boot the machine, it beeped a short series of tones through the speaker and halted. Joe thought he had counted correctly but wanted to be sure, so he powered off and on for a sixth time. He heard what he expected: one short beep, then another short beep, then three beeps. He looked up the beep pattern in a diagnostic table. The beeps indicated a CMOS checksum failure.

Joe replaced the CMOS RAM, rebooted, and got the same error message. But then he went into the ROM BIOS setup, adjusted the system settings, saved them, and rebooted again. The checksum error had disappeared.

PERFORMANCE OBJECTIVES

Upon completion of this exercise, within 5 minutes and 100% accuracy, you will be able to

1. Diagnose a computer problem from the results of a POST.

BACKGROUND INFORMATION

DEFINITION OF POST

Every time you turn on a computer, it automatically goes through a series of internal tests. These internal tests are called the **Power-On Self-Tests (POSTs).** The specific testing done by POST may vary slightly from one computer to the next. However, the essential idea of this kind of automated computer testing during power-up is the same. Understanding what the POST does on any one system will give you enough information to understand other slightly different systems. The POST program is contained in the computer's ROM. Recall that ROM is memory that retains its contents even when the computer is turned off.

TABLE 36.1 POST procedures

Test Name	Action Taken
Basic system	Checks the operation of the microprocessor as well as the area of memory that contains the POST itself and some other areas of memory, including the system buses.
Display	Checks the hardware that operates the video signals. These are the signals that operate the monitor. If there is a second monitor installed in the system, it is not checked.
Memory	Checks all the computer memory. This is accomplished by having data written into the memory and read from it. A number appears in the upper left-hand corner of the screen. This number represents the amount of memory checked at that point in the test. You can watch this number slowly increase to the full amount of memory available in the system.
DMA controller Interrupt logic Programmable timers	Check the operation of all motherboard support circuitry. All components are essential for proper operation.
Keyboard	Checks the circuits that connect the keyboard to the computer. The keyboard is also checked for stuck keys.
Disk drives	Determines which disk drives are installed. If, for any reason, the system does not have any disk drives, this part of the test is omitted.
Adapter cards	Check and configure installed adapter cards. Some of these cards may themselves contain additional ROM chips that have their own POSTs.

What POST does for you, every time you turn on your computer, is to check the major sections of the motherboard, as well as the disk drives, display, and keyboard sections. If no errors are encountered, the POST begins booting up the operating system. The components tested by POST are given in Table 36.1.

POST ERROR MESSAGES

Generally, when the POST encounters an error, it indicates the type of error by a number called a POST **error code.** The POST error codes are numbers whose values indicate the type of problem encountered. Another method used by the POST to indicate errors is an audible signal. Using the system speaker, POST can indicate the type of problem by a series of short beeps.

Table 36.2 gives the general types of indications that occur when a particular type of test encounters an error in the system.

Audible Error Codes

As you saw in Table 36.2, certain errors cause specified types and numbers of beeps when detected by POST. For example, when there is a problem with the display circuitry, one long beep and two short beeps are emitted from the system speaker.

POST also has other beep codes to indicate detected failures in other parts of the system. The audible codes for IBM-BIOS are listed in Table 36.3. Other BIOS manufacturers, such as AMI and Phoenix, use different codes.

TABLE 36.2 POST error indications

Type of Failure	Indication
Basic system	A failure in the basic system test is indicated by the system's being halted, with no visible display and no beep. The cursor is left visible on the screen.
Display	A display error causes the system to emit one long beep followed by two short ones. POST continues.
Memory	With this type of error, the screen displays a numeric error code. The value of the error code indicates where the problem is in memory.
Motherboard logic	Most errors cause a beep code to be emitted prior to a system halt.
Keyboard	A keyboard failure causes a number to be displayed on the screen. For a stuck key, the value of that key is displayed.
Disk drive	The error code 601, 1780, or 1781 is displayed on the screen.

TABLE 36.3 IBM-BIOS POST audible error codes

1 short beep	System OK
2 short beeps	POST error displayed on screen
Repeating short beeps	Power supply, system boards
3 long beeps	3270 keyboard card
1 long, 1 short beep	System board
1 long, 2 short beeps	Display adapter (MDA, CGA)
1 long, 3 short beeps	EGA
Continuous beep	Power supply, system board

UNDERSTANDING POST ERROR CODES

In almost all personal computer POSTs, the numeric error codes can be broken into two major parts: a device number and a two-digit error code. The device number followed by two zeros (such as 100 for the system board) indicates that no errors have been detected. Specific error codes should be referenced using the manuals that come with the computer system.

Table 36.4 lists the major error codes and their causes. The table uses numbers such as 4xx. This means that a 4 represents an error in the monochrome display adapter. The xx represents any value from 00 (meaning no problem) to 99, indicating the specific problem in that part of the computer system. Thus, error codes for the monochrome display adapter are between 400 and 499.

TABLE 36.4 Major error codes and their causes

Error Code	Cause
01x	Undetermined problem
02x	Power supply errors
1xx	System board errors
2xx	RAM memory errors

(Continued on the next page)

TABLE 36.4 *(Continued)*

Error Code	Cause
3xx	Keyboard errors
4xx	Monochrome display adapter errors
4xx	On PS/2 systems, parallel port errors
5xx	Color graphics adapter card errors
6xx	Floppy drive or adapter card errors
7xx	Math coprocessor errors
9xx	Parallel printer adapter errors
10xx	Alternate parallel printer adapter errors
11xx	Asynchronous communications adapter errors
12xx	Alternate asynchronous communications adapter errors
13xx	Game adapter control errors
14xx	Printer errors
15xx	Synchronous data link control; communications adapter errors
16xx	Display emulation errors (specifically 327x, 5520, 525x)
17xx	Fixed disk errors
18xx	I/O expansion unit errors
19xx	3270 PC attachment card errors
20xx	Binary synchronous communications adapter errors
21xx	Alternate binary synchronous communications adapter errors
22xx	Cluster adapter errors
24xx	Enhanced graphics adapter (EGA) errors
29xx	Color or graphics printer errors
30xx	Primary PC network adapter errors
31xx	Secondary PC network adapter errors
33xx	Compact printer errors
36xx	General-purpose interface bus (GPIB) adapter errors
38xx	Data acquisition adapter errors
39xx	Professional graphics controller errors
48xx	Internal modem errors
71xx	Voice communications adapter errors
73xx	$3^{1}/_{2}$" external disk drive errors
74xx	On PS/2 systems, display adapter errors
85xx	IBM expanded memory adapter (XMA) errors
86xx	PS/2 systems point device errors
89xx	Music card errors
100xx	PS/2 multiprotocol adapter errors
104xx	PS/2 fixed disk errors
112xx	SCSI adapter errors

FIGURE 36.1 Device Manager tab

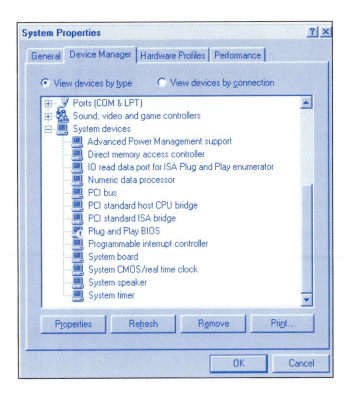

TROUBLESHOOTING TECHNIQUES

Windows 95/98 provides a way to examine what kind of hardware problems exist in the current machine configuration. Within the control panel, the Device Manager submenu of the System Properties window is used to examine all the installed hardware components recognized by Windows 95/98. Figure 36.1 shows a typical Device Manager display, indicating a problem with the plug and play BIOS.

Often, the Device Manager provides a clue to determining why the affected hardware is not working (as we have already seen in Exercise 19). If you do not have a BIOS error code reference, the Device Manager is a good substitute.

SELF-TEST

This self-test is designed to help you check your understanding of the background information presented in this exercise.

True/False

Answer *true* or *false*.

1. Every time you turn on the computer, it automatically goes through a series of internal tests.
2. The test that your computer automatically goes through checks everything except the microprocessor and math coprocessor.
3. All memory except for the memory that contains the test itself is checked by the test through which your computer goes.

Multiple Choice

Select the best answer.

4. POST stands for
   a. Power-On Self-Test.
   b. Passed-On Surface-Testing.
   c. PreprOgrammed Self-Test.
   d. None of the above.

441

5. A keyboard error encountered by POST causes
   a. The POST to stop.
   b. A number to be displayed on the screen and the testing to continue.
   c. The keyboard is not tested by POST.
   d. Only a stuck key is displayed by POST.
6. A power supply problem encountered by POST could cause
   a. No beeps at all.
   b. A continuous beep.
   c. A series of short beeps.
   d. Any of the above.

Matching

Match a problem code on the right with each type of error on the left.

7. RAM memory errors   a. 13xx
8. Keyboard errors   b. 12xx
9. Game adapter control errors c. 2xx
                               d. 3xx

Completion

Fill in the blank or blanks with the best answers.

10. The error code 8924 would represent a(n) _____ _____ error.
11. An error code of 728 means a(n) _____ _____ error.
12. Error codes of the form _____ represent problems with the graphics printer.

FAMILIARIZATION ACTIVITY

Observe and record the following error codes. Make sure you listen for audible error codes as well as look for numerical error codes.

1. Boot a normally working computer. Observe all beeps, the duration (short or long) of the beeps, and how many beeps there are. At the same time, view your screen to observe all numerical values produced by your computer's POST. Record this information here.

2. Turn off the computer and remove the keyboard plug from the computer (so the keyboard is no longer attached to the computer). Now boot the computer. Again, record all of your POST observations.

3. Turn off the computer. Reconnect the keyboard. Now, while booting the computer, hold down a key on the keyboard (this simulates a stuck key). Record your observations here (including the key you held down).

4. Turn off the computer. Remove the monitor cable from the computer case (so the monitor is no longer connected to the computer). Boot the system and record your observations of the POST.

5. If your instructor gives a demonstration of POST messages for specific problems introduced into the computer, record your observations in the following space.

QUESTIONS/ACTIVITIES

1. How long does the POST take for your computer system? (Use a digital watch.)
2. Can you tell how much computer memory your system has from any information given during the POST? Explain.
3. What happens during the POST if you hold down more than one key at the same time (simulating more than one stuck key)?
4. List all the sections that are being tested in your specific computer system during the POST. How do you know this?

REVIEW QUIZ

Within 5 minutes and with 100% accuracy,

1. Diagnose a computer problem from the results of a POST.

37 Motherboard Replacement and Setup

INTRODUCTION

One Saturday afternoon, Joe Tekk was visiting some relatives. Many of the children and young adults were involved in a heated argument about why the figures in the new computer game they were using were not moving around like they were supposed to. Joe knew none of their theories were correct, so he added his two cents to the conversation.

Joe asked, "Does your computer have an MMX processor?"

The group fell silent. Joe continued, "That software uses the new MMX instructions that are part of all of the new microprocessors. The new instructions are designed for multimedia applications." Joe examined the system and showed all who were interested how to identify the processor type. Joe was right, there was no MMX capability.

Before Joe knew what was happening, he was roped into upgrading the PC. Before he left, he determined what he would need to perform the upgrade. The next weekend he returned with all of the necessary parts.

After carefully extracting the old motherboard and inserting the new one, the procedure was a complete success. Soon after that, the screen was filled with moving game characters.

PERFORMANCE OBJECTIVES

Upon completion of this exercise, within 30 minutes and with 100% accuracy, you will be able to

1. Remove and replace a personal computer motherboard and successfully boot the computer system.
2. Identify the form factor of a motherboard.

BACKGROUND INFORMATION

In past exercises, you had the opportunity to learn about different aspects of the motherboard. You learned about the various types of microprocessors and coprocessors, types and uses of memory, and motherboard expansion slots.

In this exercise, you will discover how to remove and replace the motherboard of a personal computer properly and safely. The first time you change one should be under supervised laboratory conditions.

OVERALL SAFETY

In every case of motherboard removal and replacement, always observe electrical safety precautions. Make sure the system is turned off and the power cable is *completely* removed.

WHY REPLACE THE MOTHERBOARD?

There are many reasons why a motherboard might be replaced. The first and foremost reason is advances in technology. For someone who purchased one of the early PCs, the fast processing speeds and large hard disk space available on a Pentium would be a dream come true. For someone who already owns a Pentium, a newer Pentium or Pentium II/III with MMX technology might the desired goal. These two situations require different approaches.

Generally, the first thing to do is determine what upgrade paths are available. For someone with an old PC (386 or older), it makes sense to just buy a new system, rather than upgrade. The reason for this is simple—every piece of hardware must be replaced. This includes items like the motherboard, CPU, memory, etc. That is a complete system. Instead of building a PC from parts (unless you want the experience), simply purchase one of the new under $1000 PCs that contains a Pentium II 400 MMX, 64MB memory, 8GB disk, 32X CD-ROM, 56K modem, and so on, including the monitor.

Another common situation is when a system already contains relatively new memory, a large hard disk, super VGA graphics, a fast CD-ROM drive, and so on, but runs too slowly. These PCs may be candidates for a simple processor and/or memory upgrade. For example, a Pentium 75 can be upgraded to a Pentium 166. The various upgrade options are determined by the capabilities of the motherboard. A simple upgrade might accomplish the desired goal of increased speed at a fraction of the cost of a new PC.

The last common situation involves upgrading the PC to new technology that can still use the existing expansion cards, 72-pin SIMM memory, and disks. For example, let us take the same Pentium 75 as the starting point, but this time add the additional requirement of MMX technology as the desired goal. In this case, the motherboard can be replaced, providing a new Pentium or Pentium II microprocessor with MMX technology while at the same time using the old hard disks, memory, and all other hardware as well. At a cost of about $300 to $400, this upgrade path may provide a cost-effective solution for access to the latest technology.

The rest of this exercise is devoted to the last scenario, where the motherboard must be replaced. For example, we can assume the upgrade path is from a Pentium without MMX technology to a Pentium, Pentium Pro, or Pentium II/III using MMX technology. Each of these upgrade paths will result in the purchase of different motherboards.

MOTHERBOARD FORM FACTORS

Motherboard form factors describe the physical size, layout, and features of a particular motherboard. Currently there are several popular choices available, each of them providing different types of technology. For example, NLX motherboards provide capability for Pentium II/III processors with AGP graphics support. Table 37.1 shows some of the details for the Baby-AT, ATX, LPX, and NLX motherboard form factors.

TABLE 37.1 Common motherboard form factors

Form Factor	Size
Baby-AT	8.5" × 11"
ATX	8.5" × 11"
LPX	9" × 10.6"
NLX	Generic Riser card

TABLE 37.2 Chip set properties

Property	Types
Memory type	FPM, EDO, SDRAM, ECC, parity
Secondary cache	Burst, pipeine burst, synchronous, asynchronous
CPU type	486, P-24T, P5, P54C/P55C, Pentium Pro, Pentium II/III
Maximum memory bus speed	33, 40, 50, 60, 66, 75, 83, 100 MHz
PCI type	32 bit, 64 bit

CHIP SETS

Each motherboard provides a *chip set* designed to control the activity of the system. The chip sets are designed to provide the general features built into the motherboard. This includes items such as the memory controller, EIDE controller, PCI bridge, clock, DMA control, mouse, and keyboard controls. Table 37.2 shows some of the common chip set properties. When shopping for a motherboard, it pays to know as much as possible about your chip set options.

BIOS UPGRADES

BIOS, or the Basic Input Output System, is stored in ROM (Read Only Memory) on many of the older motherboards. In contrast, new motherboards may contain updatable BIOS memory. The ability to update BIOS may be important, as many features of the motherboard are exploited by the BIOS. For example, the new plug-and-play standard Version 1.1A is implemented in BIOS. Without the BIOS plug-and-play feature, the new plug-and-play hardware may not be recognized by the operating system. Using flash memory, a BIOS upgrade may be as close as the Internet. The actual procedure may involve the following steps:

1. Write down all current settings.
2. Download the BIOS files.
3. Create a boot floppy containing the BIOS upgrade.
4. Perform the upgrade.
5. Reset system.
6. Verify the new BIOS version.
7. Verify the settings.

Note that if the BIOS upgrade does not execute properly, the system may be left in an unbootable state. Extreme caution must be exercised when upgrading system BIOS.

REMOVAL PROCESS

Often the process of extracting the motherboard can be broken down into several steps. For example, the following steps can be used to remove the motherboard from most PCs:

1. Remove the power cord.
2. Remove all cables.
3. Remove the system cover.
4. Remove the peripheral cards.
5. Label and remove all cables from the motherboard.
    Floppy disk drive
    Hard disk drive (primary and secondary)
    CD-ROM drive

Serial and parallel
Power supply
Speaker
Indicator
Reset
Etc.

6. Remove the motherboard.
7. Clean the case, fan, and so on.

If for some reason your PC is a little different, you may need to develop a custom solution.

NEW MOTHERBOARD INSTALLATION

The installation of a new motherboard can also be broken down into several steps, as follows:

1. Insert new motherboard into the case and secure it.
2. Insert microprocessor (and cooling fan).
3. Insert memory modules.
4. Replace all *cables*.
    Floppy disk drive
    Hard disk drive (primary and secondary)
    CD-ROM drive
    Serial and parallel
    Power supply
    Speaker
    Indicator
    Reset
    Etc.
5. Insert and secure all peripheral cards.
6. Replace the cover.
7. Replace the power cord.
8. Power up the system.

TROUBLESHOOTING TECHNIQUES

This approach to changing a motherboard works very well, but sometimes there is an occasional problem, such as a connector not fitting properly or forgetting to perform one of the steps. Retrace your steps to identify and correct any problems.

It is important to note that changing the motherboard hardware is only half of the installation procedure. The hard drive still contains information about the old motherboard. For example, when booting Windows for the first time after a new motherboard has been installed, Windows will detect the new hardware and begin a process to make the necessary modifications to the Windows operating system to support the new hardware.

During this process, it may be necessary to reboot the computer several times while you view each of the new devices Windows finds. Note that you may need to supply a Windows installation CD-ROM so that Windows can copy any required files. Follow each of the instructions provided by Windows.

SELF-TEST

This self-test is designed to help you check your understanding of the background information presented in this exercise.

True/False

Answer *true* or *false*.

1. Before removing the motherboard, you should make sure all hard disks have been removed from the enclosure.

2. Whatever upgrade path is chosen, a new motherboard must be installed.
3. A Pentium can be upgraded to MMX without purchase of a new motherboard.
4. A chip set can be updated using a software procedure.
5. The form factor describes the physical size, layout, and features of a motherboard.

Multiple Choice

Select the best answer.

6. The most common motherboard form factors are
   a. ATX, LPX, and NLX.
   b. A memory controller, EIDE controller, and PCI bridge.
   c. Designed to all provide the same features.
   d. All of the above.
   e. None of the above.
7. Some types of BIOS software
   a. May be updated at a later time.
   b. May be stored in ROM.
   c. Cannot be updated.
   d. All of the above.
   e. None of the above.
8. When updating BIOS software, it is always a good idea to
   a. Reset the computer before making any changes.
   b. Write down all the BIOS settings before making any changes.
   c. Update all files directly.
   d. Get a version from the manufacturer with the best features.
9. Each chip set is designed to
   a. Address data on the hard disks directly.
   b. Coordinate all the activity on the motherboard.
   c. Stop viruses from infecting the system.
   d. Allow for use of the MMX instructions.
10. Immediately after a motherboard has been replaced, the system
    a. Boots normally and runs much faster.
    b. Does not boot and beeps three times.
    c. Detects all new hardware in one easy step.
    d. Detects all new hardware during several reboot operations.

Completion

Fill in the blank or blanks with the best answers.

11. Extreme caution must be exercised when performing a BIOS upgrade because the system may be left in a(n) _____ state.
12. The _____ includes support for plug-and-play hardware.
13. AGP graphics are used on the _____ motherboard form factor.
14. Each of the internal _____ cards must be removed prior to motherboard replacement.
15. Both the _____ and _____ disk cables must be reconnected after a motherboard is replaced.

FAMILIARIZATION ACTIVITY

Using the proper procedure, remove the motherboard on your system unit. Once the motherboard has been removed, have your instructor check all the system parts removed, including the motherboard. Then, with your instructor's approval, reinstall the motherboard in the computer system. Have your instructor check out the system before you plug it in and apply power to the system.

QUESTIONS/ACTIVITIES

1. What type of computer system was used by your lab group?
2. Outline the steps required for the removal of the motherboard from your computer system.
3. What difficulties did you encounter, if any, in the removal of the motherboard from your system?
4. Were there other computer systems used in the lab for this exercise that were different from your computer system? Explain.

REVIEW QUIZ

Within 30 minutes and with 100% accuracy,

1. Remove and replace a personal computer motherboard and successfully boot the computer system.
2. Identify the form factor of a motherboard.

38 Hard Disk Fundamentals

INTRODUCTION

Joe Tekk drove his car to a local computer store. It was a rainy, windy Saturday. He'd been all over town looking for a specific Maxtor hard drive. Everyone was sold out. By the time Joe walked into the computer store, he was soaking wet and fuming.

He walked to the rear of the store, where the hard drives were kept. Because he was wet, he did not touch anything, but he carefully searched the display cases for the hard drive he was looking for. He found it and his whole mood changed, just as a salesperson came to assist him.

"Can I help you?"

"Sure," Joe replied, his face breaking into a large grin. "I'd like this Maxtor hard drive." Joe pointed to the drive, which had the highest storage capacity of all the drives in the display.

"You are pretty happy for a guy who's soaking wet."

Joe laughed. "I don't care about being wet. I just hate having to delete files when my hard drive is full."

PERFORMANCE OBJECTIVES

Upon completion of this exercise, within 10 minutes and with 100% accuracy, you will be able to

1. Determine if the hard disk has been set up for the most efficient system operation by looking at the directory structure and the PATH command in the AUTOEXEC.BAT file.
2. Determine the type of hard drive used in your system.

BACKGROUND INFORMATION

The construction of a hard disk system is much different from that of a floppy disk system. With the floppy disk system, data is stored on each side of the disk, but in a hard disk system, there is usually more than one disk, or **platter.** Figure 38.1 shows the structure of a two-platter hard disk system, in which there are four sides for storing data.

As you can see from Figure 38.1, the four sides are labeled 0 through 3. Figure 38.2 shows how a floppy disk organizes its data in single concentric tracks, as compared with a hard disk system, which organizes its data in a combination of tracks called **cylinders.**

FIGURE 38.1 Typical hard disk structure

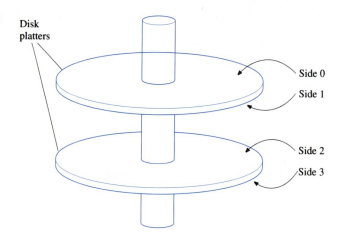

FIGURE 38.2 Floppy and hard disk organization

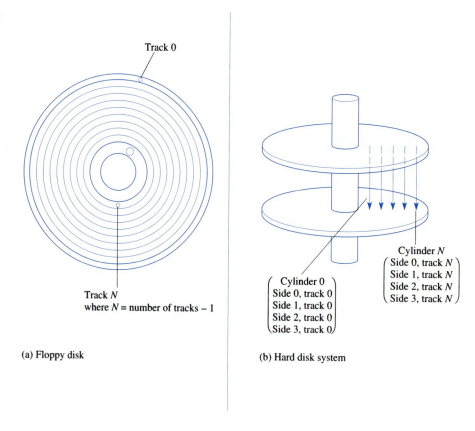

(a) Floppy disk

(b) Hard disk system

PARTITIONS

A hard disk can be formatted so that it acts as two or more independent systems. As an example, it is possible for a hard disk to operate under two entirely different operating systems, such as Windows and UNIX. Doing this is called **partitioning** the disk, as shown in Figure 38.3. This is sometimes necessary when a computer is part of a collection of computers connected together over a network.

In the early versions of DOS, no disk partition could be larger than 32MB, and no one operating system could access more than one partition. This means that no matter how much data the hard disk could hold, early versions could make use of only 32MB of the disk surface. DOS 4.0 broke the 32MB limitation on hard disks.

When a disk is partitioned, its **primary partition** (the one from which it boots) is called the C drive:, and the remaining partitions are referred to as D, E, F, and so on.

FIGURE 38.3 Single and multiple partitions of a hard disk

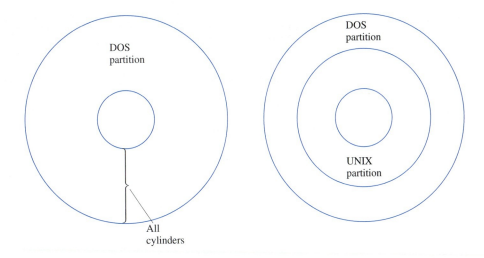

FIGURE 38.4 Windows 98 FDISK utility

Every disk may be partitioned differently, but there are a few rules that must be followed. For example, Windows 95 Version A can recognize disks as large as 2.1GB. Windows 95 Version B and above can address disks as large as 4TB (terabytes). The operating system determines the maximum size that can be handled. Updates to an operating system add capabilities for new technology.

The FDISK tool supplied by DOS and Windows is used to create partitions on the hard disk. Figure 38.4 shows the Windows 98 FDISK menu. It looks very similar to the old DOS FDISK program. FDISK is used to create, modify, or delete partitions on a hard drive. Extreme caution must be observed when working with the FDISK program.

There are also many specialty programs designed to make the disk partitioning process easier and more flexible than the FDISK program. For example, PartitionMagic by PowerQuest allows for a hard drive to be partitioned dynamically, saving time and disk space. Figure 38.5 shows the PartitionMagic main window. Information about the default drive is displayed automatically, showing the size of each partition and associated disk format. By selecting the Info Options button, the Partition Information window is displayed showing the default disk usage statistics, as shown in Figure 38.6.

The Cluster Waste tab shows the current amount of disk space that is wasted. This waste is attributed to the smallest amount of disk space that can be allocated by the operating system. For

FIGURE 38.5 PartitionMagic main window

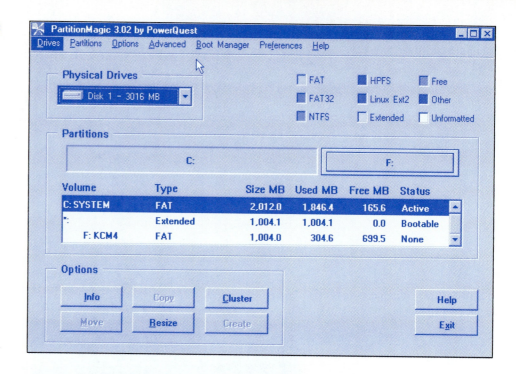

FIGURE 38.6 Disk Usage Partition Information

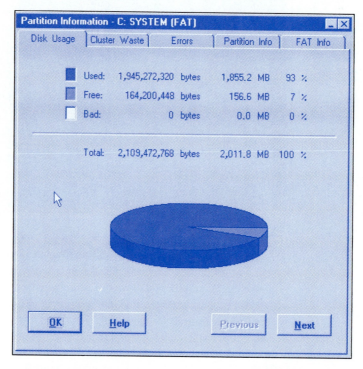

example, if we want to store one character in a file, it will be stored in a 32K chunk of space on a computer with 32K clusters (shown in Figure 38.7). The cluster size is determined by the type of file structure used on the disk, such as FAT16 or FAT32, which will be discussed shortly.

PartitionMagic can also display information about the physical layout of a partition, as shown in Figure 38.8. The first, last, and total physical sectors are displayed along with the corresponding cylinder and head information. The disk physical geometry is also indicated on the Partition tab.

Details about the FAT are available on the FAT Info tab shown in Figure 38.9. This window contains the details of the FAT structure, such as the number of FATs, root directory capacity, First FAT sector, First Data sector, and other interesting information.

FIGURE 38.7 Cluster Waste Partition Information

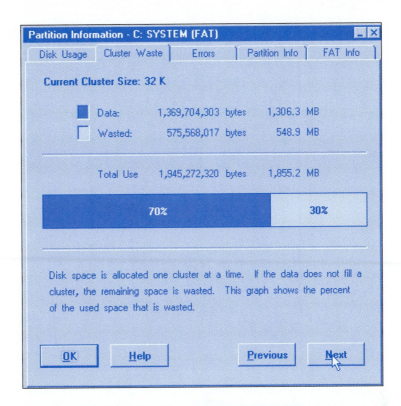

FIGURE 38.8 Physical Partition Information

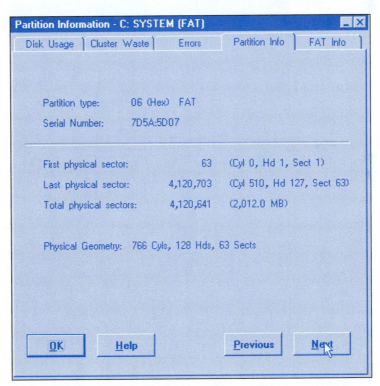

PartitionMagic can also change partition information dynamically, such as changing the cluster size. Figure 38.10 shows the common cluster sizes and associated wasted space. In this case, since the disk is close to capacity, PartitionMagic cannot recommend any type of changes; otherwise, the user may select a new cluster size and change the partition size. If it is necessary to change disk partition frequently, it may be a good idea to invest in a software package.

FIGURE 38.9 FAT Info tab

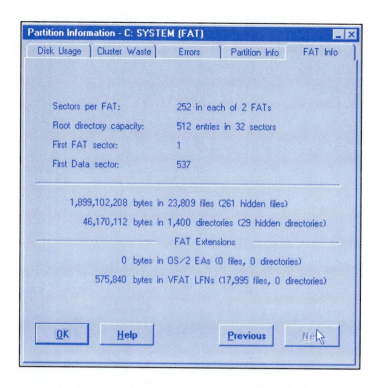

FIGURE 38.10 Cluster Analysis window

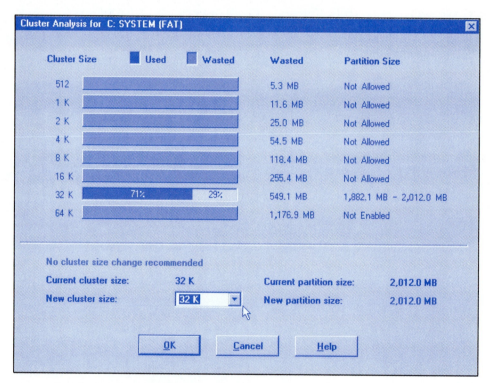

WINDOWS NT DISK ADMINISTRATOR

Windows NT provides a different method to deal with the chore of managing disks and disk partitions. The Disk Administrator utility, located in the Windows NT Administrative Tools (Common) menu, offers many advantages over running the traditional FDISK program in a DOS window. Figure 38.11 shows a graphical display of the physical partitions on a computer system. Notice how Disk 0 contains an NTFS partition at the beginning of the disk labeled "C:" and an unknown partition at the end of Disk 0 labeled "D:" with 55MB of free space remaining. Also note from the figure that Disk 1 contains an NTFS format and a

FIGURE 38.11 Windows NT Disk Administrator window

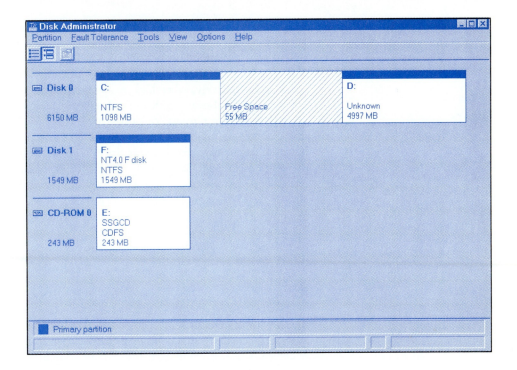

label of "F:" with no free space and the CD-ROM 0 reports information about the CD currently in the drive. On this computer, the D: disk is actually the Windows 98 operating system, but because the format is FAT32, Windows NT cannot read it. Likewise, Windows 98 cannot see the NTFS partitions.

You are encouraged to explore the capabilities of the Disk Administrator utility.

LBA

Logical block addressing (LBA), is a method to access IDE (Integrated Drive Electronics) hard disk drives. Using LBA, disks larger than 504MB (1024 cylinders) can be partitioned using FDISK or PartitionMagic. Actually, LBA has been around for quite some time now, and has been incorporated into system BIOS on most PCs. Before this, BIOS limitations prevented FDISK from using the entire drive and it was necessary to use custom software, called a Dynamic Drive Overlay. The last option was to simply stay under the limit.

LBA may be implemented in four ways. For example:

1. ROM BIOS support for INT 13h.
2. Hard disk controller support for INT 13h.
3. Use only 1024 cylinders per partition.
4. Real-mode device driver support for geometry translation.

Windows 95 and Windows 98 support the first three methods directly. The last method requires a special version of the Dynamic Drive Overlay software.

IDE disks using the ATA interface also use BIOS INT 13h services. The disk drive identifies itself to the system BIOS specifying the number of cylinders, heads, and sectors per track. The number of bytes in each sector is always 512.

FAT32

As hard drives grew in storage capacity, they quickly reached the maximum size supported by DOS and Windows (initially only 32MB, then 504MB, then 2.1GB). This limitation was based on the number of bits used to store a cluster number. The original FAT12 used a 12-bit FAT entry. FAT16 added four more bits, allowing for up to 65,536 clusters. With each cluster representing 16 sectors on the disk, and each sector storing 512 bytes, a cluster would contain 8KB. The total disk space available with 65,536 clusters of 8KB is 512MB, a small hard drive by today's standards.

One way to support larger partitions is to increase the size of a cluster. Storing 32KB in a cluster allows a 2048MB (2.048GB) hard drive, but also increases the amount of wasted space on the hard drive when files smaller than 32KB are stored. For example, a file of only 100 bytes is still allocated 32KB of disk space when it is created because that is the smallest allocation unit (one cluster). You would agree that most of the cluster is wasted space. Some disk compression utilities reclaim this wasted file space for use by other files. In general, however, large cluster sizes are not the solution to the limitation of the FAT16 file system.

FAT32 uses 32-bit FAT entries, allowing 2200GB hard drives without having to result to using large cluster sizes. In fact, FAT32 typically uses 4KB clusters, which helps keep the size of the FAT small and lowers the amount of wasted space. FAT32 is only used by Windows 95 Version B (OEM Service Pack 2) and Windows 98. Several utilities, such as PartitionMagic, are able to convert a FAT16 disk into a FAT32 disk. Windows NT has its own incompatible file system called NTFS.

NTFS

The **NT file system,** or NTFS, is used on Windows NT computers. Using NTFS, it is possible to protect individual items on a disk and therefore prevent them from being examined or copied. This is a feature commonly found on multiuser computers such as Windows NT.

HARD DRIVE INTERFACES

Many companies manufacture hard drives for personal computers. Even though each company may design and build its hard drives differently, the interface connectors on each drive must conform to one of the accepted standards for hard drive interfaces. These interface standards are illustrated in Figure 38.12.

FIGURE 38.12 Hard drive connections

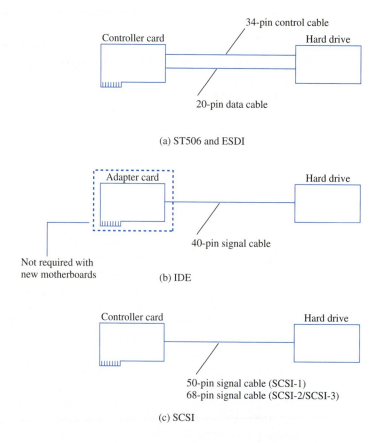

The first popular hard drive interface scheme was invented by Shugart Technologies. Called **ST506,** it requires two cables (control and data) between the controller card and the hard drive. This is shown in Figure 38.12(a). Serial data passes back and forth between the controller and hard drive over the data cable. A second hard drive is allowed; the second drive shares the control cable with the first drive, but has its own data cable. Jumpers must be set on each drive for proper operation, and the last drive needs to contain termination resistors.

An improvement on the ST506 standard was developed by Maxtor, another hard drive manufacturer. Called **ESDI** (Enhanced Small Device Interface), this interface uses the same two cables as the ST506, but allows data to be exchanged between the controller and hard drive at a faster rate. Although similar in operation to ST506, ESDI is not electrically compatible with it. Thus, ST506 hard drives require ST506 controllers, and ESDI hard drives require ESDI controllers.

The **IDE** (Integrated Drive Electronics) interface, shown in Figure 38.12(b), has virtually replaced the older ST506 standard. A single cable is used to exchange parallel data between the adapter card and the hard drive. A second hard drive uses the same cable as the first, with a single jumper on each drive indicating if it is the primary drive. A significant difference in the IDE standard is that the controller electronics are located *on the hard drive itself.* In a two-drive system, the primary drive controls itself and the other hard drive as well. The adapter card plugged into the motherboard merely connects the hard drive to the system buses, using parallel data transfers. This allows typical data transfers with an IDE hard drive of up to 8MB per second.

New motherboards have built-in **EIDE** (enhanced IDE) controllers, which provide signals for four EIDE connectors. This eliminates the need for an adapter card. One pair of connectors are the *primary* connectors, the other pair are the *secondary* connectors. Each pair can support two IDE hard drives (CD-ROM and tape backup drives as well) in a master/slave configuration. Figure 38.13 shows the pin and signal assignments for the EIDE interface, and a typical IDE hard drive.

The earlier IDE interface lacked the upper eight data lines D_8 through D_{15}. The IDE specification also limited hard drive capacity to just 504MB, a small size by today's standards. The EIDE specification increases drive capacity to more than 8GB, expands the maximum number of drives from two to four, and increases the data transfer rate to over 16MB/second.

Right-clicking on My Computer and then choosing Properties allows you to check the system properties. In Device Manager, double-click on the Hard disk controller entry and then the primary IDE controller entry. You should get a settings window similar to that shown in Figure 38.14. In Windows NT, this information is available in the Windows NT Diagnostics window.

As indicated in Figure 38.14, the primary IDE controller uses interrupt 14. A handful of I/O ports are required as well, to issue commands and read data from the controller.

The **SCSI** (Small Computer System Interface) standard, shown in Figure 38.12(c), offers more expansion capability than any of the three previously mentioned standards. Like the IDE standard, the SCSI standard uses a single cable to control the hard drive. But where all three previous standards allowed at most two hard drives, the SCSI standard allows up to *seven* devices to be daisy-chained on the single 50-pin cable. Each device can be a hard drive, if necessary. This difference is important to computer users who have large storage requirements and might require **gigabytes** (1GB = 1024MB) of hard drive capacity. SCSI is used to connect a wide variety of devices (usually 8 or 16) together on a shared bus. For example, several disk drives, a tape drive, and a scanner can be connected to one SCSI bus. Each device on the bus and the controller card itself requires an address and the end of the SCSI cable, or the last device on the bus must be terminated. Figure 38.15 shows a daisy chain of SCSI devices. The last device contains a terminator.

There are several different types of SCSI buses, each allowing for a specific cable type, transfer rate, bus width, and so on. Table 38.1 shows the different SCSI standards.

SCSI buses also have specific length requirements falling into two categories: single-ended and differential. A single-ended SCSI bus is cheap and fast over short distances. Differential SCSI can be used over longer distances. Table 38.2 shows the SCSI bus length requirements.

FIGURE 38.13 (a) Primary IDE connector pin/signal assignments and (b) typical EIDE hard drive (*photograph by John T. Butchko*)

Signal	Pin	Pin	Signal
$\overline{\text{RESET}}$	1	2	GND
D_7	3	4	D_8
D_6	5	6	D_9
D_5	7	8	D_{10}
D_4	9	10	D_{11}
D_3	11	12	D_{12}
D_2	13	14	D_{13}
D_1	15	16	D_{14}
D_0	17	18	D_{15}
Ground	19	20	Key (missing pin)
DRQ3	21	22	GND
$\overline{\text{IOW}}$	23	24	GND
$\overline{\text{IOR}}$	25	26	GND
IOCHRDY	27	28	ALE
DACK3	29	30	GND
IRQ14*	31	32	IO16
A_1	33	34	GND
A0	35	36	A_2
CS0	37	38	CS1
SLV/ACT	39	40	GND

* IRQ15 for the secondary IDE connector

(a)

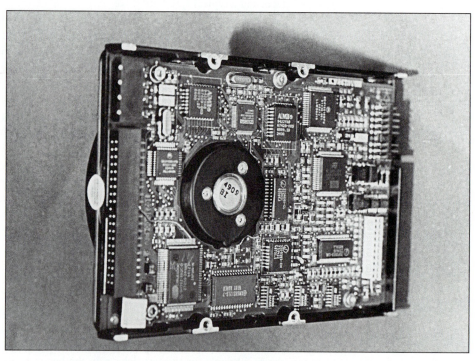

(b)

There are also different types of connectors used to connect SCSI devices together. Check the individual requirements for each SCSI device to determine the appropriate type. Note that SCSI devices are generally more expensive than non-SCSI devices, but they provide for combinations of devices not possible with standard PC technology.

FIGURE 38.14 Hardware settings for primary IDE controller

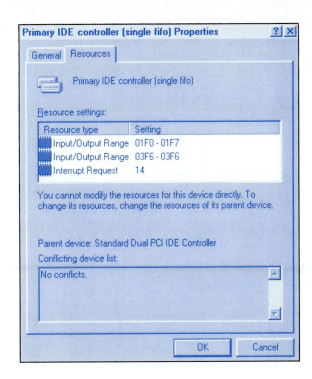

FIGURE 38.15 A daisy chain of SCSI devices

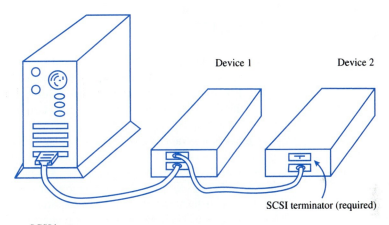

TABLE 38.1 SCSI bus standards

Standard	Bus Width	Max Transfer Rate (Mbps)	Cable Type
SCSI-1	8	4	Not specified
SCSI-2	8 16	5 10	A B
SCSI-3	16 32	10 20	P P, Q

TABLE 38.2 SCSI bus lengths

Bus Type	Single-Ended	Differential
SCSI-1	6 meters	25 meters
SCSI-2	6 meters	25 meters
SCSI-3	3 meters	25 meters

461

TABLE 38.3 Data recording techniques

Technique	Meaning/Operation
MFM	Modified Frequency Modulation. Magnetic flux transitions are used to store 0s and 1s.
RLL	Run Length Limited. Special flux patterns are used to store *groups* of 0s and 1s.
Advanced RLL	Advanced Run Length Limited. Permits data to be recorded at higher density than RLL.

DATA STORAGE

Although there are differences between IDE hard drives and SCSI hard drives (and all the other types), there is also something in common: each hard drive uses the flux changes of a magnetic field to store information on the hard drive platter surface. A number of different techniques are used to read and write 0s and 1s using flux transitions. Some of the more common techniques are listed in Table 38.3.

DISK CACHING

Because of mechanical limitations (rotational speed of the platters; movement and settling time of the read/write head), the rate at which data can be exchanged with the hard drive is limited. It is possible to increase the data transfer rate significantly through a technique called **caching.** A hardware cache is a special high-speed memory whose access time is much shorter than that of ordinary system RAM. A software cache is a program that manages a portion of system RAM, making it operate as a hardware cache. A computer system might use one or both of these types of caches, or none at all.

The main idea behind the use of a cache is to increase the *average* rate at which data is transferred. Let us see how this is done. First, we begin with an empty cache. Now, suppose that a request to the hard drive controller requires 26 sectors to be read. The controller positions the read/write head and waits for the platters to rotate into the correct positions. As the information from each sector is read from the platter surface, a copy is written into the cache. This entire process may take a few *milliseconds* to complete, depending on the drive's mechanical properties. If a future request requires information from the same 26 sectors, the controller reads the copy from the cache instead of waiting for the platter and read/write head to position themselves. This means that data is accessed at the faster rate of the cache (whose access time might be as short as 10 *ns*). This is called a cache *hit*. If the requested data is not in the cache (a *miss*), it is read from the platter surface and copied into the cache as it is outputted, to avoid a miss in the future. The cache uses an algorithm to help maintain a high hit ratio.

The same method is used for writing. Data intended for the hard drive is written into the cache very quickly (8MB/second), and then from the cache to the platter surface at a slower rate (2.5MB/second) under the guidance of the controller.

Many hard drives now come with 256KB of onboard hardware cache. In addition, a program called SMARTDRV can be used to manage system RAM as a cache for the hard drive. To use SMARTDRV, a line such as

C:\DOS\SMARTDRV.EXE 2048

must be added to your AUTOEXEC.BAT file. This command instructs SMARTDRV to use 2MB of expanded or extended memory as a cache. Small programs that are run frequently (DOS utilities stored on the hard drive) load and execute much more quickly with the help of SMARTDRV.

FIGURE 38.16 Disk sectors

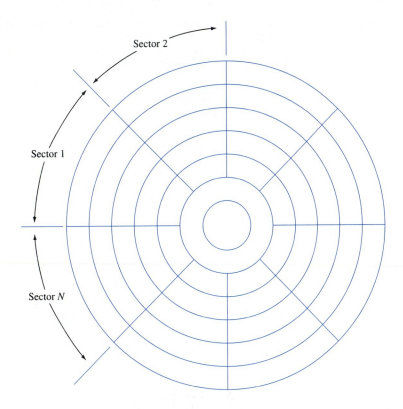

DISK STRUCTURE

The information presented here applies equally well to floppy disks as to hard disks. Figure 38.16 shows how a disk is divided into **sectors.** From the figure, you can see that a sector is a specified pie-slice area on the disk.

Disk sectors and tracks are not physically on the disk, just as data is not physically on the disk. They are simply magnetic patterns placed on the disk by electrical impulses. The number of tracks available on the disk varies. For example, a standard $3^1/2''$ floppy disk has 80 tracks, whereas a hard disk may contain 650. The number of sectors a disk has also varies. This is illustrated in Table 38.4.

DISK STORAGE CAPACITY

You can calculate the storage capacity of a disk as follows:

$$DSC = sides \times tracks \times sectors \times size$$

where

DSC = disk storage capacity
sides = number of disk sides used
tracks = number of disk tracks per side
sectors = number of disk sectors per side
size = size of each sector in bytes (usually 512)

As an example, consider a double-sided, double-density (nine-sector-per-track) disk. From Table 38.4, you can see that such a disk has 40 tracks per side and nine sectors per track, where each sector stores 512 bytes. Thus, for this type of disk, the total disk storage capacity is

$$DSC = 2 \times 40 \times 9 \times 512 = 368,640 \text{ bytes (or 360KB)}$$

Recall that each disk contains a boot sector. This boot sector is contained in sector 1, side 0, track 0, as illustrated in Figure 38.17. Table 38.5 lists the information contained in a boot sector.

TABLE 38.4. Floppy disk configurations

Disk Size	Disk Type	Tracks/Side	Total Sectors*
5.25	Single-sided—8 sectors per track	40	320
5.25	Single-sided—9 sectors per track	40	360
5.25	Double-sided—8 sectors per track	40	640
3.5	Double-sided—9 sectors per track	40	720
3.5	Quad-density—9 sectors per track	80	1440
5.25	Quad-density—15 sectors per track	80	2400

*Common sector sizes for disks are 128, 256, 512, and 1024 bytes.

FIGURE 38.17 Location of the boot sector

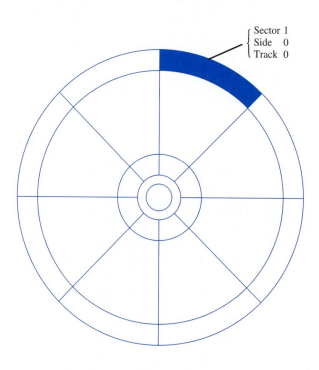

TABLE 38.5 Information in a boot sector

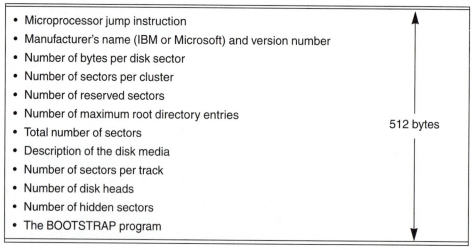

- Microprocessor jump instruction
- Manufacturer's name (IBM or Microsoft) and version number
- Number of bytes per disk sector
- Number of sectors per cluster
- Number of reserved sectors
- Number of maximum root directory entries
- Total number of sectors
- Description of the disk media
- Number of sectors per track
- Number of disk heads
- Number of hidden sectors
- The BOOTSTRAP program

512 bytes

Following the boot sector, there is a file allocation table (FAT). This table is used by DOS to record the number of disk sectors on the disk that can be used for storage as well as bad sectors that must not be used for storage. Several sectors are reserved for the FAT. All hard disks come from the factory with a certain number of bad sectors. During final product testing, these bad sectors are usually found and are usually then labeled on the hard drive unit itself.

In order to ensure reliability, each disk contains a duplicate copy of the FAT. This means that if one copy of the FAT goes bad, the backup copy is available for use.

DISK FRAGMENTATION

Disk fragmentation is the result of one or more disk files being contained in scattered sectors around the disk, as shown in Figure 38.18. Observe that the read/write heads may take more than one revolution of the disk to read all the file information scattered across the various sectors as a result of disk fragmentation.

Next, consider the file data distributed in contiguous sectors around the disk, as shown in Figure 38.19. The way the data is distributed here, it is conceivable that it could all be read in one revolution of the disk. The difference between fragmented data and contiguous data is that it takes longer to read fragmented data.

FIGURE 38.18 Disk fragmentation

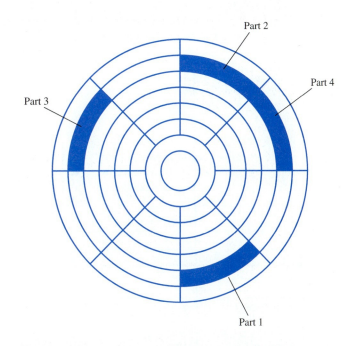

FIGURE 38.19 Contiguous file data on the same disk track

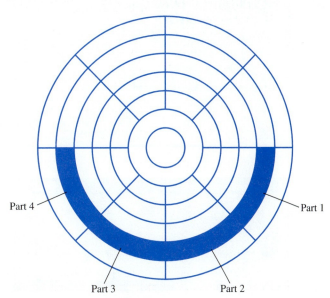

You must keep in mind that disk drives are slow when compared with the rest of the computer system. If information is fragmented over the hard disk, it takes longer to read the disk and significantly slows the entire computer system (because it takes longer to read and write the information). Disk fragmentation occurs when files are repeatedly added and deleted on a disk. Once disk fragmentation occurs (especially on the hard drive), your system will begin to run more slowly when it interacts with the hard drive. To eliminate this, you must use a hard disk utility program to defragment the disk. You will learn how to do this in the next exercise. It is important now to realize that you must do this periodically in order to maximize system performance.

The defragmentation process can be automated in Windows 95/98 using the system agent. During periods of inactivity, the system agent will periodically run the disk utility tools, keeping the disk in reasonably good condition. The defragmentation process can also be run on demand by selecting Disk Defragmenter from the System Tools submenu (located under Accessories on the Start menu). Figure 38.20 shows the initial defrag screen presented by Windows.

From this menu, it is very easy to start the defragmentation process, select another drive, check or modify the advanced parameters, or exit. Figure 38.21 shows the Disk Defragmenter Settings window.

Figure 38.22 shows the brief status of the defragmentation process. We can also view the details of the defragmentation process by clicking the Show details button as shown in

FIGURE 38.20 Selecting a disk to defragment

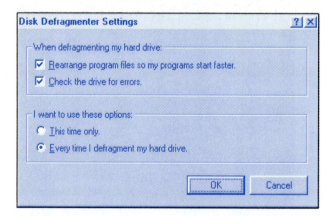

FIGURE 38.21 Disk Defragmenter Settings window

FIGURE 38.22 Defragmentation status window

FIGURE 38.23 Details of the defragmentation process

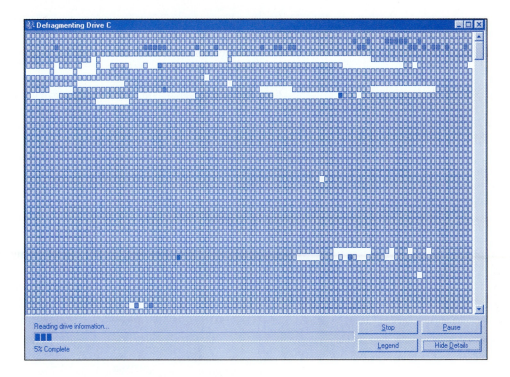

FIGURE 38.24 Defragmentation details legend

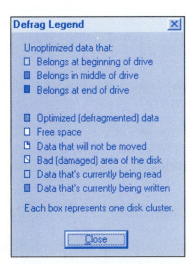

Figure 38.23. Notice how each of the disk clusters is presented on the screen. It is interesting to watch the defragmentation process. Depending on the amount of fragmentation, the process may last just a few minutes or as long as a few hours. Select Legend to view the legend shown in Figure 38.24 to help you identify the different types of disk clusters. As you can see, there are many different possible states for a disk cluster.

FILE ALLOCATION

Every time DOS has to get space on the disk for a file, it looks at the FAT for unused disk **clusters.** A cluster is a set of contiguous disk sectors. DOS will always try to minimize disk fragmentation, if possible. Every file written to the disk has a directory entry that is a 32-byte record, which contains the following information:

- Name of file
- File extension
- Attribute byte

- File time
- File date
- Number of the starting cluster
- Size of the file

DOS allocates disk space for directory entries in a special directory area that follows the FAT. This means that a disk can hold only a certain number of files, depending on the size and density of the disk. To DOS, a subdirectory is treated simply as a standard DOS file. This means that DOS stores subdirectory information in the same manner as it stores file information. Because of this, the number of subdirectories is also limited by the amount of free file space on the disk.

OPTIMIZING DISK PERFORMANCE

When setting up the structure of a hard disk, you should do so in a manner that will optimize the overall performance of the entire system. This means having your files organized to enhance maximum disk performance and to defragment the disk periodically to get rid of disk fragmentation.

For setting up the directory of a hard disk, the root directory should contain only the following:

- Operating system files
- AUTOEXEC.BAT
- CONFIG.SYS

All other files should be placed in subdirectories. For example, all the Windows files should be placed in \WINDOWS. No other files should be added to the root directory. Doing this will improve system performance.

In addition to careful directory planning, you can instruct Windows to use the hard disk in a certain way, depending on the role your machine is playing on the network. Figure 38.25 shows the File System Properties window, accessed through the Performance menu of System Properties. Currently the machine is set up as a network server, enabling a 64KB read-ahead on the hard drive to help supply data faster. Other options are desktop computer and mobile docking station.

USING THE PATH COMMAND

To maintain optimum system performance, you should carefully structure the PATH command. Recall that the PATH command specifies a series of drives and subdirectory paths that can be searched by the operating system every time a command is not found in

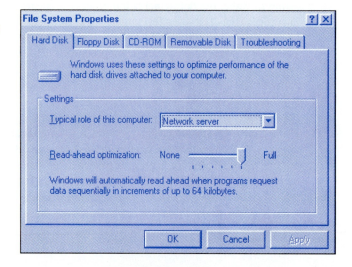

FIGURE 38.25 Windows 98 File System Properties window

TABLE 38.6 Cost of 1 bit of hard drive storage

Hard Drive	Cost	Year Purchased	Cost/bit
18MB	$2500	1981	4.29¢
30MB	$300	1985	.000953¢
212MB	$375	1989	.000168
853MB	$319	1995	.000035
1.7GB	$250	1996	.000013¢
5.1GB	$199	1998	.000004¢
8.4GB	$129	1999	.000001¢

the current directory. This command is kept in the AUTOEXEC.BAT file. Recall that the form of the PATH command is

PATH [[Drive:]path[;[drive:][path] . . .]

What happens here is that the operating system first searches in the order given by the PATH command, from left to right, for the path of a particular file that is not in the active directory. Thus, you should organize the PATH command so that the most frequently used paths appear first and the least frequently used paths appear last.

TROUBLESHOOTING TECHNIQUES

For a little historical perspective about the price of 1 bit of hard drive storage, examine the numbers in Table 38.6. Clearly, it is getting cheaper every day to buy a large hard drive, with more than a 160-fold reduction in the cost of a bit in the past ten years.

SELF-TEST

This self-test is designed to help you check your understanding of the background information presented in this exercise.

True/False

Answer *true* or *false*.

1. The construction of a hard disk system is the same as that of a floppy disk system.
2. A disk in a hard drive system is referred to as a *platter*.
3. A hard disk can be partitioned so that it has more than one operating system.
4. IDE and EIDE hard drives have the same capacity.
5. Each disk sector contains 1024 bytes.

Multiple Choice

Select the best answer.

6. In a hard disk system, a combination of tracks is called
   a. The drive.
   b. A cylinder.
   c. A cluster.
   d. The spindle.
7. In DOS versions 3.3 and earlier, the maximum size of the primary partition was
   a. 640KB.
   b. 32MB.
   c. 32KB.
   d. Limited by the size of the disk.
8. A disk *sector* is
   a. The part of the disk only written to.
   b. Two or more consecutive tracks.
   c. Used in hard disks, not in floppies.
   d. A pie-slice portion of a disk.

9. SCSI support requires
   a. A special socket.
   b. An SCSI controller card.
   c. The SCSI chip set.
   d. All of the above.
10. FAT32 file systems require
    a. 32 copies of the FAT for safety.
    b. Windows 95B.
    c. A compressed disk.
    d. All of the above.

Matching

Match a statement on the right with each DOS version on the left.

11. 3.2
12. 4.01
13. 5.0

a. Partition size may be up to 2GB.
b. Removed the 32MB limit for the primary partition.
c. Size limit of a single partition is 32MB.
d. No size limit on a partition other than the primary one.

Completion

Fill in the blank or blanks with the best answers.

14. The term _____ _____ is used to indicate that one or more disk files are contained in scattered sectors around the disk.
15. When setting up a hard disk for optimum performance, you should have as few files as possible in the _____ directory.
16. The _____ command determines the order in which directories will be searched for files not in the active directory.
17. SCSI stands for _____ _____ _____ _____.

FAMILIARIZATION ACTIVITY

1. Using the technical manual for the hard disk system in your assigned computer, determine the following:
   a. The number of platters in the hard drive system.
   b. Maximum storage capacity of the hard drive system.
   c. The type of hard disk system used.
2. Turn on your system and determine the following:
   a. Operating system version used.
   b. Structure of the directories.
   c. Structure of the PATH command.
3. Based on the information you obtained in this activity, write a recommendation of how the system performance could be improved; if you don't think it could be improved, state why.

QUESTIONS/ACTIVITIES

1. In your own words, explain the major differences between a hard disk system and a floppy disk system.
2. List and explain the different types of hard disk systems.
3. Sketch what is meant by a cylinder in a hard disk system.
4. Describe what is meant by a disk partition.
5. Determine the disk storage capacity of a hard disk system that has four sides, 160 tracks, nine sectors, and a sector size of 1024 bytes.

REVIEW QUIZ

Within 10 minutes and with 100% accuracy,

1. Determine if the hard disk has been set up for the most efficient system operation by looking at the directory structure and the PATH command in the AUTOEXEC.BAT file.
2. Determine the type of hard drive used in your system.

39 Hard Drive Backup

INTRODUCTION

Joe Tekk had just gotten back from lunch when the phone rang. It was Don, his manager. Don described a problem one of RWA's customers was having with their computer.

Don said, "Jim told me that when they turn the computer on, it displays the message 'hard disk 0 failure,' or something like that."

Joe knew this was a bad sign and immediately began wondering if someone was regularly backing up the files on the computer. Joe asked, "Don, did they perform backups?"

Don responded in a very serious tone, "I asked that question, too, and got the response you don't want to hear." He continued, "Joe, please take care of this for me."

Joe decided to make the best out of a bad situation by helping the customer implement a regular backup procedure so this problem will not happen again. Luckily, he was also able to recover many of the files from the defective hard drive. When he returned to the shop, the disk worked fine . . . for about 15 minutes.

PERFORMANCE OBJECTIVES

Upon completion of this exercise, within 10 minutes and with 100% accuracy, you will be able to

1. Back up and restore a set of files from one disk to another and read the backup log created in the process.
2. Use PKZIP and PKUNZIP to back up and restore a compressed directory of files.
3. Back up and restore a set of files from disk to tape and tape to disk.

BACKGROUND INFORMATION

In this exercise, you have the opportunity to learn how to copy the information from a hard disk onto floppies or tape, partition the hard disk, and then restore information back to the hard disk. This process needs to be done whenever a hard disk is replaced.

First, you will see how to back up a hard drive. Then you will be presented with the information on how to partition the drive using DOS or Windows. Next, you will see how to return the information from the backup disks to the hard drive. Last, you will discover the low-level formatting process used to prepare new disks for higher-level DOS formatting. Note that many hard drives *do not* require low-level formatting. They are automatically low-level formatted at the factory.

DISK BACKUP

The contents of a hard drive are very important to the overall operation of any computer system. This is where all the frequently used programs, files, and information are kept. A hard drive crash can cause the loss of some or all of this valuable information. The periodic backing up of its files is an important part of using a computer. The hard disk should be backed up periodically, as well as before a computer is shipped.

BACKUP SCHEDULES

Generally, the task of performing disk backups is taken very seriously. A strict schedule of backups should be performed regularly. Since it can take a lot of time to back up a hard disk, usually the backup process is broken down into two types of backups. First, a full backup is used to write all the information from a hard drive to a tape drive, floppy disks, or Zip disks with enough excess capacity to contain the backed-up files. After a full disk backup has been completed, incremental backups can be performed to save any files that have been modified since the last backup date. Table 39.1 shows a typical backup schedule using a combination of both full and incremental saves.

The tapes or disks containing the backed-up data may be kept in a safe place for as long as necessary. For example, a PC containing information collected at a bank, medical, or dental office may need to be saved for some specific period of time as required by law, thus requiring many tapes to be used. In contrast, disk file information maintained by a video store needs to be completely restored from the last full backup and any incremental backups. Each situation may be different, but the same question needs to be answered in each case: If the hard drive crashes, can the data be restored?

WINDOWS 95/98 BACKUP

Windows 95/98 comes with backup software used to write information to a limited number of tape devices. In order to determine whether a compatible tape device is attached, it is necessary to run the backup utility program. Figure 39.1 shows the first screen displayed by the Microsoft Backup program indicating the general steps involved in performing a backup. Generally, performing a backup involves selecting the files to back up, selecting the destination device, and starting the process. It is likely the backup software will confirm any action where data would be overwritten. Answer all questions very carefully.

Next, as shown in Figure 39.2, Backup will automatically create a full backup file set, which can be used to restore a Windows 95/98 boot disk, including all the Registry files. This file set is used only when performing a full backup and restore, and is not used for incremental backups. This backup file set is used to restore files after a catastrophic hard disk failure.

The Options window selected from the Settings pull-down menu allows the individual settings of the Backup program to be modified. The Backup utility contains components used to back up, restore, and compare files. For example, to modify Backup settings, simply select the Backup tab. Figure 39.3 shows the Settings—Options for the Backup operation.

TABLE 39.1 Typical file backup schedule

Day	Backup Type
Monday	Incremental
Tuesday	Incremental
Wednesday	Incremental
Thursday	Incremental
Friday	Full

FIGURE 39.1 Windows Backup system tool

FIGURE 39.2 Backup automatically creates a full backup file set

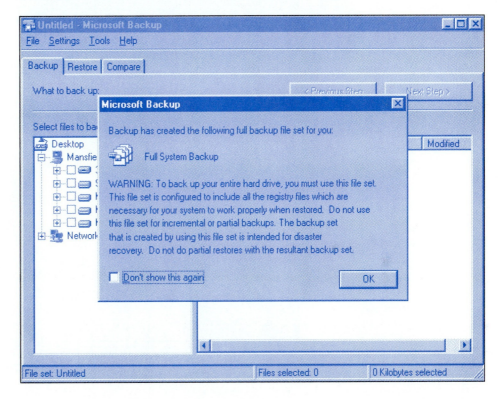

Either a full backup or an incremental backup *type* can be selected, and other options such as data compression and tape or floppy erasure settings can be modified as necessary. Care must be exercised when changing any of the Backup settings.

Just as the Backup settings may be changed, the settings of the Restore operation may also be modified as shown in Figure 39.4. Choices can be made as to where files will be restored, and what types of action Backup will take when it comes across files with dates different from those recorded on the tape. Again, care must be taken when modifying any of these options.

To select files to back up, simply click on the drive or folder and then select the Next Step button as shown in Figure 39.5. It is possible to select any combination of disks or folders.

Unfortunately, as shown in Figure 39.6, Backup did not find a compatible tape drive and the destination for the backup must be selected from the remaining devices. Sometimes,

FIGURE 39.3 Backup Settings—Options

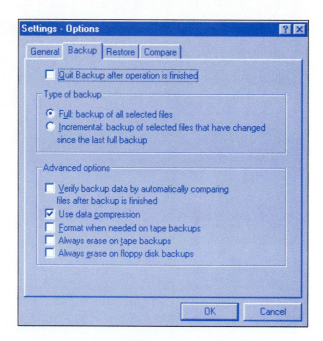

FIGURE 39.4 Restore Settings—Options

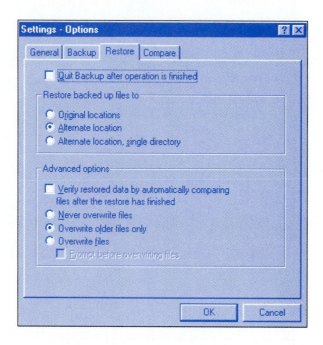

this is an adequate solution, but usually the cost to store backup information on a hard disk is too expensive when compared with the cost of a tape capable of storing 2, 4, or 8GB of information.

Many times, tape drives come with their own software and do not rely on the built-in Backup utility. The Hewlett-Packard tape drive designed for use under Windows 95/98 comes with its own software that mimics the functions of the built-in Backup utility. Figure 39.7 shows a typical backup screen using the Colorado Backup utility. The Settings Options menus are identical to the built-in Backup utility.

To begin a backup operation, simply select files to be backed up. This involves left-clicking on the appropriate disks or folders; this causes the backup software to scan for selected files to back up. (The backup software will display a window showing the status of the selection process, as illustrated in Figure 39.8.) Note the file selection operation may take several minutes to complete. Notice how the number of files selected and the size of

FIGURE 39.5 Microsoft Backup file section screen

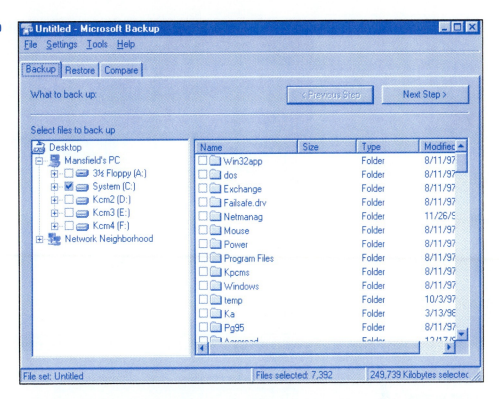

FIGURE 39.6 Backup did not find a tape drive

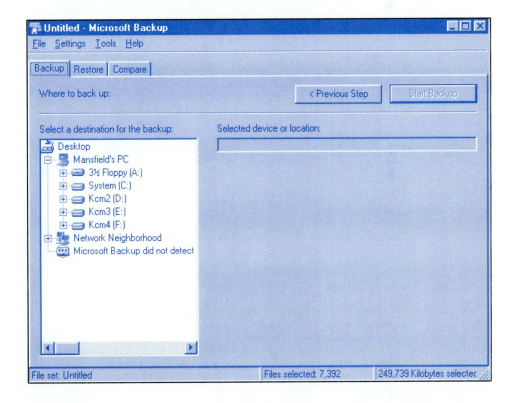

these files are displayed in the status bar at the bottom of the window. By pressing the Next button, the destination for the backup may be selected. Figure 39.9 shows the list of devices that can be selected as the destination for the backup. Notice the HP Colorado T1000e is selected in the list, and the right side of the Backup window shows the current tape's capacity and its total of free and used space. After selecting the backup device, it is again necessary to left-click the Next button to finalize any remaining options.

FIGURE 39.7 Colorado Backup file selection screen

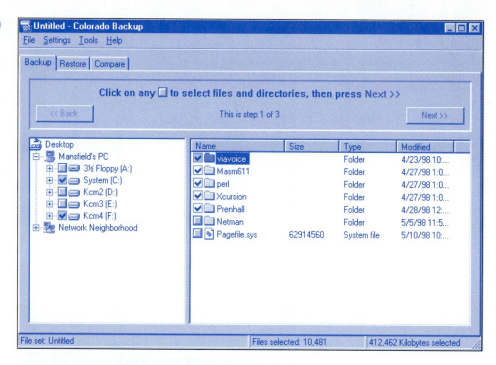

FIGURE 39.8 File Selection status window

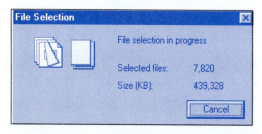

FIGURE 39.9 Selecting the destination device

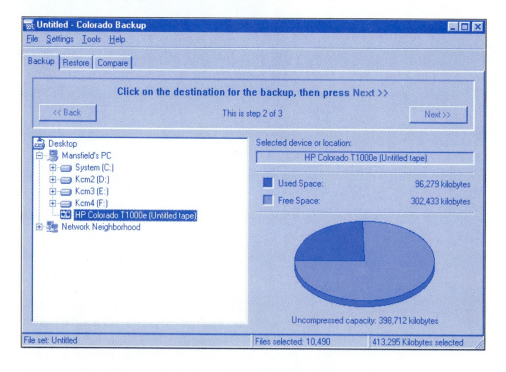

Each backup is appropriately named, usually describing the type of location of the data. For example, in Figure 39.10, the name "Files from System (C:) and other sources" is automatically displayed. The backup file set can also be password protected to guarantee privacy. The password, if selected, must be used to access the data on the tape. Without it, it

FIGURE 39.10   Backup options window

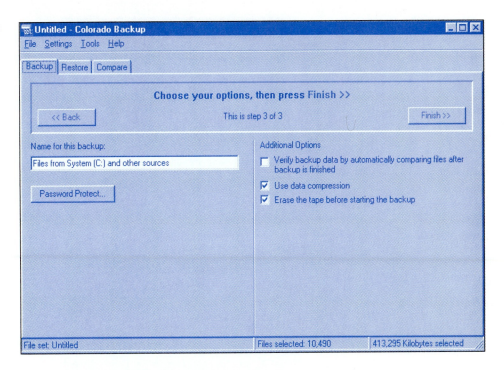

FIGURE 39.11   Remove check mark to erase the tape

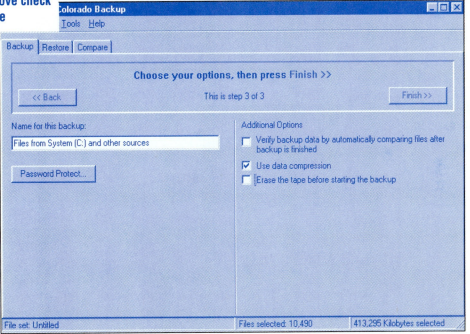

will be impossible to restore or view the contents of the tape. Obviously, care must be exercised when assigning a password to a file set or changing any of the settings. Notice how "Erase tape before starting the backup" has been deselected in Figure 39.11. This will cause the new backup information to be appended to the end of the tape. When all the settings have been verified, the backup process can be initiated by left-clicking the Finish button.

The Backup program displays the window shown in Figure 39.12 as it prepares to create the new file set. After preparation is complete, Backup will display the current status of the backup as shown in Figure 39.13. When the backup requires a new tape, it will pause until a new tape is inserted in the tape drive and then the process will continue. When the backup is complete, a pop-up window is displayed as shown in Figure 39.14.

FIGURE 39.12 Backup prepares to write information to tape

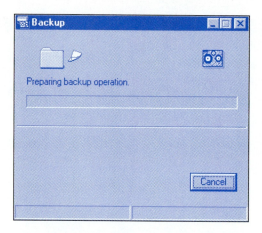

FIGURE 39.13 Backup writes information to tape

FIGURE 39.14 Confirmation of successful backup operation

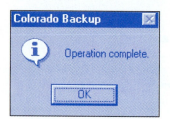

RESTORING FILES

When it becomes necessary to restore files to your system, for whatever reason, it involves selecting the appropriate tape that contains the required information. Then the process involves running the Backup program and selecting the Restore tab from the Backup window. Then it is necessary to select the appropriate device and file set. This is illustrated in Figure 39.15.

The Next button is used to display the file selection screen. Individual files, folders, and disks may be selected for restoration by simply left-clicking on each disk or folder. After the files to restore have been selected and the Next button is pressed, a few additional options shown on Figure 39.16 may be changed if necessary. It is important to use extreme caution when modifying any of these options to prevent possible loss of data.

It is a good idea to spend some time with the Backup program to become familiar and comfortable performing all types of routine backup and restore operations. You are also encouraged to experiment with the compare operation used to perform comparisons between disk and tape files.

FIGURE 39.15 Select the backup source and save set

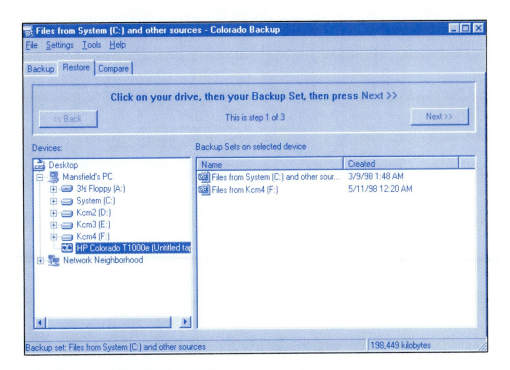

FIGURE 39.16 Selecting the appropriate Restore options

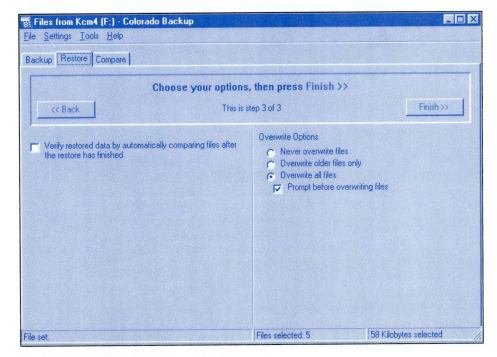

WINDOWS NT BACKUP

In a Windows NT environment, backups are usually considered serious business. A loss of data without replacement can put a company at risk of going out of business. There are two different areas to focus on in a Windows NT Domain: the backup requirements of the server itself and the backup requirements of the clients on the network. Basically, all the disks physically connected to the NT Server need to be backed up as well as all the disks on the computers in the network. Using the built-in Windows NT backup program or a program from a third party, it is easy to handle this important and otherwise difficult task. You are encouraged to explore the backup program(s) available on your Windows NT computer.

DISK BACKUP USING DOS

DOS contains a command used just for the purpose of backing up the files of a hard disk. The command has the following format:

BACKUP
[drive1:][path][filename][drive2:][/s][/m][/a][/f:size][/d:date][/t:time][/L[[drive:][path]filename]]

where: drive1 = the disk drive to be backed up (the **source disk**)
  drive2 = the disk to contain the backed-up files (the **target disk**)
  /s = a command that causes the backing up of subdirectories
  /m = a command that backs up only those files that have been changed since the last backup process
  /a = a command that causes the files that are to be backed up to be appended (as opposed to written over) on the target disk
  /f:size = a command used if the target disk is not formatted before starting the backup process. (It is important to note that the DOS *format* command must be accessible to the backup program for this command to work.)
  /d:date = a command that backs up all files that have been modified on or after *date*
  /t:time = a command that backs up all files that have been modified on or after *time*
  /L:filename = a command that causes a record to be kept in a file of all the files that have been backed up. (This file is called a **log file.** If you do not specify a file name for this log file, DOS will give it the default name of BACKUP.LOG.)

Table 39.2 lists the various values that can be used for target disks during the backup process. These values are used with the /f:size modifier when the target disk has not been formatted.

It is important to note that BACKUP does *not* copy the two hidden DOS files, IO.SYS and MSDOS.SYS. Nor does BACKUP copy the COMMAND.COM file. BACKUP allows you to place a new DOS operating system on the hard disk before restoring the backed-up files to it. Thus, the new IO.SYS and MSDOS.SYS files, as well as the COMMAND.COM file, are not overwritten during the process of transferring the backed-up files to the hard disk.

Be sure not to use an old version of DOS (such as 3.2 or earlier) to restore files backed up by DOS 4.0 or later. If you try to do this, the files will not be properly restored, which will result in the loss of data.

Remember that the backup process will erase any old files already on the target disk unless you use the /a switch with the BACKUP command.

For example, to back up an entire directory to a blank formatted disk (for a directory named \MYSTUFF), you would use

C> BACKUP \MYSTUFF A:

TABLE 39.2 Values of /f:size for unformatted target disk

Target Disk Size	Allowed Values for /f:size
160K single-sided 5"	160, 160K, 160KB
180K single-sided 5$\frac{1}{4}$"	180, 180K, 180KB
320K double-sided 5$\frac{1}{4}$"	320, 320K, 320KB
360K double-sided 5$\frac{1}{4}$"	360, 360K, 360KB
720K double-sided 3$\frac{1}{2}$"	720, 720K, 720KB
1.2M double-sided 5$\frac{1}{4}$"	1200, 1200K, 1200KB, 1.2, 1.2M, 1.2MB
1.44M double-sided 3$\frac{1}{2}$"	1440, 1440K, 1440KB, 1.44, 1.44M, 1.44MB
2.88M double-sided 3$\frac{1}{2}$"	2880, 2880K, 2880KB, 2.88, 2.88M, 2.88MB

RESTORING FILES USING DOS

Once files have been backed up, you can use the DOS RESTORE command to restore the backed-up files from the floppy disk to the hard drive. The RESTORE command has the form

RESTORE drive1:[drive2:][pathname][/s][/p][/b:date][/a:date][/e:time][/m][/n]

where: drive1 = the drive that contains the backup files (**source drive**)
        drive2 = the drive to which the backed-up files will be restored (**target drive**)
        /s = a command that lets all subdirectories be restored
        /p = a command that stops the restoring process to let you know that a file to be restored matches an existing file that has been changed since the backup process was done. (It asks if you want to replace that file with the backed-up file.)
        /b:date = a command that causes files to be restored that have been last modified on or before *date*
        /a:date = a command that causes files to be restored that have been modified on or after *date*
        /e:time = a command that causes files to be restored that have been modified at or after *time*
        /m = a command that restores only those files that have been modified since the last backup
        /n = a command that restores only those files that do not exist on the target disk

Remember that the DOS RESTORE command does not restore the system files. It is good practice to check all the restored files with a DIR command to make sure the restoring process has been completed and has accomplished what you expected. To restore the entire directory \MYSTUFF, use the command

```
RESTORE A: \MYSTUFF
```

When the RESTORE process is completed, one of the following exit codes will be displayed:

0 = normal operation
1 = no files found to be restored
2 = restoring process terminated by user
3 = restoring process terminated because of an error

USING PKZIP AND PKUNZIP FOR BACKUPS

The PKZIP and PKUNZIP utilities (available free on most computer bulletin boards) use a digital data compression algorithm to compact the amount of storage space required by a file or group of files. The algorithm is classified as *lossless,* since files must be identical to their original contents when uncompressed.

Figure 39.17 shows the contents of the GRAFX directory. The total space required by the files is 672,033 bytes.

To compress all files in the GRAFX directory, use the DOS commands:

```
C> CD \GRAFX
C> PKZIP GRAFX
```

The PKZIP program compresses each file individually and stores the entire compressed group of files in a new file called GRAFX.ZIP. When this has been done, the display looks like that shown in Figure 39.18. Notice that some files were not compressed. PKZIP decided that they could not be significantly compressed and stored them without change.

Figure 39.19 shows the new GRAFX.ZIP file in the directory listing. The GRAFX.ZIP file takes only 368,971 bytes. Recall that the original directory contained 672,033 bytes. This surely is a significant savings in disk space, and will reduce the required backup time because less information needs to be saved.

FIGURE 39.17 Contents of the GRAFX directory

```
Volume in drive D is SUPERSTUFF
Volume Serial Number is 1037-11D2
Directory of D:\GRAFX

.              <DIR>         04-19-94   9:22a
..             <DIR>         04-19-94   9:22a
DA       EXT            1,156 06-22-94   1:11p
GGEN     C             19,093 08-20-94  11:58a
GGEN     EXE           98,402 06-22-94   1:11p
GO       BAT                9 08-20-94   8:13a
OBJS     SFB          150,758 06-22-94   1:11p
PACMAP   DAT          317,049 06-22-94   1:11p
RCT      EXE           85,566 10-16-94   4:00a
         9 file(s)       672,033 bytes
                     160,186,368 bytes free
```

FIGURE 39.18 Output of the PKZIP program

```
PKZIP (R)   FAST!   Create/Update Utility   Version 1.1   03-15-90
Copr. 1989-1990 PKWARE Inc. All Rights Reserved.  PKZIP/h for help
PKZIP Reg. U.S. Pat. and Tm. Off.

Creating ZIP: GRAFX.ZIP
   Adding: DA.EXT        imploding (46%), done.
   Adding: GGEN.C        imploding (77%), done.
   Adding: GGEN.EXE      storing   ( 0%), done.
   Adding: GO.BAT        storing   ( 0%), done.
   Adding: OBJS.SFB      imploding (32%), done.
   Adding: PACMAP.DAT    imploding (66%), done.
   Adding: RCT.EXE       imploding (40%), done.
```

FIGURE 39.19 GRAFX.ZIP directory entry

```
Volume in drive D is SUPERSTUFF
Volume Serial Number is 1037-11D2
Directory of D:\GRAFX

.              <DIR>         04-19-94   9:22a
..             <DIR>         04-19-94   9:22a
DA       EXT            1,156 06-22-94   1:11p
GGEN     C             19,093 08-20-94  11:58a
GGEN     EXE           98,402 06-22-94   1:11p
GO       BAT                9 08-20-94   8:13a
OBJS     SFB          150,758 06-22-94   1:11p
PACMAP   DAT          317,049 06-22-94   1:11p
RCT      EXE           85,566 10-16-94   4:00a
GRAFX    ZIP          368,971 11-22-94   9:36a
        10 file(s)     1,041,004 bytes
                     159,793,152 bytes free
```

To get the original files back, use the following commands:

C> CD \GRAFX

C> PKUNZIP A:GRAFX

(assuming that the GRAFX.ZIP file was saved on drive A:).

Note: The WinZip utility (shareware) is designed with an easy-to-use graphical interface that zips any file dragged and dropped into its window. Download WinZip and experiment with it.

PREPARATION OF A NEW HARD DISK

To prepare a new hard disk for storing programs, you must go through what is essentially a three-step process:

1. Low-level formatting (except for IDE hard drives)
2. Hard disk partitioning
3. High-level formatting

Once you have performed these three steps, the hard drive is ready to store programs that will be used by the system in which it is installed.

A **low-level format** causes all the tracks and sectors to be developed on the hard drive. This is the process that causes the outline of these tracks to be written. You must perform this process on any new hard disk unit, except for IDE hard drives, which are low-level formatted at the factory. This step is one of the most important steps in the installation of a new hard drive. The low-level formatting process is also one of the major repair tools used to recover the system from a hard disk crash.

Part of the process of low-level formatting is called **defect mapping.** Defect mapping is the process of letting the low-level formatting software know where the defects on the hard drive are located. All new hard drives will have the locations of these defects printed on a label on the drive itself and/or in the accompanying instruction manual. You need to enter this information during the low-level formatting process in order to have the formatting program mark these sections so that nothing ever gets written to them. Basically, format makes sure that these sections of the disk are never used.

There are commercially available programs that can be used to do low-level formatting of a hard drive. Several of the manufacturers of hard disk controllers have a ROM chip installed that contains a low-level formatting program. To access this program, use the DOS DEBUG command. Once you are in DEBUG, enter the following:

$$G = C800:5$$

This is the DEBUG GO command, and it should activate the proper memory location to start the low-level formatting program. If this isn't the correct memory location for the particular controller card installed in your system, the worst that can happen is that your computer system will "hang up." This means you simply reboot and try a different memory location in the form of C800:X, where X represents a number that corresponds to your particular system (such as 5, 6, or 8).

Once the low-level formatting is complete, the next step is to partition the disk. This is the first step to perform when installing a new IDE hard drive. Partitioning is accomplished with the FDISK command. The FDISK command has the following format:

$$FDISK$$

The FDISK command configures a hard disk for use with DOS or Windows. Through the use of this command, you can do any of the following:

- Make a primary DOS partition.
- Make an extended DOS partition.
- Change the active partition.
- Show the partition data.
- If the system has more than one hard disk, select the hard disk you want.

If you reconfigure your hard disk using FDISK, you will destroy all existing files. You must make sure that you have backup files of all the hard disk files before you create a partition with this command.

When using FDISK, you must make a primary DOS partition before you can create an extended DOS partition. With DOS 4.0 and later, you will need only one DOS partition, the primary one. For making a primary DOS partition, choose the first selection on the menu; the screen display will then be as shown in Figure 39.20.

FIGURE 39.20 Screen display for primary DOS partition

```
Create DOS Partition

Current fixed disk drive: 1

   1. Create Primary DOS Partition
   2. Create Extended DOS Partition
   3. Create Logical DOS Drive(s)in
      the Extended DOS Partition

Enter choice: [1]

Press ESC to return to Fdisk Options
```

FIGURE 39.21 Screen display after partitioning

```
Create DOS Partition

Current fixed disk drive: 1

Do you wish to use the maximum size
for a DOS partition and make the DOS
partition active (Y/N).........?  [Y]

Press ESC to return to Fdisk Options
```

If you use the default settings on the display, you will be able to create a primary DOS partition of up to 2GB on the hard drive. After the partitioning is complete, the message shown in Figure 39.21 will be displayed.

After completing the instructions shown on the screen, you must then format the hard drive. Usually you will want to start your system from the hard drive. If so, use the /s option with the FORMAT command. Therefore, from drive A: (with your DOS system disk in that drive), enter:

A> FORMAT C: /S

and the high-level formatting process will be executed.

On models 386 and higher, the BIOS setup program must be executed so that new drive parameters can be specified. For example, if an older 406MB hard drive has been replaced with a newer 2.5GB hard drive, the drive parameters specifying the number of heads, cylinders, and sectors per track will surely be different. For example, a new 2.5GB hard drive uses these numbers:

Cylinders	656
Heads	128
Sectors	63

where $656 \times 128 \times 63 \times 512$ bytes per sector equals 2,708,471,808 bytes, or 2.583GB. All three parameters must be saved by BIOS so that DOS knows how to access the new hard drive correctly.

After all this, the hard drive is ready to receive programs. Remember that to boot from the hard drive you must make sure that it has the COMMAND.COM file as well as the proper CONFIG.SYS file and, usually, an AUTOEXEC.BAT file.

TROUBLESHOOTING TECHNIQUES

Many problems associated with hard drive backups are usually a result of the connection between the computer and the tape drive. The parallel port is used to communicate with tape devices (because the data can be transferred faster using parallel data lines). The parallel port in newer computers can be configured in BIOS in different ways, such as standard or EPP. If the setting is incorrect, the tape device will not work properly. It may be necessary to modify the BIOS setting to change the parallel port from EPP to standard, or vice versa. Many times, the documentation provided by the manufacturer can provide the answer to this problem as well as many other types of common problems.

Other problems and questions arise when the C: drive has been replaced. In this case, assuming we have a *good* backup, it is necessary to reinstall the operating system on the new hard disk before any tape backup files can be restored. After all the files have been restored, a reboot is necessary to complete the process.

SELF-TEST

This self-test is designed to help you check your understanding of the background information presented in this exercise.

True/False

Answer *true* or *false*.

1. The Windows built-in backup program works with all types of tape devices.
2. A full disk backup using Windows will automatically include all of the Registry files.
3. You need to use only the DOS FORMAT command on a hard disk before placing programs on it.
4. The standard method used to back up a hard drive is with the COPY or XCOPY command.
5. In the standard course of normal computer operations, you should periodically back up the programs on a hard disk.

Multiple Choice

Select the best answer.

6. You should back up a hard drive
   a. Just before shipping the computer unit.
   b. Before reformatting the hard drive.
   c. As a periodic routine of normal computer use.
   d. All of the above.
7. Before you can start the backup process with a hard drive, you must
   a. Have a set of formatted disks on which to store the information.
   b. Make sure there are no subdirectories, because these will not be backed up.
   c. Create a record of those files to be backed up, so you will be able to restore them later.
   d. None of the above.
8. In DOS 4.0 the BACKUP command will
   a. Copy the two hidden DOS files IO.SYS and MSDOS.SYS.
   b. Not copy the two hidden files IO.SYS and MSDOS.SYS.
   c. Work with the RESTORE command for DOS 3.2 and lower.
   d. None of the above.
9. A recommended backup schedule includes
   a. A complete backup of the entire system each day.
   b. An initial complete backup of the entire system, followed by daily incremental backups.
   c. Incremental backups of the hard disk followed by a complete backup.
   d. Requirements that all files be backed up monthly.
10. When modifying the Windows Backup settings
    a. There is absolutely no risk of losing information.
    b. Extreme caution must be exercised to prevent accidental loss of data.
    c. Windows will automatically prevent accidental loss of data.
    d. A special password must be used to prevent accidental loss of data.

Matching

Match an item on the right with each phrase on the left.

11. Low-level format
12. Make DOS partition
13. High-level format

a. Use DOS DEBUG
b. DOS FORMAT /S
c. DOS RESTORE
d. DOS FDISK

Completion

Fill in the blanks with the best answers.

14. It is necessary to create a(n) _____ on each hard disk before any information can be saved.
15. The settings for each hard drive are stored in the system _____ settings.
16. The DOS RESTORE command does not restore the _____ files.
17. When preparing a newly installed hard drive, the first software preparation is _____ formatting.
18. When doing a low-level format, all data on a hard disk is _____.

FAMILIARIZATION ACTIVITY

Your system may have an A: and a C: drive or an A: and a B: drive. To do this activity, you will use the A: drive as the target drive (the drive that will contain the backed-up files) and the other drive as the source drive (the drive that contains the files to be backed up).

1. Make sure both disks (source and target) are formatted.
2. Change the volume name of the source disk to SOURCE and the volume name of the target disk to TARGET.
3. Clearly label both disks: the source disk as SOURCE and the target disk as TARGET.
4. On the SOURCE disk, create the directory and file structure shown in Figure 39.22.

FIGURE 39.22 Directory and file structure for SOURCE disk

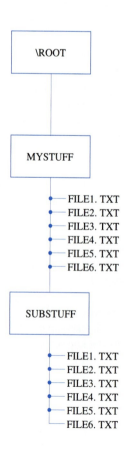

5. Place the target disk in drive A:. Place the source disk in drive B: or make drive C: the active drive.
6. Make the MYSTUFF directory the active directory on the source disk.
7. You will now back up all the files and subdirectories on the source disk to the target disk. From the source drive (with MYSTUFF as the active directory), enter

    ```
    B:> BACKUP B: A: /L/S
    ```

8. After the backup process is completed, look at the directory of the target disk and record what you see (include the volume name as well as the names of the files).
9. Try to erase the files on the target disk by using the DOS ERASE command. Record the message you get on the screen when you try this.
10. Use the DOS ATTRIB command to check the attributes of the files on the target disk. Record what you observe.
11. Go to the root directory of the source disk. Observe the new file there called BACKUP.LOG. Using the DOS TYPE command, observe and record the contents of this file.
12. Repeat this process using PKZIP and PKUNZIP.

QUESTIONS/ACTIVITIES

1. What message appeared on the screen just before the backup process started when you first backed up the source disk with the BACKUP B: A: /L/S? Explain why this message was there.
2. Did the volume name of the target disk change after the backup process? Why do you think this happened?
3. If the backup process requires more than one target disk (because of the number and size of the files to be backed up), what do you think will be the volume name of the second target disk? What do you think will be the file names on the second target disk?
4. What are the attributes of a backed-up file created by the DOS BACKUP process?
5. In what directory is the BACKUP.LOG file placed? What are the contents of this file?
6. What happens if you try to PKZIP a ZIP file? Does it compress even more?

REVIEW QUIZ

Within 10 minutes and with 100% accuracy,

1. Back up and restore a set of files from one disk to another and read the backup log created in the process.
2. Use PKZIP and PKUNZIP to back up and restore a compressed directory of files.
3. Back up and restore a set of files from disk to tape and tape to disk.

40 Hard Disk Replacement and File Recovery

INTRODUCTION

Joe Tekk examined the old computer system sitting on his desk. A client had brought it in, requesting more hard drive space. The system contained an 850MB hard drive, which was full and already compressed. Joe grabbed a spare 2GB hard drive and connected it as a slave. He powered up the system and ran the BIOS setup program to update the drive information. After the system booted up, the new 2GB drive was available as drive D:. Joe called the customer to say the computer was ready.

Later, after thinking about the installation, Joe recalled the first time he replaced a hard drive. Everything went wrong. The system would not boot with the new drive installed, and eventually a number of files were corrupted on the original drive. After much trial and error, Joe was able to get both drives operating, having learned a great deal about hard drives in the process. Each new installation teaches me something else, *Joe thought to himself.*

PERFORMANCE OBJECTIVES

Upon completion of this exercise, you will be able to do one or all of the following, depending on your lab situation:

1. Remove and replace a hard drive system, making sure the heads are parked and all the files are backed up, with 100% accuracy within 45 minutes.
2. Install a new hard drive, partitioned and formatted.
3. Use the DOS RECOVER command on a single file and on a directory of files. Show the contents of one of the recovered directory files and rename it with a descriptive name with 100% accuracy within 10 minutes.

BACKGROUND INFORMATION

In Exercises 38 and 39, you had the opportunity to learn about hard disk fundamentals and backup. In this exercise, you are shown how to replace a hard disk. You also have the opportunity to learn how to recover lost or damaged files using the DOS RECOVER command. Both of these skills are important for the repair and maintenance of microcomputer systems.

In all the following examples of a hard drive replacement, it is assumed that you have already removed the system cover. In all cases, observe electrical safety precautions, which include having the power cord disconnected from the system as well as from any electrical outlet. It is also assumed that you have completely backed up the hard drive before practicing any removal and replacement procedures. See Exercise 39 for hard drive backup procedures if you need a review. It is also assumed that you have parked the heads of the hard drive using the standard parking procedure for your system. Most hard drives (IDE hard drives, for example) automatically park their heads. This means that the drive heads land on a place on the hard drive that does not contain any data. You may need to consult with your hard drive manual or with your instructor in regard to your particular computer system.

Most hard drives are mounted in their metal cage by two or more screws on each side of the drive. It is not difficult to physically remove or install the hard drive, but a great deal of preparation is required to ensure the drive is replaced/installed properly.

PHYSICAL CONSIDERATIONS

Several factors must be considered when replacing a hard drive (or adding a new one). Is the drive an old $5\frac{1}{4}$" unit that we want to replace with a new $3\frac{1}{2}$" drive? Is a spare power connector available, along with room in the hard drive bay? Is the power supply powerful enough to support another piece of hardware? Does the motherboard have integrated-drive electronics built-in? How much money do you want to spend? Are you willing to mix one manufacturer's hard drive with another's, such as a Maxtor drive with a Connor, Seagate, or Western Digital? Does your BIOS support large hard drives (LBA performed by BIOS)? If not, do you have memory to spare for the dynamic drive overlay software that loads at boot time to control the drive?

Every user will have his or her own reasons for replacing or adding a hard drive. Let us examine some common scenarios that are typical of how systems are upgraded. Keep in mind that hard drive installations generally encounter some kind of glitch that adds more time to the job than originally planned. A wise person will set aside additional time up front, just in case.

Scenario #1: Adding a Second IDE Hard Drive (as a Slave)

Perhaps the easiest thing to do when you've run out of disk space is to buy a new drive to use as a second hard drive, a slave to the original hard drive, which is the master. An IDE hard drive contains its own onboard controller, which is disabled for slave operation. The IDE controller on the master drive controls both drives. This is illustrated in Figure 40.1.

Adding a second drive as a slave involves changing the jumpers on the hard drive to indicate slave operation. Figure 40.2 shows the master/slave jumper settings for a Maxtor IDE hard drive.

The manufacturer provides technical details on the hard drive, its jumper settings, logical parameters (such as number of cylinders, and sectors), and operational characteristics. Looking through the technical details is time well spent. You can learn useful information about BIOS, DOS, Windows, and hard drive technology from the installation directions that come with the hard drive.

To add a new, second hard drive on a machine with a BIOS that supports LBA, do the following:

1. Set the jumpers on the new drive for slave operation.
2. Connect the IDE ribbon cable to the new drive and attach power.
3. Turn on power and enter the system BIOS setup program.
4. Use the hard drive auto-detect feature to find the new drive and load its parameters. Typically the BIOS will recommend the best mode of operation (such as LBA and Normal) for the drive.
5. Save the new settings and boot the machine.
6. Use FDISK to create a partition on the new drive.
7. Reboot and format the new partition.

Note that the drive parameters can also be entered by hand, skipping auto-detect.

Be aware that adding any type of new drive may cause drive letters to change.

FIGURE 40.1 Adding a second IDE disk drive

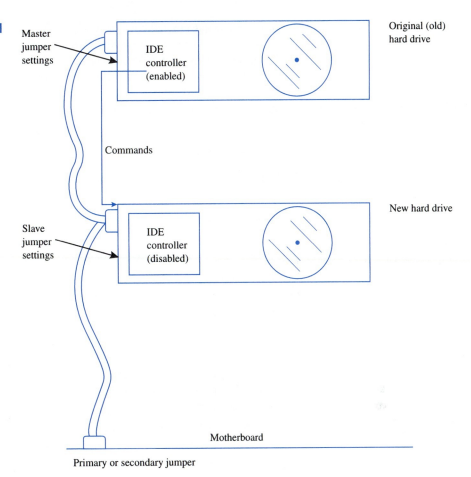

FIGURE 40.2 Disk drive jumper settings

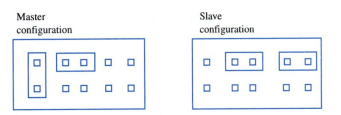

Scenario #2: Adding a Second IDE Hard Drive (as a Master)

In this scenario, the new drive becomes the master, with the old drive changing from master to slave. This will require the new drive to be formatted as a system disk, so that the operating system can boot from it at power-up.

If the old master drive contains Windows, you will have to transfer all associated files to the new drive (using XCOPY or PartitionMagic's copying feature). Alternately, you could install Windows on the new drive and leave everything else on the old drive. This option may require you to edit various setup parameters, since the old drive will be assigned a new drive letter when it is changed to slave operation. At the worst, you may need to reinstall other applications to repair the drive letter change.

Assuming that we will reinstall the operating system, here is what we need to do:

1. Set the jumpers on the new drive for master operation.
2. Connect the IDE ribbon cable to the new drive and attach power.
3. Change the old hard drive to slave mode.
4. Turn on power and enter the system BIOS setup program.
5. Use the hard drive auto-detect feature to determine the parameters for each drive.

6. Save the new settings and boot the machine.
7. Use FDISK to create a partition on the new drive.
8. Insert the Windows CD-ROM and follow the installation procedure.
9. When the installation is complete, check all applications on the old drive (which should now be drive D:) to see which ones require path and/or setting changes.
10. Delete the old operating system from the old drive.

Windows NT is able to boot from drives other than C (such as the E partition on a multiple-partition hard drive), which adds a small amount of flexibility to the installation.

Scenario #3: Replacing an Old IDE Drive

It is often the case with old hard drives that their capacity is so small compared with the new drive (200MB versus 5GB) that we do not bother to keep the old drive. This brings up an interesting problem: How can we transfer an exact copy of the old drive's data to the new drive? One way to do this would involve the following:

1. Add the second drive as a slave (as in Scenario #1).
2. Format the new drive with the /S option to transfer the operating system to it. The operating system on the new drive will match the operating system on the old drive.
3. Use the XCOPY command to copy all files from the old drive (the master) to the new drive. Try XCOPY *.* /S D:.
4. Shut the system down and remove the old drive.
5. Change the jumpers on the new drive to make it a master.
6. Power-up and set the drive parameters in the BIOS setup program.
7. Save the BIOS parameters and boot the system.

One problem with the XCOPY command is that it does not copy hidden files, so this method is not preferred. Instead, you could simply load the drive with a copy of a recent backup made of the old drive. Or, better yet, use PartitionMagic to make a *clone,* an exact copy of the old drive's partition, and store it on the new drive.

Scenario #4: Adding an SCSI Drive

When adding an SCSI drive to a PC, you also add an SCSI controller card. The SCSI controller card can be plugged into a PCI slot. The SCSI controller card can then be connected to the SCSI disk using either the internal or external connectors available on the controller card. Plug-and-play SCSI adapters are supported by Windows.

SCSI buses can support multiple devices and each device must be assigned a unique address on the bus. This is done automatically during the installation process provided by the manufacturer.

Internal drives can be mounted into any suitable location.

THE DOS RECOVER COMMAND

There are times when your disk may develop a bad sector or two. If these bad sectors are part of the contents of a file, you will lose the parts of the file that were contained on those now-bad sectors.

DOS provides a command to help restore the file. The DOS RECOVER command works by reading the file to another file and omitting the part of the file that contains the bad sector. This means that the part of the file that had the bad sector is still lost, but, if this is a text file, you can use a text editor to reinsert the lost text. If, on the other hand, this is an .EXE or .COM file, it probably will not be usable because of the data lost from the bad sector.

It should be noted that this DOS command does not recover erased files. In MS-DOS 5.0 and later, a new command called UNDELETE can be used to attempt to restore a deleted program from a disk (provided the sectors that stored this program have not been written over by another program).

The DOS command for attempting to recover a damaged file is

RECOVER[drive:][path]filename

where: drive = the drive that contains the file(s) to be recovered
        path = the path that contains the file(s) to be recovered
        filename = the name of the file (wild cards may be used) to be recovered. (Using wild cards will recover only the first matching file, not the rest.)

To use the RECOVER command to recover a file called STORY.TXT on drive A: in the root directory, you would enter

```
A> RECOVER STORY.TXT
```

The recovered file now appears in the root directory. You will find that all recovered files are *always placed in the root directory.*

If you need to recover a group of files in a directory, such as a directory called MYSTUFF, all the recovered files are placed in the root directory and named as follows:

FILE000N.REC

For example, if the files to be recovered in the MYSTUFF directory are

```
TEXT1.TXT
TEXT2.TXT
TEXT3.TXT
TEXT4.TXT
```

and you enter the DOS RECOVER command

```
A> RECOVER \MYSTUFF
```

after the recovery process, you will find the following files in your root directory:

```
FILE0001.REC
FILE0002.REC
FILE0003.REC
FILE0004.REC
```

You can also attempt to recover a whole disk of files on the A: drive, for example, by entering

```
A> RECOVER A:
```

This will place *all* the recovered files in all the directories on the disk into the root directory. Recall that the root directory can hold only a certain number of files. Because of this, you should attempt to recover only one file at a time, use a text editor to replace lost data, rename the file by a descriptive name, and place the file in a subdirectory (say, a directory called \RECOVER).

You can use the DOS CHKDSK command to check a file for bad sectors. If any are encountered, use the DOS RECOVER command.

The UNDELETE command provides a list of files it will attempt to undelete. You must supply the first letter of the undeleted file.

TROUBLESHOOTING TECHNIQUES

One of the most frightening error messages you may encounter reads like this:

Missing operating system

That simple message says it all. This error is an indication that the system cannot find a bootable partition when powered on.

If you have a startup disk, boot the system with it and use FDISK to examine the partition information for the hard drive. It is possible that the partition exists, and contains valid information (including the missing operating system), but has not been made *active*. FDISK can

FIGURE 40.3 PartitionMagic drive information

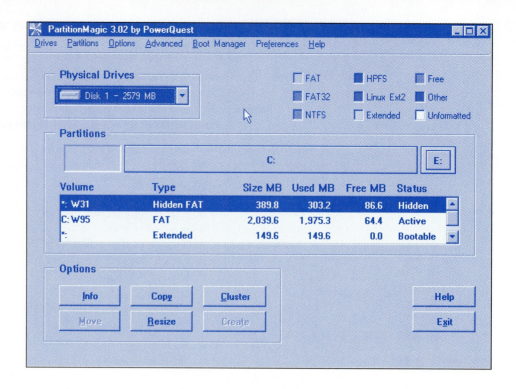

be used to set the partition active, as can any other disk utility, such as PartitionMagic. As shown in Figure 40.3, PartitionMagic clearly indicates which partitions are bootable, and which is the active partition. Always use FDISK or an equivalent utility to check for the existence of a partition before taking drastic measures, such as starting over from scratch.

SELF-TEST

This self-test is designed to help you check your understanding of the background information presented in this exercise.

True/False

Answer the following questions *true* or *false*.

1. If you lose any files on a hard drive as a result of its removal and replacement, you need only use the DOS RECOVER command to get them back.
2. Because of the DOS RECOVER command, it is no longer necessary to waste time backing up files on a hard drive.
3. Before removing a non-IDE hard drive, you should make sure the drive heads are parked.
4. When a new hard drive is added to a system, is must be the master.
5. A maximum of two IDE hard drives may be installed on a system.

Multiple Choice

Select the best answer.

6. When replacing the hard drive, you must make sure that the
   a. Drive is never dropped or mishandled.
   b. Same mounting screws are used.
   c. Power is turned off.
   d. All of the above.
7. The odd-colored stripe on the hard drive ribbon cable
   a. Represents the ground connection.
   b. Must go to the last pin of the connector.
   c. Means "do not remove."
   d. Represents pin 1 of the connector.

8. SCSI hard drives can be added to
   a. Any system.
   b. Only systems that have no IDE hard drives.
   c. Only systems that do have IDE hard drives.

Matching

Match a statement on the right with each DOS command on the left.

9. RECOVER MYFILE
10. RECOVER A:
11. RECOVER \MYSTUFF
12. UNDELETE

a. Recovers only one file.
b. Recovers a directory of files.
c. Recovers a whole disk of files.
d. Recovers every file ever deleted.
e. May recover recently deleted files.

Completion

Fill in the blanks with the best answers.

13. The DOS RECOVER command rewrites all the files except for the part contained in any bad _____.
14. Using the DOS RECOVER command for an entire disk may overload the _____ directory.
15. Recovered files from a directory are rewritten in the form of _____, where N is the file number.
16. Device IDs are used with _____ hard drives.

FAMILIARIZATION ACTIVITY

The familiarization activity for this exercise is divided into two parts. The first part deals with the removal and replacement of a hard disk drive. The second part deals with the DOS RECOVER command. Depending on your lab situation, you may not have access to a computer with a hard drive. Instead, the hard drive removal and replacement activity may be given as a classroom demonstration.

Hard Drive Removal and Replacement

Using the procedure appropriate to your computer system, remove and replace its hard drive system. Make sure you use all the proper safety procedures during this process. This procedure may involve partitioning, formatting, and reinstallation of applications.

The DOS RECOVER Command

Using your floppy disk, create a subdirectory called MYSTUFF and place five text files inside it. Make one text file in the root directory.

1. Use the RECOVER command to recover the file in the root directory. Explain how you did this and the results you got.
2. Next, use the RECOVER command to recover all the text files in the MYSTUFF directory. Again, explain how you did this.
3. What new files do you now have in the root directory? List them below.

4. Are the contents of these files the same as those of the original files? How did you determine this?

QUESTIONS/ACTIVITIES

1. Who was the manufacturer of the hard drive that you removed and replaced in the lab?
2. Explain how you made sure the heads were parked on the hard drive before you removed it.

3. Was it necessary to back up the files on the hard drive before removing them? Explain.
4. Which scenario comes closest to your lab situation?
5. Where are the files recovered using the DOS RECOVER command stored?
6. Why should you be cautious about using the DOS RECOVER command to recover a whole disk of files at the same time?

REVIEW QUIZ

Do one or all of the following, depending on your lab situation:

1. Remove and replace a hard drive system, making sure the heads are parked and all of the files are backed up, with 100% accuracy within 45 minutes.
2. Install a new hard drive, partitioned and formatted.
3. Use the DOS RECOVER command on a single file and on a directory of files. Show the contents of one of the recovered directory files and rename it with a descriptive name with 100% accuracy within 10 minutes.

41 Video Monitors and Video Adapters

INTRODUCTION

Joe Tekk had been using his new office PC for some time now. He was happy with it, but he was very curious about the new technology being marketed touting the Pentium III and AGP graphics. He decided to buy one of the new PCs for his home. He thought it might be delivered today since he chose the rush delivery option. All he could think about was the new high-speed Pentium III processor, huge hard drive, AGP video card with 8MB of video memory, and a 17" monitor.

It was just after noon when he got a phone call from Donna, the secretary. "Two big boxes were just delivered with your name on them, Joe. Can you please pick them up?" she asked. Joe quickly got the equipment cart and rushed to the office.

Within a few minutes of getting home, Joe had all of the boxes unpacked. First, he removed the system cover to examine the hardware components. Both the processor and the video card were of great interest to Joe since he had spent a lot of time researching exactly what to buy. Joe satisfied his curiosity and replaced the system cover. He was eager to see the computer in action.

Joe booted the system and began to run some of the demonstration programs that came preinstalled. While he viewed a video, he noticed immediately the high-quality output of the display. The processor and AGP graphics card displayed the full motion video file without any delay or jitter as he had seen on many other computers. For a moment, Joe forgot he was looking at a PC display.

PERFORMANCE OBJECTIVES

Upon completion of this exercise, within 10 minutes and with 100% accuracy, you will be able to

1. Identify the type of monitor and display adapter being used by a given computer system and ensure that they have been properly installed and adjusted.

BACKGROUND INFORMATION

OVERVIEW

The computer display system used by your computer consists of two separate but essential parts: the monitor and the video adapter card as shown in Figure 41.1. Note from the figure that the monitor does not get its power from the computer; it has a separate power cord and its own internal power supply.

FIGURE 41.1 The two essential parts of a computer display system

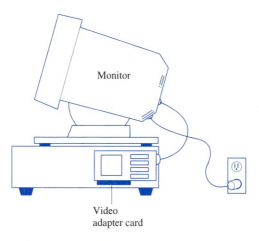

(a) Video adapter card with companion monitor

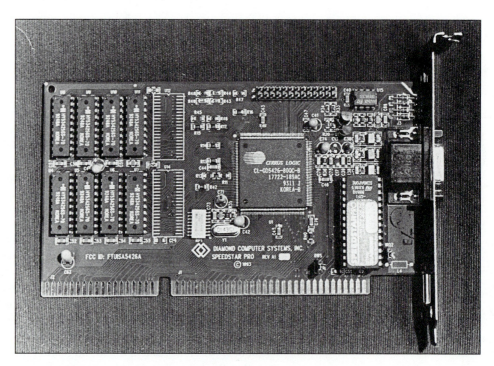

(b) SVGA graphics accelerator card (*photograph by John T. Butchko*)

The video adapter card [Figure 41.1(b)] interfaces between the motherboard and the monitor. This card processes and converts data from the computer and allows you to see all the things you are used to seeing displayed on the screen.

It is very important to realize that there are many different types of monitors and that each type of monitor essentially requires its own special video adapter card, as shown in Figure 41.2. Connecting a monitor to an adapter card not made for it can severely damage the monitor or adapter card, or both.

MONITOR SERVICING

Very seldom is the computer user expected to repair a computer monitor. Computer monitors are very complex devices that require specialized training to repair. These instruments contain very high and dangerous voltages that are present even when no power is being applied. The servicing of the monitor itself is, therefore, better left to those who are trained in this specialty.

FIGURE 41.2 Necessity of each computer monitor having its own matching adapter

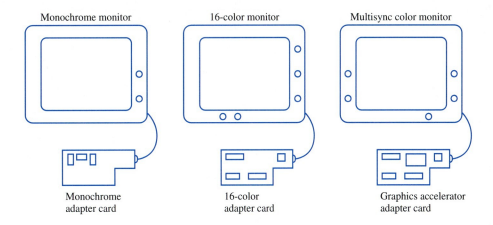

What you need to know is what kinds of monitors are available, their differences, and how they interface with the computer. Then you need to know enough about hardware and software in order to tell if a problem that appears on the monitor is in the monitor itself, its adapter card, the computer, or the monitor cabling—or is simply a result of the customer's lack of understanding about how to operate the computer.

MONITOR FUNDAMENTALS

All monitors have the basic sections shown in Figure 41.3. Table 41.1 lists the purpose of each of the major sections of a computer monitor.

MONOCHROME AND COLOR MONITORS

One of the differences between a monochrome (single-color) monitor and a color monitor is in the construction of the CRT. The differences are illustrated in Figure 41.4.

As shown in the figure, the color CRT contains a triad of color phosphor dots. Even though this consists of only three color phosphors, all the colors you see on a color monitor are produced by means of these three colors (including white, which is produced by controlling the intensity of the three colors: red is 30%, green is 59%, and blue is 11%). This process, called **additive color mixing,** is illustrated in Figure 41.5.

FIGURE 41.3 Major sections of a computer monitor

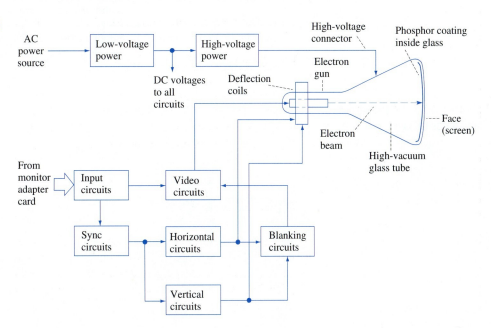

TABLE 41.1 Major sections of a computer monitor

Section	Purpose
Glass CRT	The cathode-ray tube (CRT) creates the image on the screen. It is so named because the source of electrons is called the cathode and the resulting stream of electrons is called its rays (cathode rays).
Electron gun	Generates a fine stream of electrons that are attracted toward the glass face of the CRT (the screen) by the large positive voltage applied there.
Phosphor coating	A special kind of material that emits light when struck by an electron beam.
High-voltage power source	Supplies the large positive voltage required by the CRT to attract the electrons from the electron gun.
Deflection coils	Generate strong magnetic fields that move the electron beam across the face of the CRT.
Horizontal circuits	Generate waveforms applied to the deflection coils, causing the electron beam to sweep horizontally across the face of the CRT from left to right.
Vertical circuits	Generate waveforms applied to the deflection coils, causing the horizontal sweep of the electron beam to move vertically across the face of the CRT from top to bottom and creating a series of horizontal lines.
Blanking circuits	Cause the electron beam to be cut off from going to the face of the CRT so that it isn't seen when the electron beam is retracing from right to left or from bottom to top. (This is similar to what you do when writing. You lift your pen from the surface of the paper after you finish a line and return to the left side of the paper to begin a new line just below it.)
Video circuits	Control the intensity of the electron beam that results in the development of images on the screen.
Sync circuits	Electrical circuits that help synchronize the movement of the electron beam across the screen.
Low-voltage power supply	Supplies the operating voltages required by the various circuits inside the monitor.

FIGURE 41.4 Monochrome and color CRTs

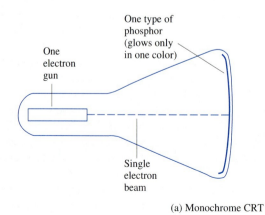

(a) Monochrome CRT

FIGURE 41.4 (continued)

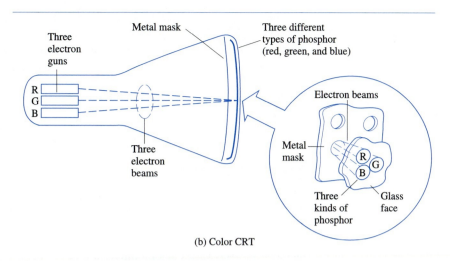

(b) Color CRT

FIGURE 41.5 Additive color mixing

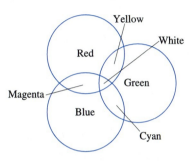

The other differences between monochrome and color monitors are the circuits inside these systems as well as their adapter cards. Some of these differences are the high voltages in a color monitor that are several times higher than those found in a monochrome monitor. Usually, these voltages are on the order of 30,000 V or more. You should note that this high voltage can be stored by the color CRT and still be present even when the set is unplugged from the AC outlet. A special probe is used to discharge the CRT.

ENERGY EFFICIENCY

Energy-efficient PCs are designed with energy efficiency in mind. The system BIOS, monitor video card, and other hardware must support either the Advanced Power Management (APM) or VESA BIOS extensions for power management (VBE/PM) standards. Some computers may support limited power management or energy saving features.

It is estimated by the U.S. Environmental Protection Agency (EPA) that the average office desktop computer or workstation uses around $105 of electrical power annually. When all desktops are considered, the total consumption adds up to around 5 percent of all electrical energy consumed in the United States. The EPA estimates that by using energy-efficient equipment, as much as $90 a year per computer can be saved.

The EPA has proposed a set of guidelines for energy-efficient use of computers, workstations, monitors, and printers. The EPA *Energy Star* program requires the computer and monitor to use less than 30 watts each when they are not being used (for a total of 60 watts including both the system unit and the monitor). Personal computers adhering to the Energy Star recommendations are also called *green* PCs.

Each computer can be set up to automatically reduce energy usage using the standby and sleep modes. The standby mode is activated after a user-specified period of inactivity. The sleep mode is automatically activated after the standby time has expired. If the computer is used during standby energy saving mode, it takes just a short period of time before the monitor is usable. Sleep mode is similar to a power-down of the monitor and requires some additional time before the monitor is usable.

The EPA Web site located at http://www.epa.gov/energy_star maintains a list of all energy-efficient computer products. Look for the Energy Star trademark on product packaging and the marketing materials supplied by most manufacturers.

VIDEO CONTROLS

Table 41.2 lists some of the major video controls and their purposes.

PIXELS AND ASPECT RATIO

Figure 41.6 illustrates two important characteristics of computer monitors. As shown in the figure, a pixel (or pel) is the smallest area on the screen whose intensity can be controlled. The more pixels available on the screen, the greater the detail that can be displayed. The number of pixels varies among different types of monitors; the more pixels, the more expensive the monitor. The aspect ratio indicates that the face of the CRT is not a perfect square. It is, instead, a rectangle. This is important to remember, especially if you are developing software for drawing squares and circles; you may wind up with rectangles and ellipses. The size of a pixel is referred to as its *dot pitch* and is a function of the number of pixels on a scan line and the distance across the display screen.

MONITOR MODES

There are two fundamental modes in which the monitor communicates with the computer: the **text mode** and the **graphics mode.** Figure 41.7 illustrates the difference.

In the text mode, the CRT display gets its information from a built-in ROM chip referred to as the **character ROM.** This may not be a separate ROM chip but part of another, larger one. This ROM contains all the characters on your keyboard, plus many more. This group of characters is known as the **extended character set** and may, among other things, be used in combination to form squares, boxes, and other shapes while your computer is still in the text mode. To get any of these extended characters on the screen (or to get *any*

TABLE 41.2 Major video controls

Control	Purpose
Contrast	A gain control for the circuits that determine the strength of the signal used to place images on the screen. It affects the amount of difference between light and dark.
Brightness	Controls the amount of high voltage applied to the CRT, which controls the strength of the beam. The higher the voltage, the stronger the beam and the brighter the picture.
Vertical size	Controls the output of the vertical circuit, changing the amount of the vertical sweep of the CRT and thus changing the vertical size of the displayed image.
Horizontal size	Controls the output of the horizontal circuit, changing the amount of horizontal sweep of the CRT and thus changing the horizontal size of the displayed image.
Vertical hold	Helps adjust the synchronous circuits so the image is stable in the vertical direction.
Horizontal hold	Helps adjust the synchronous circuits so the image is stable in the horizontal direction.

FIGURE 41.6 Pixels and aspect ratio

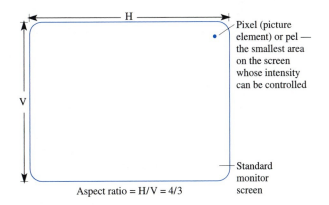

FIGURE 41.7 Text and graphics modes

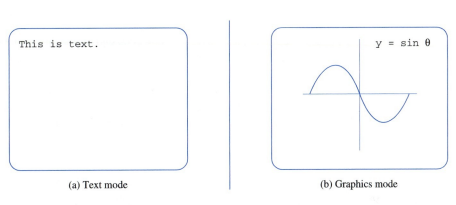

(a) Text mode (b) Graphics mode

character on the screen), simply hold down the SHIFT and ALT keys at the same time and then type in the character number. For example, to get the character ö, hold down the SHIFT and ALT keys and type in 148 on the keypad; when you lift up on the SHIFT and ALT keys, the character appears (and can also appear on the printer, depending on the type of printer).

The advantage of the text mode is that it doesn't take much memory and the visual results are predictable and easy to achieve (you need only to press a key on the keyboard). The size of the text screen is 80×25 or 40×25. The text screen is sometimes referred to as the **alphanumeric mode.**

When the monitor and its circuits are in the graphics mode, an entirely different use of memory is required. RAM is used because a program has complete control over the intensity and (in the case of color) the color of each pixel. The more pixels available on the monitor, the more memory required; the more memory required, the longer it takes to display a complete picture on the face of the CRT, which in turn means that your whole computer must be able to operate at a very high speed. In order to display detailed graphics, you must have a big and powerful machine, which means a more expensive system as well as a more expensive monitor.

Just to give you an idea of the memory requirements for graphics, if your monitor has 640 horizontal pixels and 480 vertical pixels, the total number of pixels that must be addressed by RAM is $640 \times 480 = 307,200$, which is more than a third of a megabyte for just one screen. If color is not used in the graphics mode, less memory is required (because the computer needs to store less information about each pixel).

TYPES OF MONITORS

In order to understand the differences among the most common types of computer monitors, you must first understand the definitions of the terms used to describe them. Table 41.3 lists the major terms used to distinguish one monitor from another.

Now that you know the definitions of some of the major terms used to distinguish one monitor from another, you can be introduced to the most common types of monitors in use today. Table 41.4 lists the various types of monitors and their distinguishing characteristics.

503

TABLE 41.3 Computer monitor terminology

Term	Definition
Resolution	The number of pixels available on the monitor. A resolution of 640 × 480 means that there are 640 pixels horizontally and 480 pixels vertically.
Colors	The number of different colors that may be displayed at one time in the graphics mode. For some color monitors, more colors can be displayed in the text mode than in the graphics mode. This is possible because of the reduced memory requirements of the text mode.
Palette	A measure of the full number of colors available on the monitor. However, not all the available palette colors can be displayed at the same time (again, because of memory requirements). You can usually get a large number of colors with low resolution (fewer pixels) or a smaller number of colors (sometimes only one) with much higher resolution—again, because of memory limitations.
Display (digital or analog)	There are basically two different types of monitor displays, **digital** and **analog.** Some of the first computer monitors used poor-quality analog monitors. Then digital monitors, with their better overall display quality, became more popular. Now, however, the trend is back to analog monitors because of the increasing demand for high-quality graphics, where colors and shades can be varied continuously to give a more realistic appearance.

TABLE 41.4 Common types of computer monitors

Type	Resolution*	Colors	Palette	Display
Monochrome composite	640 × 200	1	1	Analog
Color composite	640 × 200	4	4	Analog
Monochrome display	720 × 350	1	1	Digital
RGB (CGA)	640 × 200	4	16	Digital
EGA	640 × 350	16	64	Digital
PGA	640 × 480	Unlimited	Unlimited	Analog
VGA	640 × 480	256	262,144	Analog
SVGA	1280 × 1024	Varies	Varies	Digital/analog
Multiscan	Varies	Unlimited	Unlimited	Digital/analog

*In general, the higher the resolution, the higher the scan frequency. For example, the typical scan frequencies of EGA and VGA monitors are 21.5 KHz and 31.5 KHz, respectively.

Monochrome Composite Monitor

The first computers used **monochrome composite monitors,** which actually were television sets. Because of the way a standard television set works, only 40 characters could be displayed across the face of the CRT. Because of the poor resolution of this type of monitor, it is almost never used with computers today.

Color Composite Monitor

The **color composite monitor** was similar to the monochrome composite monitor except that it could show color. It had even poorer resolution than the monochrome composite monitor and, as a result, is almost never used with computers today.

Monochrome Display Monitor

The **monochrome display monitor** was the standard monitor used by IBM when the PC was first introduced. This monitor, with very good resolution and a green screen, is the single-color (monochrome) monitor you usually see with most office PC and XT systems.

RGB Monitor

The **red-green-blue (RGB) monitor** is a technologically out-of-date monitor that was one of the first popular color monitors to be used with the IBM PC. On this monitor, which had poor resolution, it was extremely difficult to read either text or any kind of graphics.

EGA Monitor

The **enhanced graphics array (EGA) monitor** is the answer to the RGB monitor. It is a very popular medium-priced color monitor. Its resolution is not quite as good as that of the monochrome display monitor.

PGA Monitor

The **professional graphics array (PGA) monitor,** which has a better resolution than that of the EGA, has not been very popular because of its high price. It is not as popular as the next color monitor, the VGA.

VGA Monitor

The **video graphics array (VGA) monitor** is one of the most popular color monitors; it provides high color resolution at a reasonable price. More and more software with graphics is making use of this type of monitor. The associated cards have a high scanning rate, resulting in less eye fatigue both in text and in graphics modes.

SVGA Monitor

Higher screen resolution and new graphics modes make the **Super VGA (SVGA) monitor** even more popular than the VGA monitor.

Multiscan Monitor

The **multiscan monitor** was one of the first monitors that could be used with a wide variety of monitor adapter cards. Since this type of monitor can accommodate a variety of adapter cards, it is sometimes referred to as the *multidisplay* or *multisync* monitor.

DISPLAY ADAPTERS

As previously stated, a computer monitor must be compatible with its adapter card. If it is not, damage to the monitor or adapter card, or both, could result.

MDA Adapter

The **MDA** or **monochrome display adapter** is an adapter card that contains both a 9-pin D-shell connector and a parallel printer connector. Figure 41.8 shows the connections for this type of adapter. This is purely a text mode adapter, offering no graphics capabilities. This adapter was used by the most popular green-screen monochrome monitors.

FIGURE 41.8   Pin diagram for MDA adapter

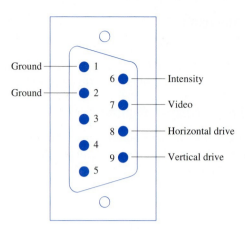

CGA Adapter

The **CGA** or **color graphics array adapter** was one of the first color adapters. It can generate 16 different colors (including black, dark gray, light gray, and white). It has several different graphic modes, with different levels of resolution. This adapter will also operate monochrome displays, but not the IBM monochrome designed for the MDA adapter. This adapter allows for three levels of resolution, the *low-, medium-,* and *high-resolution modes.* The low-resolution mode allows all 16 colors to be displayed at the same time, but the resolution is only 200×160 pixels. This results in such a poor image that this mode is not supported by any IBM system. In the medium-resolution mode, there are 200×320 pixels, which allow for only four colors to be displayed at the same time from four possible palettes. The first two palettes are red, green, brown, and black; and white, magenta, cyan, and black (or a software-selected background color). The last two palettes are identical to the first two, but with high-intensity colors.

In the high-resolution mode, you can get only one color at a time. However, the trade-off is a much higher resolution: 640×200 pixels. This adapter card comes with its own extra 16KB of memory for the displays. The pin diagram for this adapter is shown in Figure 41.9.

MGA Adapter

A small company in Berkeley, California, called Hercules Technology, created an adapter card that worked with the IBM monochrome system. This card, called the "Hercules Graphics Card," was sometimes called the **MGA** or **monochrome graphics adapter.** For the first time there was a low-cost, reliable adapter card that could provide graphics capabilities for the IBM monochrome screen. True, there was only one color, but you could now have graphics as well as text. Many other companies emulated the design of this card and the associated software that went with it. It has a resolution of 720×348 pixels, which, with its 64K of onboard memory, allows each pixel to be either ON or OFF.

FIGURE 41.9   Pin diagram for CGA adapter

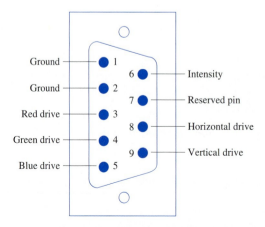

FIGURE 41.10 Pin diagram for EGA adapter

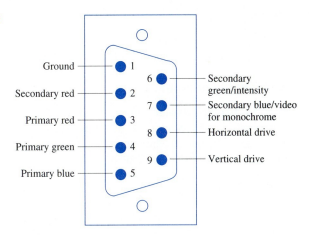

EGA Adapter

The **EGA** or **enhanced graphics adapter** can operate an RGB, EGA, or multiscan monitor. It was the first color graphics adapter put out by IBM. The pin diagram for this adapter is shown in Figure 41.10.

VGA Adapter

The **VGA** (**video graphics array card**) was the fastest-growing graphics card in terms of popularity until the SVGA card became available. The VGA adapter card uses a 15-pin high-density pin-out, as shown in Figure 41.11. The VGA 15-pin adapter can be wired to fit the standard 9-pin graphics adapter, as shown in Figure 41.12.

FIGURE 41.11 Pin diagram for VGA adapter

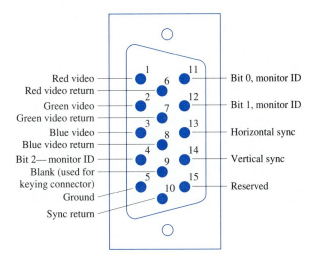

FIGURE 41.12 Nine-pin adapter cable for VGA

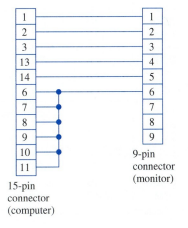

SVGA Adapter

The Super VGA graphics interface uses the same connector that VGA monitors use. However, more display modes are possible with SVGA than with VGA.

8514 Display Adapter

The **8514 display adapter** card provides a resolution of 1024 × 768 pixels and enough memory (about 0.5MB) to allow 16 colors to be used at the same time. This adapter is made for the IBM PS/2 line of computers. In order to take full advantage of this adapter, the IBM 8514 display should also be used.

MCGA Adapter

The **MCGA (multicolor graphics array) adapter** is integrated into the motherboard of some IBM PS/2 model computers. This adapter integration supports the CGA mode if a suitable analog display card is added. It has an available palette of 262,144 colors, with 64 shades of gray available at the same time. It has two graphics modes, 640 × 480 and 320 × 200 pixels.

VESA

The **VESA** (Video Electronics Standards Association) specification has been developed to guide the operation of new video cards and displays beyond VGA. New BIOS software that supports the VESA conventions is contained in an EPROM mounted on the display card. The software also supports the defined VESA video modes. Some of these new modes are 1024 × 768, 1280 × 1024, and 1600 × 1200, with up to 16 million possible colors.

GRAPHICS ACCELERATOR ADAPTERS

A graphics accelerator is a video adapter containing a microprocessor designed specifically to handle the graphics processing workload. This eliminates the need for the system processor to handle the graphics information, allowing it to process other instructions (nongraphics related) instead.

Aside from the graphics processor, there are other features offered by graphics accelerators. These features include additional video memory, which is reserved for storing graphical representations, and a wide bus capable of moving 64 or 128 bits of data at a time. Video memory is also called VRAM and can be accessed much faster than conventional memory.

Many new multimedia applications require a *graphics accelerator* to provide the necessary graphics throughput in order to gain realism in multimedia applications. Table 41.5 illustrates the settings available for supporting many different monitor types and refresh rates.

Most graphics accelerators are compatible with the new standards such as Microsoft DirectX, which provides an application programming interface, or API, to the graphics

TABLE 41.5 Common monitor support

Resolution	Colors	Memory	Refresh Rates
640 × 480	256 65K 16M	2MB 2MB 2MB	60, 72, 75, 85
800 × 600	256 65K 16M	2MB 2MB 2MB	56, 60, 72, 75, 85
1024 × 768	256 65K 16M	2MB 2MB 4MB	43 (interlaced), 60, 72, 75, 85
1280 × 1024	256 65K 16M	2MB 4MB 4MB	43 (interlaced), 60, 75, 85

TABLE 41.6 AGP graphics mode

Mode	Throughput (MB/s)	Data Transfers per Cycle
1x	266	1
2x	533	2
4x	1066	4

FIGURE 41.13 AGP configuration

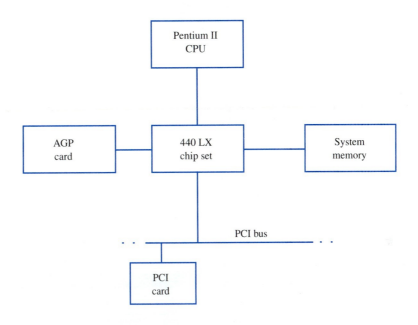

subsystem. Usually, the graphics accelerators are also compatible with OpenGL for the Windows NT environment.

AGP ADAPTER

The **Accelerated Graphics Port (AGP)** is a new interface specification developed by Intel. The AGP adapter is based on the PCI design but uses a special point-to-point channel so that the graphics controller can directly access the system main memory. The AGP channel is 32 bits wide and runs at 66 MHz. This provides a bandwidth of 266 MBps as opposed to the PCI bandwidth of 133 MBps.

AGP optionally supports two faster modes, with throughput of 533MB and 1.07GB. Sending either one (AGP 1X), two (AGP 2X), or four (AGP 4X) data transfers per clock cycle accomplishes these data rates. Table 41.6 shows the different AGP modes. Other optional features include AGP texturing, sideband addressing, and pipelining. Each of these options provides additional performance enhancements.

AGP graphics support is provided by the new NLX motherboards, which also support the Pentium II microprocessor (and above). It allows for the graphic subsystem to work much closer with the processor than previously available by providing new paths for data to flow between the processor, memory, and video memory. Figure 41.13 shows this relationship.

AGP offers many advantages over traditional video adapters. You are encouraged to become familiar with the details of the AGP adapter.

TROUBLESHOOTING TECHNIQUES

Most monitors produced today can be controlled by software. In order to take advantage of this feature, the monitor must be recognized by the operating system. Windows 95/98 will display the monitor's specific information when the Change Display Type button is selected from the Display Properties screen as shown in Figure 41.14.

FIGURE 41.14 Access to change the display type

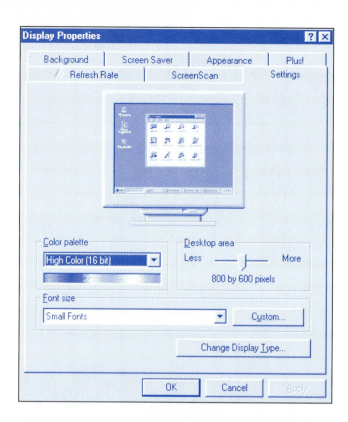

FIGURE 41.15 Setting the monitor type

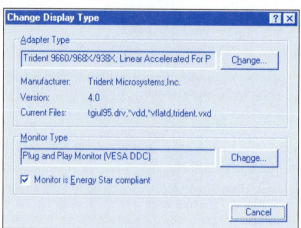

The Change Display Type window shows the current settings of the display adapter and the monitor type, as illustrated in Figure 41.15. Note the additional check box setting used to inform Windows the monitor is Energy Star compliant. If enabled, the monitor may be shut down during a period of inactivity. Be sure to verify the system, video card, and monitor can support Energy Star features before enabling them. Windows NT provides similar screens to accomplish the same tasks.

SELF-TEST

This self-test is designed to help you check your understanding of the background information presented in this exercise.

True/False

Answer *true* or *false*.

1. The standard computer monitor gets its electrical power directly from the computer.
2. A computer monitor requires special circuits such as those found on a video adapter card to process the information from the computer.

3. There are only two types of monitors used with computers: the color monitor and the monochrome monitor.
4. Any color monitor will work with any adapter card.
5. Connecting a monitor to an adapter card not made for it can severely damage the monitor.

Multiple Choice

Select the best answer.

6. Computer monitors are
   a. Easy to service.
   b. Normally serviced by computer technicians.
   c. Never in need of servicing.
   d. None of the above.
7. A computer monitor contains
   a. Very high and dangerous voltages.
   b. The same low voltages as does the computer.
   c. Easy-to-service circuits.
   d. Very few circuits, because most of the circuits are contained in the computer itself.
8. The high voltage in a computer monitor
   a. Is not dangerous because it lasts for only a short time.
   b. May still be present even when its power is disconnected.
   c. Is automatically discharged through the computer when the power is turned off.
   d. Exists only in the TV-type monitor, not the computer-type monitor.
9. The actual glass tube used in the computer monitor is referred to as the
   a. ABC.
   b. Glass tube (GT).
   c. Computer monitor (CM).
   d. Cathode-ray tube (CRT).
10. The image on the monitor is formed by
    a. Spraying the screen with electrons all at the same time.
    b. Developing a series of horizontal lines that are made across the screen from top to bottom.
    c. Developing a series of vertical lines that are made across the screen from left to right.
    d. Having the electron rays strike pixels at the proper point in time.

Matching

Match a definition on the right with each term on the left.

11. Contrast
12. Pixel
13. Aspect ratio
14. Pel
15. Text mode

a. Smallest element on the screen that can be controlled.
b. Uses a character set contained in ROM.
c. Affects the difference between light and dark on the screen.
d. The width of the monitor in comparison with the height.
e. Controls how bright the image is on the screen.
f. None of the above.

Completion

Fill in the blanks with the best answers.

16. There are basically two monitor modes, the _____ mode and the graphics mode.
17. The number of pixels available on the monitor is called the _____.
18. The _____ is the measure of the full number of colors available on the monitor.
19. The VGA display has a resolution of _____.
20. There are basically two different types of monitor displays: one is _____ and the other is analog.

Open-Ended

21. Explain what the term *multisync* means.
22. What is the fastest-growing and most popular graphics adapter card?
23. What is the resolution of the SVGA adapter card?
24. What is a green PC?
25. Why change the monitor type?

FAMILIARIZATION ACTIVITY

1. Determine the type of monitor used by your computer. How did you determine this?
2. Determine the type of display adapter card used by your computer. How did you determine this?
3. Is your monitor energy efficient? How did you determine this?
4. Does your monitor support a graphics mode? How did you determine this?
5. What is the resolution of your monitor? Are there different modes of operation of your monitor that will allow for different amounts of resolution? How did you determine this?
6. Sketch the connector used to interface your monitor and the computer. List the purpose of each connection. How did you determine this information?
7. What are the controls used on your monitor? State what each of the controls does. How did you determine this information?
8. How many different color palettes are available with your monitor? How did you determine this information?
9. How many different colors can be displayed at the same time by your monitor? How did you determine this information?
10. How much memory is installed on your display adapter? How did you determine this information?

QUESTIONS/ACTIVITIES

1. List the most popular types of monitors used by computers.
2. Explain the operation of a CRT used in a color monitor.
3. Show, by a diagram, the major sections of a monitor.
4. Explain the major differences between a monochrome CRT and a color CRT.
5. State the difference between a color monitor's palette and its ability to display several colors at the same time.

REVIEW QUIZ

Within 10 minutes and with 100% accuracy,

1. Identify the type of monitor and display adapter being used by a given computer system and ensure that they have been properly installed and adjusted.

42 The Computer Printer

INTRODUCTION

On Friday afternoon, Joe Tekk was planning to leave work a little bit early. He was just finishing up a report that he had been working on all week. *Just a few more minutes,* he thought to himself.

He sent the final copy of the document to the laser printer and waited a few minutes at his desk before he went to the shared network printer to retrieve the output. As he picked up the stack of papers, he noticed a large black streak running down the middle of the first page. He leafed through the papers—every single page had the same dark streak. *Not today,* he thought to himself.

Joe checked the printer's status display and then proceeded to open the cover to examine the components inside. He removed the toner cartridge and performed a careful visual inspection. He cleaned up a small pile of toner close to the fuser unit. He then gently shook the toner cartridge to evenly distribute the toner inside and replaced it.

As Joe put all of the pieces back together, he thought to himself, *That didn't take too much time. I still can get out early.* He reset the printer and waited for a test sheet to print to see if the problem was fixed. Everything looked fine. He went back to his desk to print another copy of his document. Again, he waited at his desk while the printer completed the print job.

When Joe picked up the papers, he studied the first sheet very carefully. It looked great. Then he leafed through the rest of the pages. A few of them were faded, and a few didn't print at all. The status indicator on the printer read "Add toner." Joe looked for a new cartridge but could not find one anywhere. When he asked Donna, the secretary, where the toner cartridges were stored, she smiled and said, "Oh, they are on backorder. Sorry!"

Joe reluctantly decided to select a different printer. When he looked at the output from the new printer, he noticed the formatting was completely different and he would have to edit the document again. He spent the rest of the afternoon making changes.

PERFORMANCE OBJECTIVES

Upon completion of this exercise, within 15 minutes and with 100% accuracy, you will be able to

1. Perform a printer self-test.
2. Produce a printer test page.
3. From the user's manual and an inspection of the printer, determine whether the printer is in need of any periodic maintenance.

> **BACKGROUND INFORMATION**

There are two fundamental types of printers used with personal computers: the **impact printer** and the **nonimpact printer.** The impact printer uses some kind of mechanical device to impart an impression to the paper through an inked ribbon. The nonimpact printer uses heat, a jet of ink, electrostatic discharge, or laser light. Nonimpact printers form printed images without making physical contact with the paper. These two types of printers are illustrated in Figure 42.1.

IMPACT PRINTERS

The most common type of impact printers is the **dot-matrix printer.** The dot-matrix printer makes up its characters by means of a series of tiny mechanical pins that move in and out to form the various characters printed on the paper.

The Dot-Matrix Printer

The dot-matrix printer, one of the most popular types of printers, uses a mechanical printing head that physically moves across the paper to be printed. This mechanical head consists of tiny movable wires that strike an inked ribbon to form characters on the paper. There are two popular kinds of dot-matrix print heads: one consists of 9 pins (the movable wires) and the other consists of 24 pins. A 9-pin print head is shown in Figure 42.2.

FIGURE 42.1 Two fundamental types of printers

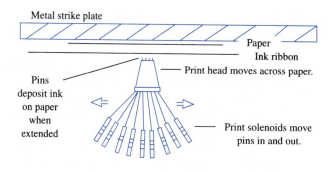

(a) Impact (dot-matrix illustration)

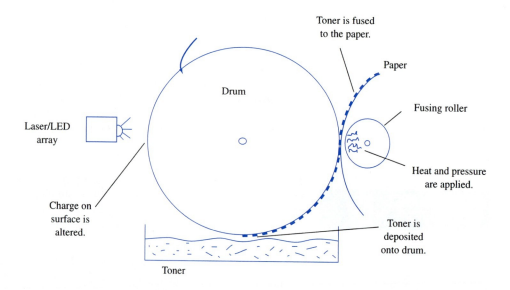

(b) Nonimpact (laser illustration)

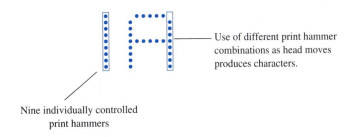

FIGURE 42.2 Nine-pin dot-matrix print head

The 24-pin dot-matrix printer is more expensive than the 9-pin model. However, because both types have modes of operation that allow for an *overstrike* of the image (with the head moving slightly and the image being struck again), the 9-pin model can produce close to what is known as letter-quality printing. The 24-pin model can produce an even sharper character when operated in the same overstrike mode. Because of the manner in which characters are formed in this type of printer, the printing of graphic images is possible.

NONIMPACT PRINTERS

The most popular nonimpact printers are the *ink-jet printer, bubble-jet printer,* and the *laser printer*. The ink-jet printer uses tiny jets of ink that are electrically controlled. The laser printer uses a laser to form characters. The laser printer resembles an office photocopying machine.

Ink-Jet and Bubble-Jet Printers

An **ink-jet printer** uses electrostatically charged plates to direct jets of ink onto paper. The ink is under pressure and is formed by a mechanical nozzle into tiny droplets that can be deflected to make up the required images on the paper. A **bubble-jet printer** uses heat to form bubbles of ink. As the bubbles cool, they form the droplets applied to the paper. Ink-jet and bubble-jet printers cost more than impact printers but are quieter and can produce high-quality graphic images.

Laser Printers

Through the operation of a laser and mirror (controlled by software), an electrical image is impressed on a photoreceptor drum. This drum picks up powdered toner, which is then transferred to paper by electrostatic discharge. A second drum uses high temperature to bond this image to the paper. The result is a high-quality image capable of excellent characters and graphics.

Because of their high-quality output, laser printers find wide application in desktop publishing, computer-aided design, and other image-intensive computer applications. Laser printers are usually at the high end of the price range for computer printers.

TECHNICAL CONSIDERATIONS

Most printer problems are caused by software. What this means is that the software does not match the hardware of the printer. This is especially true when the printing of graphics is involved. Software troubleshooting for printers is presented last in this section. When there is a hardware problem with printers, it is usually the interface cable that goes from the printer adapter card to the printer itself. This is illustrated in Figure 42.3.

Printer Cables

The interface cable is used to connect the printer to the computer. Previously, there was limited communication between the printer and computer. The computer received a few signals from the printer such as the online or offline indicator, the out-of-paper sensor, and the print buffer status. As long as the printer was sending the correct signals to the computer, the computer would continue to send data.

Advances in printer technology now require a two-way communication between the computer and printer. As a result of these changes, a new bidirectional printer cable is required to connect most new printers to the computer. The bidirectional cables may or may not adhere to the new IEEE standard for Bidirectional Parallel Peripheral Interface. The IEEE 1284 Bitronic

FIGURE 42.3 Problem areas in computer printers

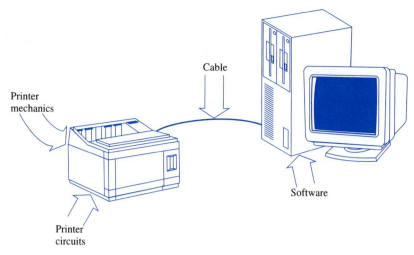

FIGURE 42.4 Typical printer cables

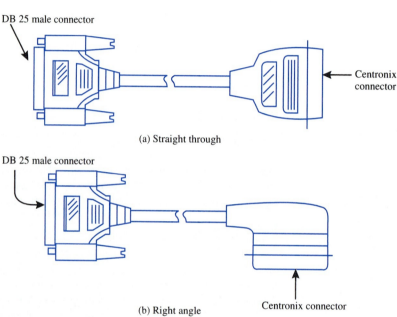

printer cable standard requires 28 AWG construction, a Hi-flex jacket, and dual shields for low EMI emissions. The conductors are twisted into pairs to reduce possible cross talk.

Check the requirements for each printer to determine the proper cable type. Figure 42.4 shows two popular parallel printer cable styles. Many different lengths of printer cables are available. It is usually best to use the shortest cable possible in order to reduce the possibility of communication errors.

Printer Hardware

A printer requires periodic maintenance. This includes vacuuming out the paper chaff left inside the printer. A soft dry cloth should be used to keep the paper and ribbon paths clean. It is a good idea to use plastic gloves when cleaning a printer, because the ink or toner is usually difficult to remove from the skin. With dot-matrix printers, be careful of the print heads. These heads can get quite hot after extended use. Make sure you do not turn the print platen rollers when the power is on because a stepper motor is engaged when power is applied. This little motor is trying to hold the print platen roller in place. If you force it to move, you could damage the stepper motor.

Laser Hardware

Essentially, laser printers require very little maintenance. If you follow the instructions that come with the printer, the process of changing the cartridge (after about 3500 copies) also performs the required periodic maintenance on the printer.

When using a laser printer, remember that such a machine uses a large amount of electrical energy and thus produces heat. So make sure that the printer has adequate ventilation and a good source of reliable electrical power. This means that you should not use an electrical expansion plug from your wall outlet to run your computer, monitor, and laser printer. Doing so may overload your system.

When shipping a laser printer, be sure to remove the toner cartridge. If you don't remove it, it could open up and spill toner (a black powder) over the inside of the printer, causing a mess that is difficult to clean up.

TESTING PRINTERS

When faced with a printer problem, first determine from the customer if the printer ever worked at all or if this is a new installation that never worked. If it is a new installation and has never worked, a careful reading of the manual that comes with the printer is usually required to make sure that the device is compatible with the printer adapter card. Table 42.1 lists some of the most direct methods for troubleshooting a computer printer.

TABLE 42.1   Printer troubleshooting methods

Checks	Comments
Check if printer is plugged in and turned on.	The printer must have external AC power to operate.
Check if printer is online and has paper.	Printers must be *online,* meaning that their control switches have been set so that they will print (check the instruction manual). Some printers will not operate if they do not have paper inserted.
Print a test page.	Select the Print Test Page option as shown in Figure 42.5. Confirm that the page printed properly. See Figure 42.6. Figure 42.7 shows the printer test page output.
Do a printer self-test.	Most printers have a self-test mode. In this mode, the printer will repeat its character set over and over again. You must refer to the documentation that comes with the printer to see how this is done.
Do a Print Screen.	If the printer self-test works, then with some characters on the computer monitor, hold down the SHIFT key and press the PRINTSCRN key at the same time. What is on the monitor should now be printed. Do not use a program (such as a word-processing program) because the software in the program may not be compatible with the printer.
Exchange printer cable.	If none of the above tests work, the problem may be in the printer cable. At this point, you should swap the cable with a known good one.
Replace the printer adapter card.	Try replacing the printer adapter card with a known good one. Be sure to refer to the printer manual to make sure you are using the correct adapter card.
Check parameters for a serial interface.	If you are using a serial interface printer from a serial port, make sure you have the transmission rate set correctly, along with the parity, number of data bits, and number of stop bits. Refer to the instruction manual that comes with the printer and use the correct form of the DOS MODE command.
Check the configuration settings.	Check all the configuration settings available on the printer.
Check the software installation.	When software is installed (such as word-processing and spreadsheet programs), the user may have had the wrong printer driver installed (the software that actually operates the printer from the program).

FIGURE 42.5 Print Test Page option

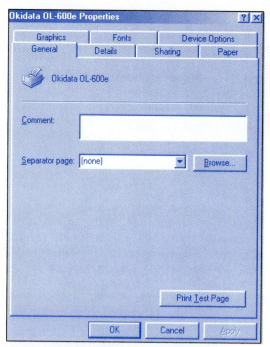

FIGURE 42.6 Printer test page confirmation

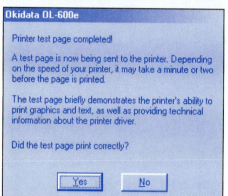

FIGURE 42.7 Printer test page output

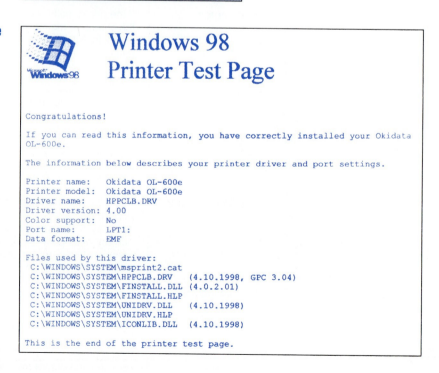

SYSTEM SOFTWARE

Many types of commands are sent to the printer while it is printing. Some of these commands tell the printer what character to print; others tell the printer what to do, such as performing a carriage return, making a new line, or doing a form feed. This is all accomplished by groups of 1s and 0s formed into a standard code that represents all of the printable characters and the other commands that tell the printer what to print and how to print it.

The ASCII Code

Table 42.2 lists all the printable characters for a standard printer. The code used to transmit this information is called the ASCII code (see Appendix A). ASCII stands for American Standard Code for Information Interchange.

As you can see in Table 42.2, each keyboard character is given a unique number value. For example, a space is number 32 (which is actually represented by the binary value 0010 0000 when transmitted from the computer to the printer). The number values that are less than 32 are used for controlling the operations of the printer. These are called **printer-control codes,** or simply **control codes.** These codes are shown in Table 42.3.

TABLE 42.2 Standard printable ASCII code

Dec	Hex	Char	Dec	Hex	Char	Dec	Hex	Char
32	20		64	40	@	96	60	'
33	21	!	65	41	A	97	61	a
34	22	"	66	42	B	98	62	b
35	23	#	67	43	C	99	63	c
36	24	$	68	44	D	100	64	d
37	25	%	69	45	E	101	65	e
38	26	&	70	46	F	102	66	f
39	27	'	71	47	G	103	67	g
40	28	(	72	48	H	104	68	h
41	29	)	73	49	I	105	69	i
42	2A	*	74	4A	J	106	6A	j
43	2B	+	75	4B	K	107	6B	k
44	2C	,	76	4C	L	108	6C	l
45	2D	-	77	4D	M	109	6D	m
46	2E	.	78	4E	N	110	6E	n
47	2F	/	79	4F	O	111	6F	o
48	30	0	80	50	P	112	70	p
49	31	1	81	51	Q	113	71	q
50	32	2	82	52	R	114	72	r
51	33	3	83	53	S	115	73	s
52	34	4	84	54	T	116	74	t
53	35	5	85	55	U	117	75	u
54	36	6	86	56	V	118	76	v
55	37	7	87	57	W	119	77	w
56	38	8	88	58	X	120	78	x
57	39	9	89	59	Y	121	79	y
58	3A	:	90	5A	Z	122	7A	z
59	3B	;	91	5B	[	123	7B	{
60	3C	<	92	5C	\	124	7C		
61	3D	=	93	5D	]	125	7D	}
62	3E	>	94	5E	^	126	7E	~
63	3F	?	95	5F	_			

TABLE 42.3 ASCII control codes

Dec	Hex	Char	
0	0	^@	NUL
1	1	☺	SOH
2	2	●	STX
3	3	♥	ETX
4	4	♦	EOT
5	5	♣	ENQ
6	6	♠	ACK
7	7	•	BEL
8	8	◘	BS
9	9	○	HT
10	A	◙	LF
11	B	♂	VT
12	C	♀	FF
13	D	♪	CR
14	E	♫	SO
15	F	☼	SI
16	10	►	DLE
17	11	◄	DC_1
18	12	↕	DC_2
19	13	‼	DC_3
20	14	¶	DC_4
21	15	§	NAK
22	16	▬	SYN
23	17	↨	ETB
24	18	↑	CAN
25	19	↓	EM
26	1A	→	SUB
27	1B	←	ESC
28	1C	∟	FS
29	1D	↔	GS
30	1E	▲	RS
31	1F	▼	US
32	20		SP

The definitions of the control code abbreviations are as follows:

ACK	Acknowledge	GS	Group separator
BEL	Bell	HT	Horizontal tab
BS	Backspace	LF	Line feed
CAN	Cancel	NAK	Negative acknowledge
CR	Carriage return	NUL	Null
DC_1–DC_4	Device control	RS	Record separator
DEL	Delete idle	SI	Shift in
DLE	Data link escape	SO	Shift out
EM	End of medium	SOH	Start of heading
ENQ	Enquiry	SP	Space
EOT	End of transmission	STX	Start text
ESC	Escape	SUB	Substitute
ETB	End of transmission block	SYN	Synchronous idle
ETX	End text	US	Unit separator
FF	Form feed	VT	Vertical tab
FS	Form separator		

The way you can enter an ASCII control code is by holding down the ALT key and the SHIFT key at the same time and using the numeric keypad to enter the ASCII control code. You will see the corresponding character on the screen when you release the ALT and

SHIFT keys. (The top row of numbers on your keyboard will not work—only the ones on the numeric keypad portion of the keyboard.)

For example, to create a text file that will cause the printer to eject a sheet of paper (a form feed), using an editor, you would press

ALT-SHIFT-12

What will appear on the screen when you release the ALT-SHIFT keys is

```
^L
```

This may not be exactly what you expected, but it is your monitor's way of interpreting what you have just entered. If you put this into a text file (say it is called FORMFED.TXT), then when it is sent to the printer, it will be interpreted as a form feed, and the printer will feed a new sheet of paper. You can then make a batch file from this form-feed text file (called FORMFED.BAT) as follows:

COPY FORMFED.TXT PRN

To test the printer's form feed, you would simply enter

```
A> FORMFED
```

and the printer (if on and ready) should then feed a sheet of paper through. You could also make up your own custom batch files to perform other types of printer tests or put your name in them (so the name appears printed on the sheet).

EXTENDED ASCII CODES

If you set up your printer to act as a graphics printer (by setting the appropriate configuration; refer to the user manual that comes with the printer), you can extend the character set to include many other forms of printable characters. These characters are shown in Table 42.4.

You can write these extended character codes to your printer by creating text files. To do this, again hold down the ALT and SHIFT keys and type the number code into the numeric keypad on your keyboard. For example, to get the Greek letter Σ, simply press ALT-SHIFT-228; when you lift up on the ALT-SHIFT keys, a Σ will appear on the monitor. If you include this in a text file (or do a PrintScreen), you can transfer it to the printer. The extended characters 176 through 223 are used for creating boxes, rectangles, and other shapes on the monitor or printer while it is still in the text mode. If you can't get these extended characters on the printer, it is either because you haven't set the printer to the graphics mode or your printer simply can't perform the functions required by this mode.

OTHER PRINTER FEATURES

Recall that most printers allow you to get different kinds of text (such as 80 or 132 characters of text across the page) or change the page orientation from portrait to landscape. You can also create batch files to test the capabilities of the printer to print the following:

- Bold text
- Underscores
- Overscores
- Superscripts
- Subscripts
- Compressed or expanded text
- Italics

For example, a new printer does not do bold text. You could move the printer to a different computer to see if the problem is in the printer, or you could have a batch file you created that will quickly test if the printer really is capable of producing bold text. If the

TABLE 42.4 Extended ASCII character set

Dec	Hex	Char	Dec	Hex	Char	Dec	Hex	Char	Dec	Hex	Char
128	80	Ç	160	A0	á	192	C0	└	224	E0	α
129	81	ü	161	A1	í	193	C1	┴	225	E1	β
130	82	é	162	A2	ó	194	C2	┬	226	E2	Γ
131	83	â	163	A3	ú	195	C3	├	227	E3	π
132	84	ä	164	A4	ń	196	C4	─	228	E4	Σ
133	85	à	165	A5	Ń	197	C5	┼	229	E5	σ
134	86	à	166	A6	a	198	C6	╞	230	E6	μ
135	87	ç	167	A7	°	199	C7	╟	231	E7	τ
136	88	ê	168	A8	¿	200	C8	╚	232	E8	φ
137	89	ë	169	A9	⌐	201	C9	╔	233	E9	θ
138	8A	è	170	AA	¬	202	CA	╩	234	EA	Ω
139	8B	ï	171	AB	½	203	CB	╦	235	EB	δ
140	8C	î	172	AC	¼	204	CC	╠	236	EC	∞
141	8D	ì	173	AD	¡	205	CD	═	237	ED	Ø
142	8E	Ä	174	AE	«	206	CE	╬	238	EE	∈
143	8F	Å	175	AF	»	207	CF	╧	239	EF	∩
144	90	É	176	B0	░	208	D0	╨	240	F0	≡
145	91	æ	177	B1	▒	209	D1	╤	241	F1	±
146	92	Æ	178	B2	▓	210	D2	╥	242	F2	≥
147	93	ô	179	B3	│	211	D3	╙	243	F3	≤
148	94	ö	180	B4	┤	212	D4	╘	244	F4	⌠
149	95	ò	181	B5	╡	213	D5	╒	245	F5	⌡
150	96	û	182	B6	╢	214	D6	╓	246	F6	÷
151	96	ù	183	B7	╖	215	D7	╫	247	F7	≈
152	98	ÿ	184	B8	╕	216	D8	╪	248	F8	°
153	99	Ö	185	B9	╣	217	D9	┘	249	F9	•
154	9A	Ü	186	BA	║	218	DA	┌	250	FA	·
155	9B	¢	187	BB	╗	219	DB	█	251	FB	√
156	9C	£	188	BC	╝	220	DC	▄	252	FC	η
157	9D	¥	189	BD	╜	221	DD	▌	253	FD	²
158	9E	₧	190	BE	╛	222	DE	▐	254	FE	■
159	9F	ƒ	191	BF	┐	223	DF	▀	255	FF	

printer can do it, the problem is probably in the software, because a new printer driver needs to be installed (which comes from the software vendor). To do this, you need to understand what printer manufacturers do in order to get their printers to create features such as **bold text,** subscripts, and superscripts.

PRINTER ESCAPE CODES

The ESC (escape) character is used by printer manufacturers as a preface. It is an easy way for them to get a whole new set of printer commands. The character ESC generally doesn't do anything by itself; what it does is to tell the printer that the character or set of characters that follows is to be treated in a special way. As an example, an <ESC> E means to begin bold text and <ESC>F means to end the bold text. The exact escape sequence is different for different printer manufacturers, and you need to find the sequence for your printer in the user's manual.

You could have a batch file calling a text file that tests for bold printing, such as

Mickey Brown's Printer test:
This is normal text.
<ESC>E
This is now bold text.
<ESC>F
This is back to normal text.

The problem in creating this kind of text file is to actually enter the ESC key into it (just pressing the ESC key doesn't do it). The secret to doing this is to enter a CTRL-V (hold down the CTRL key while pressing the V key) and then follow it with the [(left bracket). So when you see the text

```
<ESC>E
```

it really means CTRL-V[E, which will start boldface printing. Remember, for the printer you are using, the escape code may be different. All you need to do is to use the operator's manual that comes with the printer to determine the proper escape code for each printer's unique features.

MULTIFUNCTION PRINT DEVICES

It is becoming more and more common to see printers bundled with other common products, like a fax machine. For example, a fax machine usually prints any faxes received. With some modifications, it can print data received from a computer. These types of printers generally use either ink-jet or bubble-jet printer technology.

Similarly, when sending a fax, the image or text that is sent must be scanned. Again, by making some additional modifications, the scanner can provide the scanned data to a computer instead of a fax. These three features—printing, faxing, and scanning—are available on most multifunction printers. Other features such as an answering machine may also be included. Multifunction devices can save a lot of money while offering the convenience of many products in one package.

ENERGY EFFICIENCY

Like computers and monitors, printers can waste a tremendous amount of energy. This is because printers are usually left on 24 hours a day but are active only a small portion of the time. The EPA Energy Star program recommends that a printer automatically enter a sleep mode when not in use. In sleep mode, a printer may consume between 15 and 45 watts of power. This feature may cut a printer's electricity use by more than 65 percent.

Other efficiency options recommended by the EPA include printer sharing, duplex printing, and advanced power management features. Printer sharing reduces the need for an additional printer. Power management features can reduce the amount of heat produced by a printer, contributing to a more comfortable workspace and reduced air-conditioning costs. Consider turning off a printer at night, on weekends, or during extended periods of inactivity.

TROUBLESHOOTING TECHNIQUES

The most common problems with printers usually involve the quality of the output. Many problems are associated with the supply of ink or toner. Printers also contain many mechanical components that are a common point of failure. In the case of a dot-matrix printer, the ribbon may need to be replaced, the print head may need to be replaced, or the pin feeds may occasionally require some adjustment. For an ink-jet printer, the ink cartridge may become clogged with dried ink and may need to be cleaned to restore the print quality. There is no set schedule for these events to occur.

The best course of action is to be prepared for common problems that can be encountered. For example, it is a good idea to keep printer supplies on hand, so when a problem occurs, it can be remedied quickly. Table 42.5 contains a list of items that should be kept on hand. Remember, many of these items have a certain shelf life. Rotate the stock regularly.

TABLE 42.5 Common printer types and supplies

Printer Type	Supplies
Dot matrix	Ribbons, pin-feed paper
Daisy wheel	Ribbons, extra daisy wheels, pin-feed paper
Ink jet and bubble jet	Black ink cartridges, color ink cartridges, single-sheet ink-jet paper
Laser	Toner cartridges, single-sheet laser-quality paper
Color laser	Cyan, yellow, magenta, and black toner cartridges, single-sheet color laser-quality paper

SELF-TEST

This self-test is designed to help you check your understanding of the background information presented in this exercise.

True/False

Answer *true* or *false*.

1. There are basically two types of printers: impact printers and nonimpact printers.
2. A bubble-jet printer is capable of producing graphics.
3. A dot-matrix printer cannot do text, but it can do graphics.
4. An IEEE 1284 cable is used to support serial printer communications.
5. Most printer problems may be corrected by simply turning the printer off and on.

Multiple Choice

Select the best answer.

6. A nonimpact printer is the
   a. Dot-matrix printer.
   b. Ink-jet printer.
   c. Laser printer.
   d. Both b and c.
7. Most printer problems are caused by
   a. Hardware in the printer.
   b. The printer interface card.
   c. Software.
   d. The printer cable.
8. A quick way of checking a printer/computer interface is to
   a. Do a PrintScreen with some text on the monitor.
   b. Run it from the user's word-processing program.
   c. Swap the printer with a known good one.
   d. None of the above.
9. A printer test page
   a. Is performed automatically.
   b. Tests the connection to the printer.
   c. Tests the printer driver.
   d. Both b and c.
10. Printer features include:
    a. The ASCII character set.
    b. Extended ASCII characters.
    c. Various output choices.
    d. None of the above.

Matching

Match a definition on the right with each term on the left.

11. ASCII code
12. ESC
13. Extended character set

a. A printer code prefix.
b. Standard code used for communicating between the computer and printer.
c. Letters of the Greek alphabet.
d. None of the above.

Completion

Fill in the blank or blanks with the best answers.

14. For a printer to print special graph-type symbols (in text mode), it must be set to the _____ _____ mode.
15. The printer code for beginning bold text is called a(n) _____ sequence.
16. To enter an <ESC> in a text file, you must do a CTRL _____ followed by the _____ symbol.
17. Page orientation can be either _____ or _____.
18. ASCII characters less than 32 are called _____ _____.

FAMILIARIZATION ACTIVITY

Using the printer available with your system,

1. Determine whether the printer has a graphics or similar setting. Refer to the owner's manual.
2. If the printer does have such a setting, describe what is needed to put the printer into this setting (the switches and positions of these switches).
3. Using the owner's manual, determine what you must do in order to perform a printer self-test. Then actually run a printer self-test, and attach one sheet of the results to this exercise.
4. From the owner's manual, determine and list the special printing characteristics available (such as boldface), along with their <ESC> sequences.
5. Develop a batch file or files using the <ESC> sequences for your printer that will test all the printing characteristics of your particular printer.

QUESTIONS/ACTIVITIES

1. What kind of printer did you use for this exercise?
2. If you need to get a new ribbon cartridge for this printer, where would you get it and for what would you ask?
3. Does the printer you used in this exercise have the ability to print in color? Explain.
4. Does your printer use a serial or a parallel interface? How did you determine this?

REVIEW QUIZ

Within 15 minutes and with 100% accuracy,

1. Perform a printer self-test.
2. Produce a printer test page.
3. From the user's manual and an inspection of the printer, determine whether the printer is in need of any periodic maintenance.

43　Keyboards and Mice

INTRODUCTION

Joe Tekk's phone rang just as Joe walked into his office. It was his friend Ken Koder.

"Hey, Joe, you remember that problem I had with my keyboard when we were trying to use pcANYWHERE? Well, it showed up in other places. I would press a key and a different character would appear on the screen."

Joe recalled the problem. "Sure, I remember. You tried to enter '1' and kept getting an exclamation point. What did you find out?"

Ken explained, "Out of desperation I deleted my keyboard using Device Manager. When I rebooted, Windows found my keyboard and reinstalled drivers for it. Everything is fine now."

Joe thought about that approach. "I never would have tried that, Ken."

Ken laughed. "I don't know why I did it; it was just something else to try."

PERFORMANCE OBJECTIVES

Upon completion of this exercise, within 15 minutes and with 100% accuracy, you will be able to

1. Ensure that the computer keyboard has been properly connected to the system.
2. Explain how the keyboard should be maintained.
3. Ensure that the computer mouse has been properly connected and adjusted.
4. Check keyboard/mouse properties under Windows.

BACKGROUND INFORMATION

One of the most common ways of getting information into a computer is through a keyboard. A computer keyboard consists of separate keys that, when tapped, send specific codes to the computer. Essentially, such a code tells the computer that a key is depressed, what key it is that is being depressed, and when the key is no longer depressed.

Another device used for getting information into a computer is a computer **mouse.** A mouse is simply a device that moves a cursor to any desired area of the screen. The computer always knows at what position on the screen the cursor is located. On the mouse itself, there are buttons (usually three). When a button is depressed, this—along with the position of the cursor on the screen—gives the computer specific information. Usually the screen contains information as to what that particular area of the screen means to the mouse user. For instance, it could mean to begin or terminate a process. Figure 43.1 shows actions of a keyboard and a mouse.

FIGURE 43.1 Actions of a keyboard and a mouse

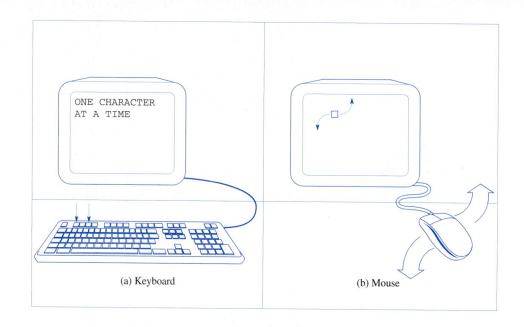

As you can see from Figure 43.1, both these devices are input devices. The major disadvantages of the keyboard are the typing skill required to use it and the need to know specific key sequences to initiate computer actions (such as DIR in order to get the directory listing of a disk).

The disadvantages of the keyboard are overcome through the use of a mouse. Using the mouse does not require any typing skills or knowledge of special key sequences (such as the DOS commands). Windows essentially requires only the use of the mouse for system interaction. Windows allows you to execute multiple programs simultaneously, and quickly change from one application to another, with a single mouse click.

THE KEYBOARD

IBM and compatible computers have gone through an evolution of keyboards. The first keyboard was the 83-key PC keyboard that was also used with some of the first XT systems. This keyboard had ten function keys along the left side, with an ESC key at the upper left and a combination numeric keypad–cursor control section on the right (see Figure 43.2).

The 84-key AT keyboard became the next in the line of keyboards and the standard for all AT-model computers. It was similar in layout to the 83-key PC keyboard. One of the changes was moving the ESC key to the top of the numeric keypad section. Three LED status lights were added to indicate when the Caps Lock, Num Lock, and Scroll Lock functions of the keyboard were enabled.

FIGURE 43.2 The original 83-key PC keyboard

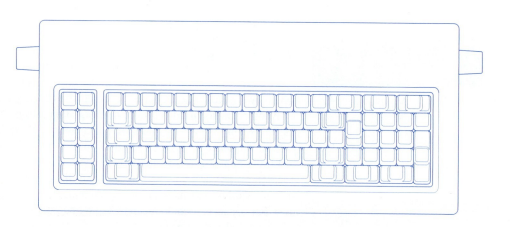

FIGURE 43.3 The 101-key enhanced keyboard

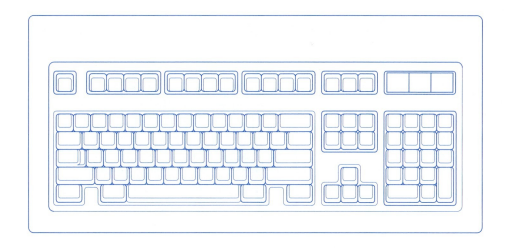

FIGURE 43.4 Four ways of identifying a key on an IBM keyboard

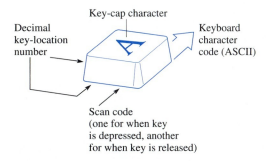

IBM later came out with an 84-key **space-saving keyboard** for some versions of IBM models 25 and 25-286. This keyboard has its function keys along the top with no numeric keypad–cursor control section on the right.

A new keyboard, used with most PCs, contains 101 keys. The function keys are located horizontally along the top of the keyboard, where there are now 12 of them. The ESC key is at the upper left. The keyboard contains duplicate cursor-movement and other similar keys. This is sometimes referred to as the **enhanced keyboard** (Figure 43.3).

Identifying Keys

There are four ways of identifying a key on an IBM keyboard: by the **character** on the cap of the key, by the **character code** associated with each key-cap character, by the scan code of the key, and by the decimal key–location number. These are illustrated in Figure 43.4.

During the POST, the first part of the keyboard scan code is displayed if there is a problem with that particular key. Figure 43.5 shows the scan codes for the IBM 101-key keyboard, and Figure 43.6 shows the key-location numbers for the same keyboard.

As shown in Figure 43.6, each key is assigned a decimal number that is used as a key-location reference on most IBM drawings. These numbers are used only as convenient guides for the physical location of the various keys and bear no relationship to the actual characters generated by the corresponding keys.

Keyboard Servicing

Outside of routine cleaning of the keyboard, there is little you can do to service it. In many cases, the keyboard assembly is a sealed unit. The major hazards to a keyboard are spilled liquids. Periodically you can use a chip puller to pull the keytops off the keyboard. (Be sure to have a similar keyboard to use as a reference when replacing these key caps.) Then hold the keyboard upside down and blow it out with compressed air.

The keyboard is connected to the computer through a cable to the **keyboard interface connector.** This connector is shown in Figure 43.7.

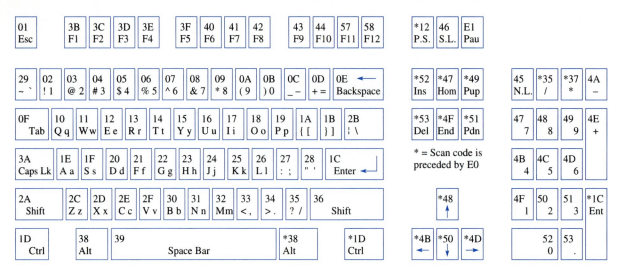

FIGURE 43.5 Scan codes for 101-key keyboard

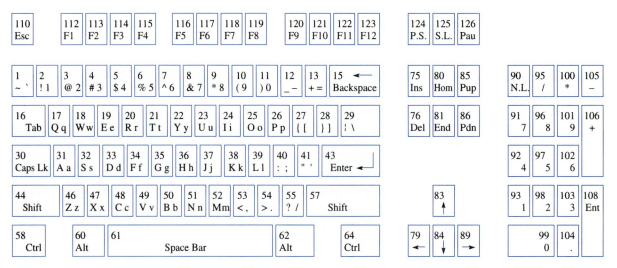

FIGURE 43.6 Key-location numbers for IBM 101-key keyboard

FIGURE 43.7 Keyboard interface connector (socket)

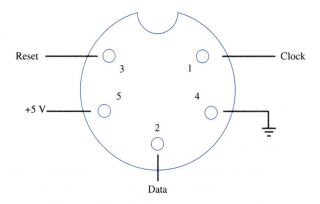

You can use a voltmeter to test the operation of the keyboard interface connector. Voltages between all pins and pin 4 of the connector should be in the range of 2 to 5.5 V DC. If any of these voltages are wrong, the problem is usually in the computer's system board. If these voltages are correct, the problem may be in the keyboard or its connector cable.

FIGURE 43.8 Typical computer mouse

Some keyboards have one or more switches (on the bottom side) to make them compatible with the computer to which they are connected. Check these switch settings as well as the documentation that comes with the keyboard.

If you find that only one key is malfunctioning, you can check the small spring on the key. Simply remove the key cap, under which you will see a small spring. Try pulling the spring slightly and then replace the key cap. You can check the cable continuity by carefully removing the bottom plate of the keyboard and observing how the cable interfaces with the computer.

Because new keyboards are so inexpensive, it is usually cheaper to replace a bad keyboard.

Today, you can purchase custom keyboards that have themes (a *Star Trek* keyboard), keyboards with infrared transmitters and no cables, or a keyboard with the keys arranged into two groups, one for each hand. You can even buy a keyboard with a scanner built in.

THE MOUSE

There are basically two types of mouse drivers used with PCs: one is a **serial mouse** and the other is a **parallel mouse.** The serial mouse interfaces with the computer through the serial port; the parallel mouse interfaces through a parallel port. In the PC, the mouse is typically connected to the 9-pin male plug on the COM1 serial port. A 9- to 25-pin adapter is available if the serial port has a 25-pin connector.

When a serial port is not available, a **bus mouse** can be interfaced with the computer using its own **bus interface** board. Figure 43.8 shows a typical computer mouse.

TRACKBALLS

A trackball is similar to a mouse except the device does not move. Instead, the user pushes a round trackball around inside its case, allowing the same movement as a mouse but not requiring a mousepad or large surface for movement. Many laptop and notebook computers have trackball mouse devices built in.

TROUBLESHOOTING TECHNIQUES

Usually the biggest problems with a mouse-to-computer connection is improper installation of the mouse software. To correct this problem, read the literature that comes with the mouse. In order for the mouse to interface effectively with the computer, the software that comes with the mouse must be installed in the system as directed by the manufacturer. If you are running DOS, you also want to make sure that the system's CONFIG.SYS and AUTOEXEC.BAT files have been properly set up so that the mouse driver is automatically installed each time the system is booted up. Again, this information is included in the literature that comes with the mouse. The important point here is that you read the literature and follow the directions.

FIGURE 43.9 Initial Mouse Properties window

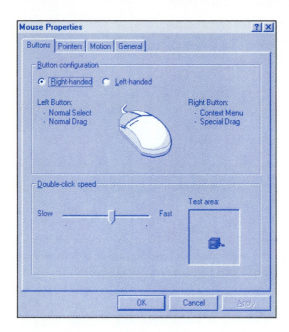

FIGURE 43.10 Mouse pointer types

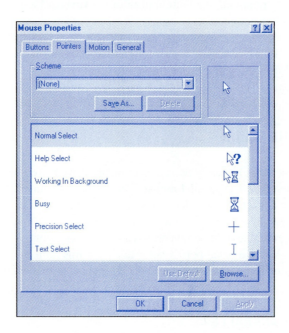

Usually, there is an optional utility program with the software that comes with the mouse that helps you test and adjust the mouse interface (by means of software). These utilities are perhaps one of the best tests of mouse performance.

Windows supplies its own mouse drivers, eliminating the need for any setup in CONFIG.SYS and AUTOEXEC.BAT, and provides a great amount of control over how the mouse appears and operates.

Clicking on the Mouse icon in Control Panel brings up the Mouse Properties window shown in Figure 43.9. Here you can adjust such important parameters as the double-click speed and left/right-handed operation. Figures 43.10 and 43.11 show two additional control windows dealing with the appearance of the mouse pointer. If the mouse is not responding correctly, it may be necessary to change its driver or hardware properties. In Control Panel, double-clicking the System icon and then selecting the Device Manager tab will allow you to double-click Mouse and check the driver information. This information is illustrated in Figure 43.12. Mouse information is available under the Devices icon in the Windows NT Control Panel. Figure 43.13 shows the associated Windows NT mouse information.

FIGURE 43.11 Additional pointer controls

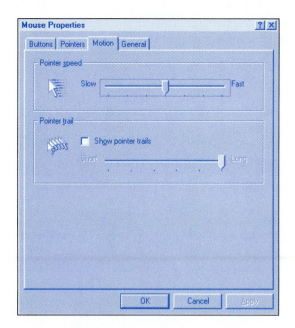

FIGURE 43.12 Examining the Mouse Driver information

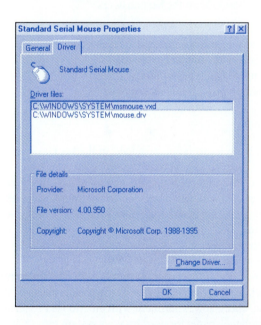

FIGURE 43.13 Windows NT Mouse Driver status

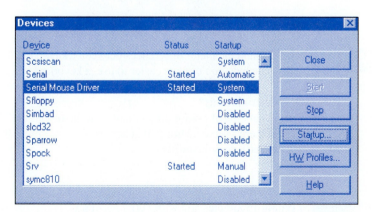

SELF-TEST

This self-test is designed to help you check your understanding of the background information presented in this exercise.

True/False

Answer *true* or *false*.

1. One of the most common ways of getting information into the computer is through the keyboard.
2. Generally speaking, a mouse is easier to use than a keyboard.
3. Scan codes are the same as ASCII codes.

Matching

Match one or more descriptions on the right with each term on the left.

4. 101-key keyboard
5. 83-key keyboard
6. 84-key keyboard

a. The enhanced keyboard.
b. Used in the PS/2 line.
c. First IBM keyboard.
d. None of the above.

Completion

Fill in the blanks with the best answers.

7. The _____ code is associated with each key-cap character on the keyboard.
8. The key-location number is used as a convenient guide to the _____ location of the actual key on the keyboard.
9. There are two types of mouse drivers: _____ and parallel.
10. Windows provides its own mouse _____.

FAMILIARIZATION ACTIVITY

1. Using the information provided in the background information of this exercise, determine the type of keyboard used by your computer.
2. Sketch the connector that the keyboard uses on the rear of the computer.
3. Where are the function keys located on your keyboard? What purposes do the keys serve?
4. Does your keyboard have any switches (usually located on the bottom of the keyboard)? If so, determine and record the purposes of these switches.
5. If your system has a mouse, what software has been added to the system so that the mouse can function?
6. If there is a utility program for checking the operation of the mouse with your system, practice using it. This is one of the requirements of the performance objectives. Otherwise, check and experiment with the Mouse Properties under Windows.
7. Sketch the connector used by your system mouse at the rear of the computer.

QUESTIONS/ACTIVITIES

1. Explain the major differences between a keyboard and a mouse.
2. Which do you think is easier to use, a keyboard or a mouse? Why?
3. If your keyboard has a Num Lock key, what purpose does this key serve? How did you determine this?
4. What type of mouse do you have—serial or parallel? How did you determine this?
5. Is there a separate utility that comes with the mouse for testing its operation? If so, what exactly does the utility do?

REVIEW QUIZ

Within 15 minutes and with 100% accuracy,

1. Ensure that the computer keyboard has been properly connected to the system.
2. Explain how the keyboard should be maintained.
3. Ensure that the computer mouse has been properly connected and adjusted.
4. Check keyboard/mouse properties under Windows.

44 Telephone Modems

INTRODUCTION

Joe Tekk was looking through his storage boxes in the attic of his parent's home. In one box he found his old 300-baud acoustic modem. He pulled it out and looked at it: shiny white plastic with black rubber couplers for the telephone handset. Toggle switches on the control panel selected different modes of operation. Joe recalled connecting to a college mainframe for the first time and endlessly dialing his old rotary telephone until the single line to the college was available.

Joe thought it was funny that back then 300 baud seemed really fast. Today, modems as small as credit cards operate almost 200 times faster and still do not seem fast enough to satisfy the growing needs of personal and business computing. Joe wondered exactly when the change had occurred.

PERFORMANCE OBJECTIVES

Upon completion of this exercise, within 20 minutes and with 100% accuracy, you will be able to

1. Successfully use a computer modem to transmit and receive a text file.
2. Explain the various MNP and CCITT standards.
3. Describe several encoding methods.

BACKGROUND INFORMATION

This exercise has to do with communications between computers. The user of one computer may interact with another computer—which may be located thousands of miles away—as if it were sitting right in the same room.

In order for computers to communicate in this manner, four items must be available, as shown in Figure 44.1.

As shown in Figure 44.1, there must be some kind of link between the computers. The most convenient link to use is the already-established telephone system lines. Using these lines and a properly equipped computer allows communications between any two computers that have access to a telephone. This becomes a very convenient and inexpensive method of communicating between computers.

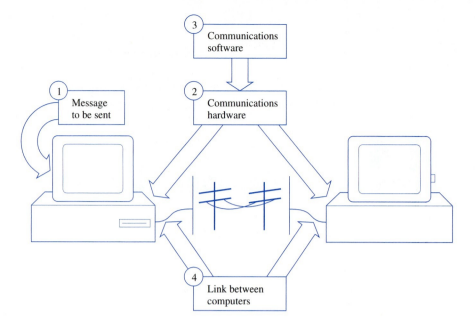

FIGURE 44.1 Four items necessary for computer communications

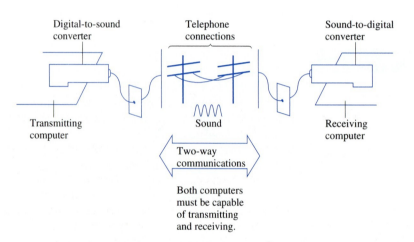

FIGURE 44.2 Basic needs for the use of telephone lines in computer communications

There is, however, one problem. Telephone lines were designed for the transmission of the human voice, not for the transmission of digital data. Therefore, in order to make use of these telephone lines for transmitting computer data, the ONs and OFFs of the computer must first be converted to sound, sent over the telephone line, and then reconverted from sound to the ONs and OFFs that the computer understands. This concept is shown in Figure 44.2.

THE MODEM

The word *modulate* means to change. Thus an electronic circuit that changes digital data into sound data can be called a *modulator*. The word *demodulate* can be thought of as meaning "unchange," or restore to an original condition. Any electronic circuit that converts the sound used to represent the digital signals back to the ONs and OFFs understood by a computer can, therefore, be called a demodulator. Since each computer must be capable of both transmission and reception, each computer must contain an electrical circuit that can modulate as well as demodulate. Such a circuit is commonly called a mo̲dulator/de̲modulator, or **modem.**

For personal computers, a modem may be an internal or an external circuit—both perform identical functions.

FIGURE 44.3 The RS-232 standard

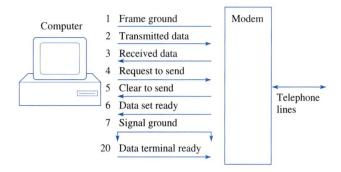

TABLE 44.1 Standard baud rates

Low Speed	High Speed
300	
600	
1200	14,400
2400	28,800
4800	33,600
9600	56K

THE RS-232 STANDARD

The EIA (Electronics Industries Association) has published the EIA *Standard Interface Between Data Terminal Equipment Employing Serial Binary Data Interchange*—specifically, EIA-232-C. This is a standard defining 25 conductors that may be used in interfacing **data terminal equipment** (DTE, such as your computer) and **data communications equipment** (DCE, such as a modem) hardware. The standard specifies the function of each conductor, but it does not state the physical connector that is to be used. This standard exists so that different manufacturers of communications equipment can communicate with each other. In other words, the RS-232 standard is an example of an interface, essentially an agreement among equipment manufacturers on how to allow their equipment to communicate.

The RS-232 standard is designed to allow DTEs to communicate with DCEs. The RS-232 uses a DB-25 connector; the male DB-25 goes on the DTEs and the female goes on the DCEs. The RS-232 standard is shown in Figure 44.3.

The RS-232 is a digital interface designed to operate at no more than 50 feet with a 20,000-bit/second bit rate. The **baud,** named after J. M. E. Baudot, actually indicates the number of *discrete* signal changes per second. In the transmission of binary values, one such change represents a single bit. What this means is that the popular usage of the term *baud* has become the same as bits per second (bps). Table 44.1 shows the standard set of baud rates.

TELEPHONE MODEM SETUP

The most common problem with telephone modems is the correct setting of the software. There are essentially six distinct areas to which you must pay attention when using a telephone modem:

1. Port to be used
2. Baud rate
3. Parity
4. Number of data bits
5. Number of stop bits
6. Local echo ON or OFF

Most telephone modems have a default setting for each of these areas. However, as a user you should understand what each of these areas means. You will have to consult the specific documentation that comes with the modem in order to see how to change any of the above settings. For now it is important that you understand the idea behind each of these areas.

Port to Be Used

The most common ports to be used are COM1 and COM2. Other possible ports are COM3 and COM4. The port you select from the communications software depends on the port to which you have the modem selected. On most communications software, once you set the correct port number, you do not need to set it again.

Baud Rate

Typical values for the baud rate are between 9600 and 56K. Again, these values can be selected from the communications software menu. It is important that both computers be set at the same baud rate.

Parity

Parity is a way of having the data checked. Normally, parity is not used. Depending on your software, there can be up to five options for the parity bit, as follows:

Space: Parity bit is always a 0.
Odd: Parity bit is 0 if there is an odd number of 1s in the transmission and is a 1 if there is an even number of 1s in the transmission.
Even: Parity bit is a 1 if there is an odd number of 1s in the transmission and is a 0 if there is an even number of 1s in the transmission.
Mark: Parity bit is always a 1.
None: No parity bit is transmitted.

Again, what is important is that both the sending and receiving units are set up to agree on the status of the parity bit.

Number of Data Bits

The number of data bits to be used is usually set at 8. There are options that allow the number of data bits to be set at 7. It is important that both computers expect the same number of data bits.

Number of Stop Bits

The number of stop bits used is normally 1. However, depending on the system, the number of stop bits may be 2. Stop bits are used to mark the end of each character transmitted. Both computers must have their communications software set to agree on the number of stop bits used.

WINDOWS MODEM SOFTWARE

Windows has built-in modem software, accessed through the Control Panel. Clicking on the Mouse icon displays the window shown in Figure 44.4. Notice that Windows indicates the presence of an external Sportster modem. To test the modem, click the Diagnostics tab. Figure 44.5 shows the Diagnostics window.

Selecting COM2 (the Sportster modem) and then clicking More Info will cause Windows to talk to the modem for a few moments, interrogate it, and then display the results in a new window, shown in Figure 44.6.

Specific information about the modem port is displayed, along with the responses to several AT commands. The *AT command set* is a standard set of commands that can be sent to

FIGURE 44.4 Modem Properties window

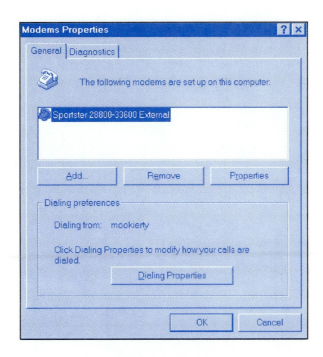

FIGURE 44.5 Modems Diagnostics window

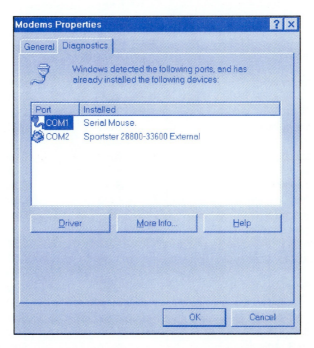

the modem to configure, test, and control it. Table 44.2 lists the typical **Hayes compatible** commands (first used by Hayes in its modem products). An example of an AT command is:

ATDT 778 8108

which stands for AT (attention) DT (dial using tones). This AT command causes the modem to touch-tone dial the indicated phone number. Many modems require an initial AT command string to be properly initialized. This string is automatically output to the modem when a modem application is executed.

TELEPHONE MODEM TERMINOLOGY

In using technical documentation concerning a telephone modem, you will encounter some specialized terminology. Figure 44.7 illustrates some of the ideas behind some basic

FIGURE 44.6 Modem diagnostic information

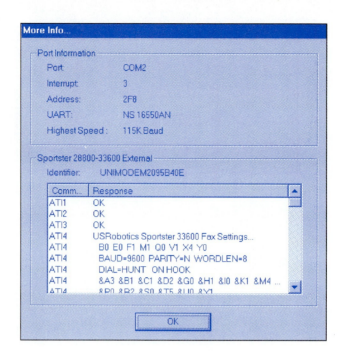

TABLE 44.2 Selected AT commands

Command	Function	Command	Function
A/	Repeat last command	Xn	Result code type
A	Answer	Yn	Long space disconnect
Bn	Select CCITT or Bell	Zn	Recall stored profile
Cn	Carrier control option	&Cn	DCD option
D	Dial command	&Dn	DTR option
En	Command echo	&F	Load factory defaults
Fn	Online echo	&Gn	Guard tone option
Hn	Switch hook control	&Jn	Auxiliary relay control
In	Identification/checksum	&M0	Communication mode option
Kn	SRAM buffer control	&Pn	Dial pulse ratio
Ln	Speaker volume control	&Q0	Communication mode option
Mn	Speaker control	&Sn	DSR option
Nn	Connection data rate control	&Tn	Self-test commands
On	Go online	&Vn	View active and stored configuration
P	Select pulse dialing	&Un	Disable Trellis coding
Qn	Result code display control	&Wn	Stored active profile
Sn	Select an S-register	&Yn	Select stored profile on power-on
Sn=x	Write to an S-register	&Zn=x	Store telephone number
Sn?	Read from an S-register	%En	Auto-retrain control
?	Read last accessed S-register	%G0	Rate renegotiation
T	Select DTMF dialing	%Q	Line signal quality
Vn	Result code form	-Cn	Generate data modem calling tone

communication methods. As you can see from the figure, **simplex** is a term that refers to a communications channel in which information flows in one direction only. An example of this is a radio or a television station.

Duplex

The **duplex** mode refers to two-way communication between two systems. This term is further refined as follows. **Full duplex** describes a communication link that can pass data in two directions at the same time. This mode is analogous to an everyday conversation between two people either face-to-face or over the telephone. The other mode, which is not

FIGURE 44.7 Some basic communication methods

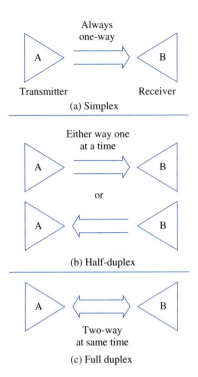

commonly available with telephone modems, is the **multiplex** mode. Multiplex refers to a communications link in which multiple transmissions are possible.

Echo

Terminology used here has to do with how the characters you send to the other terminal are displayed on your monitor screen. The term **echo** refers to the method used to display characters on the monitor screen. First, there is a **local echo.** A local echo means that the sending modem immediately returns or echoes each character back to the screen as it is entered into the keyboard. This mode is required before transmission, so that you can see what instructions you are giving the communications software. Next there is **remote echo.** Remote echo means that during the communications between two computers, the remote computer (the one being transmitted to) sends back the character it is receiving. The character that then appears on your screen is the result of a transmission from the remote unit. This is a method of verifying what you are sending. To use the remote-echo mode, you must be in the full-duplex mode. This idea is illustrated in Figure 44.8.

MODULATION METHODS

Many different techniques are used to encode digital data into analog form (for use by the modem). Several of these techniques are:

- AM (amplitude modulation)
- FSK (frequency shift keying)
- Phase modulation
- Group coding

Figure 44.9 shows how the first three of these techniques encode their digital data.

To get a high data rate (in bits per second) over ordinary telephone lines, group coding techniques are used. In this method, one cycle of the transmitted signal encodes two or more bits of data. For example, using *quadrature modulation,* the binary patterns 00, 01, 10, and 11 encode one of four different phase shifts for the current output signal. Thus, a signal that changes at a rate of 2400 baud actually represents 9600 bps!

FIGURE 44.8 Echo modes

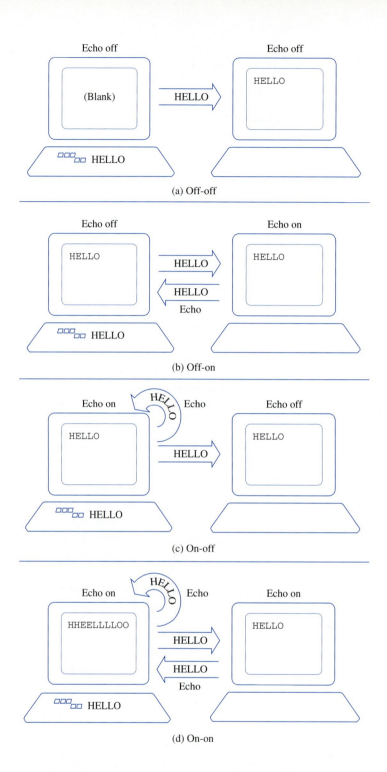

Another technique, called *Trellis modulation,* combines two or more other techniques, such as AM and quadrature modulation, to increase the data rate.

MNP STANDARDS

MNP (Microcom Networking Protocol) is a set of protocols used to provide error detection and correction, as well as compression, to the modem data stream. Table 44.3 lists the MNP classes and their characteristics.

MNP classes 4 and above are used with newer, high-speed modems. When two modems initially connect, they will negotiate the best type of connection possible, based

FIGURE 44.9 Modulation techniques

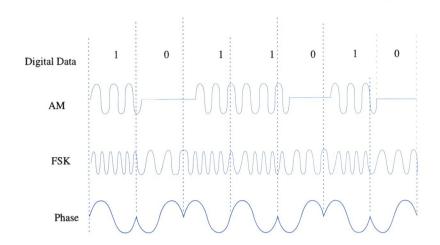

TABLE 44.3 MNP standards

Class	Feature
1	Asynchronous, half-duplex, byte-oriented
2	Asynchronous, full-duplex, byte-oriented
3	Synchronous, full-duplex, byte-oriented
4	Error correction, packet oriented
5	Data compression
6	Negotiation
7	Huffman data compression
9	Improved error correction
10	Line monitoring

Note: There is no MNP-8 standard.

on line properties, and the features and capabilities of each modem. The CCITT standards supported by the modem are also part of the negotiation. Let us look at these standards as well.

CCITT STANDARDS

CCITT (French abbreviation for International Telegraph and Telephone Consultive Committee) standards define the maximum operating speed (as well as other features) available in a modem (which is a function of the modulation techniques used). Table 44.4 lists the CCITT standards.

Earlier, low-speed standards not shown are the Bell 103 (300 bps using FSK) and Bell 212A (1200 bps using quadrature modulation). V.22 is similar in operation to Bell 212A, and is more widely accepted outside the United States.

The V.90 standard, finalized in early 1998, outlines the details of modem communication at 56K bps, currently the fastest speed available for regular modems. Fax modems have their own set of standards.

ISDN MODEMS

ISDN (Integrated Services Digital Network) is a special connection available from the telephone company that provides 64K bps digital service. An ISDN modem will typically

TABLE 44.4 CCITT standards

Standard	Data Rate (bps)
V.22	1200
V.22 bis	2400
V.32	9600
V.32 bis	14,400
V.32 terbo	19,200
V.34	28,800/33,600
V.90	56K

connect to a *basic rate ISDN* (BRI) line, which contains two full-duplex 64Kbps B channels (for voice/data) and a 16Kbps D channel (for control). This allows up to 128Kbps communication. ISDN modems are more expensive than ordinary modems, and require you to have an ISDN line installed before you can use it.

CABLE MODEMS

One of the most inexpensive, high-speed connections available today is the cable modem. A cable modem connects between the television cable supplying your home and a network interface card in your computer. Two unused cable channels are used to provide data rates in the hundreds of thousands of bits per second. For example, downloading a 6MB file over a cable modem takes less than 20 seconds (during several tests of a new cable modem installation). That corresponds to 2,400,000 bps! Of course, the actual data rate available depends on many factors, such as the speed the data is transmitted from the other end and any communication delays. But unlike all other modems, the cable modem has the capability to be staggeringly fast, due to the high bandwidth available on the cable. In addition, a cable modem is typically part of the entire package from the cable company, and is returned when you terminate service. The cost is roughly the same as the cost of basic cable service.

FAX/DATA MODEMS

It is difficult to find a modem manufactured today that does not have fax capabilities built into it. Since fax/data modems are relatively inexpensive, it does not make sense to purchase a separate fax machine (unless it is imperative that you be able to scan a document before transmission). Word-processing programs (such as WordPerfect) now support the use of a fax/data modem, helping to make the personal computer almost an entire office by itself.

PROTOCOLS

A **protocol** is a prearranged communication procedure that two or more parties agree on. When two modems are communicating over telephone lines (during a file transfer from a computer bulletin board or an America Online session), each modem has to agree on the technique used for transmission and reception of data. Table 44.5 shows some of the more common protocols. The modem software that is supplied with a new modem usually allows the user to specify a particular protocol.

TABLE 44.5 Modem communication protocols

Protocol	Operation
Xmodem	Blocks of 128 bytes are transmitted. A checksum byte is used to validate received data. Bad data is retransmitted.
Xmodem CRC	Xmodem using Cyclic Redundancy Check to detect errors.
Xmodem-1K	Essentially Xmodem CRC with 1024-byte blocks.
Ymodem	Similar to Xmodem-1K. Multiple files may be transferred with one command.
Zmodem	Uses 512-byte blocks and CRC for error detection. Can resume an interrupted transmission from where it left off.
Kermit	Transmits data in packets whose sizes are adjusted to fit the needs of the other machine's protocol.

TROUBLESHOOTING TECHNIQUES

Table 44.6 lists some of the most common problems encountered in telephone modems. As you will see, most of the problems are software related.

Other common problems encountered involve very simple hardware considerations. For example, telephone modems usually come with two separate telephone line connectors.

The purpose of the phone input is to connect a telephone, not the output line from the modem, to the telephone wall jack. The phone input is simply a convenience. It allows the telephone to be used without having to disconnect a telephone line from the computer to the wall telephone jack. If you mistakenly connect the line from the wall telephone jack to the phone input, you will be able to dial out from your communications software, but your system will hang up on you. Make sure that the telephone line that goes to the telephone wall jack comes from the *line output* and not the *phone output* jack of your modem.

Another common hardware problem is a problem in your telephone line. This can be quickly checked by simply using your phone to get through to the other party. If you can't do this, then neither can your computer.

A problem that is frequently encountered in an office or school building involves the phone system used within the building. You may have to issue extra commands on your software in order to get your call out of the building. In this case you need to check with your telecommunications manager or the local phone company.

TABLE 44.6 Common telephone modem problems

Symptom	Possible Cause(s)
Can't connect	Usually this means that your baud rates or numbers of data bits are not matched. This is especially true if you see garbage on the screen, especially the { character.
Can't see input	You are typing in information but it doesn't appear on the screen. However, if the person on the other side can see what you are typing, it means that you need to turn your local echo on. In this way, what you type will be echoed back to you, and you will see it on your screen.
Get double characters	Here you are typing information and getting double characters. This means that if you type HELLO, you get HHEELLLLOO; at the same time, what the other computer is getting appears normal. This means that you need to turn your local echo off. In this way, you will not be echoing back the extra character. With some systems *half-duplex* refers to local echo on, whereas *full duplex* refers to local echo off.

Sometimes your problem is simply a noisy line. This may have to do with your communications provider or it may have to do with how your telephone line is installed. You may have to switch to a long-distance telephone company that can provide service over more reliable communication lines. Or you may have to physically trace where your telephone line goes from the wall telephone jack. If this is an old installation, your telephone line could be running in the wall right next to the AC power lines. If this is the case, you need to reroute the phone line.

SELF-TEST

This self-test is designed to help you check your understanding of the background information presented in this exercise.

True/False

Answer *true* or *false*.

1. The most convenient link to use in order to communicate between computers at long distances is the two-way radio.
2. Telephone lines easily allow for the transmission of the computer's ON and OFF signals.
3. The word *modulate* means "to change."

Multiple Choice

Select the best answer.

4. The word *modem* comes from
    a. The first manufacturer of the device.
    b. An FCC regulation.
    c. A combination of the words "modulate" and "demodulate."
    d. None of the above.
5. The RS-232 standard
    a. Sets the transmission rate.
    b. Specifies the function of each conductor used in data communications.
    c. Was an old standard set by the FCC for communications between computers.
    d. Both b and c are correct.
6. Full-duplex mode means
    a. Simultaneous communications between two systems.
    b. One-way communication from one system to another.
    c. Communication to more than one system.
    d. None of the above.

Matching

Match a corrective action on the right with each symptom on the left.

7. Nothing appears on the screen when you type.
8. You get double characters on the screen when you type.
9. You see a bunch of incorrect characters on the screen.

a. Check baud rate and number of data bits.
b. Turn local echo on.
c. Turn local echo off.
d. None of the above.

Completion

Fill in the blanks with the best answers.

10. The purpose of the _____ connection on the modem is as a convenience for connecting a telephone.
11. The purpose of the _____ connection on the modem is for the purpose of connecting the modem to the telephone lines.

12. The _____ rate is a common unit of measurement of the number of discrete changes per second.
13. A prearranged communication method is called a(n) _____.

FAMILIARIZATION ACTIVITY

1. Read the documentation that accompanies the modem used by your system.
2. If one is not already present, create a text file on your disk that can be sent from one computer to another over the telephone modem.
3. If it has not already been done, properly connect your computer to a designated telephone outlet.
4. Your instructor will assign a telephone number for you to dial. Using your modem software, dial the assigned number and set parameters to an agreed-to setting.
5. Transmit the text file on your disk to the remote station.
6. Hang up, and then verify (by telephone) that your text file has been received.
7. Now set your modem software so that your computer is in the receive mode. An assigned remote terminal will now contact you and send you a text file. Make sure that the received text file is stored on your floppy disk.
8. By telephone, verify to the sender that the text file has been properly received.

QUESTIONS/ACTIVITIES

1. Explain, through the use of diagrams, what is needed in order to have two computers communicate with each other over standard voice telephone lines.
2. What is a modem? What does it do? What is the origin of the term "modem"?
3. State what is meant by the RS-232 standard.
4. Define baud rate. How is this term used?
5. Explain how the echo mode is properly used between computers.
6. What is the purpose of a modem protocol?
7. Check your modem properties under Windows.

REVIEW QUIZ

Within 20 minutes and with 100% accuracy,

1. Successfully use a computer modem to transmit and receive a text file.
2. Explain the various MNP and CCITT standards.
3. Describe several encoding methods.

45 CD-ROM and Sound Card Operation

INTRODUCTION

Joe Tekk was in his office at RWA Software, speaking slowly and clearly into a microphone. "Notepad . . . notepad . . . notepad," he was saying as Don, his manager, walked in. Don waited until Joe was finished speaking and asked, "What are you doing?"

"I'm training a speech recognizer to understand a bunch of different words. I want to use it to let me do simple things in Windows, such as open Notepad or Calculator, or shut down automatically whenever I say, 'Goodnight.'"

"Why did you say each word three times?"

Joe turned and picked up the speech board manual lying open on his desk. "It says you have to say each word more than once to build an accurate recognition envelope." Joe handed the manual to Don. "Pretty cool, huh?"

"Just beautiful, Joe," Don replied. "I think I'd like to teach it a few words of my own."

PERFORMANCE OBJECTIVES

Upon completion of this exercise, within 10 minutes and with 100% accuracy, you will be able to

1. Discuss what is required to operate a CD-ROM drive and a sound card.
2. Explain what it means to be MPC compliant.
3. Discuss the various CD-ROM formats.

BACKGROUND INFORMATION

The term *multimedia* is now generally applied to personal computers equipped with CD-ROM drives and sound cards. Entire encyclopedias are now available on CD-ROM. Access the subject of spacecraft, and you get live-action video of an *Apollo* moon shot, complete with the accompanying audio. Hundreds of software packages are available on CD-ROM, with more appearing every day.

A computer is said to be **MPC (Multimedia PC) compliant** if it contains the following hardware:

- 386SX-16 processor (or better)
- 4MB of RAM (or more)
- 40MB hard drive (or more)
- A color VGA display (or better)
- A mouse
- A single-spin CD-ROM (or faster)

Single-spin CD-ROM drives transfer data at a maximum rate of 150KB/second. A double-spin CD-ROM drive transfers data at 300KB/second, and so on.

A computer is **MPC-2 compliant** if it contains the following updated hardware:

- 486SX-25 processor (or better)
- 8MB of RAM (or more)
- 40MB hard drive (or more)
- A color VGA display (or better)
- A mouse
- A double-spin CD-ROM (or faster)

Note that current technology far exceeds the minimum requirements established by MPC-2.

In this exercise, we will see what is necessary to install a CD-ROM drive and sound card in a machine.

CD-ROM OPERATION

A CD-ROM stores binary information in the form of microscopic *pits* on the disk surface. The pits are so small that a CD-ROM typically stores more than 650MB of data. This is equivalent to more than 430 1.44MB floppies. A laser beam is shone on the disk surface and either reflects (no pit) or does not reflect (pit), as you can see in Figure 45.1.

These two light states (reflection and no reflection) are easily translated into a binary 0 and a binary 1. Since the pits are mechanically pressed into a hard surface and only touched by light, they do not wear out or change as a result of being accessed.

PHYSICAL LAYOUT OF A COMPACT DISK

Figure 45.2 shows the dimensions and structure of a compact disk. The pits previously described are put into the reflective aluminum layer when the disk is manufactured. Newer recordable CDs use a layer of gold instead of aluminum so that they can be written to using a low-power laser diode.

THE HIGH SIERRA FORMAT

The High Sierra format specifies the way the CD is logically formatted (tracks, sectors, directory structure, file name conventions). This specification is officially called ISO-9660 (International Standards Organization).

FIGURE 45.1 Reading data from a CD

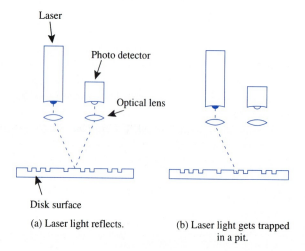

FIGURE 45.2   Compact disk

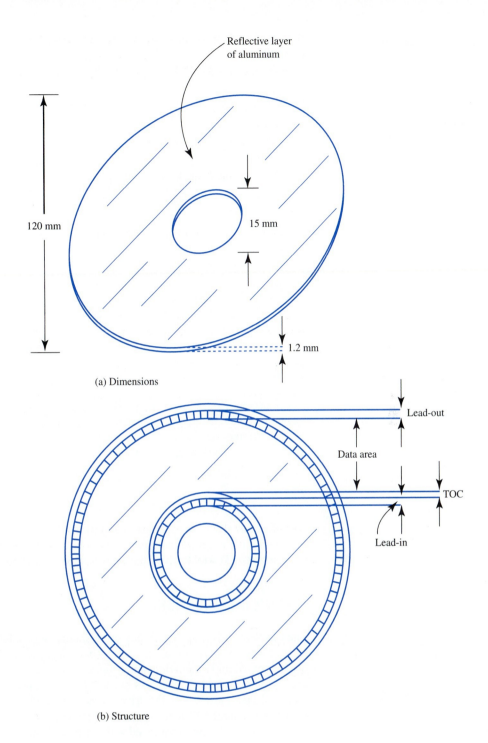

CD-ROM STANDARDS

The evolution of the CD-ROM is documented in several *books*. These are described in Table 45.1. The red book describes the method used to store digital audio on the CD-ROM. Pulse code modulation (PCM) is used to sample the audio 44,100 times/second with 16-bit sampling.

The green book specifies the CD interactive format typically used by home video games. Text, audio, and video are interleaved on the CD-ROM, and the MPEG-1 (Motion Picture Entertainment Group) video encoding method requires special hardware inside the player to support real-time video. The CD-ROM-XA (extended architecture) format is similar to CD-I.

The orange book provides the details on recordable CD-ROM drives. Gold-based disks are used to enable data to be written to the CD-ROM, up to 650MB. A *multisession*

TABLE 45.1 CD-ROM standards

Book	Feature
Red	CD-DA. Digital audio. PCM encoding.
Green	CD-I. Interactive (text, sound, and video). ADPCM and MPEG-1 encoding.
Orange	CD-R. Recordable.
Yellow	CD-ROM. Original PC CD-ROM format. 150Kbps transfer rate.

CD-ROM allows you to write to the CD more than once, and requires a multisession CD-ROM drive. Compact disks that only allow one recording session are known as WORM drives, for write once, read mostly.

The yellow book describes the original PC-based CD-ROM format (single spin), which specifies a data transfer rate of 150Kbps. 2x CD-ROMs transfer at 300Kbps. 4x CD-ROMs transfer data at 600Kbps. Currently there are 48x CD-ROM drives on the market, with faster ones coming.

PHOTO CD

Developed by Kodak, the photo CD provides a way to store high-quality photographic images on a CD (using recordable technology) in the CD-I format. Each image is stored in several different resolutions, from 192×128 to 3072×2048, using 24-bit color. This allows for around 100 images on one photo CD.

ATAPI

ATAPI stands for AT Attachment Packet Interface. ATAPI is an improved version of the IDE hard drive interface, and uses *packets* of data during transfers. The ATAPI specification supports CD-ROM drives, hard drives, tape backup units, and plug-and-play adapters.

CD-ROM INSTALLATION

The CD-ROM drive, like the floppy drive, mounts with a few screws in an empty drive bay and is connected to the computer by a ribbon cable plugged into an adapter card. A second cable reserved for left and right audio signals is also connected between the CD-ROM drive and the adapter card. This allows music CDs to be played with the CD-ROM drive. As a matter of fact, most CD-ROM drives come with application software that turns the color display into a large CD control panel for a stereo.

The CD-ROM drive is equipped with the same 4-pin power connector found on hard and floppy drives.

There are two styles of CD-ROM drives: those that have manual cartridge loading and those that have automatic cartridge loading. As shown in Figure 45.3, both drives contain an activity indicator, an audio jack, and a volume control on the front panel. In the manual drive, the user must pull the cartridge out and push it in. In the automatic drive, the user presses a button to load or eject a small platter that holds the CD.

Once the CD-ROM drive has been installed, two pieces of software must be loaded to make the drive operational. The first is a manufacturer-specific driver that manages the hardware interface between the CD-ROM drive and the computer. The second required program, which is the same for all CD-ROM drives, is the MSCDEX.EXE (Microsoft CD-ROM Extensions) file that comes with DOS and Windows. This program is used to access the CD-ROM drive as if it were an ordinary floppy drive, with sectors, tracks, a file allocation table, and so on. The first program needs to be loaded from the CONFIG.SYS file.

FIGURE 45.3 Two styles of CD-ROM drives

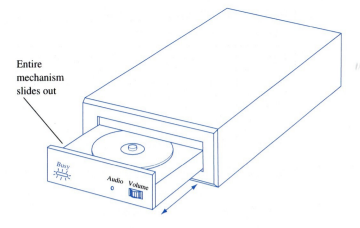

(a) Manual cartridge opening/closing performed by user

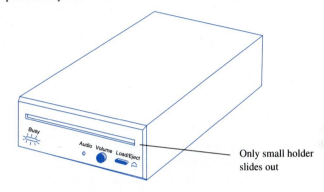

(b) Automatic cartridge opening/closing when user presses Load/Eject button

MSCDEX.EXE gets loaded during processing of the AUTOEXEC.BAT file. Most CD-ROM drive manufacturers provide an automated software installation program with their drives that makes the necessary changes in both CONFIG.SYS and AUTOEXEC.BAT, and copies the necessary files to the hard disk.

During the software installation, you are usually offered a choice between software transfer and DMA transfer of CD-ROM data. The DMA transfer option is faster than software transfer, but requires a spare DMA channel and an unused IRQ line. Generally, the setup parameters of the CD-ROM drive must be tweaked to get the optimal performance out of the drive.

A sample line from CONFIG.SYS for a Mitsumi CD-ROM drive is as follows:

DEVICE=C:\MTMCDS.SYS /D:CDROM-1 /L:R /P:300 /A:1 /M:1 /I:3

where: MTMCDS.SYS = the name of the software transfer driver
/D:CDROM-1 = the device name of the CD-ROM drive
/P:300 = the base address of the adapter card
/A:1 = the audio play mode
/M:1 = the number of memory buffers to use
/I:3 = the IRQ signal used by the CD-ROM drive

The corresponding line from AUTOEXEC.BAT is as follows:

C:\DOS\MSCDEX.EXE /D:CDROM-1 /L:R

where: MSCDEX.EXE = the required DOS interface program
/D:CDROM-1 = the same device name used in the CONFIG.SYS file
/L:R = the drive letter (R) of the CD-ROM

CD-ROM PROPERTIES IN WINDOWS

Right-clicking on the CD-ROM icon in the My Computer window displays the Properties window shown in Figure 45.4. The label box indicates that there is a disk in the CD-ROM drive (the Windows CD-ROM). Note that there is no free space indicated in the pie chart.

SOUND CARD OPERATION

Along with CD-ROM drives, sound cards for PCs have also increased in popularity. Currently, 16-bit sound cards are available that provide multiple audio channels and FM-quality sound, and are compatible with the MIDI (Musical Instrument Data Interface) specification.

The basic operation of the sound card is shown in Figure 45.5. Digital information representing samples of an analog waveform are inputted to a *digital-to-analog* converter, which translates the binary patterns into corresponding analog voltages. These analog voltages are then passed to a *low-pass filter* to smooth out the differences between the individual voltage samples, resulting in a continuous analog waveform. All of the digital/analog signal processing is done in a custom **digital signal processor** chip included on the sound card.

Sound cards also come with a microphone input that allows the user to record any desired audio signal.

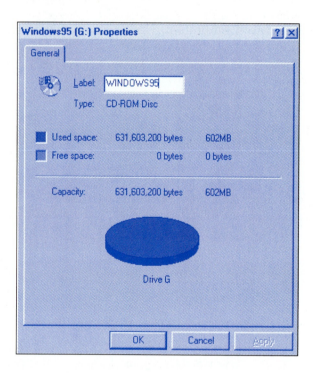

FIGURE 45.4 CD-ROM Properties window

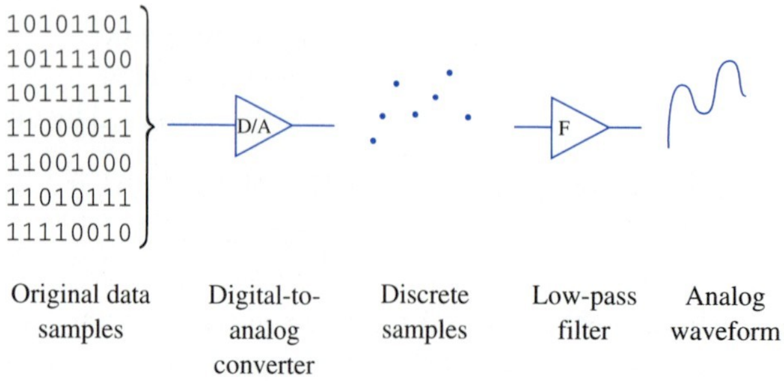

FIGURE 45.5 How binary data is converted into an analog waveform

FIGURE 45.6 Note envelope

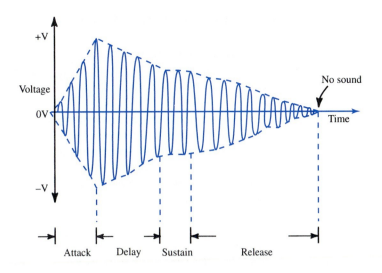

MIDI

MIDI stands for Musical Instrument Digital Interface. A MIDI-capable device (electronic keyboard, synthesizer) will use a MIDI-in and MIDI-out serial connection to send messages between a *controller* and a *sequencer*. The PC operates as the sequencer when connected to a MIDI device. MIDI messages specify the type of note to play and how to play it, among other things. Using MIDI, a total of 128 pitched instruments can generate 24 notes in 16 channels. This can be accomplished in a PC sound card by using frequency modulation or *wave table synthesis,* the latter method utilizing prerecorded samples of notes stored in a data table. The output of a note is controlled by several parameters. Figure 45.6 illustrates the use of attack, delay, and release times to shape the output waveform envelope. Each of the four parameters can be set to a value from 0 to 15.

SOUND CARD INSTALLATION

Once the sound card has been plugged into the motherboard, it is necessary to connect a set of external speakers to it. Sound cards do not use the internal speaker of the PC.

Like the CD-ROM drive, sound cards require a software driver program in order to work properly. Once again, the manufacturer of the sound card usually provides an automated setup program that will make the necessary changes in the file. Even so, it is necessary to pick the correct combination of I/O ports and IRQ lines so that both the CD-ROM drive and the sound card work together. A connector for a CD-ROM is provided on the sound card, which helps simplify installation and setup.

The CONFIG.SYS lines for a Creative Labs 16 MultiCD Sound Blaster are as follows:

DEVICE=C:\SB16\DRV\CTSB16.SYS /UNIT=0 /BLASTER=A:220 /I:7 /D:1 /H:5
DEVICE=C:\SB16\DRV\CTMMSYS.SYS

where: CTSB16.SYS and CTMMSYS.SYS = the device drivers
/BLASTER=A:220 = the base port address
/I:7 = IRQ 7
/D:1 = the low-DMA channel 1
/H:5 = the high-DMA channel 5

System DMA channels are used to transfer data to and from the sound card, and must be properly picked so that they do not conflict with DMA channels being used by other devices (such as the CD-ROM drive).

No software is loaded from the AUTOEXEC.BAT file, but a few environment variables are defined as follows:

> SET SOUND=C:\SB16
> SET BLASTER=A220 I7 D1 H5 P330 T6
> SET MIDI=SYNTH:1 MAP:E

These environment variables are examined by the Sound Blaster software when it needs to know vital settings after the machine has been booted.

The software supplied with sound cards is capable of using many different types of audio file formats. These include the .WAV files used by Windows, as well as .VOC (voice), .CMF (Creative Music File), and MIDI file formats.

When combined with a CD-ROM drive, the sound card provides a complete multimedia environment on the personal computer.

TROUBLESHOOTING TECHNIQUES

One of the most common reasons a new CD-ROM drive or sound card does not work has to do with the way its interrupts and/or DMA channels are assigned.

Figure 45.7 shows the location of the sound card in the hardware list provided by Device Manager. The AWE-32 indicates that the sound card is capable of advanced wave effects using 32 voices.

Figure 45.8 shows the interrupt and DMA assignments for the sound card. Typically, interrupt 5 is used (some network interface cards also use interrupt 5), as well as DMA channels 1 and 5. If the standard settings do not work, you need to experiment until you find the right combination.

SELF-TEST

This self-test is designed to help you check your understanding of the background information presented in this exercise.

True/False

Answer *true* or *false*.

1. Music CDs can be played on a CD-ROM drive.
2. Only MSCDEX.EXE is needed to operate a CD-ROM drive.
3. The CONFIG.SYS file loads the CD-ROM device driver.
4. The CONFIG.SYS file loads the sound card device driver.
5. The CD-ROM drive and sound card work with any IRQ line.

Multiple Choice

Select the best answer.

6. In the term *MPC compliant,* MPC stands for
   a. Master Peripheral Controller.
   b. Multi-Purpose CD-ROM.
   c. Multimedia Personal Computer.
7. Digital information is stored on the surface of a CD as
   a. Microscopic bar codes.
   b. Microscopic pits.
   c. Microscopic 1s and 0s.
8. Sound cards
   a. Require their own device driver.
   b. Share a device driver with the CD-ROM drive.
   c. Do not require a device driver.

FIGURE 45.7 Selecting the sound card

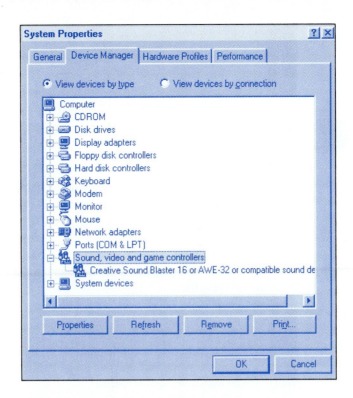

FIGURE 45.8 Sound card settings

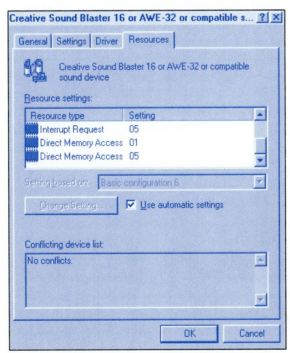

9. Which format supports recordable CDs?
    a. CD-I.
    b. CD-XA.
    c. CD-R.

Completion

Fill in the blank or blanks with the best answers.

10. Digital data is stored through the use of _____ on the CD surface.
11. A double-spin CD-ROM drive transfers data at the maximum rate of _____.

12. Sound cards use a _____ chip to generate audio signals.
13. AWE stands for _____ _____ _____.
14. A standard musical serial interface is called _____.

FAMILIARIZATION ACTIVITY

1. Put a software CD into a CD-ROM drive and do a complete directory listing of its contents with the DOS command

   ```
   DIR /S
   ```

   How many files are stored on the CD? How many bytes are stored? Does the directory listing indicate the free space on the CD? Should there be any free space?
2. Put an audio CD into a CD-ROM drive. Try to do another directory listing. What type of error do you get?
3. Use whatever sound card software you have to record three .WAV files: one that is 1 second long, one that is 5 seconds long, and one that is 10 seconds long. Compare the lengths of these files.

QUESTIONS/ACTIVITIES

If a sound card is capable of recording speech from a microphone, then it should be possible to issue spoken commands to your computer. How do you think that a computer program might be used to recognize speech?

REVIEW QUIZ

Within 10 minutes and with 100% accuracy,

1. Discuss what is required to operate a CD-ROM drive and a sound card.
2. Explain what it means to be MPC compliant.
3. Discuss the various CD-ROM formats.

46 Multimedia Devices

INTRODUCTION

Joe Tekk went with a friend—Marilyn Jayne, a photographer—to the camera store to look for a new digital camera. She invited Joe because she knew he was familiar with the lingo associated with digital cameras and with computers in general. She was interested in professional-style equipment that would allow her to perform high-quality work on her PC.

She decided to make the switch to digital because many of her competitors were already using the new technology. She was impressed with the quality of their work.

The salesman showed them every camera in the store, commenting on each special feature. Some of them use DOS-formatted disks and store the images as .JPEG files.

She settled on a reasonably priced model that was easy to use and had enough features to offer tremendous flexibility.

As they left the store, Joe replied, "I love to spend other people's money!"

PERFORMANCE OBJECTIVES

Upon completion of this exercise, within 10 minutes and with 100% accuracy, you will be able to

1. Identify different types of multimedia devices.
2. Recognize how each device is interfaced to the personal computer.
3. Discuss a method of troubleshooting multimedia devices.

BACKGROUND INFORMATION

Multimedia devices are part of a growing portion of the computer industry. New technology is making it possible to interface more and more electronic devices to the personal computer. Recall that the new MMX instructions are specially designed to speed up multimedia applications. Multimedia devices include scanners, bar code readers, cameras, and television and video add-on cards, just to name a few. Other multimedia applications build on the multimedia devices already installed in the system, such as voice recognition software, which uses the existing sound card and microphone as input to the voice recognition application software.

Multimedia devices involve three major areas of product development:

- Hardware
- Interface
- Software

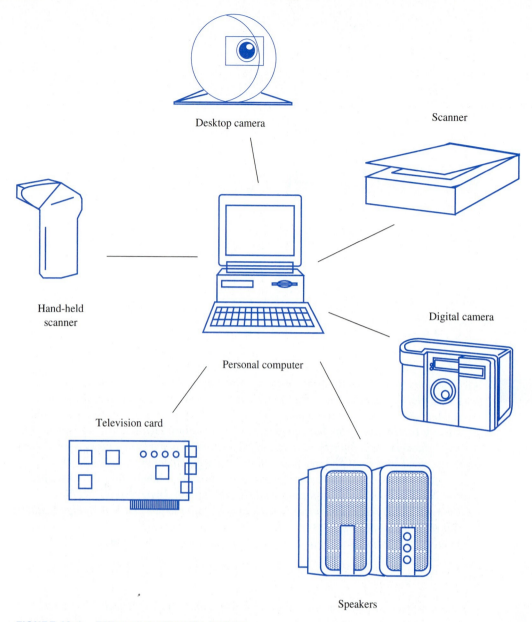

FIGURE 46.1 Different multimedia devices

Generally, advances in multimedia devices come about from advances in related technological areas, such as changes in data compression algorithms, data storage devices, data transports, and new interfaces, such as the universal serial bus (USB). Figure 46.1 shows several different types of multimedia devices. Let us examine a few of them.

CAMERAS

The digital camera stores images digitally (in the camera memory) as opposed to recording them on film. After a picture has been taken, it can be downloaded to a computer and manipulated with a graphics program. Unlike film, which has high resolution, digital cameras are limited by the amount of available memory and the resolution of the digitizing mechanism. Ultimately, the resolution is determined by the output device on which it is displayed, for example, a graphics display at 800×600, or a laser printer with a resolution of 600×600.

A digital camera uses either charged-coupled devices (CCDs) or CMOS chips. More expensive cameras use the CCD method, whereas the CMOS chips are found in cheaper cameras. In either case, the big advantage of digital cameras is the reduced cost of obtaining the images, and the speed, because there is no traditional film processing.

There are many other types and styles of digital cameras. Some of them are designed to sit on top of a monitor, so a picture of the computer user can be stored, displayed on the screen, or sent out on the Internet.

SCANNERS

Scanners are a common multimedia device used to work with both text and graphical images. There are many different types of scanners available on the market, such as flatbed, image, and the single-sheet feed scanners commonly found on multipurpose machines. Most devices are Twain compliant (the de facto interface standard for scanners), meaning that they are compatible with standard Twain device drivers and are supported by all Twain compatible software packages.

OCR, or optical character recognition, is performed by examining a scanned image that contains text. The software intensively examines the scanned image to match characters with the graphical representation. There is always an element of error introduced, depending on the quality of the input source, orientation, and quality of the hardware. Most software packages perform a spell check of any information that has been read to identify where the scanner had difficulty identifying the characters. Figure 46.2 shows OmniPage Pro recognizing some scanned text.

TELEVISION CARDS

One of the newest multimedia add-on cards is the television adapter. With the addition of cable television wire, television signals can be displayed directly on the desktop. An alternate source of programming is offered through the Web.

FIGURE 46.2 Using OCR to recognize text

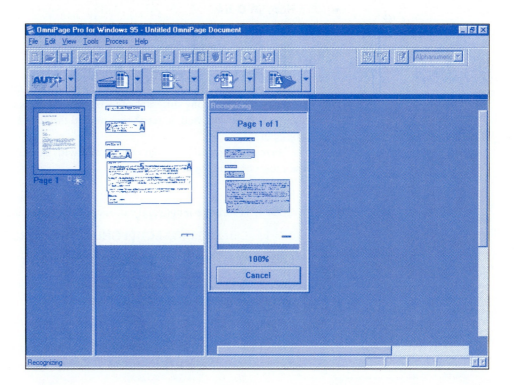

TABLE 46.1 DVD storage capacity

Prerecorded DVD		
Format	Capacity	Sides/Layers
DVD-5	4.7 GB	1/1
DVD-9	8.5 GB	1/2
DVD-10	9.4 GB	2/1
DVD-18	17.0 GB	2/2
Recordable DVD		
Format	Capacity	
DVD-R	3.8 GB/side	
DVD-RAM	2.6 GB/side	

Television cards support the various recognized standards, such as NTSC, PAL, and SECAM, and many of them offer full-motion MPEG capture capability. Optional features include MTS stereo, remote control, and radio features. The tuners can usually display many channels on the screen simultaneously and switch between channels very easily.

DVD

Digital Versatile Disk (DVD) CDs and CD-ROM drives make use of advancements in laser technology to significantly increase data storage capacity. Compare an ordinary CD's 650MB with a DVD CD's 4.7 to 17GB of storage. Typically, two disks are bonded back-to-back, and may contain two data layers on each side.

DVD CDs support digital audio (surround-sound format), up to four hours of video (MPEG-2 video compression) per side, and the DVD-ROM file system. Table 46.1 shows the various DVD prerecorded and recordable formats.

Do not confuse DVD with DIVX (Digital Video Express), a different video technology used in pay-per-view systems. DIVX disks are not DVD disks, but a DIVX player will play a video DVD disk.

BAR CODE READERS

A bar code reader is one of the most common multimedia devices. Bar codes are used to easily identify many different objects, such as books in a library, pieces of equipment that must be inventoried, letters being mailed through the postal service, or a host of other applications.

There are many different bar codes, each with a different purpose or capability. Some of the most common bar codes are:

- Code 39

Code 39 is variable-length and can encode the following characters:

0123456789ABCDEFGHIJKLMNOPQRSTUVWXYZ-.*$+%

- Code 93

Code 93 is variable length and can encode the complete 128 ASCII character set. It also provides a higher character density than code 39.

- Code 128

Code 128 is a variable-length, high-density code capable of 106 different bar and space patterns, each of which can have one of three different meanings.

- POSTNET

    POSTNET (POSTal Numeric Encoding Technique) is a five-, nine-, or 11-digit numeric code used by the U.S. Postal Service to encode ZIP code information. POSTNET is unlike other bar codes because the data is encoded in bar height instead of bar width and spaces. Most standard bar code readers cannot decode the POSTNET code.

- UPC codes

    UPC codes fall into three categories: UPC-A, UPC-E, and UPC supplemental. UPC-A codes consist of 11 data digits, whereas the UPC-E can encode only six digits. A supplemental two- or five-digit number can be appended to the main code.

- EAN (European Article Numbering, also called JAN in Japan)

    EAN coding uses the same size requirement and coding mechanism as the UPC codes, changing only the meaning of the digital data. For example, EAN-13 is similar to UPC-A, except for the addition of a country code, which is encoded into positions 12 and 13.

- Codabar

    The Codabar code is variable length, encoding the following characters:

    0123456789-$:/.+ABCD

- PDF417

    PDF417 is a high-density, two-dimensional code capable of encoding the entire ASCII character set. The PDF417 code can encode as many as 2725 characters in a single bar code.

INTERFACING MULTIMEDIA DEVICES

Every type of multimedia device must be interfaced to the PC. Usually, the interface is in the form of a cable connection to a serial, parallel, or USB port, but connections can also be of the wireless nature, using infrared or radio signals.

For communication to occur between the PC and the multimedia device, it is necessary for the interfacing to be working properly. If problems occur, all three items are suspect: the multimedia device, the cable, and the PC interface. Check the cabling and the status indicators on all equipment to help isolate any problems or errors.

DEVICE DRIVERS

Each multimedia device will probably contain its own device driver to control each device. There are many times when a device driver is out of date or contains an error that causes a problem. In such cases, it is best to make sure the most recent version of the device driver is on hand. All manufacturers provide access to the latest set of device drives for their products on the World Wide Web. Search their Web site, or check the documentation provided with each device for a specific way to obtain the latest drivers.

Note that multimedia vendors constantly fix minor problems and enhance their products. You may want to visit the manufacturer's device drivers Web page just to see if there are any reported problems. If there are, they can be fixed by downloading the latest driver. Be sure to follow all directions provided with the software.

OTHER MULTIMEDIA APPLICATIONS

Other multimedia applications, such as voice recognition software, use the existing hardware already installed on the personal computer, such as a sound card, speakers, and a

FIGURE 46.3 Headset with earphone and microphone

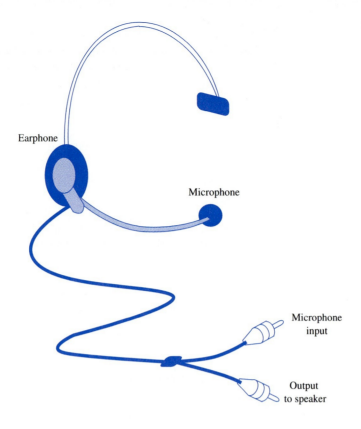

microphone. If a computer does not have a sound card, for example, it must be purchased separately before the voice recognition application will run. Some of the voice recognition products include a headset for use with the software as shown in Figure 46.3. The voice recognition software must be *trained* to recognize the speech patterns of each user. Commonly a set of words or sentences is read aloud and the computer stores the characteristic information.

Video editing is another multimedia application commonly being performed on a PC. Most products require special hardware to save and compress the video in real time. During playback, no special hardware is required to display the output.

TROUBLESHOOTING TECHNIQUES

Many different problems can be encountered with every multimedia device. Since there are three potential sources of any problem, it may be difficult to determine the cause. Use the following list of techniques when trying to resolve specific problems:

- Make sure your system meets all the requirements described by the product documentation.
- Make sure all cable connections are secure.
- Restart your computer to see if the problem is still present.
- Make sure you have the latest device drivers.
- Uninstall and reinstall the software.
- Refer to the customer support offered by the product manufacturer.

SELF-TEST

This self-test is designed to help you check your understanding of the background information presented in this exercise.

True/False

Answer *true* or *false*.

1. Digital cameras require traditional photo processing techniques.
2. OCR is used to store the image in a digital camera.

3. Every multimedia device may be interfaced to the PC.
4. Multimedia applications take advantage of MMX instructions.
5. Every bar code reader can read every type of bar code.

Multiple Choice

Select the best answer.

6. Multimedia devices commonly involve three areas of product development:
    a. Hardware, software, and the interface.
    b. Input, output, and indicator lights.
    c. Microphones, speakers, and pointing devices.
7. Bar code readers can be used to
    a. Differentiate between many different items.
    b. Identify multiple objects with the same code.
    c. Both a and b.
8. A device driver is used to
    a. Specify a multimedia device operation.
    b. Make sure all cable connections are secure.
    c. Both a and b.
9. A headset contains:
    a. A speaker and a microphone.
    b. A speaker only.
    c. A microphone only.
10. CCD cameras are:
    a. More expensive than CMOS cameras.
    b. Cheaper than CMOS cameras.
    c. Approximately the same price.

Matching

Match a description of bar code on the right with each item on the left.

11. Code 128   a. High density, two-dimensional code encoding 2725 characters.
12. Code 93   b. Encodes the complete 128-character ASCII code.
13. EAN   c. High density, using 106 different bar and space patterns.
14. POSTNET   d. Encodes five, nine, or 11 digit numeric codes.
15. PDF417   e. Encodes a country code.

Completion

Fill in the blank or blanks with the best answers.

16. Each multimedia device requires a(n) _____ _____ to control the multimedia device.
17. _____ communication uses infrared light and radio signals.
18. Before installing a multimedia device, make sure your computer meets the _____ requirements as specified by the manufacturer.
19. The latest version of the _____ _____ can usually be found on the manufacturer's Web site.
20. Voice recognition software requires _____ in order to increase the recognition accuracy.

FAMILIARIZATION ACTIVITY

1. Take a trip to a local office-supply store (such as Staples or OfficeMax) and examine the multimedia equipment available. Check the prices on the following:
    - Color scanner
    - Color camera (with internal floppy)
    - Bar code reader/printer
    - Polaroid picture scanner

- Business card scanner
- Video telephone

2. Visit a local desktop publishing house and ask for a demonstration of its multimedia software.

QUESTIONS/ACTIVITIES

1. What multimedia devices are missing from this exercise?
2. What multimedia devices would you most like to have and use? Why?
3. Why put a TV card in your computer when you can just watch a real television?

REVIEW QUIZ

Within 10 minutes and with 100% accuracy,

1. Identify different types of multimedia devices.
2. Recognize how each device is interfaced to the personal computer.
3. Discuss a method of troubleshooting multimedia devices.

UNIT VI Application Software

EXERCISE 47 Word Processors
EXERCISE 48 Spreadsheets
EXERCISE 49 Databases
EXERCISE 50 Presentation Software
EXERCISE 51 Web Development
EXERCISE 52 Science and Technology

47 WORD PROCESSORS

INTRODUCTION

Joe Tekk finished installing the latest version of Microsoft Word in the RWA Software corporate training center. Joe thought about how many people would be using the computers during the next few days. A corporate training session, a tour for a local scout troop, and a group of high school students coming in to write their resumes would all be running the new version of Word.

Joe also thought about the company-wide documentation project that had been going on for several months. All the company documents were being entered into Word and then exported in HTML for use on RWA's internal network. Joe never realized how much time was spent doing word processing.

PERFORMANCE OBJECTIVES

Upon completion of this exercise, within 10 minutes and with 100% accuracy, you will be able to:

1. Enter and leave Microsoft Word.
2. Resize and move the Word window.
3. Create a new document.
4. Save a document.
5. Open and modify existing documents.
6. Spell check your document.
7. Use bold and underline format tools.
8. Print documents.
9. Use online help.

BACKGROUND INFORMATION

Using Microsoft Word, all types of word-processing chores can be accomplished. Aside from the basic editing features necessary to create letters, memos, mailing labels, reports, and essays, Word provides several advanced features, such as mail merge, table editing, drawing tools, word art, and WWW page editing, to name just a few. Word provides a graphical WYSIWYG (what-you-see-is-what-you-get) window that is powerful and easy to learn. Figure 47.1 shows a typical Word editing window.

As you can see from Figure 47.1, some of the text displayed on the screen is bold, other text is justified, there are bulleted lists, and the text is displayed using different font sizes.

FIGURE 47.1 Microsoft Word editing window

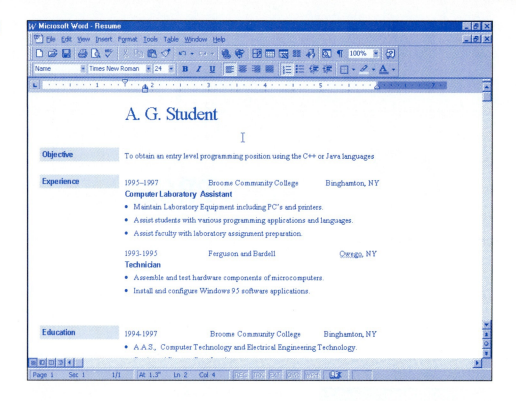

WYSIWYG means that what you see in Figure 47.1 is what you will get when you print the Word document.

All the controls for these WYSIWYG features are contained in toolbars or pull-down menus. Both new and experienced users can work efficiently with a few simple keystrokes or mouse clicks. This exercise will give you the Word experience necessary to create your own Word documents, such as technical reports, resumes, correspondence, and other written assignments.

The title bar located at the top of the Word window in Figure 47.1 shows the current document name on the left and the common Windows application options on the right: minimize window, maximize window, and close. The pull-down menus show the category names that provide access to all of Word's features. The status bar at the bottom of the Word window shows the status of the word processing session (page 1 of 1, cursor located at line 2, column 4).

By selecting a pull-down menu, all the options available on the menu are displayed. Figure 47.2 shows the View pull-down menu. The View menu provides a number of ways

FIGURE 47.2 Pull-down menu for View

FIGURE 47.3 Additional menu items in the Toolbars submenu

to view the document while it is being prepared. For example, you can preview the actual page as it will look on the printer. Several small icons at the lower left of the Word window in Figure 47.1 provide shortcuts to change the view without having to open the View menu.

A pull-down menu can be selected with the mouse by left-clicking on the desired menu option, or by using keyboard commands such as the Alt-F to access the File menu. Arrows shown to the right of items on the pull-down menu indicate that more selections are available for those specific items. Figure 47.3 shows the View/Toolbars pull-down menu. Simply moving the mouse to the Toolbar option automatically displays the additional menu options.

The toolbars allow the user to easily select and use the various features of Word by pressing one button. We will examine both pull-down menus and toolbars later in this exercise.

CREATING A NEW DOCUMENT

Microsoft Word is a member of the Microsoft Office product suite and can be accessed through the Microsoft Office options available using the Start menu or the Start Programs menu. As shown in Figure 47.4, Word can be accessed from the Start menu by selecting the New Office Document option. Figure 47.5 shows the New Office Document selection screen. This screen shows options for the complete Microsoft Office suite. For now, we will select the Blank Document icon available on the General tab to begin the Word editing session. To do this, select the Blank Document icon using the mouse and then press the OK button, or simply double-click on the Blank Document icon itself.

Microsoft Word can also be accessed directly by selecting Microsoft Word from the Programs menu or a Programs submenu. As you become more familiar with Microsoft Word and other Microsoft Office products, you will probably use a combination of these methods to open Word.

Once started, Word presents the user with a blank editing screen. Now it is just a matter of entering the text. We can begin to get comfortable with Word by entering a simple letter of correspondence. Enter the text shown in Figure 47.6.

You may notice several different things as you type on the Word screen. First, Word is constantly updating the current page information and examining each character as it is typed. If a word is misspelled, Word will underline it in red. If Word finds a grammatical or

FIGURE 47.4 Accessing Word from the Start menu

FIGURE 47.5 New Office Document screen

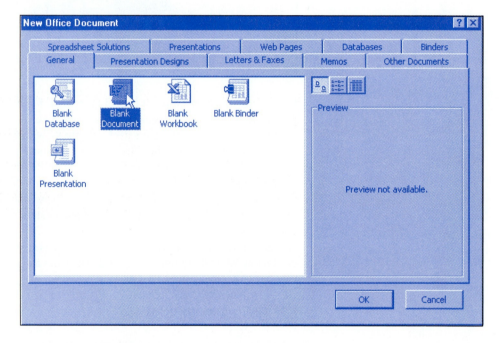

format error in a sentence, such as too many spaces between words in a sentence or problems with tense, Word will underline that in green. You are encouraged to become familiar with Word status indicators as you enter some text into the Word window.

Experiment with the bold, italic, centering, and font controls to format the text. This can be done to text as it is being entered, or after it has been entered. For example, to make a block of text bold, you could:

- Select the text and then left-click the bold button.
- Left-click the bold button (to turn bold on), enter the text, and then left-click the bold button again (to turn bold off).

Selecting text after it has been entered can be easily done with the mouse or keyboard. Position the mouse over the first character or line to be selected, left-click and hold, then drag the mouse over the last character or line to select. The selected text will be highlighted to make it stand out from the rest of the text. You can select text with the keyboard by positioning the cursor at the beginning of the text and holding the Shift key down while moving

FIGURE 47.6 Sample letter

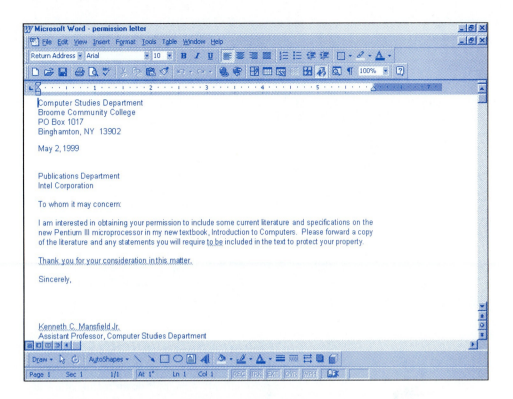

to the end of the selected text with the arrow keys. To deselect a block of text, left-click the mouse anywhere inside the editing window.

Many editing features require that a block of text be selected first, so it is worth the time invested to master this technique.

One last important point: do not forget about the Undo button (a small, blue counter-clockwise arrow). There will be times when you have made one or more errors while editing a document and the results are not what you intended. Left-clicking the Undo button will back up the editing process and undo your editing changes one by one.

EXITING FROM WORD

To exit from Word, two methods are available:

- Select Exit from the File pull-down menu, or
- Select the close option available on the Word window by clicking the box with an X in it located on the top right-hand corner of the screen.

If the current document in the editing window has not been saved yet, Word will ask you to choose whether to save the changes in your file. This dialog box is shown in Figure 47.7. If you want to save your document, select the Yes option and then verify the File name displayed in the File name field on the Save As dialog box, shown in Figure 47.8. Word will automatically fill in File name with the first words you entered into the document. If this name is suitable, select the Save option. If you want to change the name, simply enter a new name in the File name field and left-click the Save option.

FIGURE 47.7 Close dialog box

FIGURE 47.8 Saving a file

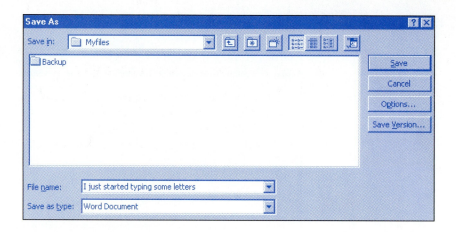

FIGURE 47.9 Standard Word toolbar

PULL-DOWN MENUS AND TOOLBARS

As we have already seen, pull-down menus allow access to the specific features of Word. These features are grouped into the following categories: File, Edit, View, Insert, Format, Tools, Table, Window, and Help. Examine each of the pull-down menus.

In addition to the pull-down menus, Word also uses icons to allow direct access to some of the most common features, such as file operations such as save, open, and print; format operations such as text justification; and text attributes like bold, underline, and italics. The icons provide one-button access to features that were previously several mouse clicks or keystrokes away using the pull-down menus. The standard toolbar menu of icons is shown in Figure 47.9.

Many other toolbar menus are available and are probably already displayed on your Word screen. The toolbars are selected by choosing the Toolbars option from the View pull-down menu. This was previously illustrated in Figure 47.3.

The Formatting toolbar shows most of the common formatting options such as font size and selection; character attributes such as bold, italic, and underline; paragraph justification choices; and list and indentation options. As you view the toolbar, items that are currently selected show up as buttons that are depressed. For example, if left margin justification is selected, that button will appear to be depressed. Similarly, as a word or sentence is bolded or underlined, these buttons will also appear to be depressed. These options are selected by left-clicking on them.

You can also customize the individual items that are displayed on each toolbar menu using the Customize option. You are encouraged to become familiar with all of the toolbar choices.

MOVING THE WINDOW

If you want to move the Word window on the Windows desktop, simply

1. Place the mouse cursor on the Microsoft Word window title bar located at the top of Word screen.
2. Press *and hold* the left mouse button.
3. Drag the window to its new place on the desktop. The window may be outlined as it is being dragged.
4. Release the left mouse button.

RESIZING THE WINDOW

You can change the size of the Word window, making it as large or small as Windows will allow. This is accomplished in the following way:

1. Place a mouse cursor on an edge or corner of the window. Notice that the mouse pointer changes when you are in this position.
2. Press *and hold* the left mouse button.
3. Drag the window edge or corner until the window is the desired size.
4. Release the mouse button.

To make the window as large as the entire screen, simply left-click on the maximize icon displayed at the top right-hand corner of the Word window.

SCROLLING THROUGH A DOCUMENT

Whenever a document window is resized, a scroll bar may be added or removed from the Word window. A scroll bar is present on a window whenever there is more information than can be displayed on the screen. On the Word screen, there will be two scroll bars: one to display the complete document from top to bottom and one to display the contents of the document from left to right. The mouse is used to control the scroll bars.

In addition to scrolling, the entire display can be updated by paging through the document. The Page Down and Page Up buttons on the keyboard do not actually move an entire page in either direction. Instead, they move the cursor a few inches forward or backward. Pressing and holding the Ctrl (Control) key while pressing Page Up or Page Down will move the cursor to the beginning of the next, or previous, page. Pressing Ctrl and End will take you to the end of the document. Pressing Ctrl and Home returns you to the beginning of the document.

SAVING THE DOCUMENT

As you are typing text into Word, it is always a good idea to save your work frequently to prevent accidental loss of data. To save your document, simply press the Save icon (the Floppy Disk) on the Standard Toolbar. The Save option is also located on the File pull-down menu. An AutoSave feature is built in that allows Word to save your document at predefined intervals such as every 10 minutes. Look in the Options submenu of the Tools pull-down menu. The Save tab contains all the parameters associated with the AutoSave feature.

If you have not named your document yet, you will be presented with the Save As dialog window previously shown in Figure 47.8. If your document has already been saved, Word will simply update the file already stored on the disk.

OPENING AN EXISTING DOCUMENT

There are several ways to open a Word Document. First, on the Start Menu, the "Open Office Document" selection will present a list of all Office documents. This is illustrated in Figure 47.10. Each of the files shown with a .doc extension is a Word document. To open one of these documents, simply double-click on the document, or select the document with a single click and use the Open option.

Another way is to open Word directly and use the Open option on the Toolbar or the File menu or by selecting one of the document names listed at the bottom of the File menu. This list contains the most recent files accessed by Word.

The opened document is displayed beginning with the first page. New material may be added to the document by left-clicking at the desired insertion point.

FIGURE 47.10 Opening a file in Word

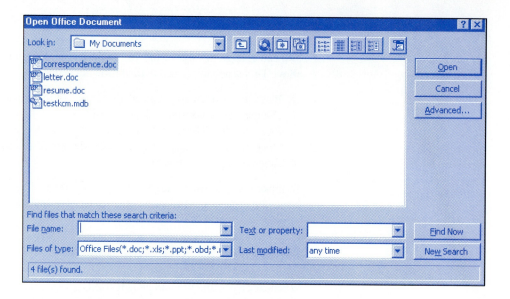

PRINTING DOCUMENTS

The Print option is available on the Standard Toolbar and the File pull-down menu. Using either one of these methods will cause Word to print the current document using the Windows default printer. The Print Preview option will display a graphic image of how the hard copy output should look after it is printed. This can help save paper and produce better looking output.

Word will print the entire document, a selected range of pages, or just the current page. Word can even print to a file instead of the printer.

OTHER WORD-PROCESSING APPLICATIONS

Another popular word-processing application is WordPerfect. WordPerfect began as a DOS application, and has grown into a powerful Windows application, comparable with Microsoft Word.

If you do not want to spend any money, two built-in word processors are available in Windows. They are Notepad and Wordpad. Notepad is text-based. Wordpad, like Microsoft Word, allows text formatting and other WYSIWYG features, but is not as powerful as Word.

TROUBLESHOOTING TECHNIQUES

The Word help facility available from the pull-down menu or from the Office Assistant provides instant access to the features of Word. Figure 47.11 shows the help options available. You are even allowed to enter a question, such as "How do I insert page numbers," and Word help will look up the associated topic.

FIGURE 47.11 Word Help pull-down menu

SELF-TEST

This self-test is designed to help you check your understanding of the background information presented in this exercise.

True/False

Answer the following questions *true* or *false*.

1. Word displays documents in WYSIWYG format.
2. Word checks spelling and grammar as the text is entered into the document.
3. Word menu options can be accessed only using keyboard commands.
4. The Word toolbars can be customized by the user.
5. You can forget to save your document and leave Word, losing all your changes.

Multiple Choice

Answer the following questions by selecting the best answer.

6. The Word Office Assistant is used
    a. To access Word help features.
    b. To control how Word automatic features are configured.
    c. By Word to manage the user interface.
7. A submenu is indicated by
    a. The underscored character in the submenu category.
    b. The underscored character in the submenu category and the triangle located on the right margin in the pull-down menu.
    c. An icon shown next to the pull-down category.
8. To go to the beginning of a new page, press
    a. Page Down.
    b. Shift-Page Down.
    c. Ctrl-Page Down.
9. Word documents have the extension
    a. .doc.
    b. .txt.
    c. .wrd.
10. To make an existing block of text bold you must first
    a. Select it.
    b. Format it.
    c. Highlight it.

Completion

Fill in the blanks with the best answers.

11. The Thesaurus is located on the _____ pull-down menu.
12. The spelling and grammar check can be activated using the _____ icon.
13. The buttons in the lower left corner of the Word document window are used to change the _____.
14. Print _____ shows a graphic display of what the printed page will look like.
15. Sentences underlined in green have _____ errors.

FAMILIARIZATION ACTIVITY

1. Open a new Word Document. Enter your name on the first line, and the title "Word Processing Experience" on the second line. Beginning on the third line, write a small paragraph explaining your experience using a word processor. Save the document.
    a. Select the name and title portion of the text and then try each of the formatting tools to see how the format of the text changes.
    b. Try the same operations on the paragraph portion of the text.
2. Experiment with tables. Try to duplicate the one shown in Table 47.1.

TABLE 47.1 **Sample table**

Input	Output	Memory	CPU
64-bit digital	64-bit digital	256 MB RAM	Pentium III
Four 16-bit analog	Two 16-bit, four 8-bit analog	20 GB HD	500+ MHz

3. Insert an image into your Word document. This is done by selecting Insert/Picture, Insert/File, or Insert/Object. You can even put the image on the clipboard and paste it into your Word document.
4. Save a Word document as an HTML file and then view the file with a browser. Does it look the same?

QUESTIONS/ACTIVITIES

1. Locate information in Word about a mail merge. List the number of subtopics.
2. Enter a few words and deliberately spell a few of them wrong. Does Word underline them in red? What happens if you right-click the mouse over one of the incorrect words?
3. Run the Resume Wizard (File/New/Other Documents) to create a resume of your work experience and employment goals.

REVIEW QUIZ

Within 10 minutes and with 100% accuracy,

1. Enter and leave Microsoft Word.
2. Resize and move the Word window.
3. Create a new document.
4. Save a document.
5. Open and modify existing documents.
6. Spell check your document.
7. Use bold and underline format tools.
8. Print documents.
9. Use online help.

48 Spreadsheets

INTRODUCTION

Joe Tekk was just putting the finishing touches on a spreadsheet he had been working on. Joe thought he was spending too much money and had decided to set up a budget. He didn't have any experience with spreadsheets before, but since RWA Software had standardized on the Microsoft Office product, he used the Excel application program (which had never been run except for a quick look when Microsoft Office was originally installed).

Joe was quite impressed with how easy it was to set everything up. Now it was just a matter of keeping track of receipts, entering the information, and looking at the results. As Joe thought about the spreadsheet, he wondered what he would be doing when he found out where all of his money was going.

PERFORMANCE OBJECTIVES

Upon completion of this exercise, within 10 minutes and with 100% accuracy, you will be able to

1. Understand the terminology associated with spreadsheets.
2. Navigate around the spreadsheet window.
3. Create a spreadsheet.
4. Use several built-in spreadsheet functions.

BACKGROUND INFORMATION

An electronic spreadsheet is the equivalent of a traditional accounting worksheet. Although both of these documents (the spreadsheet and the worksheet) organize the data into a matrix of rows and columns, the electronic worksheet is much more flexible and easier to work with than its paper counterpart. The flexibility of a spreadsheet is demonstrated with the ability to apply the spreadsheet to every area of technology. For example, an electrical engineering student might use a spreadsheet to store data for a range of values, such as voltage or current, that are tracked over a period of time. A mechanical engineering student may be interested in storing data about the placement of components on a printed circuit board where the tolerance measurements are critical and must be reviewed frequently to ensure a high level of quality control. An engineering science student may use a spreadsheet to implement a formula in physics to determine the results of an equation, thereby eliminating the need to write a computer program.

FIGURE 48.1 Microsoft Excel spreadsheet window

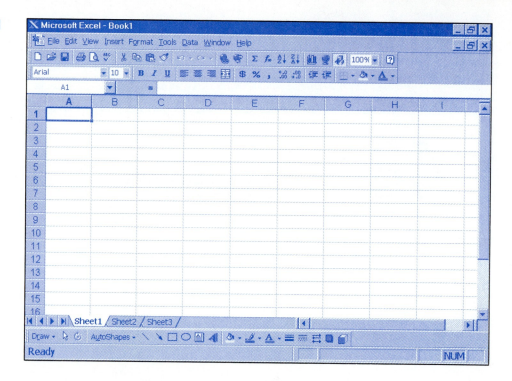

Before we get involved in the actual development of a spreadsheet of our own, it is necessary to learn about the terminology of a spreadsheet. Figure 48.1 shows a Microsoft Excel spreadsheet. The Microsoft Excel window contains several different components such as the title bar, pull-down menus, toolbars, status bar, and, lastly, the actual worksheets where all of the activities actually take place in the spreadsheet.

The smallest element in a spreadsheet is called a **cell.** A cell is defined by the intersection of one row and one column. As you can see from Figure 48.1, each column in an Excel spreadsheet is identified by a letter and each row is identified by a number. Therefore, specifying one column letter and one row number uniquely identifies each spreadsheet cell.

Each cell in a spreadsheet may contain text, a number, or a formula. Any text in a cell is commonly referred to as a label. Each label describes the contents of one or more cells. Examples of some common cell labels include months of the year, days of the week, or a list of any other set of items or objects being tracked in the spreadsheet. Numbers in a spreadsheet cell may be integers or floating-point numbers. Formulas in a spreadsheet may contain hard-coded values, references to other cells in the spreadsheet, built-in formulas, or custom calculations.

SPREADSHEET PLANNING

The process of building a spreadsheet consists of several different steps. These steps are similar to the steps used to develop a database application or the development of a computer program. In general, the basic steps are as follows:

1. State the purpose of the spreadsheet.
2. Design a model of the spreadsheet.
3. Build the electronic version of the model.
4. Test the spreadsheet.
5. Document the spreadsheet.

First, the purpose of the spreadsheet identifies what the spreadsheet is supposed to do, what inputs are required, and what type of output is required from the spreadsheet such as reports or graphs. With a clear purpose defined for how the spreadsheet should work, it becomes a simpler task to get started in the actual spreadsheet development.

Second, it is necessary to develop an outline or model indicating what the spreadsheet should look like and how the spreadsheet should function. The model should identify the basic form of the spreadsheet, indicating the number of cells necessary to hold the data. The model should also identify the different types of calculations that must be performed. With the simple model written on paper, the actual process of coding the spreadsheet can begin.

Third, the actual development of the spreadsheet is performed in steps. Each part of the spreadsheet must be tested to verify that the individual elements are correct and working properly. By testing each part of the spreadsheet individually as you go, problems are eliminated as they are encountered, without having to worry about fixing all the errors at the end of the project. In a small spreadsheet, it may not be necessary to perform the work in steps, but as the size of a project grows, testing in stages becomes more and more important.

Testing of the spreadsheet involves entering test data and determining whether the results are correct. The results are compared with manually calculated test data, usually developed when creating the spreadsheet model. When the results of the spreadsheet agree with the calculated test data, the spreadsheet can be documented as necessary.

The documentation process involves preparing a list of the instructions necessary to work with the spreadsheet or actually describing how the spreadsheet operates. Most spreadsheet software allows for comments to be added to individual cells to indicate what decisions were made and why. With these basic steps in mind, let us examine the details of the spreadsheet window to see how these development steps can be followed.

BUILDING A SPREADSHEET

As an example, let us build a spreadsheet to compute the distance that a body falls in feet per second for the first 10 seconds of free fall given by the equation:

$$S = \tfrac{1}{2} a t^2$$

where
S = the distance in feet
a = acceleration due to gravity (32 ft/sec^2)
t = time in seconds

The purpose of the spreadsheet is to determine the distance in feet that the body falls during the first 10 seconds of free fall. The spreadsheet model therefore must contain all the elements necessary to calculate the free-fall distance. Although this may seem like a simple problem, it is necessary to calculate a few of the values for S that we expect to see when the data is displayed using the spreadsheet. Using a pocket calculator, the constant value for a and the individual values of t, the actual values, can be computed very easily.

Let us begin the actual development of the spreadsheet using Sheet1 in the Excel workbook by entering the spreadsheet title and the column headings for both the time and the distance traveled. As this information is entered, it is displayed in the selected cell as well as the formula bar. We will discuss the formula bar in detail soon. The heading information is shown in Figure 48.2. The spreadsheet designer may choose any location on the spreadsheet to contain the heading information. In general, all the decision-making regarding the placement of information such as titles, headings, and data are made so that the user can enter spreadsheet data easily and at the same time make the spreadsheet visually appealing.

Next, it is necessary to supply the initial value for time, a zero, which is entered into the appropriate cell location under the time column heading as indicated in Figure 48.3. Notice that all that is necessary to enter the zero in cell C5 is to use the arrow keys or mouse to select the cell, and then to press the zero key and then the Enter key. Notice that when a particular cell is selected, the border around the cell is thicker and darker than the other cells in the spreadsheet.

To supply the remaining values for the time, the user may simply enter the remaining digits 1 through 10 in the appropriate cell locations C6 through C15, although a better choice would be to create a formula that performs the same job. To create a formula, it is

FIGURE 48.2 Adding a title and heading to the spreadsheet

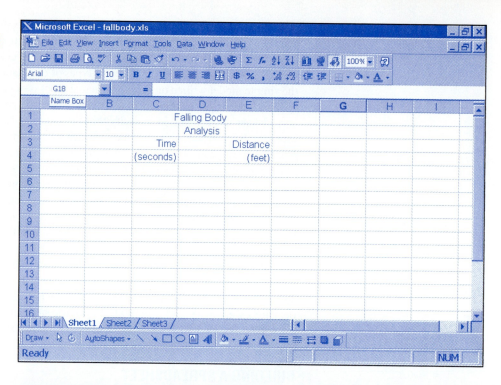

FIGURE 48.3 Entering the initial value for time

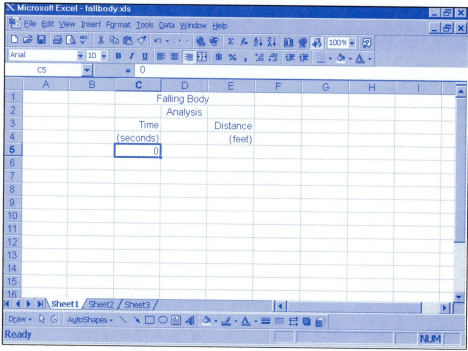

necessary to select the cell C6, which is where the formula is to be stored. Then, it is necessary to enter an = (equal sign) at the beginning of the formula followed by the specific formula to execute. A formula to increase the cell size by 1 is accomplished by specifying the formula

$$=C5 + 1$$

in cell location C6. This formula causes the cell C6 to look at the value stored in location C5 and then add 1 to it. Figure 48.4 illustrates this process. Notice that the contents of the cell are displayed in the formula toolbar simultaneouly. Also notice that the size of the formula toolbar display is much longer than the size of a cell and allows for very long formulas to be examined. Note that it is also possible to create and edit formulas using the formula toolbar.

FIGURE 48.4 Entering a formula for the time values

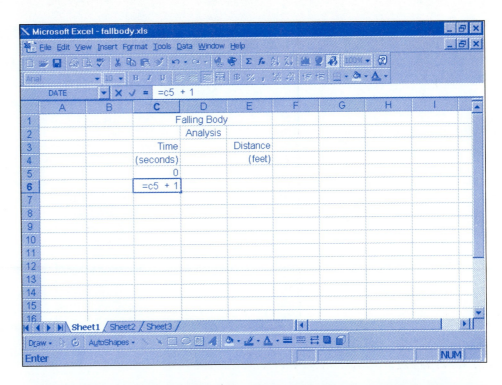

FIGURE 48.5 Value in cell C6 produced by the formula

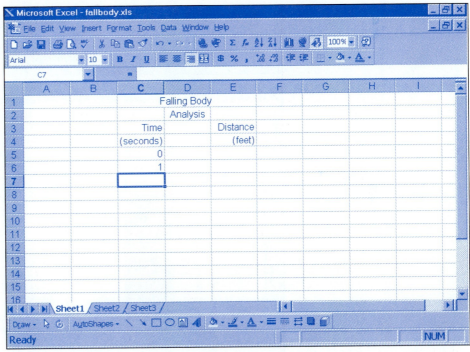

After the formula is entered, it is necessary to press the Enter key. This causes Excel to execute the formula and display the results inside of the specified cell, as is indicated in Figure 48.5. Notice that a 1 is now displayed in cell C6. After the formula in cell C6 is executed, cell C7 automatically becomes the next selected cell.

To continue the list of numbers in the time column, it is necessary to enter formulas in the remaining cells C7 through C15. Fortunately, it is not necessary to reenter the formula in each of the cells. Instead, the formula in cell C6 can be copied and then pasted into the remaining cells C7 through C15. To perform the copy process, it is necessary to select the formula in cell C6 and copy its contents to the Windows clipboard. For example, a right-click of the mouse on cell C6 displays the menu shown in Figure 48.6, where the copy option is selected. Following the copy process, the cell C6 is displayed in an outline box.

583

FIGURE 48.6 Preparing to copy a formula

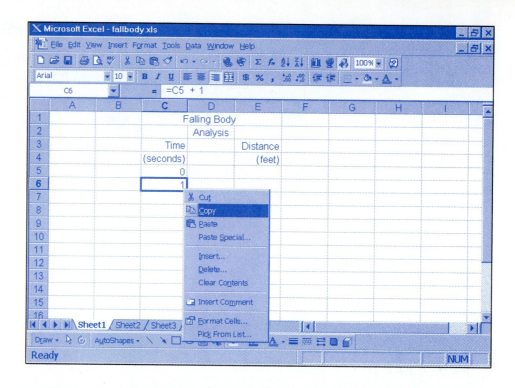

FIGURE 48.7 Copying the contents of a cell

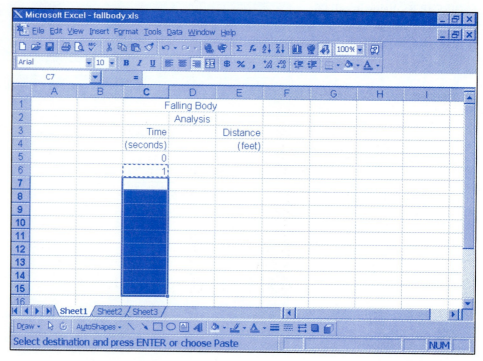

The last step in the copy process involves selecting the cells that the formula should be copied to. These cells, C7 through C15, must be selected as shown in Figure 48.7. The first selected cell, C7, is displayed using the thicker lines around the cell and the remaining cells C8 through C15 are displayed in black. Notice that the status bar displays the simple instructions "Select destination and press ENTER or choose Paste." Following these instructions and pressing the Enter key causes the formula to be pasted into the selected cells as illustrated in Figure 48.8. Aside from the cell location C5, which contains a zero, the rest of the cells C6 through C15 use a formula to determine their values.

Each cell in the distance column where a number will be displayed must contain the distance S using the formula $S = \frac{1}{2} at^2$. We can begin by selecting the first cell location in the distance column where the new formula is to be entered, which is E5. The elements of the

FIGURE 48.8 Results of the pasted formula are displayed

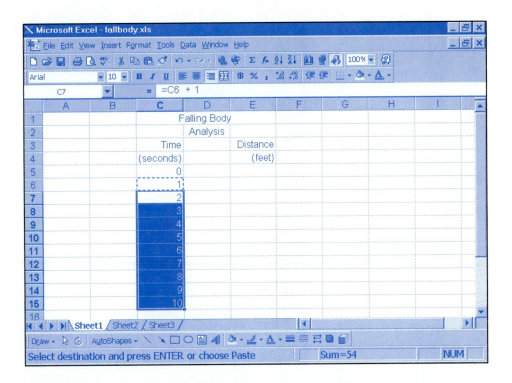

FIGURE 48.9 Entering the distance formula

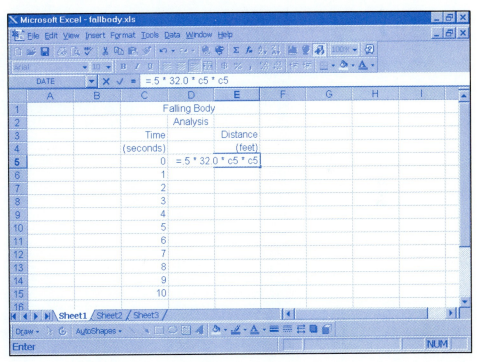

formula consist of constant values (0.5 and 32.0) and the variable *time,* which is located in the adjacent cell C5. Using this information, the formula

=0.5 * 32.0 * C5 * C5

is used to compute the first distance calculation, which is shown in Figure 48.9. After the Enter key is pressed, the results of the calculation are displayed and the next cell in the column becomes the currently selected cell, as indicated by the window displayed in Figure 48.10. Note that the value 0 displayed in cell E5 is correct since the value of time is zero. Now, all that remains is the process to copy the formula from cell E5 to the other cells in the distance column E6 through E15. The steps to perform this process are the same as we used to copy the formulas in the cells C6 through C15. First, the cell containing the formula, E5,

FIGURE 48.10 Distance formula calculation results

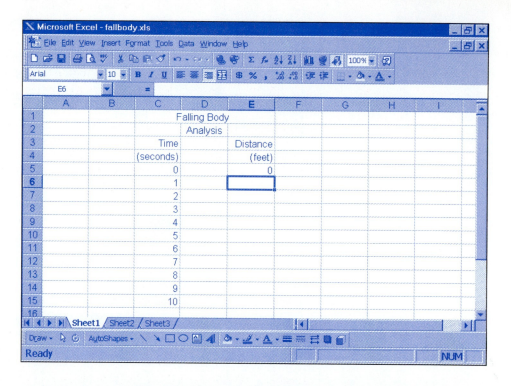

FIGURE 48.11 Completed spreadsheet display

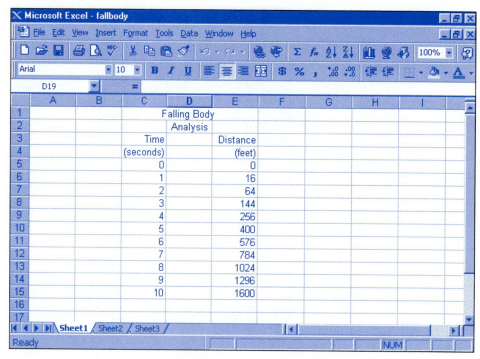

must be copied to the clipboard. Then copy the selected formula to the new cells E6 through E15. After the formulas have been pasted, notice that the new values for the distance traveled are displayed in the appropriate cell locations. Figure 48.11 shows the completed worksheet. Now it is necessary to verify that the values are correct by choosing a few sample elements in the list and manually performing the same calculations that were implemented in the formula. If the results are the same, the spreadsheet is working properly.

The last step in the development of a spreadsheet involves producing any required documentation and/or instructions that might be helpful to the user. For this example, no additional documentation or instructions would be necessary.

VIEWING THE OUTPUT FROM A SPREADSHEET

After a spreadsheet has been developed, it is common to print a hard copy of the results; this is accomplished by selecting the Print option from the File menu or by pressing the Print icon on a toolbar. Microsoft Excel also provides several chart and graph options to view the spreadsheet data. Continuing with the previous example, a graph can be created by selecting the Chart Wizard icon located on the toolbar. Figure 48.12 shows the first window of four displayed by the Chart Wizard. As you can see from the figure, many different chart options exist. First, it is necessary to choose the chart type from the list of items displayed along the left side of the display. The defaults selected by the Chart Wizard for "Column" are an appropriate selection for the Falling Body Analysis. You are encouraged to experiment with all the various charting options available. By selecting the Next button, the second step is displayed by the Chart Wizard.

In step 2, it is necessary to select the range of data to be displayed in the chart. The default selection

includes the numbers we want to examine. Microsoft Excel has also chosen the Column button for the series option. You may notice that the data displayed inside the Chart Wizard Window shown in Figure 48.13 is actually using the data from the worksheet. By selecting the next button, we can move to step 3.

The next step involves labeling the various parts of the graph. We have the option of entering a chart title, category for the X-axis, and the value for the Y-axis. This process is illustrated in Figure 48.14. Excel does not provide any default values for these options and they must be entered manually. After the necessary text has been entered, press the Next button to move to the last step.

The last step in the Chart Wizard process determines where the chart should be displayed. Figure 48.15 indicates that there are two choices: as a new sheet or as an object in an existing sheet. The default option selected by Word is as an object in Sheet1. Rather than choose the default value, it is desirable to select the option to create the chart in a new sheet. By selecting the Finish button, the Chart Wizard creates the chart as a new sheet in the spreadsheet, as shown in Figure 48.16.

FIGURE 48.12 Select the chart type using the Chart Wizard

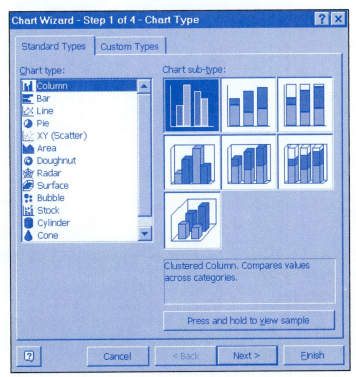

FIGURE 48.13 Choose the chart data source

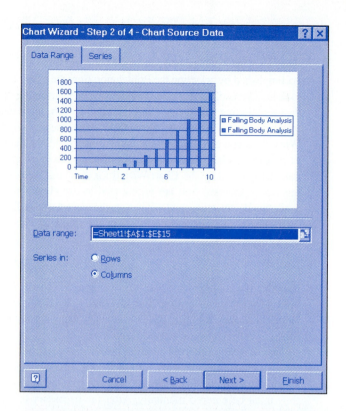

FIGURE 48.14 Specify the Chart options

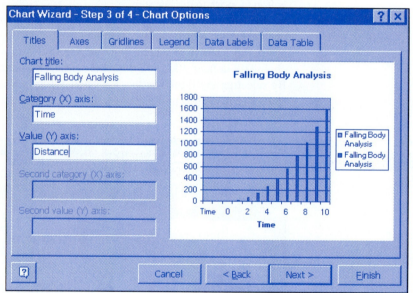

FIGURE 48.15 Select the location to store the chart

FIGURE 48.16 New sheet for the spreadsheet

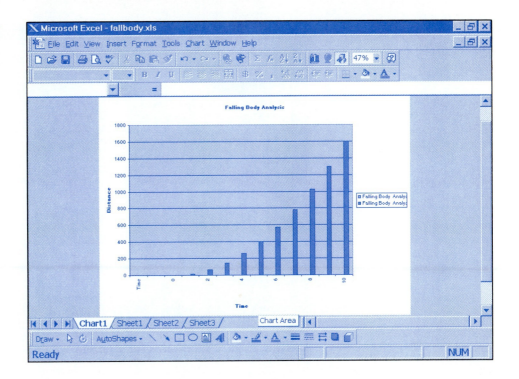

Many of the options available when using the Chart Wizard can be used to create very useful output. The old saying that a picture is worth a thousand words applies to spreadsheets too. Countless options are provided to the spreadsheet user to allow various outputs to be created.

Many other output-formatting options are also available, including page orientation, headers, and footers.

FORMULAS

In the previous spreadsheet example, it was not necessary to develop sophisticated formulas to solve the problem. Many times, however, it is necessary to develop formulas that are somewhat sophisticated. Fortunately, Excel is well-equipped to handle these situations by using any of the built-in functions. Figure 48.17 shows the Paste Function window, which lists each function by category. Each category contains a list of related functions. A

FIGURE 48.17 Partial list of Microsoft Excel built-in functions

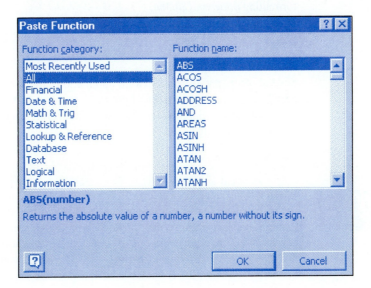

FIGURE 48.18 Cell format options

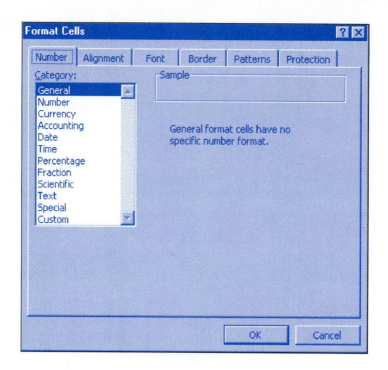

description of the selected function is provided as well, indicating what numbers or variables are needed to use the function. When developing a spreadsheet, it is useful to know about the different types of built-in functions. You are encouraged to examine each of the function categories and each of the individual functions within the category.

SPREADSHEET APPEARANCE

The appearance of a spreadsheet is an important part of the development process. Microsoft Excel provides the ability to format each cell or a range of cells in many different ways. For example, it might be necessary to center a heading over a range of cells, or center the data in the cells. The Format pull-down menu allows access to change the format of a cell, row, column, or the entire sheet. Figure 48.18 illustrates the variety of cell formatting options. Notice that if the cell contains a number, any of the items displayed in the list may be selected. As a particular item is selected, a sample display is also provided to show what the data will look like. Once again, you are encouraged to examine each of the various formatting options available.

OTHER SPREADSHEET SOFTWARE

Several other spreadsheet software applications are also available in addition to Microsoft Excel, such as Quattro Pro and Lotus 123, to name just a few. Each of these packages provides features similar those contained in Excel. You can perform the same operations using any of these packages.

TROUBLESHOOTING TECHNIQUES

When troubleshooting problems in a spreadsheet, it is necessary to compare the results of the spreadsheet with calculations that have been manually calculated. It is a mistake to assume that just because the spreadsheet contains some numbers, they are correct. It is necessary to check at least one type of each calculation and verify its correctness. As discrepancies are discovered between the spreadsheet and the manual calculations, it is necessary to review the formulas as well as to review the manual calculations to determine whether they are correct.

SELF-TEST

This self-test is designed to help you check your understanding of the background information presented in this exercise.

True/False

Answer *true* or *false*.

1. The smallest element in a spreadsheet is called a cell.
2. Cells may only contain numbers.
3. The value of a cell may be used in an equation.
4. It is not necessary to verify the calculations performed by a spreadsheet.
5. A1:B5 represents a range of cells.

Multiple Choice

Select the best answer.

6. The best way to write a spreadsheet is
    a. By scratch, composing it as you go.
    b. To make a plan of the design.
    c. To revise an existing spreadsheet.
7. What formula is required to add 15 to the value of cells A1 and B2 and store the result in a new cell?
    a. A1 + B2 + 15 =
    b. 15 = A1 + B2
    c. = A1 + B2 + 15
8. Charts are displayed
    a. On new sheets.
    b. As objects on existing sheets.
    c. Both a and b.
9. ABS stands for
    a. Absolute value.
    b. Average-based spreadsheet.
    c. Arithmetic bias.
10. After a formula is entered, the value of its cell
    a. Remains the same.
    b. Is updated based on the formula.
    c. Is set to zero by default.

Completion

Fill in the blanks with the best answers.

11. A cell's border is thicker and darker when it is _____.
12. A spreadsheet may contain one or more _____.
13. Graphs can be easily created using the Graph _____.
14. ABS, AND, and ATAN are examples of _____.
15. The first cell on a spreadsheet is _____.

FAMILIARIZATION ACTIVITY

1. Duplicate the Falling Body Analysis spreadsheet. Extend the time to 25 seconds. Use the spreadsheet to determine how long it takes the object to fall 10,000 feet.
2. Design a spreadsheet to keep track of the score in a bowling game.
3. Design a spreadsheet that analyzes a three-resistor series circuit. Figure 48.19 provides the required equations.
4. Use a spreadsheet to graph the data shown in Table 48.1.

FIGURE 48.19 For Familiarization Activity step 3

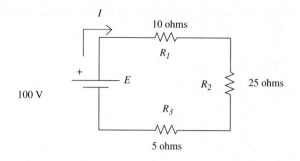

(a) Schematic of three-resistor series circuit

$$R_T = R_1 + R_2 + R_3$$

$$I = \frac{E}{R_T}$$

$$V_1 = I \cdot R_1 \quad V_2 = I \cdot R_2 \quad V_3 = I \cdot R_3$$

(b) Equations

TABLE 48.1 For Familiarization Activity step 4

Load (K ohms)	Power (mW)
1	23
2	45
3	89
4	121
5	167
6	126
7	98
8	61
9	39
10	27

QUESTIONS/ACTIVITIES

1. What limitations does the electronic spreadsheet have, if any?
2. Explain how a spreadsheet could be used to design a new part.
3. What is the advantage of using several sheets instead of one large sheet?

REVIEW QUIZ

Within 10 minutes and with 100% accuracy,

1. Understand the terminology associated with spreadsheets.
2. Navigate around the spreadsheet window.
3. Create a spreadsheet.
4. Use several built-in spreadsheet functions.

49 Databases

INTRODUCTION

Joe Tekk was frustrated. His manager, Don, had put him in charge of the electronics lab after one of his co-workers had retired. Unfortunately, Joe was having difficulty trying to keep all the necessary electrical components and other various parts in stock. Every time Joe turned around, someone was asking him for some part that turned out to be out of stock. This usually meant that the lab project needed to be put on hold because it typically took between seven and 14 days to complete the order process and get the stocks replenished.

Joe knew that he needed to get a better handle on the situation. As he explored his options, he soon determined that a database could be used to keep track of the various components and also automate the reorder process at the same time. Joe began to read some of the documentation about the Microsoft Access program, which was included with the Microsoft Office Professional package that everyone at RWA Software used. After just a few minutes, Joe realized Microsoft Access included a Database Wizard, which could help him develop the database with a minimum of effort.

As Joe began the database development, he identified all the electronic components and other parts that he had to keep track of. He then proceeded to use a Database Wizard to develop the tables, create the forms necessary to enter all the data, and produce the reports that he planned to use to stay on top of the lab inventory. Before he knew it, the job was done.

PERFORMANCE OBJECTIVES

Upon completion of this exercise, within 10 minutes and with 100% accuracy, you will be able to

1. Understand the terminology associated with databases.
2. Navigate around the different components of a database.
3. Create a database using the Database Wizard.
4. Use a Report Wizard to create a new custom database report.

BACKGROUND INFORMATION

The database is a revolutionary way to store and access data. A database may contain information applicable to any field of technology. A more appropriate and descriptive term for a database is a "relational database." The database designer determines what relationships exist between each element of data. Traditionally, a database is used in the field of business technology to keep track of customers, maintain the inventory in the company warehouse, maintain credit accounts for a company's customers, and keep track of the employees of a company. The applications for databases in technology include inventory control, task scheduling, and service call management. The applications for a database are limited only by the imagination of the database designer.

Before we are able to explore the wide variety of database applications, it is necessary to develop a basic knowledge of database terminology. The terminology of databases begins with a general definition of what a database actually is. A database is a collection of related information. The data contained in the database is organized into a hierarchy consisting of tables, records, and fields. Within the database, the data is grouped into categories called tables. Within each table are the records, which contain the individual data fields. The smallest element that can be referenced in a database is a field.

To determine what relationships exist in the data contained in the database, it is necessary to indicate what fields in the database are designated as *key fields*. A relational database contains both primary and secondary keys. The keys in a database allow for specific records of information to be selected. Consider a part number as a key to a database. By entering a part number in the key field, all data associated with the particular part may be displayed.

DATABASE DESIGN

Whenever a new database must be developed, it is necessary to follow a set of steps that are designed to specify all the necessary requirements. In general, the steps that can be followed when building a database are:

1. State the purpose of the database.
2. Design a model of the database on paper.
3. Build the electronic version of the paper model.
4. Test the database.
5. Document your work.

Before a database can be built, it is important to understand what is expected to be accomplished. It is necessary to identify each of the elements or fields that will be included in the database and what relationships, if any, exist between the database elements. It is also helpful to produce a model of what the database will look like.

For example, a database needed to keep track of the individual components in an electronics laboratory inventory begins with the identification of the components. A review of the lab inventory might reveal that the components fall into several standard categories: resistors, capacitors, integrated circuits, and one other category, miscellaneous items. In addition to the actual components, it is also necessary to keep track of the component supplier information, purchasing information, and shipping information. Once a plan is in place, the actual entry of the database elements can begin.

USING AN EXISTING DATABASE

To use a database, a user simply runs a database program. One of the most popular database applications for the personal computer is Microsoft Access, which is purchased separately or as a part of the Microsoft Office Professional suite of products. When the Access database is started, a window is displayed similar to Figure 49.1. The user can select a database that already exists (the default selection), or the user can create a new database. The default selection as shown in Figure 49.1 lists the names of the databases that have been accessed. An existing database that has never been opened before can be located using the More Files selection option.

To open the Inventory Control database, simply select it and press the OK button. This causes Access to begin executing that database and display the Main Switchboard as shown

FIGURE 49.1 Opening an existing database

FIGURE 49.2 Inventory Control Main Switchboard

in Figure 49.2. The Main Switchboard is called a *form*. A form contains either menu options or data entry screens. The Main Switchboard contains options to Enter/View Products, Enter/View Other Information, Preview Reports, Change Switchboard Items, and Exit this database. When selected, each of these buttons will cause another form to be displayed, either containing another menu or a data entry screen.

For the Inventory Control database, the items being tracked are called *products*. When a user selects the Enter/View Products option from the Main Switchboard, the Products window is displayed as illustrated in Figure 49.3. The first product, R100, a 100-ohm resistor, is automatically displayed. The ProductID is a unique number automatically assigned by Access. It is one of the keys to the database. The program user can enter data into the other fields on the form. Each product is given a name, description, category, lead time, and a reorder level. Purchase order information is also maintained for each product. Notice

FIGURE 49.3 The Enter/View Products form

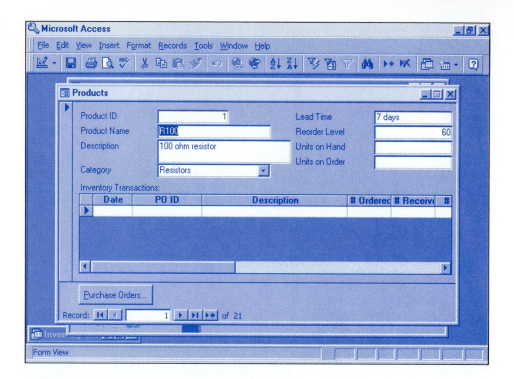

FIGURE 49.4 Adding a new product

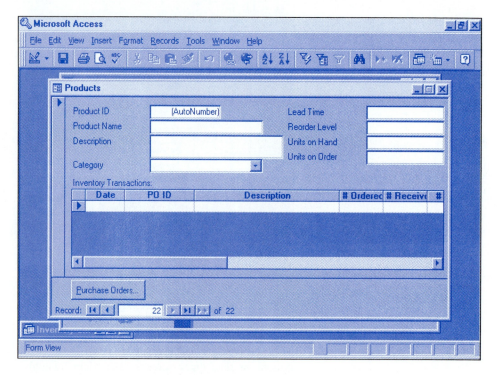

that R100 is record number 1 in the Products database. This is shown in the bottom left corner of Figure 49.3, along with a few control buttons. Using these buttons, the user can move back and forth between all of the products very easily. Two of the buttons (the arrowheads with bars next to them) provide access to the beginning of the list and the end of the list, respectively, with the click of one button. Two other buttons (the arrowheads without the bars) move one record at a time. All new records are entered at the end of the list.

To get to the end of the list and enter a new record, all that is necessary is to press the button that contains the right arrowhead with the asterisk next to it; this causes the screen shown in Figure 49.4 to be displayed. Notice that the Product ID field contains the text

FIGURE 49.5 Other information contained in the database

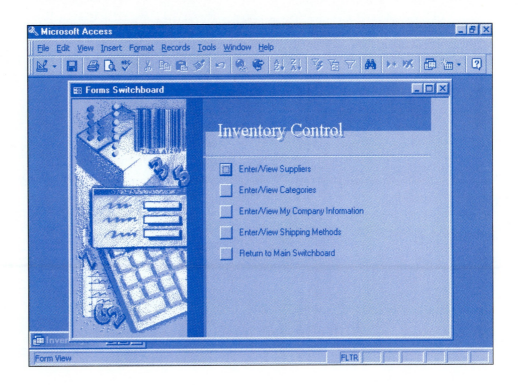

message (AutoNumber), which is how the sequential record numbers are assigned, and the record number is set to a value of 22, the next number that was available. The database user can then enter the necessary new product information into the database. After the products have been updated, or added, the window is closed and the Main Switchboard is redisplayed.

As previously stated, in an electronics laboratory inventory database, in addition to the components and miscellaneous parts, it is also necessary to keep track of the component suppliers, component categories, shipping information, and other company data. Selecting the Enter/View Other Information option, which is located on the Main Switchboard, provides access to enter or view each of these items. Figure 49.5 shows the Forms Switchboard menu. Data about suppliers, categories, company information, and shipping methods can be added to or modified as required using steps similar to those used when working with the Products form.

RELATIONSHIPS BETWEEN DATA ELEMENTS

With all the data elements contained in the Inventory Control database, it is possible for many different relationships to exist between the various data elements. For example, products are associated with a particular category of component, and purchase orders are associated with different suppliers and shipping methods. Figure 49.6 shows the relationship between the various elements in the database. Each of the boxes shown in the Relationships window is a data table. The data in a database is stored in tables. The lines between the data elements indicate what relationships exist.

In order to keep the database as simple as possible, relationships are created between two tables using data elements of the same name in both tables. For example, a supplier identification number stored with the purchase order is the same supplier identification number used in the supplier table. Relationships between the data elements are classified as one-to-many, many-to-many, and one-to-one. Notice that the links in Figure 49.6 indicate what type of relationship exists by specifying next to each table either a 1 or an infinity sign ∞.

The one-to-many relationship is the most common type of relationship. In a one-to-many relationship, a record in Table 1 can have many matching records in Table 2, but a

FIGURE 49.6 Different relationships between data elements

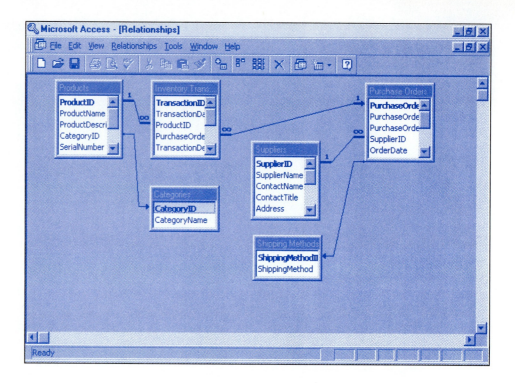

record in Table 2 has only one matching record in Table 1. In a many-to-many relationship, a record in Table 1 can have many matching records in Table 2, and a record in Table 2 can have many matching records in Table 1. In a one-to-one relationship, each record in Table 1 can have only one matching record in Table 2, and each record in Table 2 can have only one matching record in Table 1. This type of relationship is not common, because information related in this way would usually be defined using only one table. The kind of relationships that can be created depends on how the fields are defined. For example,

- A one-to-many relationship is created if only one of the related fields is a primary key or has a unique index.
- A many-to-many relationship is actually two one-to-many relationships that require use of a third table whose keys consist of the keys from the original two tables.
- A one-to-one relationship is created if both of the related fields are primary keys or have unique indexes.

When the definition for each field is correct, relationships are created by simply selecting an element in one table and dragging the link to the corresponding element in the other table.

DATABASE INTERNAL STRUCTURE

Each of the tables, which are used to store the database information, is maintained using the Inventory Control database. Notice that the Inventory Control database is the minimized window located in the bottom left corner of Figure 49.2 through Figure 49.5. When this window is restored, all the internal structure of the database can be examined or modified. The Inventory Control database window is shown in Figure 49.7. Each of the tabs displayed in the window are used to access the specific database elements such as tables, queries, forms, and reports.

The Tables tab lists all eight of the tables associated with the Inventory Control database. The program user can choose from three different options: open a table, design a table, or create a new table. Figure 49.8 shows the result of selecting the Open option. The Datasheet View lists each record of data that is stored in the table in a format

FIGURE 49.7 Inventory Control database window

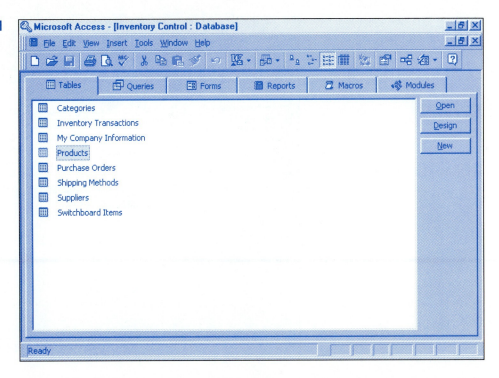

FIGURE 49.8 Datasheet View of the Products Table

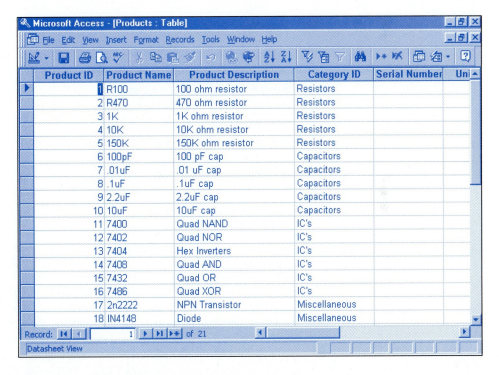

similar to a spreadsheet. Using the Datasheet View, data may be entered, modified, or deleted.

Choosing the Design option, the user is presented with the Products Table design window as shown in Figure 49.9. Notice that each of the field names is associated with a data type and a description. The common data types are listed in Table 49.1. In addition, each field is associated with different types of properties.

For example, the field properties for the ProductID specified by the AutoNumber data type include field size, new values, format, caption, and indexed indicator. You are encouraged to explore the field properties for each of the various data types available.

FIGURE 49.9 Products Table element details

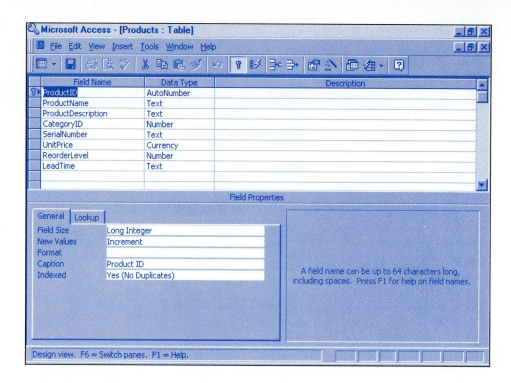

TABLE 49.1 Common data types

Data Type	Field Contents
Text	Text or numbers
Memo	Text or numbers
Number	Numbers only
Date/Time	Date/Time values
Currency	Currency values
AutoNumber	System assigned values
Yes/No	Yes or No

DATABASE FORMS

Database forms are a convenient way to enter and review information in a database. Figure 49.10 shows all the different forms associated with the Inventory Control database. The Switchboard form contains the menus, and the others provide access to the data. The Forms window provides the same options as the Tables window. By selecting the Open option, the form is opened and the data or a menu is displayed similar to Figures 49.2 through 49.5. The Design option allows for the form to be modified. Figure 49.11 shows the Products Form being edited. Notice how each of the data fields on the form is identified by a descriptive label. This helps the user to begin working with the different forms in the database very quickly. The form details are a very important part of the development process.

CREATING A DATABASE

As you can see from the existing Inventory Control example database, much thought, planning, and development time are spent creating a database. Recall from Figure 49.1 that there are two options available when creating a new database. The first option is to

FIGURE 49.10 Inventory Control database forms

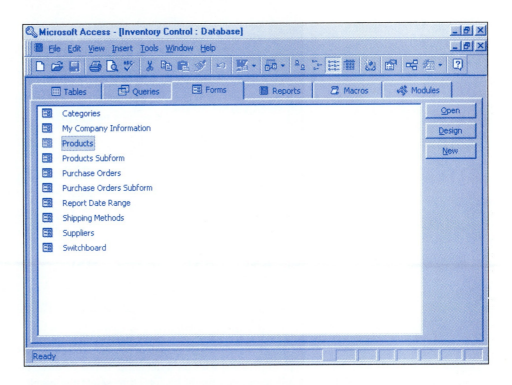

FIGURE 49.11 Editing the Products Form

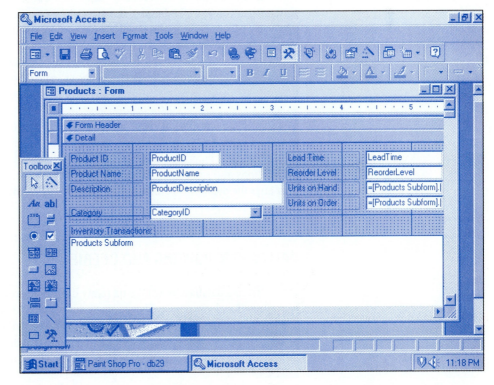

create a blank database. In a blank database, the development of each of the individual database components (tables, queries, forms, and reports) are handled by the database designer. The development of a blank database takes place in stages. It begins with the definition of all of the table elements and properties. Next, all the relationships between the various tables are defined. What remains is the creation of queries, forms, and reports that are required. These can be written in any order.

The second option available to the database designer is to use the Database Wizard to develop all the required components for standard types of database applications. Figure

FIGURE 49.12 Database Wizard custom application categories

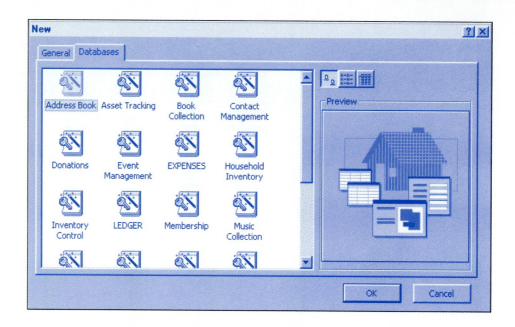

49.12 shows the Databases tab on the New database window. This list contains many of the typical personal uses for a database. If one of these databases does not fit your particular application, you might be forced to select the blank database and build it from scratch. Rather than resort to the blank database option immediately, review each of the database applications provided in the Database Wizard to see if one of them comes close. If so, it may be possible to modify the database produced by the wizard to meet many other applications not included on the list. For example, many of the items in the list refer to some type of collection, whether it is books, music, or a video collection. It is possible that one of these existing applications might be adapted to other types of collections as well.

When choosing one of the Database Wizard applications, Access will automatically create all the tables, forms, and reports that may be required. All that is necessary by the user is to begin entering the data. You are encouraged to review all the Database Wizard applications and sample databases. Note that Access can also provide a sample set of test data after the database is created. This provides an example of what the database looks like and what the reports look like.

DATABASE REPORTING AND QUERIES

Output from a database may take on many different forms. Traditionally, a database report was printed out on paper. The paper reports could then be distributed to anyone who needed a copy. In an effort to save resources, it is now more common to view a report right on the screen. This saves the cost of the paper but might also make it more convenient and up-to-date. Note that it is also possible to produce a report for distribution using the Internet.

Figure 49.13 shows the reports that are included in the Inventory Control database. As their titles suggest, they provide information about cost comparisons, product purchases by supplier, a product summary report, and a product transaction detail report. Other reports can be added as required using the one of the wizards. Figure 49.14 illustrates choosing the Chart Wizard to help create a new report that will be displayed as a graph. As you can see from the list, many different reporting capabilities are available.

Queries in a database provide access to a subset of the information contained in the database based on parameters selected by the user. When a query is executed, only the records that meet the query specifications are selected. For example, in the Inventory

FIGURE 49.13 Inventory Control report selection screen

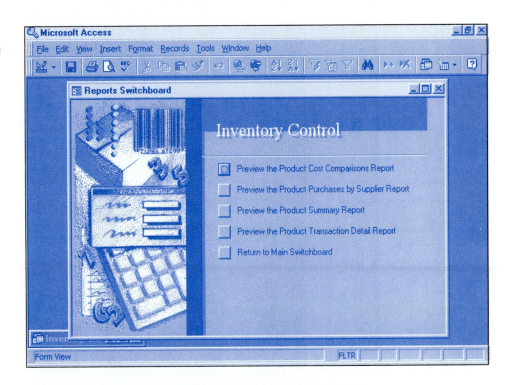

FIGURE 49.14 Select the Chart Wizard to graph supplier data

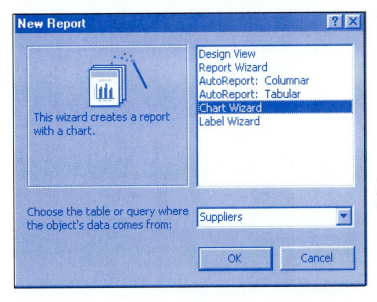

Control database, a query could be used to select only products in the database that need to be ordered. A query may be created manually or using one of the wizards.

OTHER DATABASE SOFTWARE APPLICATIONS

Many different database application programs are available for use on the PC. Reasons to choose one product over another are based on many different types of decisions. Some of the items that might be considered include cost, license fees, additional resource requirements such as hard disks and memory, transaction speed, and data security. Here is a short list of several popular products:

- Foxpro
- DBASE
- Oracle

TROUBLESHOOTING TECHNIQUES

The best type of troubleshooting advice for a database is to plan the database well. For example, it can sometimes be difficult to add new elements to a database after the database file already contains data. Occasionally, a process to migrate the data from one file format to another is required.

When each data element is created, review the field properties to ensure that each of the items contain appropriate values and that the key field settings contain the appropriate values. Remember that the relationships between the data elements and tables are determined by the key field values.

SELF-TEST

This self-test is designed to help you check your understanding of the background information presented in this exercise.

True/False

Answer *true* or *false*.

1. Databases organize information by tables.
2. The smallest element in a database is a key.
3. The keys for two different items can be identical.
4. Information in a database is organized into a hierarchy.
5. Database reports must be created using a wizard.

Multiple Choice

Select the best answer.

6. Relationships are created between
    a. Keys.
    b. Tables.
    c. Fields.
7. Which is not a relationship?
    a. One-to-many.
    b. Many-to-many.
    c. Many-to-none.
8. The AutoNumber feature
    a. Is applied to the whole database.
    b. Is applied to each table.
    c. Is applied to a field.
9. A blank database contains
    a. Empty tables connected in a standard way.
    b. No tables at all.
    c. Tables containing test data with no relationships.
10. The smallest element in a database that can be referenced is a
    a. Table.
    b. Record.
    c. Field.

Completion

Fill in the blank or blanks with the best answers.

11. The Main _____ is displayed when a database is opened.
12. Two different kinds of keys in a database are _____ and _____.
13. A database is also called a _____ database.
14. Each record of data stored in a table is displayed in the _____ view.
15. Information is entered into a database using a _____.

FAMILIARIZATION ACTIVITY

1. Design a database that will keep track of IP address assignments for a small network.
2. Create a database using the Database Wizard. Select one of the built-in database templates.

QUESTIONS/ACTIVITIES

1. Ask a professional from a local business to visit your class and explain how their company uses a database.
2. Search the Web for database products and/or applications. What are the common features of each?

REVIEW QUIZ

Within 10 minutes and with 100% accuracy,

1. Understand the terminology associated with databases.
2. Navigate around the different components of a database.
3. Create a database using the Database Wizard.
4. Use a Report Wizard to create a new custom database report.

50 Presentation Software

INTRODUCTION

Joe Tekk walked into the reception room of a large banquet hall. His friend and manager at RWA Software, Don Beers, was getting married. Joe quietly set up his laptop on a table near the door and started a PowerPoint presentation. Within 10 minutes, most of the guests were watching the presentation with delight, laughing and joking with each other. Joe had scanned pictures of Don, given to him by friends and relatives, and placed the images in a PowerPoint presentation. Then he added the many quotes and stories that went along with the pictures.

Don, watching with Joe, smiled happily, but mumbled to Joe under his breath, "I'm glad no one else is going to see this."

Joe glanced at Don and shrugged. "Sorry, Don, I saved the presentation in HTML format and posted it on your Web page."

PERFORMANCE OBJECTIVES

Upon completion of this exercise, within 15 minutes and with 100% accuracy, you will be able to

1. View an existing PowerPoint presentation.
2. Create your own PowerPoint presentations (from scratch and from a template).
3. Print out a PowerPoint presentation (outline and slide formats).

BACKGROUND INFORMATION

Presentation software is used to create professional presentations of text, graphics, and even multimedia audio and video files. Any kind of report can be improved by the powerful features of presentation software.

MICROSOFT POWERPOINT

Microsoft PowerPoint is a powerful application that allows you to make dynamic presentations using a variety of resources. The presentation format can be tailored so that the information is presented in the style envisioned by the presenter.

PowerPoint presentations consist of *slides,* graphical objects that may contain text, images, or other types of objects (almost anything that can be cut and pasted). Presentations can be manually stepped through by the presenter, or automatically sequenced in a slide-show format. Individuals who do not have PowerPoint can still view presentations that are specially saved as stand-alone slide shows.

Let us examine many of the basic features of PowerPoint.

OPENING AN EXISTING PRESENTATION

When PowerPoint is started, the dialog box shown in Figure 50.1 is displayed. If you are creating a new presentation, the AutoContent Wizard is a good place to start. We will use the AutoContent Wizard later in this exercise. For now, we will examine an existing presentation. Left-clicking the radio button to Open an existing presentation and then left-clicking OK will produce the Open dialog box in Figure 50.2. Like any other file open dialog, you can navigate to the directory of your choice to locate your presentation.

Note the series of PowerPoint files in the file display. The one selected (*convert*) has its first slide displayed in the viewing window at the right. This is a nice way to find the presentation you are looking for.

Double-clicking on the file name, or left-clicking Open after selecting a file, opens the presentation. Depending on how the presentation was last saved, it may come up in outline form or in slide form. Figure 50.3 shows the Outline view of the *convert* presentation. A nice feature of the Outline view is that it can be printed. Then everyone watching the show has a detailed outline they can make notes on during the presentation.

FIGURE 50.1   Initial PowerPoint dialog box

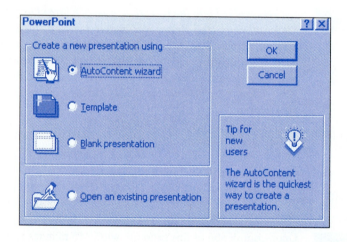

FIGURE 50.2   Opening a PowerPoint presentation

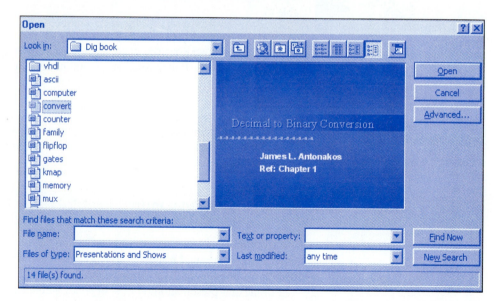

FIGURE 50.3 Outline view of a presentation

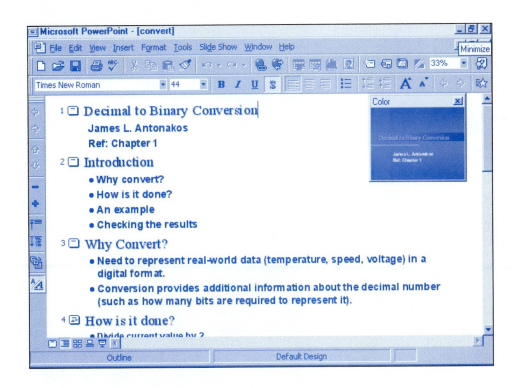

VIEWING A PRESENTATION

The Outline view shown in Figure 50.3 contains a number of features. Note the small color window to the right. It displays a reduced version of the current slide. The current slide is the slide where the cursor is positioned.

The small icons next to each slide number are also significant, as they indicate the type of slide presented. Notice that slide 4's icon is different from the other three slides. This is because slide 4 contains a graphic object as well as text. The other three slides only contain text.

The font of the outline is the same font that is used in the slides. So, if text appears bold or italic in the outline, it will appear bold or italic in the slide as well. The same goes for the font size and style.

To change the presentation view, use the View pull-down menu illustrated in Figure 50.4, or left-click one of the small view buttons at the lower left corner of the PowerPoint window. The Slide format replaces the text-based Outline view with the actual graphic slide. Figure 50.5 shows the Slide view of slide 1 in the *convert* presentation. The current slide is indicated in the status box at the lower left of the PowerPoint window. The total number of slides (6) in the presentation is also indicated.

The slide can be edited by left-clicking inside it and then making the necessary changes. Individual slides are accessed via the vertical scrollbar on the right of the display.

A window showing a small black-and-white version of the slide may also be displayed. It is enabled through the View pull-down menu. Its purpose is to show what the slide would look like when printed on a black-and-white printer.

To view the entire presentation, select the Slide Show option from the View pull-down menu or the View Show option in the Slide Show pull-down menu. Slides can be advanced during the slide show by pressing Space or Enter, or by left-clicking anywhere on the screen. Built-in tools for generating the timing of the slide show are included for presenters who require a specific pace for their presentations.

Right-clicking during a slide show will bring up the context-sensitive menu shown in Figure 50.6. From this menu you can move forward or backward in the slide show, adjust options, and even end the show.

FIGURE 50.4 View pull-down menu

FIGURE 50.5 Slide view of a presentation

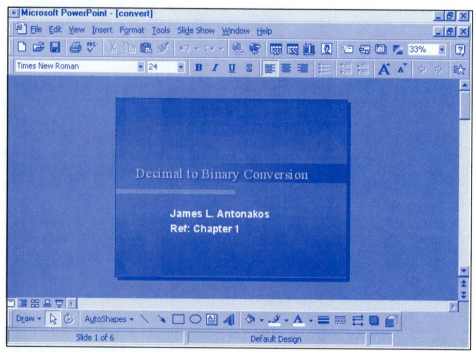

FIGURE 50.6 Slide show control menu

ADDING SLIDES TO A PRESENTATION

Slides are easily added to a presentation by selecting New Slide from the Insert pull-down menu. The slide is inserted after the current slide.

A large variety of slide types is available. One must be selected when you insert a new slide. Figure 50.7 shows the New Slide window. Left-clicking a slide icon causes a brief description to be displayed so that you know what you are getting. Note the wide variety of slide formats: text-only, object-only, text and object, multiple objects, and so on.

After choosing a slide, the new slide is displayed and set up for editing. For slides that contain objects, it may be better to edit the slide in Slide view than Outline view.

Figure 50.8 shows the new slide, which clearly indicates what area of the slide can be edited. The number of the slide is displayed in the status box (5 of 7). Other parameters, such as the font style and size (Times New Roman, 24 pt.), justification (left), and text formatting (bold), are also indicated.

To add text for the title, left-click inside the rectangular title box and begin entering text. Figure 50.9 shows the slide with its new title. Note that the title box is outlined differently once it has been selected. Any edge or corner of the box can be dragged to change the size of the box.

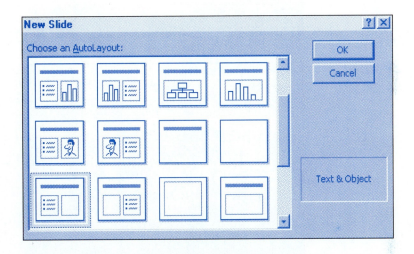

FIGURE 50.7 Choosing a new slide format

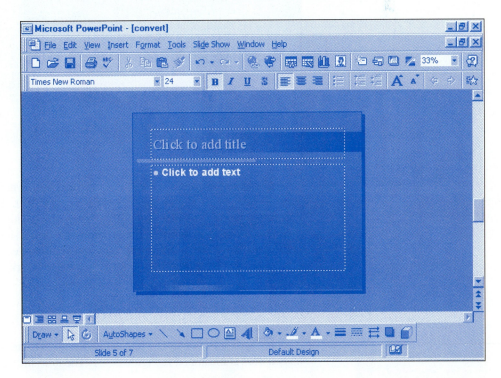

FIGURE 50.8 New text-only slide

FIGURE 50.9 Adding the title to the new slide

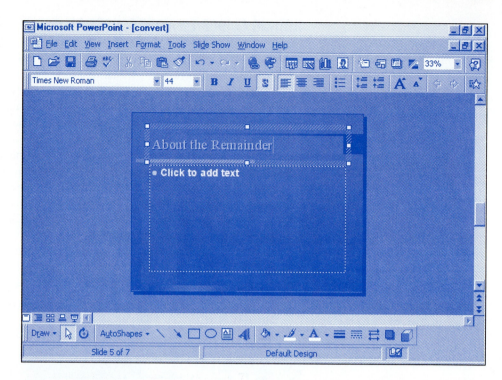

FIGURE 50.10 Adding text to the new slide

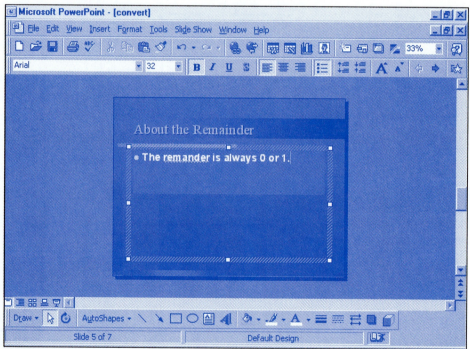

Text is added by left-clicking inside the text box, which may also be resized as necessary. Figure 50.10 shows the first line of new text. PowerPoint has underlined the misspelled word *remander* (should be *remainder*). This spell-checking ability is a nice feature and one more boost in the creation of a professional presentation. When multiple lines of text are added to a text box, they can be formatted as an ordinary paragraph of text, or as ordered or unordered lists by selecting the appropriate icons on the toolbar.

INSERTING OBJECTS INTO A SLIDE

A slide that contains one or more objects will indicate where to double-click to add an object. This is illustrated in Figure 50.11. The new slide contains a text box and an object

FIGURE 50.11 Slide containing text and object elements

FIGURE 50.12 Insert Object dialog box

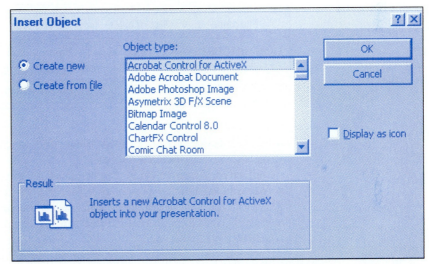

box. As usual, you can resize the object just as you can the text. Double-clicking the object icon brings up the window shown in Figure 50.12. Note that you can insert an existing object that has been saved in a file, or you can create a new object to insert. After the object has been inserted into the slide, double-clicking it will bring up the application associated with the object. This is useful if the object needs to be edited at a later time.

Figure 50.13 shows the finished slide containing an object created with an equation editor. Virtually any type of object, from clip art to animation files, can be inserted into a slide.

SAVING A PRESENTATION

PowerPoint lets you save your presentation in many different ways. Typical formats are Presentation, PowerPoint Show, and HTML (Web-based). PowerPoint Shows are presentations that always open up as slide shows.

Another useful save option is Pack-and-Go. In this option a viewer program (PPVIEW32.EXE) is packaged with the presentation (which is compressed). Individuals who do not have PowerPoint can view the presentation with PPVIEW32. The file extension

FIGURE 50.13 Slide containing text and a graphic object

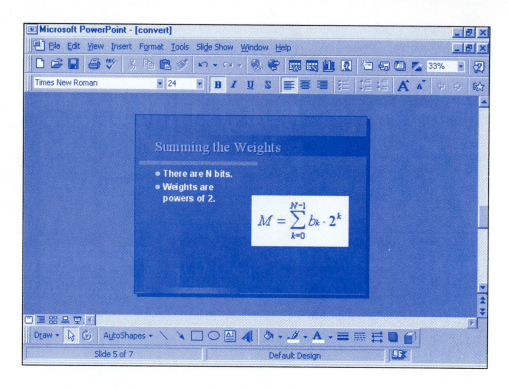

TABLE 50.1 File sizes for various Save options

File name	Size (K bytes)	File Type
CONVERT.PPT	63.5	Presentation
CONVERT.PPS	62.5	PowerPoint Show
PRES0.PPZ	890	Pack-and-Go

for this save option is .PPZ. Examine Table 50.1, which shows all versions of the *convert* presentation saved in different formats.

Once saved, the presentation may be placed on a floppy disk to hand out, or posted on the Web (as an ordinary presentation file or after conversion to HTML format). The PRES0.PPZ file is saved to a floppy with a small setup program that extracts the viewer and presentation when they are installed on a different computer. Its large increase in size is due to the PPVIEW32.EXE code.

PowerPoint also supports three other file formats: WMF (Windows Metafile) for graphic slides, RTF (Rich Text Format) for outlines, and POT (PowerPoint Template).

PRINTING A PRESENTATION

As previously mentioned, PowerPoint presentations can be printed in a variety of formats. These formats include:

- Slides (which can be scaled)
- Handouts (two or more slides per page)
- Outline

The outline format is nice for handing out at the beginning of a presentation. This allows the audience members to take notes without being rushed.

CREATING A PRESENTATION

Recall from Figure 50.1 that we can use the AutoContent Wizard to create a new presentation. This is done when PowerPoint is started, but may also be initiated at any time by selecting the

FIGURE 50.14 Choices for new presentations

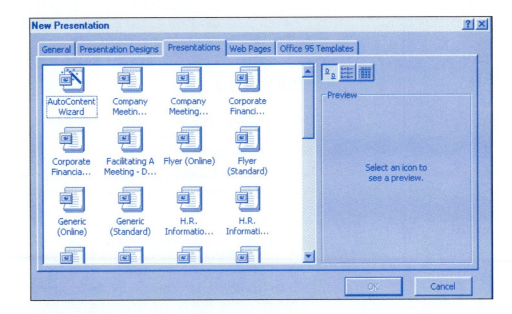

FIGURE 50.15 Initial AutoContent Wizard window

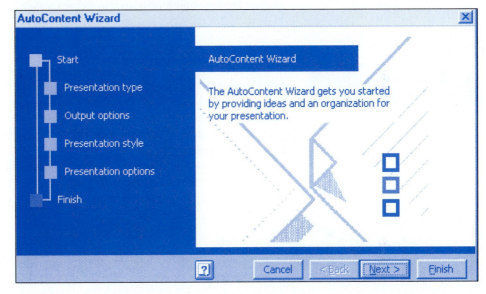

New option from the File pull-down menu, and then choosing AutoContent Wizard in the Presentations section of the New Presentation window shown in Figure 50.14. Note the additional predesigned presentations. These are called *templates,* and contain fully developed presentations, with specific wording (such as "Company Meeting Title" and "Presenter") included to indicate where to customize the presentation by entering your own information.

The AutoContent Wizard starts with the screen shown in Figure 50.15. The first thing to do is choose a presentation type, such as General, Corporate, Personal, or Sales/Marketing. This selects a template to use for the presentation.

Then the following output options must be specified:

- On-screen presentation
- Black-and-white slides
- Color slides
- 35-mm slides
- Handouts

And these presentation options:

- Title
- Name
- Additional information

Once all the information has been entered, the AutoContent Wizard builds the presentation.

615

A new presentation can also be started by left-clicking the New icon in the Standard toolbar. Here there are no choices except for what type of slide will be used for slide 1.

OTHER SOFTWARE

Many companies offer presentation software products. It is worth the time spent browsing the Web for presentation software and multimedia authoring tools. Here is a short list of several popular products:

- Multimedia Toolbook (www.asymetrix.com): Multimedia authoring tool. Create "books" that contain active objects (such as buttons that can be clicked). Include audio, video, and animation.
- Corel Presentations (www.corel.com): An easy-to-use presentation software program with many of the same features that are found in PowerPoint.
- Harvard Graphics (www.harvardgraphics.com): PowerPoint will convert Harvard Graphics presentations into its own format.
- Lotus Freelance (www.lotus.com): Powerful presentation package containing support for Web publishing, networking, database access, and group presentations. PowerPoint will convert Freelance presentations into its own format.

TROUBLESHOOTING TECHNIQUES

There may come a time when you are editing a slide in outline format and cannot perform some simple editing step such as left-clicking to select an insertion point. Whatever the difficulty, it may help to switch display modes and try the same step when viewing the graphical slide. This is especially true for slides that contain objects and text. The outline does not make it clear what needs to be done to access the object, but it is very obvious on the graphical version.

Also, when saving your presentation for others to view, bear in mind that they may not have the same fonts installed on their machine as you do. You may want to have the fonts embedded in your presentation to avoid this situation.

SELF-TEST

This self-test is designed to help you check your understanding of the background information presented in this exercise.

True/False

Answer *true* or *false*.

1. PowerPoint presentations are strictly text-based.
2. The Slide Show format is used to display slides larger than normal for editing purposes.
3. Pack-and-Go presentations contain their own viewers.
4. PowerPoint presentations can be saved in HTML format.
5. Title and text areas of a slide may be resized as necessary.

Multiple Choice

Select the best answer.

6. The outline view of a presentation
    a. Can be printed.
    b. Contains all text and graphics in the presentation.
    c. Neither a nor b.
7. Which presentation file has the extension .PPT?
    a. Presentation.
    b. PowerPoint Show.
    c. Pack-and-Go.

8. Which presentation file is the largest if the same presentation is saved many different ways?
   a. Presentation.
   b. PowerPoint Show.
   c. Pack-and-Go.
9. New slides are inserted
   a. Before the current slide.
   b. After the current slide.
   c. At the end of the presentation.
10. Which file extension is not supported by PowerPoint?
    a. .RTF.
    b. .PPT.
    c. .PCX.

Completion

Fill in the blanks with the best answers.

11. PowerPoint presentations consist of one or more _____.
12. A printed presentation containing two or more slides per page is called a(n) _____.
13. PowerPoint performs _____ checking on text entered into a slide.
14. New presentations are easily started using the _____ Wizard.
15. Fully developed presentations that require customization are called _____.

FAMILIARIZATION ACTIVITY

1. Design a PowerPoint presentation that explains how to add two three-digit decimal numbers. Show an example, such as 143 plus 893 equals 1036.
2. Prepare a PowerPoint presentation on a subject of your own choice. Include several different slide styles.
3. Search the Web for PowerPoint presentations posted by others. Select three that you find interesting and merge them into a single presentation.

QUESTIONS/ACTIVITIES

1. What information is not shown in the Outline form of a presentation?
2. Compare the file sizes of a presentation when saved as a Presentation and as a PowerPoint Show.

REVIEW QUIZ

Within 15 minutes and with 100% accuracy,

1. View an existing PowerPoint presentation.
2. Create your own PowerPoint presentations (from scratch and from a template).
3. Print out a PowerPoint presentation (outline and slide formats).

51 Web Development

INTRODUCTION

Joe Tekk and Don, his manager, were browsing the Web. They decided to check some information on the RWA Software home page. Joe clicked on the link to RWA's page and was dismayed to find that it took longer than 30 seconds to load.

"Don, this is really bad. No one wants to wait that long for a page to load."

"I agree," Don replied. "What can we do about it?"

One hour later, Joe called Don and told him to look up the RWA Software home page again. Don started his browser and clicked on RWA's link.

The new page loaded in two seconds.

PERFORMANCE OBJECTIVES

Upon completion of this exercise, within 15 minutes and with 100% accuracy, you will be able to

1. Discuss the basic concepts of Web development.
2. Use an HTML editor and a text editor to create an HTML document.
3. Describe the two graphical image formats that are used on the Web.
4. Develop a new image from scratch.

BACKGROUND INFORMATION

The changes in the computer industry brought about by the World Wide Web are nothing short of revolutionary. Everyone with the slightest interest in computers knows about the World Wide Web. With a few new skills, everyone is given the ability to publish information on the World Wide Web. In this exercise, we will examine how a Web page is planned and created.

PLANNING THE PAGE

The process of creating a Web page consists of several elements. These elements include planning the Web page layout, creating any required graphical images or buttons, and HTML coding. The Web page designer usually begins planning by producing a rough sketch of what the page should like when it is complete. As an example, Figure 51.1 shows the design of a Web page for RWA Software. Using the sketch as a blueprint, the process to actually create the Web page may begin. Let us begin by creating the new images that will be displayed on the RWA Software home page.

FIGURE 51.1 Web page design plan

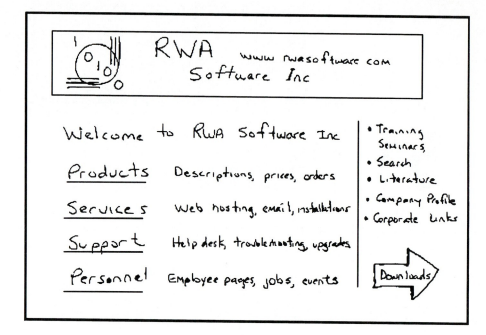

IMAGES

A Web page that includes graphical images provides a different viewing experience from that of viewing a Web page that contains only text. Two different image formats, GIF and JPG (short for JPEG), are supported for use on the Web.

The benefits of using images on a Web site are many (recall the old adage that a picture is worth a thousand words). One of the problems with images is that the limited bandwidth available to many Internet users causes much time to be wasted while waiting for images or graphics to download. Therefore, it is important for the Web page developer to keep the size of every image to a minimum. This allows the Web page to load as quickly as possible.

When choosing between the two graphics formats, the following points must be considered. The size of a GIF image is typically much smaller than the same image stored in a JPG format. The reduced file size is a result of a GIF image being limited to a total of 256 colors. A GIF image has two other advantages not offered by a JPG image. A GIF image may be used to create animation files on the Web, and it may contain a transparent background. An animation file contains a series of GIF images that are put together in a slide show format. A GIF image saved with a transparent background blends seamlessly into a Web page.

Unlike a GIF image, a JPG image may contain up to 16 million colors. The increased color capability available in the JPG format allows for high-quality images to be distributed over the Web. The increase in the number of colors accounts for the large size associated with JPG images. As a rule of thumb, use a GIF image whenever possible and use a JPG image only as necessary such as when a high-quality color image is required.

Creating graphical images for use on a Web page is straightforward, although there are many details to consider. The process begins by choosing image-editing software. The Paint Shop Pro application produced by Jasc Software is a popular image editing and animation product. Using Paint Shop Pro, images may be created from scratch or modified as necessary. The initial Paint Shop Pro window is shown in Figure 51.2. The Tip of the Day window (displayed by many Windows application programs) provides useful tips on how to use Paint Shop Pro. The Tip of the Day window and the two windows that it covers (Controls and Layers) can safely be closed. The File Associations window is displayed the first time Paint Shop Pro is run. The associations that are selected in the File Associations window will cause Paint Shop Pro to automatically be used as the default image editor for that particular file type.

FIGURE 51.2 Paint Shop Pro development window

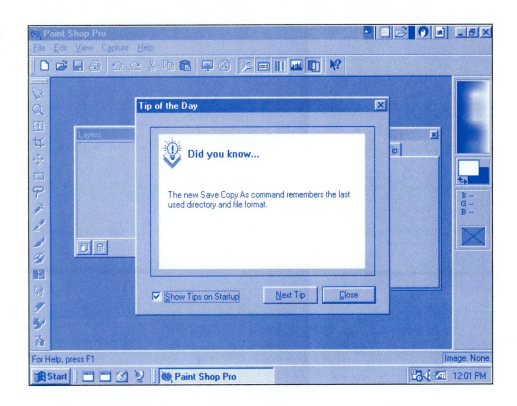

Paint Shop Pro automatically displays many of the toolbars that are available. The Tool Bar, Tool Palette, Status Bar, and Color Palette toolbars are usually displayed. The Tool Bar toolbar is located directly under the pull-down menu and contains the most common Paint Shop Pro functions such as file, clipboard, and other toolbar operations. The Tool Palette toolbar located along the left side of the window allows for quick selection of the common editing tools. Because no image is currently active, none of the icons are available for use. The Status Bar located along the bottom of the screen displays useful information to the image editor. The Color Palette toolbar is displayed along the right-hand side of the Paint Shop Pro window. At the top of the toolbar is the color selection panel. As the mouse is moved over the color selection panel, the RGB box located under the selection area indicates the amount of red, green, and blue to use in the current color. Between the color selection panel and the RGB box are the current foreground and background colors. Initially these colors are black (background) and white (foreground). The foreground and background colors can be swapped by clicking on the L-shaped arrow that points to both colors.

New Paint Shop Pro users are encouraged to explore each of the pull-down menus and examine the context-sensitive help messages that are available for each of the toolbar icons. Install a copy of Paint Shop Pro from the companion CD before continuing.

CREATING IMAGES

To create a new image using Paint Shop Pro simply select the new file icon from the Paint Shop Pro editing screen as shown in Figure 51.3. The New Image window presents the program user with several choices, which are divided into two categories. First, the image dimensions consisting of width, height, and resolution must be specified. Second, the image characteristics, which include the background color and the image type, must also be chosen. The default image dimensions of 300-by-300 pixels at a resolution of 72 pixels per inch displayed using up to 16.7 million colors requires an image size of 263.7KB. To create the new logo image for RWA Software, most of these values need to be modified.

Referring to the Web page sketch in Figure 51.1, it appears that the RWA logo is much wider than it is high. Using a typical VGA monitor, the screen resolution is 640 pixels wide and 480 pixels high. Using those screen dimensions, it would appear that the RWA logo

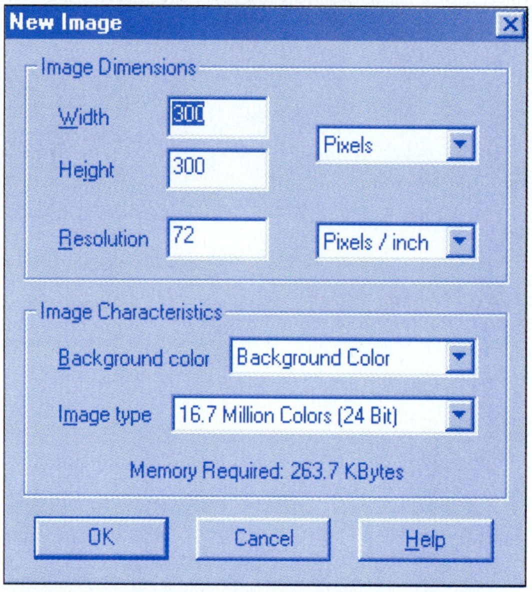

FIGURE 51.3   Default New Image properties

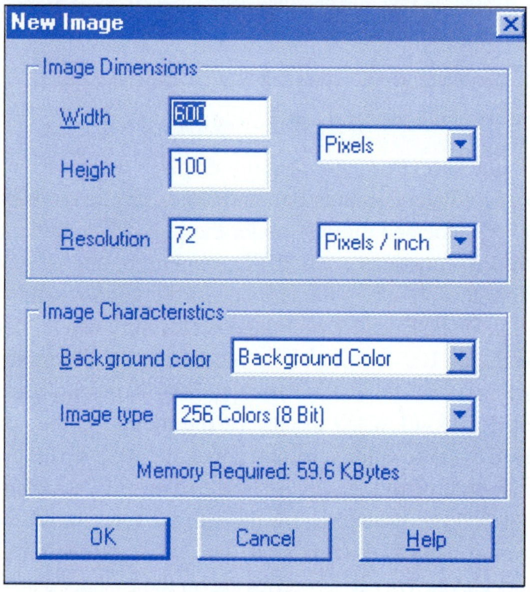

FIGURE 51.4   New Image information for RWA Software logo

should be set to dimensions of 600 pixels wide by 100 pixels tall. Figure 51.4 shows the New Image window with these specifications. Notice that by choosing a 256-color image type, the memory required for this image is only 59.6KB.

Selecting the OK button causes the new image to be displayed as illustrated in Figure 51.5. Also notice that the Tool Palette toolbar has been moved from the left-hand side of the window to below the Tool Bar. This is because several of the important palette icons were not displayed due to the limited resolution of the computer monitor. The new image title bar shows the name of the image (Image1), current zoom setting (1:1), and Background, indicating that the new image was created using the background color setting, which is currently set to black. The first time the image is saved, the user will be given the opportunity to change its name.

Let us begin by changing the black background of the image to white. Select the Flood Fill Tool Palette icon and then left-click on the black background of the image. The color is changed to the foreground color, white. This result of this operation is shown in Figure 51.6. Now we are ready to begin the actual image development. The RWA Software logo consists of two parts: text and image elements. Once again, using the Web page sketch as a model, the text can be placed on the logo. On the Web, using color is an easy way to make your text stand out. To add colored text to the logo, we must first choose the color of the

FIGURE 51.5 New image displayed in the background color

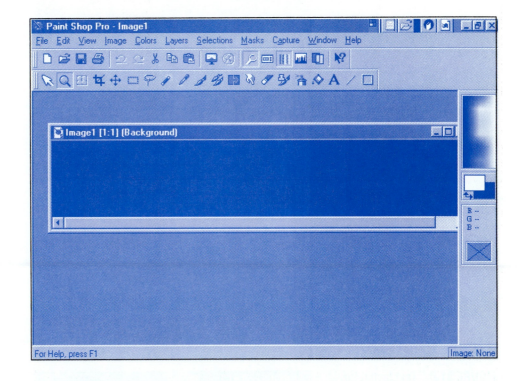

FIGURE 51.6 New image window

FIGURE 51.7 Selecting a new color

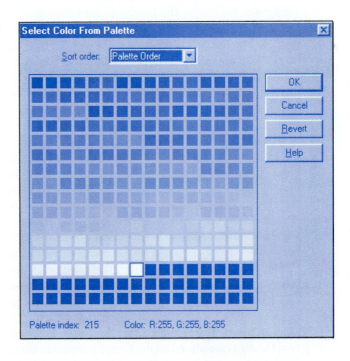

text. Left-click on the foreground color (white) in the Color Palette toolbar and choose a different color from the many that are available on the Select Color From Palette menu, as illustrated in Figure 51.7. When a selection has been made, select OK and the new color will now be displayed in the foreground color box.

FIGURE 51.8 Add Text menu options

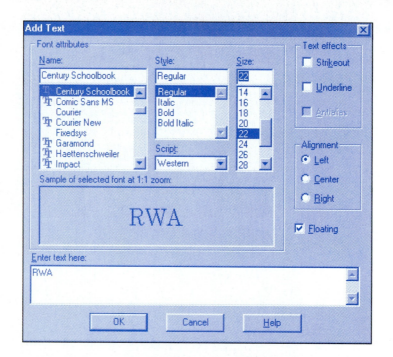

FIGURE 51.9 Text layout for the RWA logo

Next, select the Text icon (the icon with A in it) from the Tool Palette toolbar and left-click inside the image. This causes the Add Text dialog window shown in Figure 51.8 to be displayed. Choose the font, style, size, and any other attributes and enter the text to be displayed in the Enter text here box. To keep the text display interesting, it will be added in stages to allow for different colors, fonts, and sizes to be mixed together. The first text added to the logo is "RWA," using a large size of the Century Schoolbook font. The Text icon must be selected several times in order for the remainder of the text components to be added. New text colors must be selected before placing new text. The completed text layout for the RWA Software logo is shown in Figure 51.9.

The second stage of the RWA Software logo is a custom icon that will be displayed to the left of the text display. Rather than perform the actual development of the icon in the logo window, it is better to open a new document, create the image, and then paste it into the existing image that contains the text. The icon is created using a 100-by-100, 256-color image.

The development of the icon is also performed in steps similar to the steps used in creating the text. Each of the stages is shown in Figure 51.10. Starting with a white background and selecting the color blue, we begin to create the icon by drawing a circle. To create a circle, it is necessary to select the shape tool (Rectangle) on the Tool Bar and then select the Toggle Control Palette icon on the Tool Bar. This causes the Controls window to be displayed, which allows for the circle shape to be selected. Various shape controls are also available. Figure 51.10(a) shows the result of creating the circle shape using an outline width of 4.

Next, after selecting the color green, the four binary digits are added using the Text tool. The last binary digit is added using the color red. Figures 51.10(b) and 51.10(c) illustrate how the digits are placed on the icon. To complete the icon, a set of vertical and horizontal lines will be added along the bottom and right-hand side of the icon. To add a line, it is necessary to select the Line tool on the Tool Palette. The line is drawn by left-clicking to set the beginning point and then dragging the line to the correct ending location. Figure 51.10(d) shows the first line being added using the Line tool. The icon displayed in Figure 51.10(e) is the completed RWA Software icon.

FIGURE 51.10 Development stages of the RWA software logo

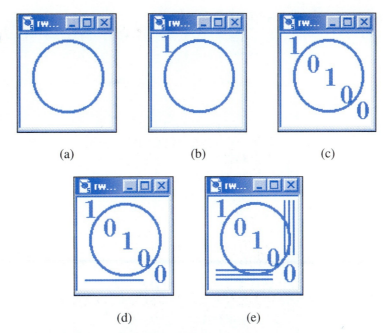

FIGURE 51.11 Final version of the RWA logo

To complete the development of the logo, all that remains is to select the icon, copy it to the clipboard, and then paste it (as a transparent image) into the proper position in the logo. Figure 51.11 shows the final version of the RWA Software logo. The image for the right-pointing Downloads arrow (see the sketch in Figure 51.1) is created using the Rectangle tool, the Line tool, and the Flood Fill tool.

HTML

HTML (hypertext markup language) code can be entered directly using a text editor or an HTML editor. When using a text editor, it is necessary for the Web page developer to be familiar with the specific details of HTML. Using an HTML editor, such as HoTMetaL by SoftQuad, provides access to the HTML language by simply clicking on the toolbar icons. Both of these development options will be explored in this exercise.

CREATING A WEB PAGE

Now that the image files for the Web page have been created, it is time to generate the HTML code for the actual Web page. When the Web was new, people simply used Notepad or some other text editor to write the HTML code for their Web pages. As the interest grew, a market for Web-page editing software was created. Today there are many different graphical Web-page editors. It is almost unnecessary to know the specifics of HTML, because the graphical interfaces provided by the software typically display the page under construction in WYSIWYG format.

For the remainder of this exercise we will be examining how one editor, HoTMetaL PRO, can be used to create a Web page. Install a copy of HoTMetaL PRO from the companion CD to experiment with your own Web page.

FIGURE 51.12 Starting a new page in HoTMetaL PRO

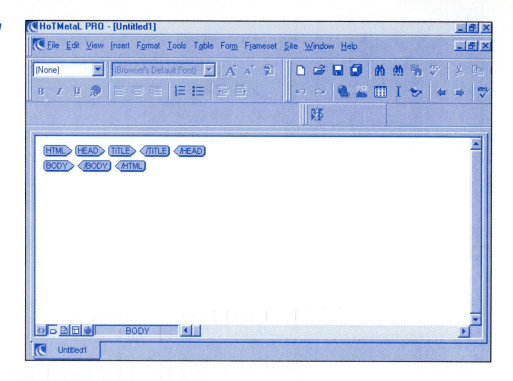

HOTMETAL PRO

Refer to the sketch of the RWA Software home page in Figure 51.1. We will try to make the actual Web page look as similar to the sketch as possible. To do so, we will use many of the basic features provided by the HoTMetaL PRO editor.

Figure 51.12 shows the initial display of a new Web page. The graphical tag icons can be turned off, but keeping them displayed helps choose where to left-click to set insertion points for new material. HoTMetaL PRO generates the initial set of matching tags required by any Web page (<HTML> ... </HTML>, etc.). Note that almost all the buttons in the toolbars are inactive. We need to left-click somewhere inside the Web page to make them active.

The first thing to do is to enter the title of the Web page. This is important, since the title is identified by the browser when your page is displayed. In addition, the contents of the title are used by search engines to locate information on your page. So choose a title that reflects what the page is used for. Figure 51.13 shows the Web page with its new title. Also, the title bar of the HoTMetaL window indicates that the Web page has been saved as RWA. The actual file name is RWA.HTM. File names ending with .html are also used.

To begin entering the contents of the Web page, left-click between the <BODY> tags. This is where everything else in the page goes.

ADDING AN IMAGE

The first thing we want to place on the page is the RWA Software logo image created with Paint Shop Pro. To control the placement and alignment of the image on the Web page, the image should be placed inside a paragraph. To create a set of paragraph tags, left-click on the down arrow in the leftmost selection box (underneath the File, Edit, and View pull-down menus). Select Paragraph from the list. Figure 51.14 shows the list of elements that may be inserted. Normally, once we have created a paragraph we insert text into it. But this paragraph will have an image inserted into it, and then the paragraph will be centered, which will center the image (the RWA Software logo) on the Web page.

To select an image, left-click the Image button (small mountain with moon, next to the Table button). This will bring up the Image Properties window shown in Figure 51.15. There are two ways to specify the image file. One is to enter the full path name to the file in the Image File box. Left-clicking the Choose button instead brings up the window shown in

FIGURE 51.13 Adding the title

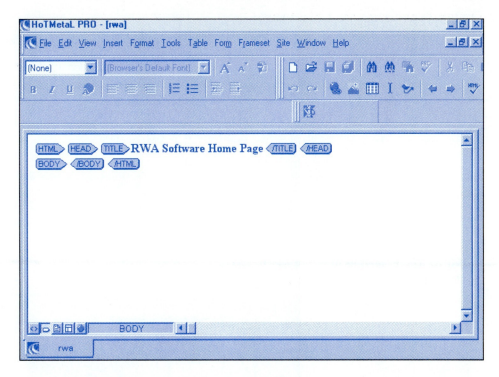

FIGURE 51.14 Choosing a style element

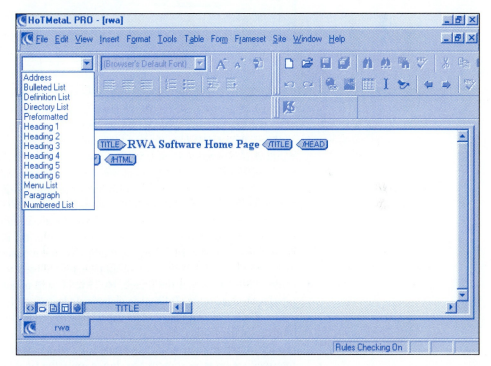

FIGURE 51.15 Initial Image Properties window

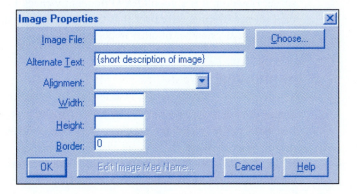

FIGURE 51.16 Choosing an image

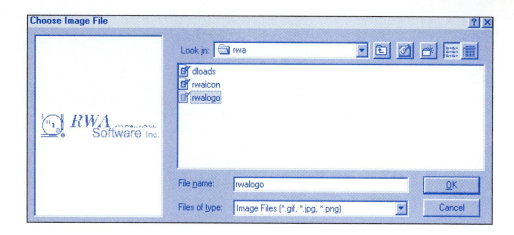

FIGURE 51.17 Properties of the selected image

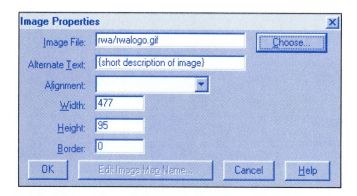

Figure 51.16. Using the Choose Image File controls, we can browse images in a small view window while we look for the one we need. This is very useful if you do not remember what you named the image file and there are a lot of image files in the directory.

The RWALOGO file is selected as the image. When we return to the Image Properties window, illustrated in Figure 51.17, many of the boxes are now filled with image information. The Width and Height values are the number of pixels in each direction. The Alternate Text box can hold an identification string for the image for individuals doing text-only browsing. Left-click inside the box and enter "RWA Software logo." Then left-click the OK button. This will place the image on the Web page, as indicated in Figure 51.18. By default the image is aligned with the left margin. To center the image, select it by left-clicking between the <P> and <IMG> tags and dragging the mouse to the other set of tags (to the left of </P>). The image element should be highlighted when it is selected. Left-click the Center button to center the image. HoTMetaL will update the display with the image centered to the best of its ability (the graphical tags affect the WYSIWYG appearance).

To check your Web page in an actual browser, left-click the Browser button in the lower toolbar (underneath Undo). If you have not recently saved the Web page, you will be prompted before the browser is loaded. For the purpose of this presentation, none of the intermediate browser screens are shown, only the final Web page. This is done to conserve space. However, it is a good idea to frequently check the appearance of your Web page in the browser. You may discover that the text is too small, or the colors do not look good together, or any number of other imperfections.

ADDING TABLES

Tables provide a very powerful way to format and present information. Tables with invisible borders are useful for making multicolumn layouts. Tables within tables are also useful, as we shall see.

To insert a table, left-click the Table button (underneath the binoculars). Be sure the current cursor position on the Web page is between the </P> and </BODY> tags. The Insert

FIGURE 51.18 Adding the logo image

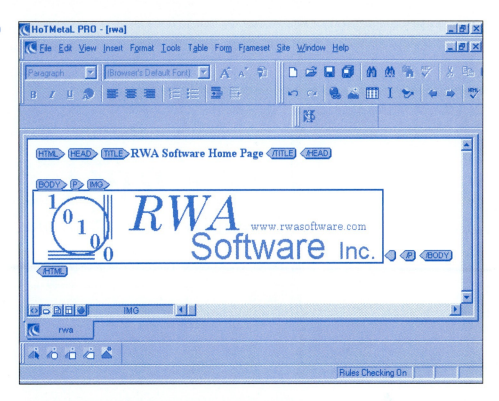

FIGURE 51.19 Setting the table properties

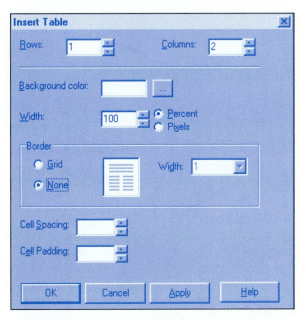

Table window shown in Figure 51.19 allows all the properties of the table to be adjusted before it is inserted. The table properties in Figure 51.19 indicate one row, two columns, 100% width, and no border. This invisible table is the basis for the two-column layout indicated in the original Web page sketch shown in Figure 51.1.

The new table is outlined in Figure 51.20. As its properties in Figure 51.19 indicated, its width is 100% of the available width (the whole screen in this case). The <nbsp> tags stand for "nonbreaking space," special characters inserted by HoTMetaL PRO when a table is created that forces the borders to be drawn if there is no content in the table cell (and borders are not set to zero). These tags will be replaced with other information.

Left-click inside the first column of the table. This will set the cursor position. Enter the text "Welcome to RWA Software, Inc." Select it, and make its font size larger by left-clicking the large capital A button with the up arrow (underneath the Form pull-down menu). Figure 51.21 shows the new table. Notice that the width of the first column (containing

FIGURE 51.20 Adding a table

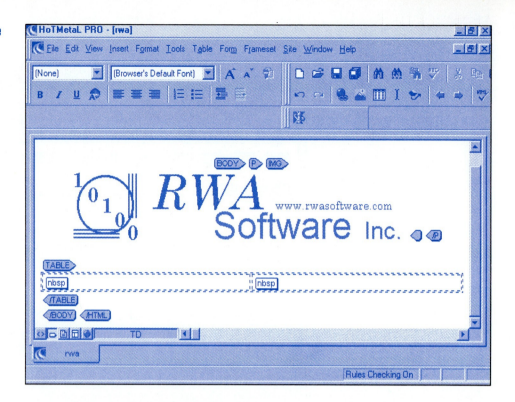

FIGURE 51.21 Adding a greeting

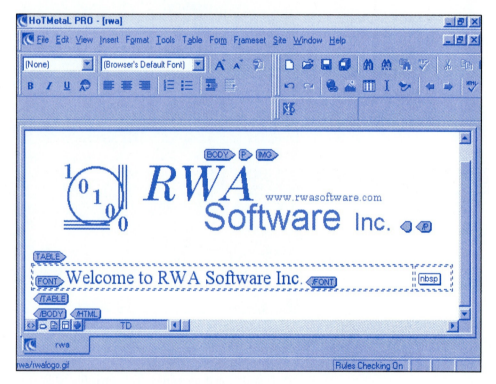

the text) is larger than the second column (currently empty). The browser will automatically resize the table columns depending on what is contained within them. The appearance of the Web page may change as more information is added.

Left-click to the right of the </FONT> tag in the first column of the table. Another table will be inserted here, a table within a table, which will contain links to other pages. Insert an invisible table (border equals zero) with four rows and two columns. The width should once again be 100%, although the 100% is now relative to the width of the first column in the original table.

ADDING LINKS

Left-click inside the first row and column of the new table. To insert a link to another page or file, left-click the Anchor button (to the left of the Image button). Figure 51.22 shows the resulting dialog box. Enter "products.htm" into the File or URL text box. This specifies a local Web page as the object of the link. Many other types of links, such as FTP, e-mail, and Telnet are possible. Left-click the down arrow to view a list of types.

After left-clicking the OK button, the link is entered into the table. Figure 51.23 shows the new link, which is selected for editing by default when created. This allows us to change the name of the link. Enter "Products" and increase the size of the font a few point sizes.

To add a description of the link, left-click in the second column of the inner table's first row. Enter the text "Descriptions, prices, ordering information…" into the table. The table should look like that shown in Figure 51.24. Once again, it is important to use a browser to display the page, so that you can make adjustments to the size of the fonts and other formatting changes.

Three other links are entered the same way as the Products link, completing the left half of the two-column layout presented in Figure 51.1. The right half must contain several

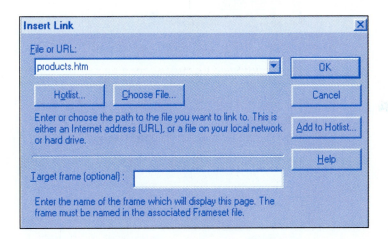

FIGURE 51.22 Entering the link

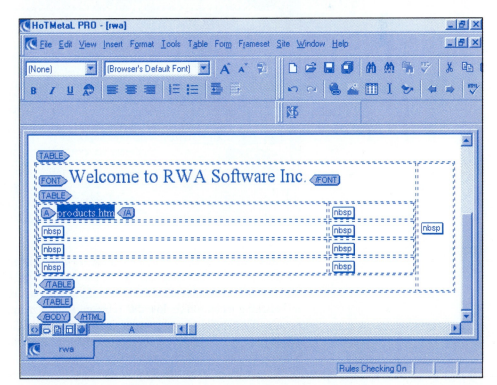

FIGURE 51.23 Default link name

FIGURE 51.24 Link and description table entries

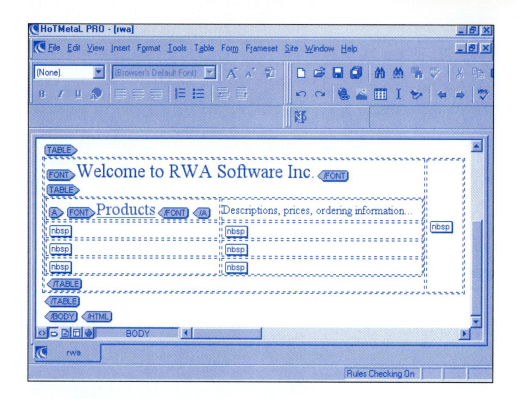

FIGURE 51.25 Starting a bulleted list

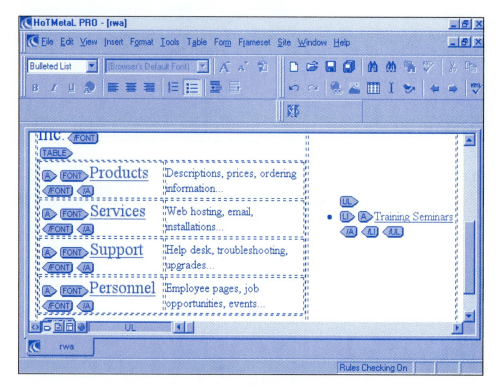

additional links (no descriptions) and the Download image (which will be made clickable). An unordered list (bulleted list) will be used to store the links.

Left-click inside the second column of the original table. Then start an unordered list by left-clicking its button (below the Table pull-down menu). Enter a link in the first list element to TRAIN.HTM and rename the link "Training Seminars." Figure 51.25 shows the added list element. Notice that the table elements have all been resized. This is a good time to view the page in a browser to verify formatting.

FIGURE 51.26 Adding the Downloads image

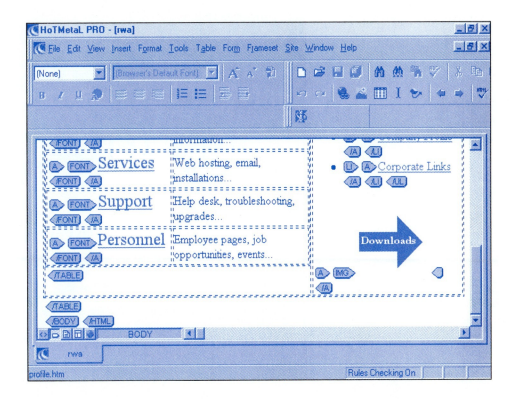

To enter another list element, left-click between the </A> and </LI> tags and press Enter. A new list element will appear. A similar method can be used to enter the additional links to SEARCH.HTM, LIT.HTM, PROFILE.HTM, and LINKS.HTM.

To complete the RWA Software home page we only need to add the Downloads image and make it clickable. The image must be inserted after the </UL> tag. Once inserted, the image itself can be made clickable (an anchor) by selecting it and left-clicking on the Anchor button. Figure 51.26 shows the final HoTMetaL view of the Web page. There are many more features in HoTMetaL than were presented here. If you are interested in the Web, invest time in learning about the additional features so that your Web pages will reflect your skill and imagination.

THE FINAL WEB PAGE

Figure 51.27 shows the final RWA Software home page. It is worthwhile pointing out a few features of the page:

- The background is white. This helps for readability and printing.
- There is no background image. This also helps improve readability and printing, as well as page loading time (one less image to load).
- Images are used sparingly (and contain few colors). This decreases the page loading time.
- There are no Java scripts or other time-consuming add-ons.
- The user does not have to scroll to view the entire contents of the page.
- All links are clearly labeled. The user does not have to wonder where a particular link may go.

Anyone who has spent time browsing the Web has discovered that some pages seem to take forever to load, or are cluttered with images or loaded with Java scripts that tie up processing power. Although you may want to make a page splashy with animations, music, or other objects, it is a good idea to keep in mind that users grow impatient quickly and may not hang around long enough to view the page if it takes more than a few seconds to load.

FIGURE 51.27 Final RWA Software home page

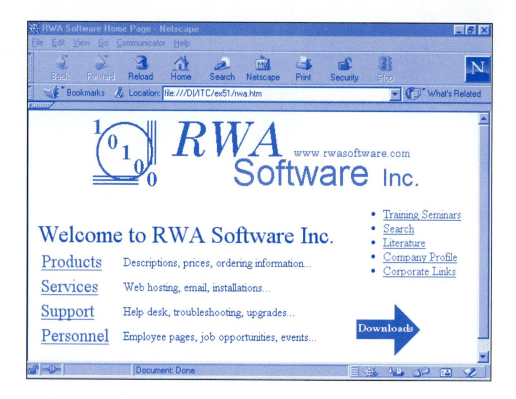

MAINTAINING A WEB PAGE

Once the main body of HTML code is in place for a Web page, making small updates to the page can be easily done using an ordinary text editor. Figure 51.28 shows the RWA Software home page opened using Notepad. One good item to include on a Web page is a message indicating when the page was last updated. Adding this message to the RWA Software home page can be done inside Notepad by left-clicking on the desired insertion point and adding a statement similar to the following:

```
<P ALIGN="CENTER"><I>Last updated on 5/15/1999</I></P>
```

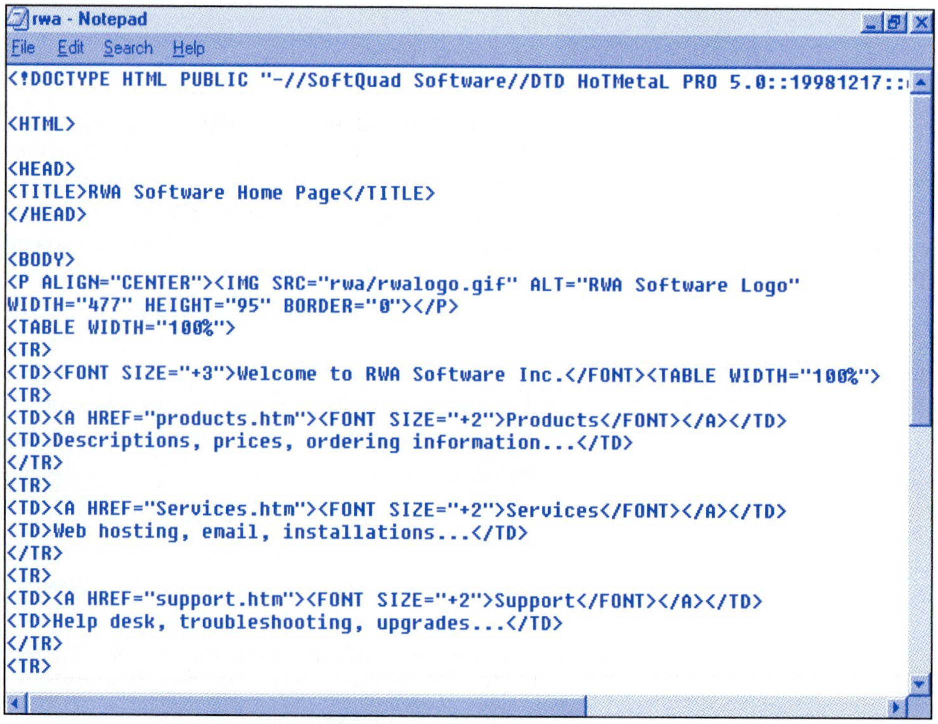

FIGURE 51.28 Viewing the HTML source code using Notepad

Once the page is saved on the Web server, the message

Last updated on 5/15/1999

will be displayed, providing the user with an indication of how up-to-date the Web page is.

OTHER SOFTWARE

Many companies produce software used to streamline the development of material (both HTML and graphics) for distribution on the Web. For HTML development, in addition to SoftQuad's HoTMetaL product, other popular HTML products include:

- Microsoft Front Page (www.microsoft.com)
- Adobe Pagemill (www.adobe.com)

Note that some popular word-processing programs such as Microsoft Word and Corel WordPerfect are able to read and save documents in HTML format as well, offering a convenient way to produce Web documents without having to learn a new software package.

For graphics and image processing, in addition to Jasc Software's Paint Shop Pro, other popular graphics and image editing programs include:

- Adobe Photoshop (www.adobe.com)
- Corel Web Graphics (www.corel.com)
- Micrographics Simply3D (www.micrographics.com)

Each of these programs basically perform similar tasks; the difference between them is in the user interface. Fortunately, many of these Web-based software development tools can be evaluated before purchase by simply downloading a demonstration copy from the company Web site.

TROUBLESHOOTING TECHNIQUES

A few tips to keep in mind regarding Web development are as follows:

- The advances in Web technologies progress at an astounding rate. For the Web developer, this means that it is necessary to invest additional time to *try* to keep up with the changes. Several different versions of HTML have been used on the Web. The older software written using the old standards are not able to take advantage of the new features now available. HTML editors typically provide rule checking to make sure the HTML code meets the required standard.
- When a graphical image is added to a Web page, it causes the page to take longer to download. Sometimes just playing with the image type or properties can reduce the number of bytes required to store the image. For example, a JPG image, saved as a GIF, may require less memory without sacrificing much quality. Reducing the color count in an image also saves memory, with only slight changes in coloring in many cases.

SELF-TEST

This self-test is designed to help you check your understanding of the background information presented in this exercise.

True/False

Answer *true* or *false*.

1. Image size (in bytes) is not a factor in Web page design.
2. Only GIF and JPG image files are used in Web pages.
3. HTML stands for hypertext transfer meta language.
4. The <P> tag does not require a matching </P> tag.
5. The border on a table may be invisible.

Multiple Choice

Select the best answer.

6. Links are also called
   a. Anchors.
   b. Jump points.
   c. Targets.
7. The size (in bytes) of an image depends on
   a. The number of pixels.
   b. The number of colors.
   c. Both a and b.
8. HoTMetaL's graphical interface is
   a. WDSIWDS.
   b. WFDFAW.
   c. WYSIWYG.
9. The <TITLE> of a Web page is used by
   a. Search engines.
   b. E-mail applications.
   c. HTML editors.
10. Nbsp stands for
    a. New binary space partition.
    b. Nonbreaking space.
    c. Negative breaking space.

Completion

Fill in the blank or blanks with the best answers.

11. A bulleted list is also called a(n) _____ list.
12. A GIF image is limited to _____ colors.
13. During the planning stage, a _____ is made of the new Web page.
14. Image colors are stored in a _____.
15. HTML source code is stored in a file whose extension is _____ or _____.

FAMILIARIZATION ACTIVITY

1. Draw a sketch of a home page layout for a company that sells computer supplies.
2. Develop several images to use on the computer supply home page. In addition, browse the Web for related images and save one to edit and use on the home page.
3. Create the computer supply home page.

QUESTIONS/ACTIVITIES

1. Visit several Web sites from a variety of subjects: business, technology, education, and entertainment. Compare the styles and load times of the pages.
2. Visit a local ISP (or the computer center at your school) and find out what is required to have them host your Web page.

REVIEW QUIZ

Within 15 minutes and with 100% accuracy,

1. Discuss the basic concepts of Web development.
2. Use an HTML editor and a text editor to create an HTML document.
3. Describe the two graphical image formats that are used on the Web.
4. Develop a new image from scratch.

52 Science and Technology

INTRODUCTION

Joe Tekk was listening to a voice-mail message from Don, his manager. Don was describing the slow network performance that a customer reported, asking Joe for help. Joe packed his laptop and drove to the customer's office. After connecting his laptop to the customer's network, Joe ran a protocol analyzer to see what kind of traffic existed on the network. The protocol analyzer indicated a significant amount of broadcast traffic coming from a print server. Joe discovered that the print server was a new machine, just installed on the network, and improperly configured.

Joe showed the customer how the protocol analyzer worked and explained how to download a trial version.

PERFORMANCE OBJECTIVES

Upon completion of this exercise, within 15 minutes and with 100% accuracy, you will be able to

1. Capture and display packets using LanExplorer.
2. Set up and simulate a simple circuit using Electronics Workbench.
3. Summarize the various science- and technology-related applications available.

BACKGROUND INFORMATION

In this exercise we examine several science- and technology-related applications that are useful in many different educational settings (as well as practical applications). The applications included are not meant to be exhaustive. Only a few examples of popular, useful applications are presented.

LANEXPLORER

Computer networks play a large role in society today. Often it is necessary for a network engineer or a technology student to troubleshoot problems that arise on the network using a software tool called a protocol analyzer. Using a protocol analyzer, the network interface card in a computer is put into promiscuous mode, allowing it to see all the traffic that is transmitted on the local network. This potentially causes a network security risk because it is possible to capture data that is considered to be confidential, such as passwords, social security numbers, and salaries. Therefore, extreme caution and good judgment should be exercised when using a protocol analyzer.

FIGURE 52.1 Initial LanExplorer window is displayed

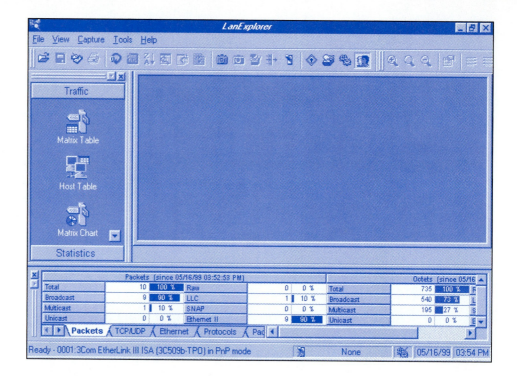

Traffic Monitoring

A typical use for a protocol analyzer is to collect baseline network traffic data. The baseline historical data is then used to compare against network data collected at a different time. The baseline data can be used as an early warning detection system because it is possible to identify several possible harmful situations. These situations include identification of network capacity issues, DHCP errors, duplicate IP assignments, and network utilization trends. In some situations, it is possible to identify a piece of faulty network equipment that has been causing excessive collisions. The process of investigating a network begins with monitoring all the network traffic to develop the baseline data. The baseline may be taken over the course of several hours, days, or weeks using a product such as the Intellimax Systems LanExplorer (www.intellimax.com).

LanExplorer is one of the most popular protocol analyzer packages available. Note that a demonstration copy of LanExplorer is included on the companion CD. The first time LanExplorer is started, it is necessary to select the default network adapter. Since there is usually only *one* network card in a PC, there is only *one* choice available. When more than one network card is present, one of them must be selected. The first screen displayed by LanExplorer after selecting a default adapter is shown in Figure 52.1. The LanExplorer screen is divided into three areas. First is the Task Panel, which contains two items, the Traffic option and the Statistics option. The Traffic option allows for quick access to built-in displays. The Statistics option provides access to Distribution and Rate information, which is automatically set. Selections that can be made in the Traffic Task Panel are displays of the Matrix Table, Host Table, Matrix Chart, Host Chart, and Alarm Log. As the items are selected, the corresponding data is displayed in the panel to the right of the Task Panel. At the bottom of the LanExplorer display is the Console Panel, which is used to display a breakdown of the monitored network traffic. Note that as soon as LanExplorer is started, it begins to monitor all the traffic that is present on the network. Let us begin by discussing the various traffic display options.

When the Matrix Table display option is selected by left-clicking on the icon, the panel on the right displays a line of data for each packet that is examined. The information available for examination that is displayed on this screen include:

- Address (host name or host number)
- Octet ratio (bytes of data)
- Total octets

- Total packets
- Duration of the network activity
- Octets in, packets in
- Octets out, packets out
- Broadcast and multicast message counts
- IP packet type
- Time stamp information

Figure 52.2 shows a typical Matrix display containing Internet Protocol information. By using the scroll bars, all of the various data elements may be examined.

Moving down the list of Traffic Task Panel options, the Host Table is displayed, as in Figure 52.3. As the traffic is monitored, statistics are gathered for each of the hosts that is transmitting data on the network. The graphical bar chart displayed in Figure 52.3 indicates the hosts with the highest ratio of traffic.

The Traffic Matrix Chart shows the same information that was listed in the table format, but now is displayed graphically using a pie chart as illustrated in Figure 52.4. The graphical display shows a breakdown of the traffic data, by both a count and a percentage. The

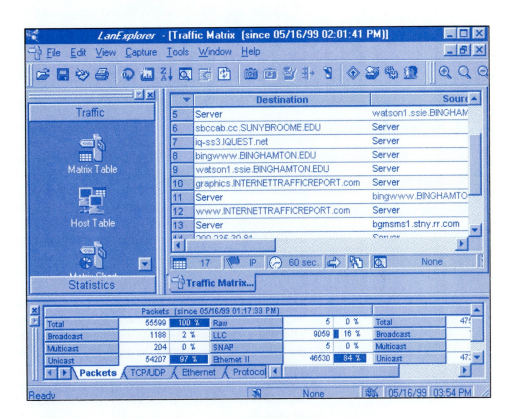

FIGURE 52.2 LanExplorer Traffic Matrix display

FIGURE 52.3 LanExplorer Host Table display

FIGURE 52.4 LanExplorer Traffic Matrix Chart

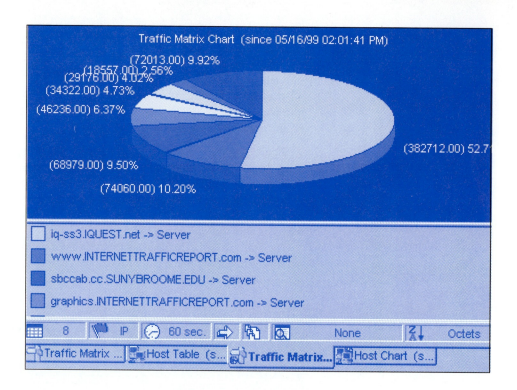

name of the hosts with the highest traffic is displayed in ascending order as space permits. As the traffic patterns change, the pie chart is automatically updated.

The remaining item in the Traffic Task panel is the Alarm Log. The Alarm Log is used to keep track of all security-related issues such as plain-text password transmissions, DHCP address issues, duplicate IP network assignments, and many other important items. You are encouraged to check the items on the Alarm Log frequently because many potential security issues may be identified.

The LanExplorer Console window located at the bottom of the display contains information about all the traffic being monitored. Figure 52.5 shows the various information displayed in the Console window. The first tab contains information about the Packet Statistics, which is displayed in Figure 52.5(a). The total number of packets as well as the packet type is identified. Figure 52.5(b) shows a breakdown of the TCP/UDP Internet Protocol packets. Figure 52.5(c) shows all of the Ethernet statistics. Figure 52.5(d) shows what type of protocols have been monitored on the network, and Figure 52.5(e) completes the list of available information by providing a breakdown of the various sizes of the packets that have been transmitted.

Each of the categories presented in the Console window display can be very important when developing a baseline of activity for a network or when performing troubleshooting.

Packet Capture

Aside from the process of passively monitoring the network traffic, LanExplorer can also capture network traffic. Simply by selecting the Start option from the Capture pull-down menu or using the toolbar icon, LanExplorer will keep a copy of each packet of information transmitted. When the capture process is "in progress," the screen shown in Figure 52.6 is displayed. The number of packets captured, number of octets captured, packets seen, octets seen, elapsed time, filter information, and buffer usage is all displayed in real time.

When the buffer is full, or the Stop Capture command is executed, the user is presented with a screen similar to Figure 52.7. Although it looks very much like the Traffic Matrix display discussed earlier, packets that are captured may be decoded. To decode a packet, the program user simply double-clicks the specific line in the display. This causes the Protocol Decode window to automatically be displayed as shown in Figure 52.8. The Protocol Decode window contains two areas. First, the raw data window is displayed at the top of the screen. Second, the protocol specific breakdown is displayed at the bottom. The

(a) Packet statistics

	Packets (since 05/16/99 01:17:33 PM)						Octets (since 05/16		
Total	56768	100 %	Raw	5	0 %	Total	47625478	100 %	
Broadcast	1250	2 %	LLC	9070	16 %	Broadcast	124077	0 %	
Multicast	209	0 %	SNAP	5	0 %	Multicast	30188	0 %	
Unicast	55309	97 %	Ethernet II	47688	84 %	Unicast	47471213	100 %	

Packets / TCP/UDP / Ethernet / Protocols / Pac

(b) TCP/UDP statistics

	Packets (since 05/16/99 01:17:33 PM)							
FTP	37988	67 %	NNTP	0	0 %	FTP		3
Telnet	0	0 %	NetBIOS	816	1 %	Telnet		
SMTP/POP3/IMAP4	0	0 %	SNMP	0	0 %	SMTP/POP3/IMAP4		
HTTP(S)	6874	12 %	Others	660	1 %	HTTP(S)		

Packets / TCP/UDP / Ethernet / Protocols / Pac

(c) Ethernet statistics

	Transmit (since 05/16/99 01:17:33 PM)					Receive (since 05/16/99 01:17:33 PM)		
OK	4230	Error	1	OK	52244	Error		1
1 Collision	22	Collision	0			No Buffer		1
1+ Collision	28	Late Collision	0			CRC		0
Deferral	611	Underrun	1			Alignment		0

Packets / TCP/UDP / **Ethernet** / Protocols / Pac

(d) Protocol statistics

	Packets (since 05/16/99 01:17:33 PM)						Octets (since 05/16/99 01:17:3		
NetBIOS	8887	15 %	AppleTalk	0	0 %	NetBIOS	7834077	16 %	AppleTalk
IP	51756	85 %	SNA	6	0 %	IP	40110014	84 %	SNA
IPX	237	0 %	Vines	0	0 %	IPX	34478	0 %	Vines
XNS	0	0 %	DEC	0	0 %	XNS	0	0 %	DEC

Packets / TCP/UDP / Ethernet / **Protocols** / Pac

(e) Packet size statistics

	Packet Size (since 05/16/99 01:17:33 PM)					
64	19567	32 %	256-511	2509	4 %	
65-127	6949	11 %	512-1023	1505	2 %	
128-255	1102	2 %	1024-1518	29450	45 %	

TCP/UDP / Ethernet / Protocols / **Packet Size**

FIGURE 52.5 LanExplorer Console window display information

FIGURE 52.6 LanExplorer packet capture status display

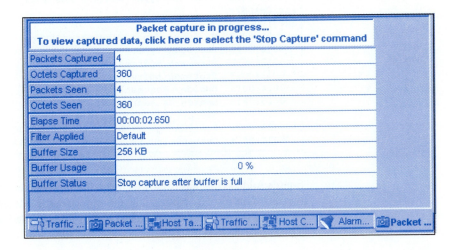

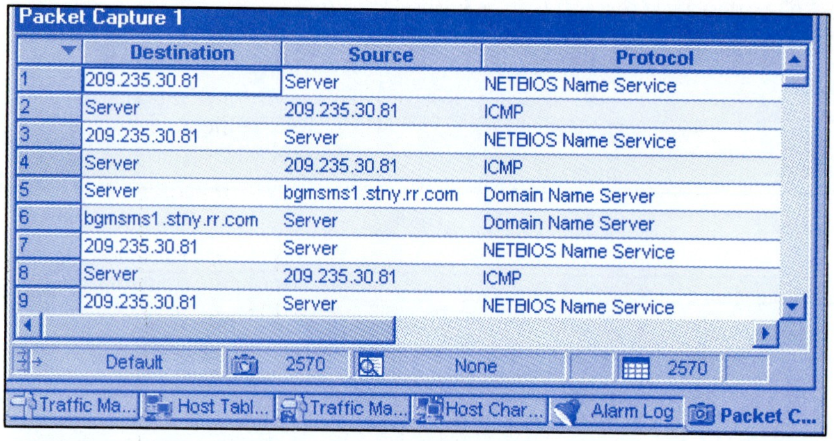

FIGURE 52.7 LanExplorer captured packet display

FIGURE 52.8 LanExplorer Protocol Decode display window

FIGURE 52.9 Decoded LanExplorer packet saved in text format

```
Capture 1:Packet 88

Destination Source       Protocol      Summary  Size      Time Tick
--------------------------------------------------------------------------
24.95.142.229    24.95.142.1    ARP   Request    60    05/16/99 15:27:33.767

Addr.  Hex. Data                                          ASCII
0000:  FF FF FF FF FF FF 08 00 3E 02 07 8D 08 06 00 01    ........>.......
0010:  08 00 06 04 00 01 08 00 3E 02 07 8D 18 5F 8E 01    ........>...._..
0020:  00 00 00 00 00 00 18 5F 8E E5 30 38 02 01 00 04    ......_..08....
0030:  18 5F 8E F8 03 03 A9 90 00 00 00 00                ._..........

802.3 [0000:000D]
  0000:0005  Destination Address: FFFFFFFFFFFF (Broadcast)
  0006:000B  Source Address: 08003E02078D
  000C:000D  Ethernet Type: Address Resolution Protocol (ARP) (for IP and
for CHAOS)
ARP [000E:0029]
  000E:000F  Hardware Type: Ethernet (10Mbps)
  0010:0011  Protocol Type: DOD Internet Protocol (IP)
  0012:0012  Hardware Address Length: 6
  0013:0013  Protocol Address Length: 4
  0014:0015  Opcode: Request
  0016:001B  Source HW Address: 08003E02078D
  001C:001F  Source IP Address: 24.95.142.1
  0020:0025  Destination HW Address: 000000000000
  0026:0029  Destination IP Address: 24.95.142.229
```

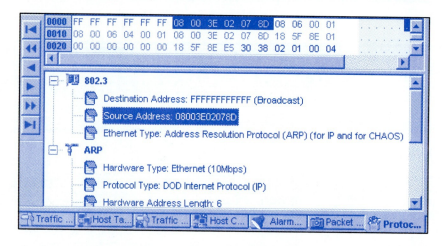

Source Address that has been selected in the Protocol area has a corresponding selection in the raw data display. Notice that the same source address is highlighted in both areas. Choosing the Save option creates a copy of this decoded information in a text file. Figure 52.9 shows the data that is written to the text file.

Packet Capture Filters

Many times when monitoring or troubleshooting a network, it is necessary to monitor only a small portion of the data being transmitted. For example, LanExplorer can be used to look for network traffic from a specific host computer or look for a specific type of protocol. This is accomplished by setting up a packet capture *filter*. Packet filter selections are made from either the network layer or address. Figure 52.10(a) shows the check box options available when selecting Layer 3+ IP/ARP and TCP/UDP categories. Each item that contains a check mark is identified for capture. Figure 52.10(b) shows the Address packet capture filter. Notice that the address may be a MAC address or an IP address. Addresses in the Known Addresses box may be dragged down to the Address Filter list as required. The Filter Mode is used to specify whether the addresses listed in the address filter list are inclusive or exclusive, allowing for unlimited filter choices.

By using a filter, much of the networking traffic is eliminated from the buffer, thereby saving only the traffic that is desired. Note that it is also possible to set a trigger event, which will cause LanExplorer to begin the packet capture. This helps guarantee that the

FIGURE 52.10 Packet Capture Filter configuration options

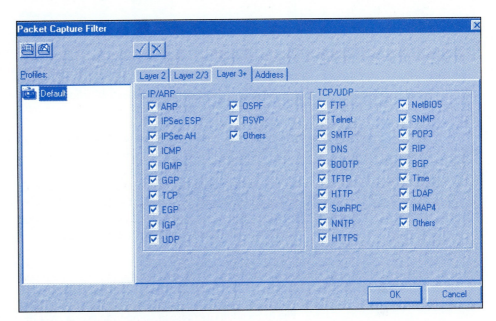

(a) Layer 3+ filter selection options

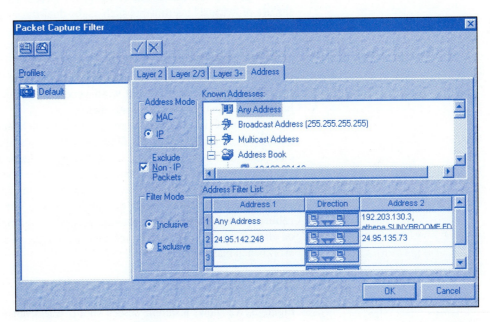

(b) Address filter selection options

FIGURE 52.11 Packet Capture Trigger events

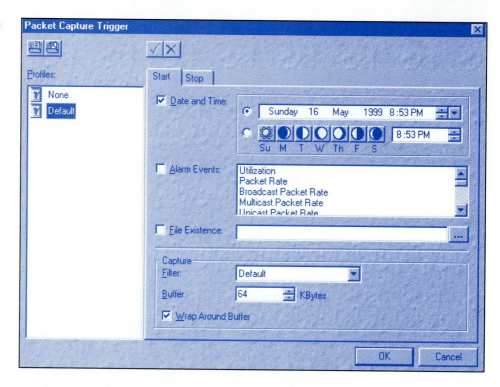

network traffic that is captured follows the triggering event. This makes investigating network problems much easier by isolating the information. Notice from Figure 52.11 that it is possible to start or stop capturing network traffic using trigger events based on a date and/or time, a specific network event, or by the existence of a specific file.

Rather than pay thousands of dollars for a hardware-based protocol analyzer, spend a fraction of the amount on a software-based protocol analyzer such as LanExplorer. Your networking knowledge and experience will increase rapidly.

ELECTRONICS WORKBENCH

The tremendous increase in computing power over the past several years has paved the way for computer-based simulation of electronic circuitry (both analog and digital), including the use of "virtual" instruments such as the dual-trace oscilloscope, Bode plotter, function generator, and power supply. Electronics Workbench (www.interactiv.com) provides these capabilities, and many more, with an easy-to-learn-and-use environment.

Nothing can replace the hands-on experience of actually setting up a real circuit on a lab bench. But learning how to use the virtual instruments of Electronics Workbench, including its logic analyzer and word generator, requires the same correct thinking and handiwork as in using the real instruments.

Furthermore, instructors can create a circuit with a fault introduced into it for students to analyze. The exact nature of the fault can be hidden and could be anything you might typically find in a circuit, such as a shorted or open component or faulty logic gate. Students can gain troubleshooting experience by searching for the fault and determining a repair.

Using a computer to run lab experiments has many advantages. It costs less to purchase one computer than a full bench of equipment. The virtual instruments will not go out of calibration or blow up if connected improperly. A complicated circuit setup can be saved on disk. These are just a few of the advantages of computer simulation. Electronics Workbench makes it all possible.

The Main Window

Figure 52.12 shows a sample screen shot of Electronics Workbench in operation. A two-resistor series circuit is being simulated. The ammeter and voltmeters indicate that the

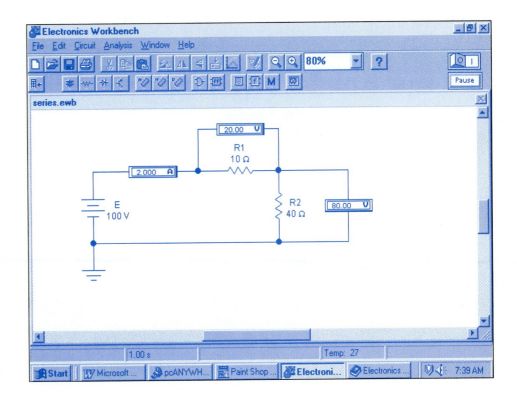

FIGURE 52.12 Two-resistor series circuit simulation

current is 2 amperes and the voltage drops across the resistors are 20 V and 80 V (which add up to 100 V, as they should).

This workspace is where all your work is performed. Circuits are constructed by dragging components from the selected parts bin into the workspace. Instruments are selected by dragging them "off the shelf" and into the workspace. Components are wired together by dragging wires from one component lead to another. Entire circuits can be loaded or saved using the File menu. Both DC and AC analyses are performed. Analog and digital circuits may be mixed together. The pull-down menus, types of instruments, and the contents of each parts bin are detailed in the next three sections.

The Pull-Down Menus

Electronics Workbench has six pull-down menus: File, Edit, Circuit, Analysis, Window, and Help. Left-click to select a menu, or press Alt followed by the underlined letter in the menu name. Left-click on the desired option to choose it. Left-click anywhere in the circuit window to close the menu without choosing an option.

The Instruments

Electronics Workbench provides seven virtual instruments for making measurements and generating signals. These instruments are the multimeter, function generator, oscilloscope, Bode plotter, word generator, logic analyzer, and logic converter. To use an instrument, drag it from the "shelf" into the workspace. Then connect to the instrument terminals as you do any other connection. To see the instrument's details (controls, displays, connector names), double-click on the instrument to open it. To remove an instrument from the workspace, simply delete it.

Sample Circuits

Several sample circuits are provided so that you can begin using Electronics Workbench as soon as it is installed. Figure 52.13 shows the Open Circuit File dialog with all the sample circuit files listed. Double-clicking on the 2M-OSCIL.EWB file brings up the workspace shown in Figure 52.14. The Description window explains what the circuit does. The schematic indicates that the oscilloscope is used to view the output waveform. Double-click the oscilloscope icon to see the full-sized instrument panel.

FIGURE 52.13 Sample circuits included with Electronics Workbench

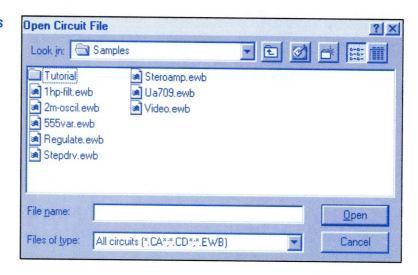

FIGURE 52.14 Sample circuit file (Colpitts oscillator)

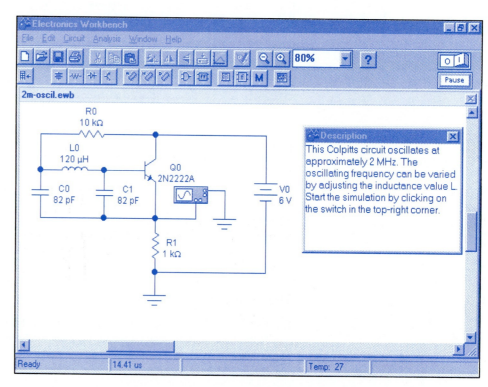

FIGURE 52.15 Oscilloscope display of Colpitts oscillator output

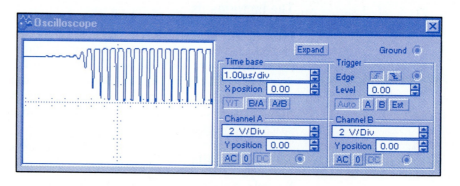

To simulate the circuit, left-click the power switch located at the top right corner (above Pause). The oscilloscope display will look similar to that shown in Figure 52.15. Note that the values indicated on the oscilloscope controls allow us to make accurate measurements of time and voltage.

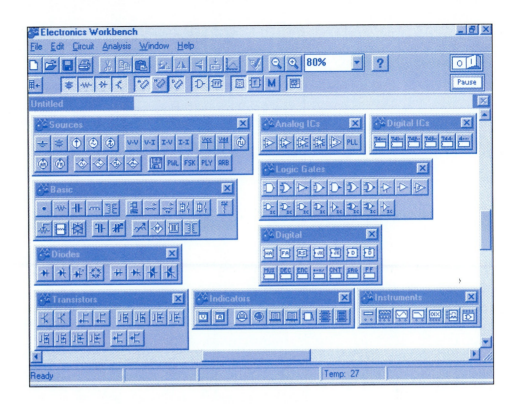

FIGURE 52.16 Many of the components and instruments available in Electronics Workbench

The Parts Bins

If you are not going to use an existing circuit, then you must design one yourself and save it. The components you require in your circuit must come from the *parts bins*. Electronics Workbench provides 12 different parts bins for building your circuits (as shown in Figure 52.16). In order, they are: Sources, Basic, Diodes, Transistors, Analog ICs, Mixed ICs, Digital ICs, Logic Gates, Digital, Indicators, Controls, and Miscellaneous. To select a bin, click on it once. To move a part into the workspace, drag it from the parts bin to the workspace. To delete a part, click on it once and press Delete.

Spend some time experimenting with the demonstration version of Electronics Workbench included on the companion CD. You will find that complex circuits are easy to set up and analyze.

MATLAB

MATLAB (www.mathworks.com) stands for Mathematics Laboratory, a powerful graphing and matrix algebra tool. Figure 52.17 shows the initial MATLAB window. The user enters commands at the >> prompt. To see a demonstration of what MATLAB can do, just type "demo" and press Enter.

Meta Files

Calculations and other operations can be performed through the command prompt, or they can be saved in a MATLAB *meta* file. An example of a simple meta file is PEAKS50.M, which contains the following statements:

```
% Peaks50.m: Graph of peaks function

z = peaks(50);surf(z);
colormap(hsv);
title(`Graph of peaks(50)
using hsv colormap');
```

The statements in a meta file are processed whenever the name of the meta file is entered at the command prompt. The meta file must be in the current directory (typically \work in the

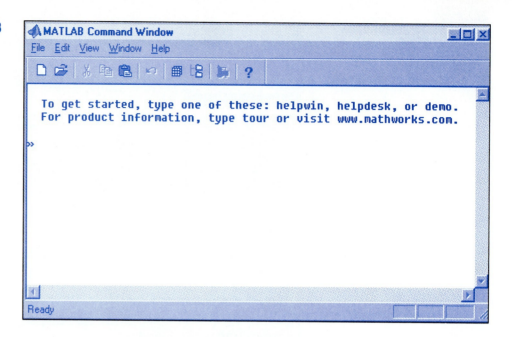

FIGURE 52.17 Initial MATLAB window

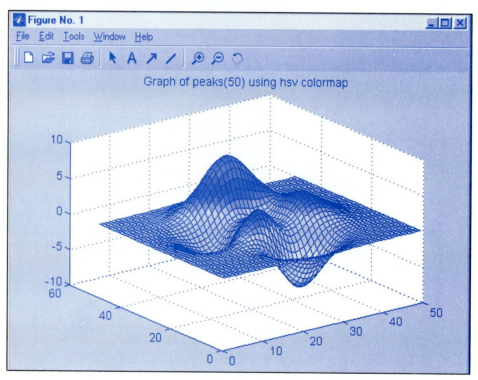

FIGURE 52.18 Graphing the built-in peaks function

main MATLAB directory). Keying "peaks50" and pressing Enter will produce the graph shown in Figure 52.18.

3-D Graphs

Let us examine each statement in the PEAKS50 meta file. The line beginning with the % symbol is a comment. Next, the statement

```
z = peaks(50);
```

uses the built-in peaks function to generate 50 points of data. Then,

```
surf(z);
```

is used to graph a surface plot of the data in z. The types of graphs available are mesh, surface, contour, quiver, and slice.

FIGURE 52.19 Graph of a complex function

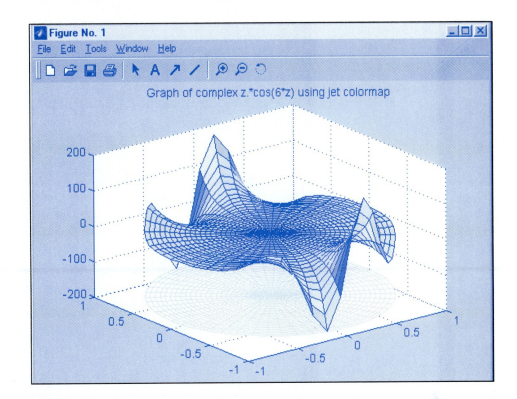

The color scheme used with the graph is chosen from a set of *colormaps*. The statement

```
colormap(hsv);
```

loads the hsv (hue-saturation-value) colormap. Other colormaps are jet, cool, hot, gray, pink, and copper.

Figure 52.19 shows the graph of a complex variable function (real and imaginary components). The corresponding meta file is called COSCURVE.M, and contains these statements:

```
% coscurve.m: Graph of z.*cos(6*z) where z is complex

z = cplxgrid(25);
cplxmap(z, z.*cos(6*z));
colormap(jet);
title('Graph of complex z.*cos(6*z) using jet colormap');
```

The .* operator is used to multiply z and cos(6*z) on an element-by-element basis.

2-D Graphs

When a 3-D graph is not required, there are several 2-D graphing functions available that support line, bar, stair, polar, and stem graphs. One way to graph the data would be to enter all the (X,Y) pairs. Another way would be to generate the Y-axis data using the X-axis values and an equation.

Suppose the equation we need to graph is as follows:

$$P = \left(\frac{80}{5 + R_L}\right)^2 \cdot R_L$$

Simply plugging in all the required values of R_L will allow us to calculate the respective values of P (power). Examine the meta file MAXPOWER.M, which implements the power equation.

```
% Maxpower.m: Plot of load power versus load resistance

xvals = [0,0.5,1,1.5,2,2.5,3,3.5,4,4.5];
xvals = [xvals,5,5.5,6,6.5,7,7.5,8,8.5,9,9.5,10];
```

FIGURE 52.20 Bar graph of maximum power transfer

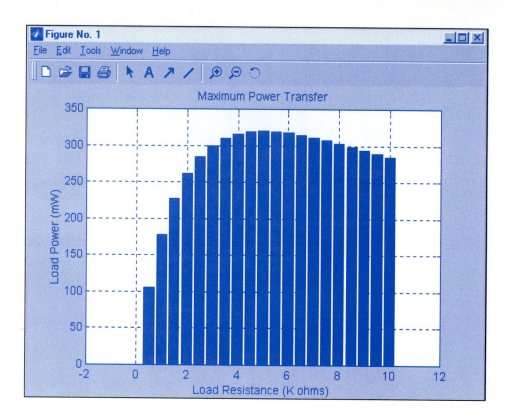

```
tempa = 5 + xvals;
tempb = 80 ./ tempa;
tempb = tempb .^ 2;
yvals = tempb .* xvals;
bar(xvals,yvals);
grid;
title('Maximum Power Transfer');
xlabel('Load Resistance (K ohms)');
ylabel('Load Power (mW)');
```

Since there are so many values of load resistance, the xvals array is built up in separate statements.

The bar graph of the power formula is shown in Figure 52.20.

Matrix Operations

Figure 52.21 shows a number of matrix operations being performed in MATLAB. In this example, a 3-by-3 matrix is entered and inverted. The original matrix is then multiplied by its inverse, producing the identity matrix as expected. Note the use (or lack of) a semicolon at the end of each statement. When the semicolon is not used, the results of the operation are displayed. The elements of the inverse matrix are not displayed because the semicolon is included in the inverse statement.

MATLAB is an easy-to-use but powerful mathematical tool. Visit the MathWorks home page to see what other packages are available for use with MATLAB, such as the Image Processing and Signal Processing Libraries.

VISIO

Visio (www.visio.com) is an electronic drawing application that uses graphical *stencils* similar to the plastic stencils used to trace letters and shapes. Figure 52.22 shows a Visio drawing being developed using the Flowchart stencil. Visio will import files from other drawing applications, such as AutoCAD and Excel, and save files in a variety of formats.

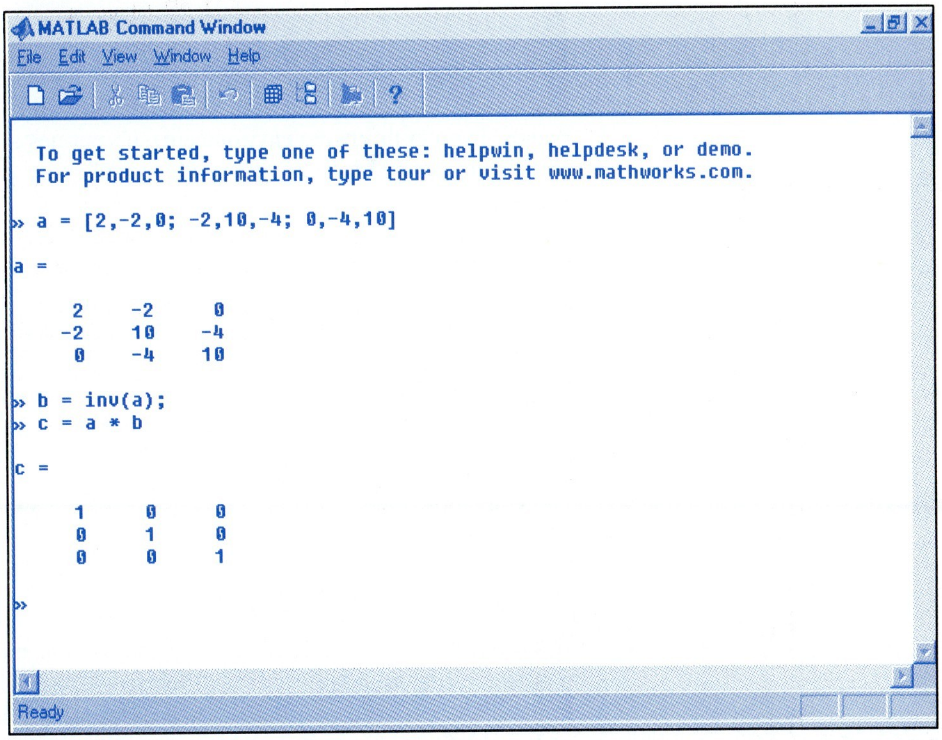

FIGURE 52.21 Performing matrix calculations

FIGURE 52.22 Developing a flowchart using Visio

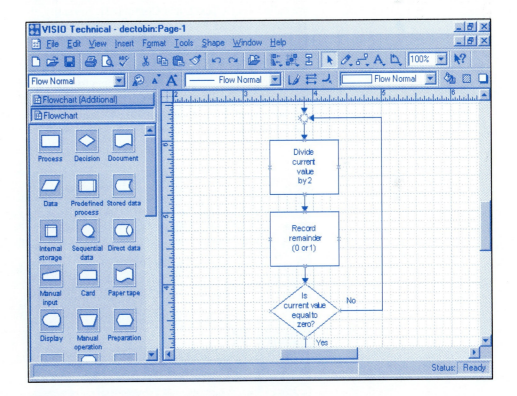

VISUAL BASIC

Visual BASIC is Microsoft's version of the BASIC programming language adapted for use in the Windows environment. Like Visual C/C++, Visual BASIC is also an object-oriented programming language. This requires a change in thinking about how a program is structured, as well as a change in how a program is written.

Procedural Versus Event-Driven

In a procedural programming environment, such as QBASIC, the main program calls subroutines in a particular order and determines when the user is allowed to enter data or commands. In an event-driven environment, subroutines are called as needed to handle *events* that occur during execution, such as the user clicking on a Convert button to perform a conversion.

Event-driven programs utilize *objects,* such as the display window shown during execution (the user interface). The programmer adjusts the properties of the object as needed by the application. For example, some of the properties associated with a window are the name displayed in the title bar, the number and type of menu options, the font style used to display text, and its color.

After setting the properties, the code associated with the object is written.

In summary, the steps in writing a Visual BASIC program are as follows:

1. Design the user interface.
2. Set all properties.
3. Write the event-handling code.

Designing the User Interface

A Visual BASIC application typically uses one or more *forms* to provide the graphical interface to the user. A form is a fancy name for a window displayed during execution. Figure 52.23 shows a simple form in development. The form contains a number of *controls,* predefined objects that add functionality to the form. The three controls in the form are called Text1, Text2, and Command1. In reality, these three names refer to the objects associated with each control.

Setting Properties

The properties of the controls placed on a form (and the form's properties itself), must be adjusted as part of the programming process. In Figure 52.23 some of the Form1 properties are shown. Note in particular the Caption property. This is where the name in the title bar is entered.

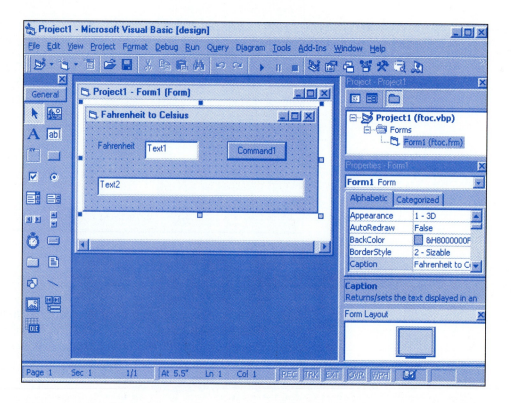

FIGURE 52.23 Visual BASIC form under development

Double-clicking on a control brings up the property list associated with its object. Each control must have its properties adjusted before continuing. For example, the label on the Command1 button will be set to "Convert." This is done by keying "Convert" into the Caption entry in the properties list.

Writing the Code

Each control placed on the form may have one or more subroutines associated with it, depending on the features desired by the programmer. For example, a button control may have event handling subroutines for mouse-click and mouse-over events.

When a control is placed onto a form, a stub procedure is created for it whose name is based on the type of event specified. For example, clicking the Convert button will activate its Command1_Click() procedure. The Visual BASIC code shown here uses the Command1_Click() procedure to convert the Fahrenheit input temperature (Text1.Text) into Celsius.

```
Private Sub Command1_Click()
 Dim Cel As Single
 On Error GoTo Notnum
 Cel = (Text1.Text - 32) * 5 / 9
 Text2.Text = "The Celsius temperature is " + Str(Cel)
 Exit Sub
Notnum:
 If Err.Number = 13 Then
  Text2.Text = "Cannot convert!"
 End If
End Sub
```

Notice that Text2.Text is assigned the value of the result. This causes the output to appear in the Text2 box. Also note that error handling is provided to guarantee that the user enters a valid number in the Text1 box.

The only way that Command1_Click() gets called is if the user clicks on the Convert button. This is the essence of event-driven programming.

Creating an Executable File

The Make option under the File menu is used to make an executable file from your Visual BASIC project. A sample execution of the temperature conversion application (FTOC.EXE on the companion CD) is shown in Figure 52.24. The user enters a Fahrenheit temperature in the Text1 box (a default value of 212 is placed in the box at startup). After clicking Convert, the result, or an error message, is displayed in the Text2 box.

VISUAL C++

Microsoft Visual C++ brings the same event-driven, object-oriented environment found in Visual BASIC to the C++ world. Figure 52.25 shows a program under development. Visual C++ supports forms and will generate 32-bit WIN32 code, rather than a DOS executable.

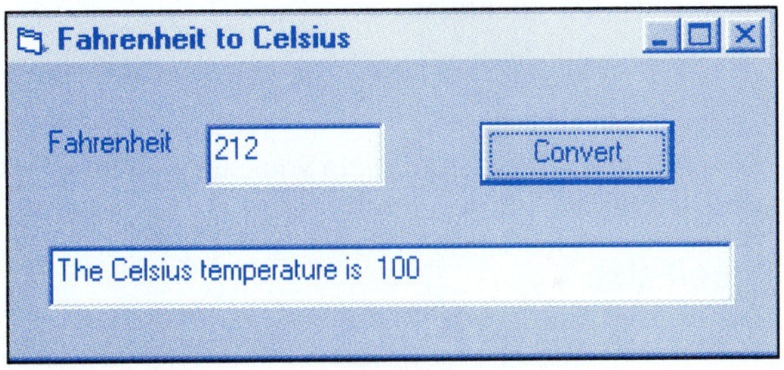

FIGURE 52.24 Sample execution

FIGURE 52.25 Visual C++ program under development

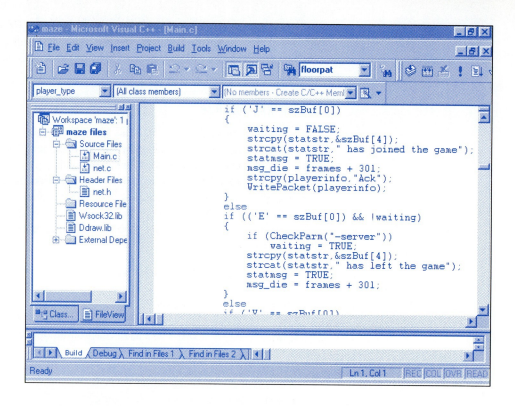

OTHER SOFTWARE

Many of the following additional applications are free, fully working versions that can be downloaded from the indicated sites. Others are commercial applications.

- DJGPP (www.delorie.com): DJGPP is a 32-bit protected-mode C/C++ compiler for DOS applications. Many commercial virtual-reality PC games have been developed using DJGPP. A graphical integrated development environment is available.
- NASM (www.web-sites.co.uk/nasm): NASM is the Netwide Assembler, a free 80x86 assembler for developers using Intel 80x86 assembly language.
- POV-Ray (www.povray.org): POV-Ray is the Persistence of Vision Ray Tracer, a free application for generating photographic-quality images. Scenes are mathematically ray traced, allowing stunning 3-D appearance of complex objects.
- Mathematica (www.wri.com): Excellent package for performing numerical, symbolic, and graphical calculations; solving calculus problems and equations; and differential equations. Both full and student versions are available for purchase.

TROUBLESHOOTING TECHNIQUES

It is important to remember that simulation software is not perfect. There are always some situations where the simulation differs from real life. For example, the oscilloscope waveform in Electronics Workbench does not show the noise associated with real electronic signals.

The benefits of computer simulation, especially with the fast machines available today, far outweigh the slight imperfections. Even so, always attempt to verify the correctness of the results of a simulation.

SELF-TEST

This self-test is designed to help you check your understanding of the background information presented in this exercise.

True/False

Answer *true* or *false*.

1. Duplicate IP assignments are included in LanExplorer's Alarm Log.

2. The LanExplorer Console Panel is where the Host Chart and Matrix Chart are displayed.
3. Electronics Workbench only simulates analog circuits.
4. MATLAB produces only 2-D graphs and charts.
5. Visio uses electronic drawing templates called stencils.

Multiple Choice

Select the best answer.

6. Monitoring network traffic is a performed to:
   a. Develop a baseline of traffic activity.
   b. Check for potential problems on the network.
   c. Both a and b.
7. A trigger is used to:
   a. Filter the packets on the network.
   b. Start and/or stop packet captures.
   c. Generate network traffic.
8. A packet capture filter can be applied to
   a. A specific host computer.
   b. A specific networking protocol.
   c. Both a and b.
9. Which is not a virtual instrument in Electronics Workbench?
   a. Oscilloscope.
   b. Logic Converter.
   c. Logic Manipulator.
10. Visual C++ generates
    a. DOS executables.
    b. WIN32 executables.
    c. Both a and b.

Completion

Fill in the blanks with the best answers.

11. The NIC is put into _____ mode when used by a protocol analyzer.
12. The address specified for a filter may either be a _____ address or an _____ address.
13. The color scheme used for a MATLAB graph is chosen from a set of _____.
14. A Visual BASIC application uses one or more _____ to provide the graphical user interface.
15. To run the demo in MATLAB, enter _____ at the command prompt.

FAMILIARIZATION ACTIVITY

1. Use LanExplorer to study the traffic on the network at your school or place of employment. What is the predominant protocol used? On average, how many bytes/second are transferred over the network?
2. Duplicate the Electronics Workbench circuit shown in Figure 52.26.

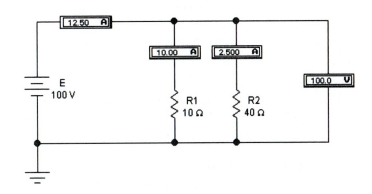

FIGURE 52.26 Two-resistor parallel circuit

QUESTIONS/ACTIVITIES

1. Search the Web for these additional science- and technology-related applications:
   - Chemical modeling (build a molecule, view its 3D model).
   - Physical systems (crash two balls together, see where they go).
   - Printed circuit board design.
   - The solar system (track planets, view constellations).
   - Artificial intelligence.
   - Biological systems (disect a virtual animal, explore a DNA strand).
2. Explain the purpose of a meta file.

REVIEW QUIZ

Within 15 minutes and with 100% accuracy,

1. Capture and display packets using LanExplorer.
2. Set up and simulate a simple circuit using Electronics Workbench.
3. Summarize the various scienc/d technology-related applications available.

UNIT VII Selected Topics

EXERCISE 53 An Introduction to Intel Microprocessor Architecture
EXERCISE 54 Computer Languages
EXERCISE 55 Hardware and Software Interrupts
EXERCISE 56 The Advanced Intel Microprocessors
EXERCISE 57 A Detailed Look at the System BIOS
EXERCISE 58 Windows Internal Architecture
EXERCISE 59 Computer Viruses
EXERCISE 60 A Typical Computer Center

53 An Introduction to Intel Microprocessor Architecture

INTRODUCTION

Joe Tekk was sitting at his desk, drawing on a sheet of paper. Don, his manager, asked him what he was drawing.

Joe replied, "I'm making a diagram of the internal registers in the Pentium processor. I have to meet with a group of Boy Scouts in the networking seminar room in 15 minutes. They are are visiting local high-technology companies to learn about careers."

Don had some free time to spare. "Do you need any help with them?"

"Sure, Don. That would be great."

Don turned to leave. "I'll go get a motherboard to show them." Joe finished sketching the register model. Then he scanned it and saved the scanned image of his register model as a .JPG image file. He copied the image file to his Web directory, and created a Web page for it that displayed the file and a short description of the register model. Don arrived with a motherboard just as Joe was finishing. He looked at what Joe had done in the six minutes he was gone.

"Nice work, Joe. What took you so long?"

PERFORMANCE OBJECTIVES

Upon completion of this exercise, within 15 minutes and with 100% accuracy, you will be able to

1. Outline the software model of the 80x86 microprocessor family.
2. Discuss the various instruction groups and addressing modes that are available.
3. Explain how a 20-bit address is formed from two 16-bit numbers.

BACKGROUND INFORMATION

Beginning with the 8088 and 8086 microprocessors, Intel started a microprocessor family that grew to include the 80186, 80286, 80386, 80486, and finally the new Pentium series. Since all processors in the series run the same basic instructions, they are collectively referred to as the 80x86 family. In this exercise, we examine the real-mode architecture of the 80x86. In Exercise 56 we will look at another mode of operation, called protected mode, where the full 32-bit power of the processor is available.

The introduction of the 8088 and 8086 into the arena of microprocessors came at a time when we were reaching the limits of what an 8-bit machine could do. With their restricted instruction sets and addressing capabilities, it was obvious that something more powerful was needed. The 80x86 contains instructions previously unheard of in 8-bit machines, a very large address space, many different addressing modes, and an architecture that easily lends itself to multiprocessing or multitasking (running many programs simultaneously).

First, let us look at the types of numbers we must work with when we deal with the microprocessor at its own level.

BINARY NUMBERS

Working with microprocessors requires a good grasp of the binary and hexadecimal number systems. Both are related to the number system we use every day, the decimal number system.

A decimal number is composed of one or more digits chosen from a set of 10 digits (0, 1, 2, 3, 4, 5, 6, 7, 8, 9), as shown in Figure 53.1. Each digit in a decimal number has an associated *weight* that is used to give the digit meaning. For example, the decimal number 357 contains three 100s, five 10s, and seven 1s. The weight of the digit 3 is 100, the weight of the digit 5 is 10, and the weight of the digit 7 is 1. The weight of each digit in a decimal number is related to the *base* of the number. Decimal numbers are base-10 numbers. Thus, the weights are all multiples of 10. Look at our example decimal number again:

Digits:	3	5	7
Weights as powers of 10:	10^2	10^1	10^0
Actual weight values:	100	10	1
Components of number:	300	50	7

Notice that the weights are all powers of 10, beginning with 0. The components of the number are found by multiplying each digit value by its respective weight. The number itself is found by adding the individual components. This technique applies to numbers in *any* base.

A **binary number** is a number composed of digits (called **bits**) chosen from a set of only two digits (0, 1), as shown in Figure 53.2. Base 2 is used for binary numbers because there are only two legal digits in a binary number. This means that the weights of the bits in a binary number are all multiples of 2. Consider the binary number 10110. The associated weights are as follows:

Bits:	1	0	1	1	0
Weights as powers of 2:	2^4	2^3	2^2	2^1	2^0
Actual weight values:	16	8	4	2	1
Components of number:	16		4	2	

The components are again found by multiplying each bit in the number by its respective power of 2. The individual components add up to 22. Thus, 10110 binary equals 22 decimal. We now have a technique for determining the decimal value of any binary number.

Going from one base to another requires a *conversion*. As we just saw, going from 10110 to 22 required us to perform a **binary-to-decimal** conversion. How do we go the

FIGURE 53.1 Three-digit decimal number

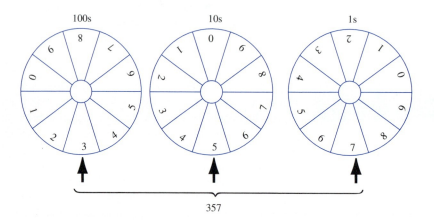

FIGURE 53.2 Five-bit binary number

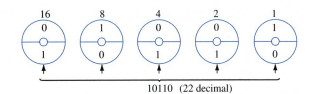

10110 (22 decimal)

other way? For example, what binary number represents the decimal number 37? This requires a **decimal-to-binary** conversion. One way to do this conversion is as follows:

```
37/2 = 18 with 1 left over
18/2 = 9 with 0 left over
 9/2 = 4 with 1 left over
 4/2 = 2 with 0 left over
 2/2 = 1 with 0 left over
 1/2 = 0 with 1 left over
```

The number is repeatedly divided by 2 and the remainder recorded. When we get to "1/2 = 0 with 1 left over," we are done dividing. The binary result is found by reading the remainder bits from the bottom up. So, 37 decimal equals 100101 binary. To check this result, we use the binary-to-decimal conversion technique:

```
 1    0    0    1    0    1
32   16    8    4    2    1
───────────────────────────
32              4         1
```

The sum of 32, 4, and 1 is 37—the original number.

Here are some common binary and decimal numbers:

Binary	Decimal
1010	10
1111	15
1100100	100
10000000	128
11111111	255
1111101000	1000

Clearly, a good understanding of the various powers of 2 is a valuable aid in performing conversions. The first 20 powers of 2 are shown in Table 53.1.

HEXADECIMAL NUMBERS

It is not easy to remember large binary numbers. For instance, examine the 20-bit binary number shown below for five seconds. Then close your eyes and try to repeat it.

10101111011010110011

TABLE 53.1 The first 20 powers of 2

Power	Value	Power	Value
2^0 =	1	2^{10} =	1024
2^1 =	2	2^{11} =	2048
2^2 =	4	2^{12} =	4096
2^3 =	8	2^{13} =	8192
2^4 =	16	2^{14} =	16,384
2^5 =	32	2^{15} =	32,768
2^6 =	64	2^{16} =	65,536
2^7 =	128	2^{17} =	131,072
2^8 =	256	2^{18} =	262,144
2^9 =	512	2^{19} =	524,288

Were you able to do it? Most people cannot, because their short-term memory is not capable of storing so many bits of information. It is possible, however, to remember shorter sequences of characters. Try this character sequence:

AF6B3

Were you able to remember it? Those who have difficulty with the 20-bit example can usually remember this five-character example easily. Here is where the trick comes in: the 20-bit binary number *has the same value* as the five-character sequence. The sequence AF6B3 is a **hexadecimal number** (base 16). This number is derived from the 20-bit binary number as follows:

1. Separate the number into groups of 4 bits each.

    1010 1111 0110 1011 0011

2. Find the individual decimal equivalent of each group.

    1010 1111 0110 1011 0011
     10   15   6   11   3

3. Replace each decimal value from 10 to 15 with the corresponding letter from A to F. Thus, the 10 becomes an A, the 11 becomes a B, and the 15 becomes an F.

    1010 1111 0110 1011 0011
     10   15   6   11   3
     A   F   6   B   3

This technique makes it possible to work with large binary numbers by using their hexadecimal equivalents.

The word *hexadecimal* refers to "6" and "10," or "16." The hexadecimal system contains numbers composed of digits and letters chosen from the set (0, 1, 2, 3, 4, 5, 6, 7, 8, 9, A, B, C, D, E, F). The decimal numbers 10, 11, 12, 13, 14, and 15 are represented in hexadecimal as A, B, C, D, E, and F, respectively. Each digit or letter in a hexadecimal number represents 4 binary bits (as we have seen). The binary patterns associated with the 16 hexadecimal symbols are shown in Table 53.2.

It is much more convenient (and easier on the memory) to use two hexadecimal symbols than to use 8 binary bits. For instance, 3EH (the H stands for Hex) means 00111110B (B for Binary). Larger binary numbers prove this point even better (as our AF6B3H example showed). Since microprocessor address and data lines commonly use 8, 16, or even 32 bits of data, the two-, four-, or eight-symbol hexadecimal equivalents are easier to deal with.

This brief discussion should have familiarized you with the types of numbers that we encounter when dealing with microprocessors. Now let's see what microprocessors do with them.

THE REAL-MODE SOFTWARE MODEL OF THE 80x86

The 80x86 microprocessor (operating in real mode) contains four data registers, referred to as AX, BX, CX, and DX. All are 16 bits wide, and may be split up into two halves of 8 bits

TABLE 53.2 Binary equivalents of hexadecimal symbols

Hex	Binary	Hex	Binary
0	0 0 0 0	8	1 0 0 0
1	0 0 0 1	9	1 0 0 1
2	0 0 1 0	A	1 0 1 0
3	0 0 1 1	B	1 0 1 1
4	0 1 0 0	C	1 1 0 0
5	0 1 0 1	D	1 1 0 1
6	0 1 1 0	E	1 1 1 0
7	0 1 1 1	F	1 1 1 1

FIGURE 53.3 Real-mode software model of the 80x86

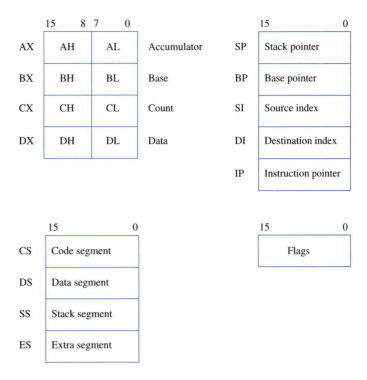

each. Figure 53.3 shows how each half-register is referred to by the programmer. Five other 16-bit registers are available for use as pointer or index registers. These registers are the *stack pointer* (SP), *base pointer* (BP), *source index* (SI), *destination index* (DI), and *instruction pointer* (IP). None of the five may be divided up in a manner similar to the data registers. AX, BX, CX, DX, BP, SI, and DI are referred to as **general-purpose registers.**

A major difference between the 80x86 and many other CPUs on the market has to do with the next group of registers, the *segment registers.* Four segment registers are used by the processor to control all accesses to memory and I/O, and must be maintained by the programmer. The *code segment* (CS) is used during instruction fetches, the *data segment* (DS) is most often used by default when reading or writing data, the *stack segment* (SS) is used during stack operations such as subroutine calls and returns, and the *extra segment* (ES) is used for anything the programmer wishes. All segment registers are 16 bits long.

Finally, a 16-bit *flag register* is used to indicate the results of arithmetic and logical instructions. Included are zero, parity, sign, and carry flags, plus a few others. Together, these 14 registers make an impressive set.

80x86 PROCESSOR REGISTERS

The 16-bit registers just introduced in Figure 53.3 are combined in an interesting way to form the necessary 20-bit address required to access memory. If you recall, there were no 20-bit registers shown in the software model. How then does the 80x86 generate 20-bit real-mode addresses?

Segment Registers

The four segment registers, CS, DS, SS, and ES, are all 16-bit registers that are controlled by the programmer. A segment, as defined by Intel, is a 64KB block of memory, starting on any 16-byte boundary. Thus, 00000, 00010, 00020, 20000, 8CE90, and E0840 are examples of valid segment addresses. The information contained in a segment register is combined with the address contained in another 16-bit register to form the required 20-bit address. Figure 53.4 shows how this is accomplished. In this example, the code segment register contains A000 and the instruction pointer contains 5F00. The 80x86 forms the 20-bit address A5F00 in the following way: first, the data in the code segment register is shifted 4 bits to the left. This has the effect of turning A000 into A0000. Then the contents

FIGURE 53.4 Generating a 20-bit address in the 80x86

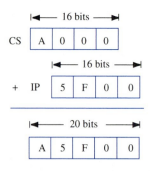

of the instruction pointer are added, yielding A5F00. All external addresses are formed in a similar manner, with one of the four segments used in each case. Each segment register has a default usage. The 80x86 knows which segment register to use to form an address for a particular application (instruction fetch, stack operation, and so on). The 80x86 also allows the programmer to specify a different segment register when generating some addresses.

General-Purpose Registers

The seven general-purpose registers available to the programmer (AX, BX, CX, DX, BP, SI, and DI) can be used in many different ways; they also have some specific roles assigned to them. For instance, the accumulator (AX) is used in multiply and divide operations and in instructions that access I/O ports. The count register (CX) is used as a counter in loop operations, providing up to 65,536 passes through a loop before termination. The lower half of CX, the 8-bit CL register, is also used as a counter in shift/rotate operations. Data register DX is used in multiply and divide operations and as a pointer when accessing I/O ports. The last two registers are the source index and destination index (referred to as SI and DI, respectively). These registers are used as pointers in string operations.

Even though these registers have specific uses, they may be used in many other ways simply as general-purpose registers, allowing for many different 16-bit operations.

Flag Register

Figure 53.5 shows the nine flag assignments within the 16-bit flag register. The flags are divided into two groups: **control flags** and **status flags.** The control flags are IF (*interrupt enable flag*), DF (*direction flag*), and TF (*trap flag*). The status flags are CF (*carry flag*), PF (*parity flag*), AF (*auxiliary carry flag*), ZF (*zero flag*), SF (*sign flag*), and OF (*overflow flag*). Most of the instructions that require the use of the ALU affect these flags.

80x86 DATA ORGANIZATION

The 80x86 microprocessor has the capability of performing operations on many different types of data. In this section, we will examine what some of the more common data types are and how they are represented and used by the processor. The 80x86 contains instructions that directly manipulate single bits and other instructions that use 8-, 16-, and even 32-bit numbers. By common practice, 8-bit binary numbers are referred to as **bytes.** Processor register halves AL, BH, and CL are examples of where bytes might be stored and utilized.

Sixteen-bit numbers are known as **words** and may require an entire processor register for storage. Registers DX, BP, and SP are used to hold word data types. In register DX, DH contains the upper 8 bits of the number and DL holds the lower 8 bits.

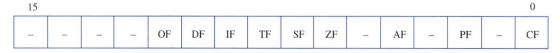

FIGURE 53.5 Real-mode 80x86 flag register

FIGURE 53.6 Word storage using Intel byte-swapping

Some instructions (particularly multiply and divide) allow the use of 32-bit numbers. These data types are called **double words** (or long words). In this case, the 32-bit number is stored in registers DX and AX, with DX holding the upper 16 bits of the number.

It is important to keep track of the data type being used in an instruction, because incorrect or undefined types may lead to incorrect program assembly or execution.

One of the differences between the Intel line of microprocessors and those made by other manufacturers is Intel's way of storing 16-bit numbers in memory. A method that began with the 8080 and has been used on all upgrades, from the 8085 to the Pentium, is a technique called **byte-swapping.** This technique is sometimes confusing for those unfamiliar with it, but it becomes clear after a little exposure. When a 16-bit number must be written into the system's byte-wide memory, the low-order 8 bits are written into the first memory location and the high-order 8 bits are written into the second location. Figure 53.6 shows how the 2 bytes that make up the 16-bit hexadecimal number 2055 are written into locations 18000 and 18001, with the low-order 8 bits (55) going into the first location (18000). This is what is known as byte-swapping. The lower byte is always written first, followed by the high byte. This method of storing numbers is also called **little-endian** format.

Reading the 16-bit number out of memory is performed automatically by the processor with the aid of certain instructions. The 80x86 knows that it is reading the lower byte first and puts it in the correct place. Programmers who manipulate data in memory must remember to use the proper practice of byte-swapping or discover that their programs do not give the correct, or expected, results.

80x86 INSTRUCTION TYPES

The 80x86 instruction set is composed of six main groups of instructions. Examining the instructions briefly here will provide a good overall picture of the capabilities of this processor.

Data Transfer Instructions

Fourteen data transfer instructions are used to move data among registers, memory, and the outside world. Also, some instructions directly manipulate the stack, whereas others may be used to alter the flags.

The data transfer instructions are:

MOV	Move byte or word (2 bytes) to register/memory
PUSH	Push word onto stack
POP	Pop word off stack
XCHG	Exchange byte or word
XLAT	Translate byte
IN	Input byte or word from port
OUT	Output byte or word to port
LEA	Load effective address
LDS	Load pointer using data segment
LES	Load pointer using extra segment
LAHF	Load AH from flags
SAHF	Store AH in flags
PUSHF	Push flags onto stack
POPF	Pop flags off stack

Arithmetic Instructions

Twenty instructions make up the arithmetic group. Byte and word operations are available on almost all instructions. A nice addition are the instructions that multiply and divide. Previous microprocessors did not include these instructions, forcing the programmer to write subroutines to perform multiplication and division when needed. Addition and subtraction of both binary and binary-coded decimal (BCD) operands are also allowed.

The arithmetic instructions are:

ADD	Add byte or word
ADC	Add byte or word plus carry
INC	Increment byte or word by 1
AAA	ASCII adjust for addition
DAA	Decimal adjust for addition
SUB	Subtract byte or word
SBB	Subtract byte or word and carry
DEC	Decrement byte or word by 1
NEG	Negate byte or word
CMP	Compare byte or word
AAS	ASCII adjust for subtraction
DAS	Decimal adjust for subtraction
MUL	Multiply byte or word (unsigned)
IMUL	Integer multiply byte or word
AAM	ASCII adjust for multiply
DIV	Divide byte or word (unsigned)
IDIV	Integer divide byte or word
AAD	ASCII adjust for division
CBW	Convert byte to word
CWD	Convert word to double word

Bit Manipulation Instructions

Thirteen instructions capable of performing logical, shift, and rotate operations are contained in this group. Many common Boolean operations (AND, OR, NOT) are available in the logical instructions. These, as well as the shift and rotate instructions, operate on bytes or words. No single-bit operations are available.

The bit manipulation instructions are:

NOT	Logical NOT of byte or word
AND	Logical AND of byte or word
OR	Logical OR of byte or word
XOR	Logical Exclusive-OR of byte or word
TEST	Test byte or word
SHL	Logical shift left byte or word
SAL	Arithmetic shift left byte or word
SHR	Logical shift right byte or word
SAR	Arithmetic shift right byte or word
ROL	Rotate left byte or word
ROR	Rotate right byte or word
RCL	Rotate left through carry byte or word
RCR	Rotate right through carry byte or word

String Instructions

Nine instructions are included to specifically deal with string operations. String operations simplify programming whenever a program must interact with a user. User commands and responses are usually saved as ASCII strings of characters, which may be processed by the proper choice of string instruction.

The string instructions are:

REP	Repeat
REPE (REPZ)	Repeat while equal (zero)
REPNE (REPNZ)	Repeat while not equal (not zero)
MOVS	Move byte or word string
MOVSB (MOVSW)	Move byte string (word string)
CMPS	Compare byte or word string
SCAS	Scan byte or word string
LODS	Load byte or word string
STOS	Store byte or word string

Program Transfer Instructions

This group of instructions contains all jumps, loops, and subroutine (called **procedure**) and interrupt operations. The great majority of jumps are **conditional,** testing the processor flags before execution.

The program transfer instructions are:

CALL	Call procedure (subroutine)
RET	Return from procedure (subroutine)
JMP	Unconditional jump
JA (JNBE)	Jump if above (not below or equal)
JAE (JNB)	Jump if above or equal (not below)
JB (JNAE)	Jump if below (not above or equal)
JBE (JNA)	Jump if below or equal (not above)
JC	Jump if carry set
JE (JZ)	Jump if equal (zero)
JG (JNLE)	Jump if greater (not less or equal)
JGE (JNL)	Jump if greater or equal (not less)
JL (JNGE)	Jump if less (not greater or equal)
JLE (JNG)	Jump if less or equal (not greater)
JNC	Jump if no carry
JNE (JNZ)	Jump if not equal (not zero)
JNO	Jump if no overflow
JNP (JPO)	Jump if no parity (parity odd)
JNS	Jump if no sign
JO	Jump if overflow
JP (JPE)	Jump if parity (parity even)
JS	Jump if sign
LOOP	Loop unconditional
LOOPE (LOOPZ)	Loop if equal (zero)
LOOPNE (LOOPNZ)	Loop if not equal (not zero)
JCXZ	Jump if CX equals zero
INT	Interrupt
INTO	Interrupt if overflow
IRET	Return from interrupt

Processor Control Instructions

This last group of instructions performs small tasks that sometimes have profound effects on the operation of the processor. Many of these instructions manipulate the flags.

The processor control instructions are:

STC	Set carry flag
CLC	Clear carry flag
CMC	Complement carry flag
STD	Set direction flag

CLD	Clear direction flag
STI	Set interrupt enable flag
CLI	Clear interrupt enable flag
HLT	Halt processor
WAIT	Wait for TEST in activity
ESC	Escape to external processor
LOCK	Lock bus during next instruction
NOP	No operation

80x86 ADDRESSING MODES

The 80x86 offers the programmer a wide number of choices when referring to a memory location. Many people believe that the number of **addressing modes** contained in a microprocessor is a measure of power. If that is so, the 80x86 should be counted among the most powerful processors. Many of the addressing modes are used to generate a **physical address** in memory. Recall from Figure 53.4 that a 20-bit real-mode address is formed by the sum of two 16-bit address registers. One of the four segment registers will always supply the first 16-bit address. The second 16-bit address is formed by a specific addressing mode operation. We will see that there are several different ways in which the second part of the address may be generated. An acceptable notation that represents both 16-bit halves of the address in Figure 53.4 is A000:5F00. We will make use of this addressing format in a subsequent exercise.

Real-Mode Addressing Space

All addressing modes eventually create a physical address that resides somewhere in the 00000 to FFFFF real-mode addressing space of the processor. Figure 53.7 shows a brief memory map of the 80x86 addressing space, which is broken up into 16 blocks of 64KB each. Each 64KB block is called a **segment.** A segment contains all the memory locations that can be reached when a particular segment register is used. For example, if the data segment contains 0000, then addresses 00000 through 0FFFF can be generated using the data segment. If, instead, register DS contains 1800, then the range of addresses becomes 18000 through 27FFF. It is important to see that a segment can begin on *any* 16-byte boundary. So 00000, 00010, 00020, 035A0, 10800, and CCE90 are all acceptable starting addresses for a segment.

Altogether, 1,048,576 bytes can be accessed by the processor. This is commonly referred to as 1 **megabyte.** Small areas of the addressing space are reserved for special operations. At the very high end of memory, locations FFFF0 through FFFFF are assigned the role of storing the initial instruction used after a RESET operation. At the low end of memory, locations 00000 through 003FF are used to store the addresses for 256 interrupt vectors (although not all are commonly used in actual practice). This declaration of addressing space is common among processor manufacturers, and may force designers to conform to specific methods or techniques when building systems around the 80x86. For instance, EPROM is usually mapped into high memory, so that the starting execution instructions will always be there at power-on.

Addressing Modes

The simplest addressing mode is known as **immediate addressing.** Data needed by the processor is actually included in the instruction. For example:

```
MOV CX,1024
```

contains the immediate data value 1024. This value is converted into binary and included in the code of the instruction.

When data must be moved between registers, **register addressing** is used. This form of addressing is very fast, because the processor does not have to access external memory (except for the instruction fetch). An example of register addressing is:

```
ADD AL,BL
```

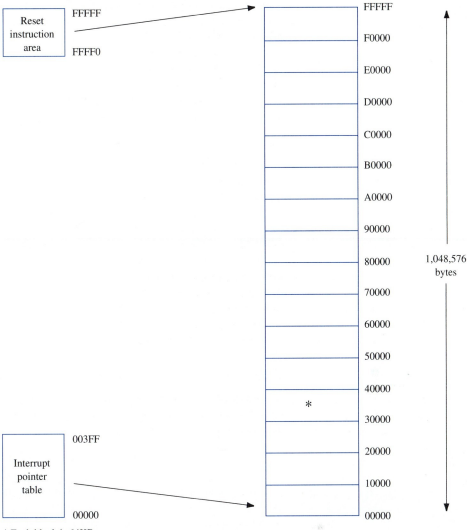

FIGURE 53.7 Real-mode addressing space of the 80x86

* Each block is 64KB.

where the contents of registers AL and BL are added together, with the result stored in register AL. Notice that both operands are names of internal 80x86 registers.

The programmer may refer to a memory location by its specific address by using **direct addressing.** Two examples of direct addressing are:

MOV AX,[3000]

and

MOV BL,COUNTER

In each case, the contents of memory are loaded into the specified registers. The first instruction uses square brackets to indicate that a memory address is being supplied. Thus, 3000 and [3000] are allowed to have two different meanings. The second instruction uses the symbol name COUNTER to refer to memory. COUNTER must be defined somewhere else in the program in order to be used in this way.

When a register is used within the square brackets, the processor uses **register indirect addressing.** For example:

MOV BX,[SI]

instructs the processor to use the 16-bit quantity stored in the SI (source index) register as a memory address. A slight variation produces **indexed addressing,** which allows a small offset value to be included in the memory operand. Consider the instruction:

MOV BX,[SI + 10]

The location accessed by this instruction is the sum of the SI register and the offset value 10.

When the register used is the base pointer (BP), **based addressing** is employed. This addressing mode is especially useful when manipulating data in large tables or arrays. An example of based addressing is:

```
MOV CL,[BP + 4]
```

Including an index register (SI or DI) in the operand produces **based-indexed addressing.** The address is now the sum of the base pointer and the index register. An example might be:

```
MOV [BP + DI],AX
```

When an offset value is also included in the operand, the processor uses **based-indexed with displacement addressing.** An example is:

```
MOV [BP + SI + 2]
```

It is easily seen that the 80x86 intends the base pointer to be used in many different ways. Other addressing modes are used when string operations must be performed.

The 80x86 processor is designed to access I/O ports, as well as memory locations. When **port addressing** is used, the address bus contains the address of an I/O port instead of a memory location. I/O ports may be accessed in two different ways. The port may be specified in the operand field, as in

```
IN AL,80H
```

or indirectly, by means of the address contained in register DX, as in

```
OUT DX,AL
```

Using DX allows a port range from 0000 to FFFF, or 65,536 individual I/O port locations. Only 256 locations (00 to FF) are allowed when the port address is included as an immediate operand.

TROUBLESHOOTING TECHNIQUES

A great amount of material was presented in this exercise regarding the Intel 80x86 architecture. It would be helpful to really commit some of the most basic material about the 80x86 to memory. As a minimum, you should be able to do the following without much thought:

- Name all the processor registers, their bit size, and whether they can be split into 8-bit halves.
- Be familiar with several architectural features, such as the processor's addressing space (1MB in real mode), interrupt mechanism, and I/O mechanism.
- Show what is meant by Intel byte-swapping.
- List the names and meanings of the most common flags, such as zero, carry, and sign.
- Describe the method used to form a 20-bit address in real mode (combining a segment register with an offset).
- Show how an instruction is composed of a operation, a set of operands, and a particular addressing mode.

Knowing these basics thoroughly will assist you in further exploration of the Intel 80x86 architecture.

SELF-TEST

This self-test is designed to help you check your understanding of the background information presented in this exercise.

True/False

Answer *true* or *false*.

1. Binary numbers must always contain at least one zero.
2. Only binary numbers can be represented with hexadecimal notation.

3. A four-symbol hexadecimal number has an equivalent 16-bit binary number.
4. Only segment registers are used to make an address.
5. The 80486 runs the same basic instructions as the Pentium.

Multiple Choice

Select the best answer.

6. The general-purpose registers are
   a. AX, BX, CX, and DX.
   b. CS, DS, ES, and SS.
   c. AX, BX, CX, DX, BP, SI, and DI.
7. The processor registers are able to hold
   a. 8-bit numbers.
   b. 16-bit numbers.
   c. Both 8- and 16-bit numbers.
   d. 20-bit numbers.
8. An example of immediate addressing is
   a. MOV AL,BL.
   b. AND AL,2FH.
   c. NOP.

Completion

Fill in the blanks with the best answers.

9. CS, DS, SS, and ES are called _____ registers.
10. A(n) _____ contains 2 bytes of information.
11. Reversing the order of the upper and lower bytes of a 16-bit number during memory access is called _____.
12. 80x86 processors can operate in either _____ mode or _____ mode.

FAMILIARIZATION ACTIVITY

1. Examine the following short list of instructions:

```
AND  AL,7
ADD  CX,5000
STC
INT  20H
RET
MOV  BL,[SI]
MOV  [DI+4],DX
```

What addressing modes are used for both source and destination operands (if any) in each instruction? What type of immediate data is used (if any), and are 8 or 16 bits needed for the data?

QUESTIONS/ACTIVITIES

1. How many different instructions are there in the 80x86 instruction set?
2. If an instruction allows four addressing modes for its destination operand and six addressing modes for its source operand, how many different instructions are possible?
3. What kinds of instructions might you like to see in an upgrade of the 80x86? Can you describe them?

REVIEW QUIZ

Within 15 minutes and with 100% accuracy,

1. Outline the software model of the 80x86 microprocessor family.
2. Discuss the various instruction groups and addressing modes that are available.
3. Explain how a 20-bit address is formed from two 16-bit numbers.

54 Computer Languages

INTRODUCTION

Don, Joe's manager, handed Joe Tekk his next assignment. "I'm sorry to do this to you, Joe, but the customer wants his custom digital camera interfaced to a parallel input/output port. You'll have to write assembly language code to control the data transfer between the camera and the port." Don paused, and then continued in a nervous voice, "He needs it in one week."

Joe took the customer's specification and examined it for a few moments. "One week?" he asked. Then he smiled at Don and said, "Thanks, Don! This is so cool!" He left the room in a hurry.

Don sat in his chair, wondering what Joe was so happy about. Thirty minutes later Joe called Don and said, "Come to my office. I got the camera working."

Don recalled a moment from Joe's job interview when Joe mentioned that he knew "a little assembly language."

That boy does not like to brag, he thought to himself.

PERFORMANCE OBJECTIVES

Upon completion of this exercise, within 15 minutes and with 100% accuracy, you will be able to

1. Enter an 80x86 assembly language source file.
2. Assemble the source file using MASM.
3. Link the object file using LINK.
4. Run the resulting 80x86 program.
5. Discuss the various statements in a C++ source file.
6. Describe the difference between an assembler and a compiler.
7. Explain the basic difference between an executable program and an interpreted BASIC program.

BACKGROUND INFORMATION

In order to use the maximum power of the personal computer, it is necessary to understand how to control and use the hardware on the motherboard and the software capabilities of the processor. In this exercise, we will examine how the 80x86 microprocessor is programmed in its own unique language, which is called **assembly language.** This exercise will expose you to assembly language and give you an appreciation for what happens inside the PC. To learn more about the details of assembly language programming, search the Web or read any of the large number of books on the subject. In addition, we will look briefly at two popular high-level languages, BASIC and C++, and see how they differ from each other and from assembly language.

MACHINE LANGUAGE VERSUS ASSEMBLY LANGUAGE

Our spoken language is one of words and phrases. The processor's language is a string of 1s and 0s. For example, the instruction

```
ADD   AX,BX
```

contains the word ADD, which means something to us. Apparently, we are adding AX and BX, two of the general-purpose registers. So, even though we might be unfamiliar with the details of the 80x86 instruction set, ADD AX,BX means something to us.

If instead we were given the binary string

```
0000 0001 1101 1000
```

or the hexadecimal equivalent

```
01 D8
```

and asked its meaning, we might be hard-pressed to come up with anything. We associate more meaning with ADD AX,BX than we do with 01 D8, which is the way the instruction is actually represented. All programs for the 80x86 are simply long strings of binary numbers.

Because of the processor's internal logic, different binary patterns represent different instructions. Here are a few examples to illustrate this point:

```
01 D8      ADD   AX,BX      ;add BX to AX, result in AX
29 D8      SUB   AX,BX      ;subtract BX from AX, result in AX
21 D8      AND   AX,BX      ;AX equals AX and BX
40         INC   AX         ;add 1 to AX
4B         DEC   BX         ;subtract 1 from BX
8B C3      MOV   AX,BX      ;copy BX into AX
```

Can you guess the meaning of each instruction just by reading it? Do the hexadecimal codes for these instructions mean anything to you? What we see here is the difference between **machine language** and **assembly language.** The machine language for each instruction is represented by the hexadecimal code. This is the binary language of the machine. The assembly language is represented by the wordlike terms that mean something to us. Putting groups of these wordlike instructions together is how a program is constructed. The format of an assembly language instruction is basically as follows:

```
<opcode>   <destination-operand>,<source-operand>
```

where <opcode> is an instruction from the 80x86 instruction set, and both <destination-operand> and <source-operand> are register names or numbers representing data or memory addresses. For example, in

```
ADD   AX,BX
```

the <opcode> is ADD, the <destination-operand> is register AX, and the <source-operand> is register BX. An instruction may have zero, one, or two operands.

Let us see how an assembly language program is written, converted into machine language, and executed.

THE NUMOFF PROGRAM

When the PC is first turned on, instructions in the start-up software turn the NUM-LOCK indicator on. This indicator is located near the NUM-LOCK button on the keyboard. It is annoying to have to push NUM-LOCK manually every time the PC is turned on (or even rebooted). Luckily, there is a single bit stored in a specific memory location used by DOS that controls the state of the NUM-LOCK indicator. We are about to see that it is possible to write a program called NUMOFF to manipulate the NUM-LOCK status bit.

Using a word processor or text editor, enter the following text file exactly as you see it. Save the ASCII text file under the name NUMOFF.ASM.

```
NUMOFF  SEGMENT  PARA 'CODE'
        ASSUME   CS:NUMOFF,DS:NOTHING
START:  MOV      AX,40H              ;set AX to 0040H
        MOV      DS,AX               ;load data segment with 0040H
        MOV      SI,17H              ;load SI with 17H
        AND      BYTE PTR [SI],0DFH  ;clear NUM-LOCK bit
        MOV      AH,4CH              ;terminate program function
        INT      21H                 ;exit to DOS
NUMOFF  ENDS
        END      START
```

These 10 lines of code constitute a *source file,* the starting point of any 80x86-based program. Thus, NUMOFF.ASM is a source file.

To convert NUMOFF.ASM into a group of hexadecimal bytes that represent the corresponding machine language, we make use of two additional programs: MASM and LINK. MASM is a **Macro Assembler,** a program that takes a source file as input and determines the machine language for each source statement. As illustrated in Figure 54.1, MASM creates two new files—the *list file* and the *object file.* The list file contains all the text from the source file, plus additional information, as we will soon see. The object file contains only the machine language.

To assemble NUMOFF.ASM, enter the following command at the DOS prompt:

```
MASM NUMOFF,,;
```

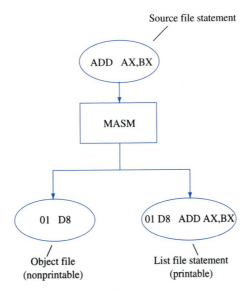

FIGURE 54.1 Operation of the MASM program

This instructs MASM to assemble NUMOFF.ASM and create NUMOFF.LST (the list file) and NUMOFF.OBJ (the object file). The list file created by MASM looks like this:

```
0000                    NUMOFF  SEGMENT PARA 'CODE'
                                ASSUME  CS:NUMOFF,DS:NOTHING
0000  B8 0040           START:  MOV     AX,40H           ;set AX to 0040H
0003  8E D8                     MOV     DS,AX            ;load data segment with 0040H
0005  BE 0017                   MOV     SI,17H           ;load SI with 17H
0008  80 24 DF                  AND     BYTE PTR [SI],0DFH ;clear NUM-LOCK bit
000B  B4 4C                     MOV     AH,4CH           ;terminate program function
000D  CD 21                     INT     21H              ;exit to DOS
000F                    NUMOFF  ENDS
                                END     START
```

In this file it is obvious that MASM has determined the machine language for each source statement. The first column is the set of memory locations where the instructions are stored. The second column is the group of machine language bytes that represent the actual 80x86 instructions.

The object file created by MASM is *not* executable. Its internal structure is used by another program, called **LINK,** to create an executable file. LINK uses NUMOFF.OBJ to create NUMOFF.EXE. The DOS command to do this is:

LINK NUMOFF,,;

The LINK program creates the executable NUMOFF.EXE file so that it conforms to a structure that DOS knows how to work with and that can be run from the DOS prompt. Recall that any .BAT, .EXE, or .COM file can be run from the DOS prompt.

Right now we have our first working 80x86 program, NUMOFF.EXE. To test it, press the NUM-LOCK button on the PC's keyboard until the NUM-LOCK light goes on. Then execute NUMOFF.EXE by entering:

NUMOFF

at the DOS prompt. The NUM-LOCK light should go off. This is what NUMOFF.EXE does. If you want the NUM-LOCK indicator to be turned off automatically each time your PC is powered up (or rebooted), add the statement

NUMOFF.EXE

to your AUTOEXEC.BAT file.

MASM and LINK have kept up with Intel's advancements in microprocessor architecture and now handle instructions for all machines up through the Pentium. However, for large programming applications, it is usually better to use a *high-level language,* such as C/C++, to write application programs. When the program is compiled, each statement is converted into one or more machine language instructions, and an executable file is generated that contains all the instructions. Thus, everything depends on the machine language instructions fed to the processor. Let us now examine two high-level languages, BASIC and C++.

HIGH-LEVEL LANGUAGES

Rather than using assembly language, it would be easier to give instructions and get answers back using symbols that we already understand. The easiest set of characters that we can use is the English alphabet. It would be very easy to program if you could give an instruction to the computer such as

> Figure out my income tax for me and let me know when you're finished so I can tell you where to mail it.

FIGURE 54.2 Different levels of computer languages

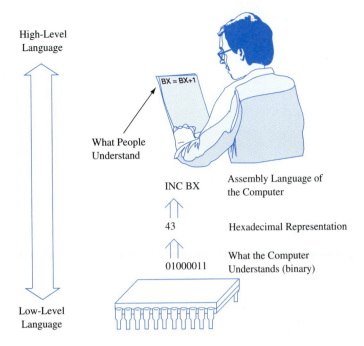

Although computers can't quite yet follow those instructions, a program can be written in words that people easily understand. When the computer is programmed using symbols and letters that the operator can read, this computer language is called a **high-level language.** The language presented here, BASIC, is a high-level language.

Figure 54.2 illustrates the different levels of computer languages. Any language that is higher in level than machine language requires a program to convert that language into machine language. For example, the next level above the 1s and 0s understood by the microprocessor is the **hexadecimal number system,** a number system containing 16 symbols: the digits 0–9 and the letters A–F. Using the hexadecimal number system, the instruction

```
01000011
```

could be entered into the computer as

```
43
```

—two keystrokes instead of eight!

Programming a computer this way is still called machine-language programming by many programmers. However, it is very difficult to write complete programs in machine language. It is easier to enter a BASIC statement, such as

```
BX = BX + 1
```

than to try and remember the correct machine language to enter. For this reason, a special program called a *BASIC interpreter* is used to automatically convert the high-level BASIC language into the low-level machine language understood by the microprocessor. The BASIC interpreter looks at each BASIC statement and converts it into the machine language required by the microprocessor. Luckily, most desktop computers come with a BASIC interpreter. If you want to program the computer in another language, such as C++, you'll need to purchase a special program called a *compiler,* which will convert the C++ statements into the 1s and 0s that the microprocessor understands. The main difference between an interpreter and a compiler is that the interpreter is used every time the program is executed, whereas the compiler is used only once, to create an *executable* machine-language program. The compiler creates an entirely new program containing only machine language. The BASIC interpreter must determine the correct machine language for every

FIGURE 54.3 Action of an interpreter

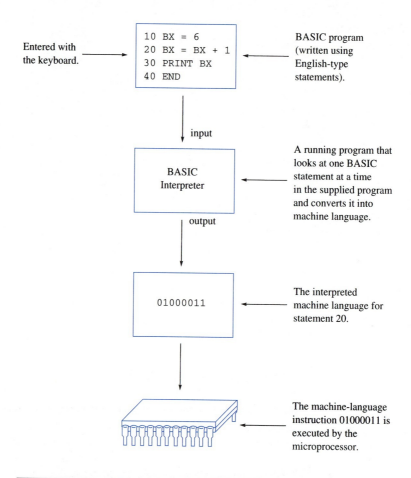

TABLE 54.1 Some programming languages and their advantages

Programming Language	Reason for Use
APL	(A Programming Language) A very powerful language for scientific work.
Assembly language	A low-level programming language that allows direct control of the microprocessor.
BASIC	Easy to learn.
COBOL	(Common Business-Oriented Language) For business-oriented programming.
LOGO	A simple programming language originally designed to teach children how to use computers.
Pascal	Teaches good programming habits.
FORTRAN	(Formula Translation) To solve mathematical formulas.
C/C++	More control over computer than other high-level languages, and built-in machine-language interface.

BASIC statement each time the program is executed. It is for this reason that compiled programs run much faster than interpreted programs.

The action of an interpreter is illustrated in Figure 54.3.

BASICA, GWBASIC, and QBASIC are three examples of interpreted BASIC. A package called Visual BASIC is capable of creating executable machine-language programs, and thus performs the actions of a compiler.

There are many different programming languages available. Table 54.1 lists some of the more common languages and their particular strengths.

FIGURE 54.4 Computing circle area

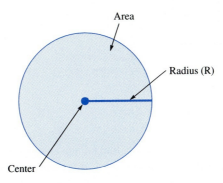

FIGURE 54.5 Output of circle area program

```
This program finds the area of a circle.
Enter the value of the radius and
the program will do the rest.
? 6
The area in square units is 113.0973
```

SAMPLE BASIC PROGRAM

The program presented here will compute the area of a circle for a given radius. You will recall that the formula for the area of a circle is

$$\text{Area} = \pi R^2$$

where Area = The area of the circle in square units.
π = The constant pi, which is approximately equal to 3.14159.
R = The radius of the circle.

This is illustrated in Figure 54.4. The BASIC program that solves for the circle area is as follows:

```
10 REM Area of a Circle Program
20 LET PI = 3.14159
30 CLS
40 PRINT "This program finds the area of a circle."
50 PRINT "Enter the value of the radius and"
60 PRINT "the program will do the rest."
70 INPUT R
80 LET Area = PI * R * R
90 PRINT "The area in square units is "; Area
100 END
```

This program will compute the area of a circle for a given radius. When the program is executed, the output on the screen will appear as shown in Figure 54.5.

What the Program Does

As you can see from Figure 54.5, when the program is executed it clears the screen and tells the program user what the program will do and how it is to be used. Then the program waits for the user to enter a number that represents the radius of a circle for which the user wants to find the area. When the user enters the value of the radius and presses the Enter key, the program computes the area of the circle and displays the result.

Look again at the program. You see that it consists of 10 lines, each line beginning with a number. Each line contains a BASIC statement. This idea is illustrated in Figure 54.6.

FIGURE 54.6 Typical BASIC statement structure

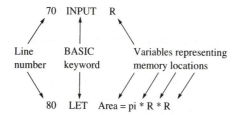

In BASICA and GWBASIC, line numbers are required for each BASIC statement in the program. This helps the editor keep the program statements in order. The line number specified by the programmer tells the editor where the statement belongs in the program.

In QBASIC, line numbers are not required, but may be used if desired. The built-in editor available with QBASIC is called a screen editor, which does not require line numbers to keep statements in order. Editing is performed by using the arrow keys (or mouse) to position the cursor at the location where a new statement will be inserted, or a current statement edited or deleted. In general, statements in QBASIC execute in the order that they appear on the screen. Table 54.2 lists the definitions of each of the BASIC statements used in the Area program.

TABLE 54.2 BASIC statements in the Area program

BASIC Statement	What It Does	Use in the Program
REM	This is a remark. It does not cause the BASIC program to do anything.	The REM statement is used by the programmer for comments and notes within the program.
LET	This is the LET statement. It signifies the beginning of a computation or logic statement.	The LET statement sets the value of the variable pi to the value of 3.14159. Now, whenever pi is used in the program, it will have the value of 3.14159.
CLS	This statement clears the monitor screen and starts the cursor at the upper left corner of the monitor.	The CLS clears the BASIC program itself from the screen. This is done to allow only the program output to appear.
PRINT	This statement causes the characters enclosed inside double quotes to be displayed (printed) on the screen when the program is executed.	The PRINT statements cause the instructions to be displayed on the monitor as well as the results of the calculation.
INPUT	This statement causes the program to wait for the user to input a value and press the Enter key.	The INPUT statement causes the program to display a question mark and wait for the user to input the value of the radius of the circle.
*	The asterisk (*) indicates multiplication.	The * is used to indicate multiplication. In this case, it is used to calculate the area of a circle: Area = πR^2 Area = pi * R * R
LINE NUMBERS	Indicates the order in which each BASIC statement will be executed. Program execution goes from the lowest line number to the largest. Line numbers may be in the range from 0 to 65529.	The program starts with line 10 and continues sequentially to line 100. Usually BASIC line numbers are numbered in units of 10. This is done in case the program needs modification and a new statement is required between two existing ones.

The C++ Environment

The C++ environment contains an **editor, compiler, include files, library files, linker,** and much more. The functions of these components are as follows:

- **Editor**   Allows you to enter and modify your C++ source code.
- **Compiler**   A program that converts the C++ program you have developed into a code understood by the computer.
- **Include Files**   Files that consist of many separate definitions and instructions that may be useful to a programmer in certain instances.
- **Library Files**   These are previously compiled programs that perform specific functions. These programs can be used by you to help you develop your C++ programs. For example, the C++ function that allows you to display text on your screen (the cout function) is not in the C++ language. Instead, its code is in a library file. The same is true of many other functions, such as graphics, sound, and working with the disk and printer. You can also create your own library files of routines that you use over and over again in different C++ programs. By doing this, you can save hours of programming time and prevent programming errors.
- **Linker**   Essentially, the linker combines all of the necessary parts (such as library files) of your C++ program to produce the final executable code. Linkers play an important and necessary role in all of your C++ programs. In larger C++ programs, it is general practice to break the program down into smaller parts, each of which is developed and tested separately. The linker will then combine all of these parts to form your final executable program code.

You will also need some kind of a disk operating system to assist you in saving your programs. The main parts of the C++ environment are shown in Figure 54.7.

Elements of C++

This section lays the ground rules for the fundamental elements of all C++ programs. In this section you will see many new definitions. These definitions will set the stage for the example program that follows.

- **Characters**   To write a program in C++, you use a set of characters. This set includes the uppercase and lowercase letters of the English alphabet, the 10 decimal digits of the Arabic number system, and the underscore (_) character. Whitespace characters (such as the spaces between words) are used to separate the items in a C++ program, much the same as they are used to separate words in this book. These whitespace characters also include the tab and carriage return, as well as other control characters that produce white spaces.
- **Tokens**   In every C++ source program, the most basic element recognized by the compiler is a single character or group of characters known as a **token.** Essentially, a token is source program text that the compiler will not break down any further—it is treated as a fundamental unit. As an example, in C++ main is a token; so is the required opening brace ({) as well as the plus sign (+).
- **C++ Keywords**   Keywords are predefined tokens that have special meanings to the C++ compiler. Their definitions cannot be changed; thus, they cannot be used for anything else except the intended action they have on the program in which they are used. The most common keywords are as follows:

```
auto     double    int       struct    break     else      public
long     switch    case      enum      register  typedef   private
char     extern    return    union     const     float     protected
short    unsigned  continue  for       signed    void
default  goto      sizeof    volatile  do        if
static   while     cin       cout      class     virtual
```

- **Types of Data**   The C++ language allows three major types of data: numbers, characters, and strings. A **character** is any item from the set of characters used by C++. A **string** is a combination of these characters.
- **Numbers Used by C++**   C++ uses a wide range of numbers. Numbers used by C++ fall into two general categories: **integer** (whole numbers) and **float** (numbers with decimal points). These two main categories can be further divided as shown in Table 54.3.

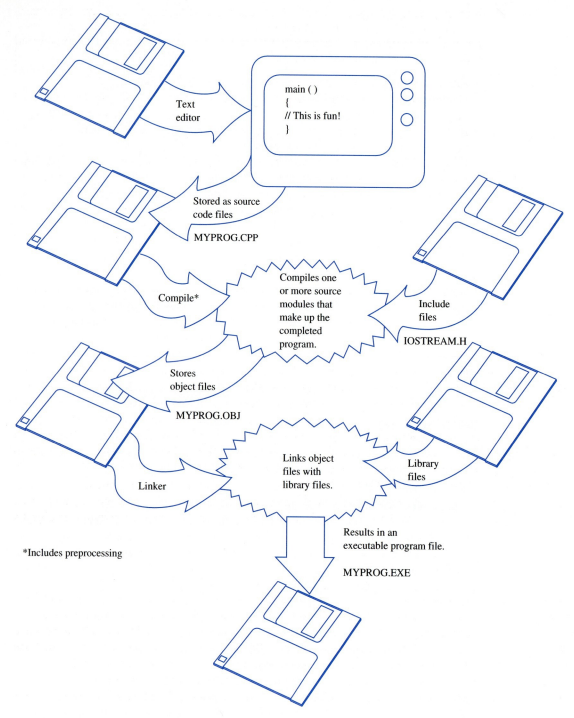

FIGURE 54.7 The C++ environment

As you can see in Table 54.3, C++ offers a rich variety of **data types.** Generally speaking, the larger the value range of the data type, the more computer memory it takes to store it. As a general rule, you want to use the data type that conserves memory and still accomplishes the desired purpose. As an example, if you needed a data type for counting objects—such as the number of resistors for a parts order—the type `int` would probably do the job. However, if you were writing a program for as much precision as possible, you might consider the type `double`, which can give you 15-digit accuracy.

Examine the sample C++ program shown in Figure 54.8. Can you spot the different tokens and keywords used by the programmer?

TABLE 54.3 Subdivisions of C++ data types

Type Identifier	Meaning	Range of Values
char	character	−128 to 127
int	integer	−32,768 to 32,767
short	short integer	−32,768 to 32,767
long	long integer	−2,147,483,648 to 2,147,483,647
unsigned char	unsigned character	0 to 255
unsigned	unsigned integer	0 to 65,535
unsigned short	unsigned short int	0 to 65,535
unsigned long	unsigned long int	0 to 4,294,967,295
enum	enumerated	0 to 65,535
float	floating point	3.4E +/− 38 (7 digits)
double	double floating point	1.7E +/− 308 (15 digits)
long double	long double floating point	1.7E +/− 4932 (40 digits)

OBJECT-ORIENTED PROGRAMMING

One of the main differences between C and C++ is the support of *objects* in the C++ environment. An object contains both code and data, and interacts with the C++ environment in certain, predefined ways. Objects may be duplicated, hidden from each other, and protected from access by other objects. Objects, unlike ordinary functions and variables, have *lifetimes*, or periods of time during execution where they are created (instantiated), used, and then destroyed (in the interest of memory management). AREA is the name of the class in the sample program that is used to instantiate the mycircle object.

In general, working with objects in a C++ program allows the programmer to skip many of the cumbersome chores of data management that would be required in an ordinary C program. The same steps are now performed automatically during an object's lifetime.

TROUBLESHOOTING TECHNIQUES

You may think it premature to discuss troubleshooting techniques when we have been exposed to so little of the assembly language and high-level programming. Even so, we have already seen a number of places where errors can occur, and it would be worthwhile to discuss them. For example, the NUMOFF.ASM source file could have contained one or more *typographical* errors, such as a misspelled instruction (MVO for MOV), or a missing comma, or a comma where a semicolon was expected. Generally, when errors such as these are present in a source file, the assembler will report them with a brief error message.

Even if the source file does not have any typographical errors, we could still run into trouble. We could enter the command to invoke MASM or LINK incorrectly, or not use the correct options.

When the source file correctly assembles and links, and an executable program has been created, there is still the possibility of a *run-time error* in the program. Run-time errors are typically caused by incorrect sequences of instructions and incomplete or faulty logicial thinking. The same problems may be encountered when working with BASIC and C++ progams.

To avoid a loss of time and effort, it is good to keep these common stumbling blocks in mind. Paying attention to the details will really pay off as you learn to create a working program with a minimum of time and effort.

Preprocessor Directives
```
#include <iostream.h>
#include <math.h>
#define PI 3.14159
#define SQUARE(x) ((x) * (x))
```

Programmer's Block
```
/*
    Program:    Circle Area
    Developed by: A. G. Programmer

    Description:    This program will solve for the area of a circle.  The
                    programmer need only enter the radius.  Value returned
                    is in square units.

    Variables:

        radius       = Radius of the circle.
        area         = Area of the circle.

    Constants:

        PI = 3.14159
*/
```

Class Definition
```
// Class Definition                        Explanation of member
                                           variables and functions
class AREA
{
public:
        float radius;                      // Radius of the circle.
        float area;                        // Area of the circle.

        AREA(void);                        // Explain program to user.
        void get_value(void);              // Get radius from user.
        void circle_area(void);            // Compute area of circle.
        ~AREA();                           // Display results.
};
```

Main Function
```
void main()
{
        AREA mycircle;                     // Instantiate object.

        mycircle.get_value();              // Get radius from user.
        mycircle.circle_area();            // Compute the circle area.
}
```

Constructor
```
void AREA::AREA()                          // Explains the program.
{
    cout << "This program calculates the area of a circle.\n";
    cout << "Just enter the value of the radius and press -RETURN-\n";
    cout << "\n";                          // Put in a blank line.
}
```

Member Functions
```
void AREA::get_value()                     // Gets radius from user.
{
    cout << "Value of the radius ==> ";
    cin >> radius;
}

void AREA::circle_area()                   // Compute the circle area.
{
    area = PI * SQUARE(radius);
}
```

Destructor
```
void AREA::~AREA()                         // Display the answer.
{
    cout << "\n\n";                        // Print two blank lines.
    cout << "The area of the circle is " << area << " square units.\n";
}
```

FIGURE 54.8 **Structure of a C++ program**

SELF-TEST

This self-test is designed to help you check your understanding of the background information presented in this exercise.

True/False

Answer *true* or *false*.

1. Machine language and assembly language are the same thing.
2. Only MASM is needed to make an executable file.
3. Both the list and object files may be printed.
4. Executable programs can be written only in assembly language.
5. BASIC is a compiled language.

Multiple Choice

Select the best answer.

6. The extension reserved for assembly language files is
   a. MLF (machine language file).
   b. SRC (source file).
   c. ASM (assembly).
7. The two files created by the assembler are
   a. MASM and LINK.
   b. Source and target.
   c. Source and object.
   d. List and object.
8. The instruction in the NUMOFF program that actually clears the NUM-LOCK bit is the
   a. MOV instruction.
   b. AND instruction.
   c. INT instruction.
9. PRINT (in BASIC) and `cout` (in C++) are both used to
   a. Input data from the user.
   b. Output data to the user.
   c. Perform input and output operations.

Completion

Fill in the blanks with the best answers.

10. A text file written in assembly language is called a(n) _____ file.
11. The linker creates an executable file using information from the _____ file.
12. Executable programs may also be written in a(n) _____ language such as C/C++.
13. A(n) _____ converts C++ statements into executable code.

FAMILIARIZATION ACTIVITY

1. Use EDIT to enter the NUMOFF.ASM source file. Deliberately misspell one of the MOV instructions (use MVO instead).
2. Assemble NUMOFF using MASM, by means of

   ```
   MASM  NUMOFF,,;
   ```

3. Did MASM discover the error? How can you tell? To see the details of the error, type out the contents of NUMOFF.LST.
4. Edit NUMOFF.ASM so that the instructions are all correct, and reassemble the program.
5. Use the DIR command to view all of the NUMOFF files.
6. Create the NUMOFF executable with the linker using this command:

   ```
   LINK NUMOFF,,;
   ```

7. Use the DIR command to verify that NUMOFF.EXE has been created.

8. Run NUMOFF a few times to verify that it actually turns the NUM-LOCK indicator off.
9. Verify the operation of the BASIC sample program using your own BASIC system.
10. Search the Web for free C++ compilers. Are any available?

QUESTIONS/ACTIVITIES

1. Examine the contents of the NUMOFF.LST file. Do the machine codes for all the MOV instructions have anything in common?
2. Can you think of a way to do the opposite of NUMOFF? That is, can you change NUMOFF so that it turns NUM-LOCK on?
3. Can you think of a way to toggle the NUM-LOCK indicator? (This would cause NUM-LOCK to alternate between ON and OFF.)
4. Explain why an assembler can be thought of as a language translator.
5. Explain why interpreted programs run more slowly than compiled programs.

REVIEW QUIZ

Within 15 minutes and with 100% accuracy,

1. Enter the following assembly language source file:

```
MTXT    SEGMENT PARA 'DATA'
MSG     DB   '80x86 assembly language!$'
MTXT    ENDS
PGM     SEGMENT PARA 'CODE'
        ASSUME CS:PGM,DS:MTXT
START:  MOV  AX,MTXT      ;set up message segment
        MOV  DS,AX
        LEA  DX,MSG       ;set up pointer to message
        MOV  AH,9         ;display string function
        INT  21H          ;DOS call
        MOV  AH,4CH       ;terminate program function
        INT  21H          ;exit to DOS
PGM     ENDS
        END  START
```

2. Assemble the source file using MASM.
3. Link the object file using LINK.
4. Run the resulting 80x86 program.
5. Discuss the various statements in a C++ source file.
6. Describe the difference between an assembler and a compiler.
7. Explain the basic difference between an executable program and an interpreted BASIC program.

55 Hardware and Software Interrupts

INTRODUCTION

Joe Tekk was busy with the installation of a network interface card. However, every time he booted the machine, the network software failed to initialize. All the other hardware in the machine, including the sound card, worked fine.

Eventually, after trial and error, Joe discovered that removing the sound card and commenting out the sound drivers in the CONFIG.SYS and AUTOEXEC.BAT files allowed the network software to load. Joe read through the manual for the sound card and found that its preassigned interrupt was the same as the interrupt used by the network interface card.

The sound card manual explained how to change the interrupt number. Joe made the necessary changes and rebooted the machine. The conflict had disappeared.

PERFORMANCE OBJECTIVES

Upon completion of this exercise, within 15 minutes and with 100% accuracy, you will be able to

1. Explain how hardware and software interact through interrupts.
2. Discuss the operation of the interrupt vector table.
3. Use DEBUG to view the interrupt vector table.
4. Find the address associated with DOS INT 21H.

BACKGROUND INFORMATION

The operation of the personal computer involves cooperation between the hardware of the system and the software running it. Essentially, there are software events that cause the hardware to respond, and hardware events that trigger a response from the software. For example, Figure 55.1 shows how the execution of the DIR command causes the disk drive to turn on so that directory information can be read.

The DIR command routine in this example uses a *software interrupt* to activate the file I/O routines. There are many of these software interrupts reserved for use by BIOS and DOS. All of them operate through the processor's INT instruction. For example, a very useful DOS interrupt is INT 21H, which is capable of performing disk, keyboard, and display I/O; memory management; time/date, printer, and directory functions; and more. We will examine the INT instruction shortly.

FIGURE 55.1 How software affects hardware

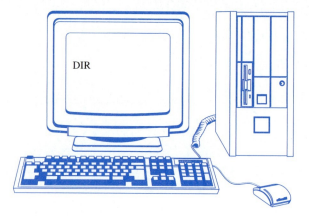

(a)   User enters DIR command.

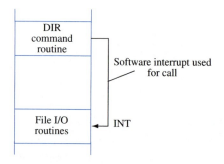

(b)   DIR subroutine calls for file input.

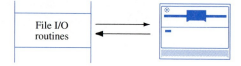

(c)   File routines activate hardware on the disk drive.

Figure 55.2 shows how a hardware event causes a response from the software. Every time a key is pressed on the keyboard, its scan code is transmitted to keyboard logic on the motherboard. The keyboard logic generates a *hardware interrupt* to signal the processor that the key code needs to be read. The hardware interrupt causes the processor to stop whatever it is doing and run the keyboard input routine (called an *interrupt handler* or *interrupt service routine*). The processor takes this action because the hardware interrupt is translated into the software interrupt INT instruction. BIOS INT 9 is used to handle the keyboard on the PC. Note that similar events occur in the Windows environment, where interrupts are used and managed much like their DOS counterparts.

THE INTERRUPT VECTOR TABLE

All types of interrupts, whether hardware or software generated, point to a single entry in the processor's interrupt vector table. This table is a collection of 4-byte addresses (2 for CS and 2 for IP) that indicate where the 80x86 should jump to execute the associated interrupt service routine. Since 256 interrupt types are available, the interrupt vector table is 1024 bytes long.

FIGURE 55.2 How hardware triggers a software response

(a) User presses a key on the keyboard.

(b) Keyboard transmits key code to keyboard logic on motherboard.

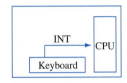

(c) Keyboard logic issues a hardware interrupt.

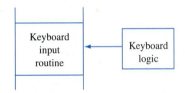

(d) Keyboard input subroutine reads key code.

The 1KB block of memory reserved for the table is located in the address range 00000 to 003FFH. BIOS and Windows (or DOS) automatically initialize the vector table at boot time.

Figure 55.3 shows the organization of the interrupt vector table. Each 4-byte entry consists of a 2-byte IP register value followed by a 2-byte CS register value. Notice that some of the vectors are predefined. Vector 0 has been chosen to handle division-by-zero errors. Vector 1 helps to implement single-step operations. Vector 2 is used when NMI is activated. Vector 3 (breakpoint) is normally used when troubleshooting a new program. Vector 4 is associated with the INTO instruction. Vectors 5 through 31 are reserved by Intel for use in their products. This does not mean that these interrupt vectors are unavailable to us, but we should refrain from using them in an Intel machine unless we know how they have been assigned.

BIOS and DOS, as well as Windows, use specific interrupt numbers when performing their respective operations. Some of the more common BIOS interrupts are listed in Table 55.1. Table 55.2 lists the more common DOS INT 21H interrupt functions. A programmer using assembly language has a great deal of power available through the use of these interrupts.

VIEWING THE INTERRUPT VECTOR TABLE WITH DEBUG

The contents of the interrupt vector table can be displayed in hexadecimal format through the use of a DOS utility program called **DEBUG.** DEBUG contains a command called

FIGURE 55.3 Interrupt vector table

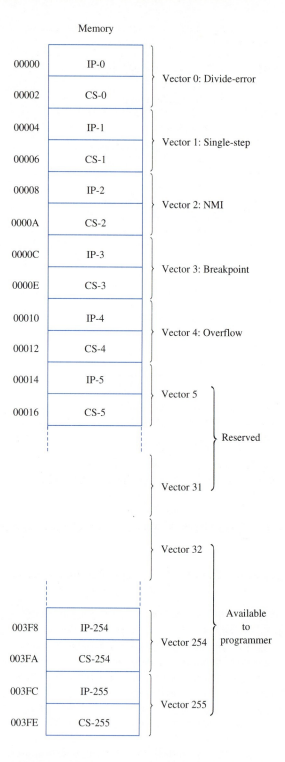

TABLE 55.1 Common BIOS interrupts

Interrupt	Function
INT 09H	Read keyboard
INT 10H	Video services
INT 13H	Disk services
INT 14H	Serial communication services
INT 16H	Keyboard services
INT 17H	Parallel printer services
INT 1AH	Time-of-day services

TABLE 55.2 Some of the DOS INT 21H interrupt functions

Function	Operation
01H	Console input with echo
02H	Display output
05H	Printer output
08H	Console input without echo
09H	Display string
1CH	Get drive information
25H	Set interrupt vector
2AH	Get date
2CH	Get system time
30H	Get DOS version number
31H	Terminate and stay resident
35H	Get interrupt vector
39H	Create subdirectory
3BH	Set current directory
3CH	Create file
3DH	Open file with handle
3EH	Close file with handle
3FH	Read from file
40H	Write to file
41H	Delete file
48H	Allocate memory
49H	Free allocated memory
4AH	Modify allocated memory blocks
4CH	Terminate program

dump that displays a range of memory locations on the screen. The user specifies the address of memory where the display will begin. To start DEBUG, enter

C> DEBUG

DEBUG responds with its own prompt, a dash (–). Enter the DEBUG command

-D 0:0

which instructs DEBUG to display the first 128 locations of memory, beginning at segment 0, address 0. The result will look similar to this:

```
0000:0000   FE 1A 4B 1E F4 06 70 00-16 00 C9 06 F4 06 70 00   ..K...p.......p.
0000:0010   F4 06 70 00 15 00 A4 15-4C E1 00 F0 6F EF 00 F0   ..p.....L...o...
0000:0020   46 0C 41 0B 23 19 B9 DF-6F EF 00 F0 6F 00 C9 06   F.A.#...o...o...
0000:0030   4D 11 62 E7 82 0E 41 0B-B7 00 C9 06 96 08 91 D6   M.b...A.........
0000:0040   FA 05 62 E7 4D F8 00 F0-41 F8 00 F0 3C 0F 17 09   ..b.M...A...<...
0000:0050   C6 0F 17 09 66 0F 17 09-2E E8 00 F0 01 00 A4 15   ....f...........
0000:0060   D0 E3 00 F0 19 10 17 09-6E FE 00 F0 EE 06 70 00   ........n.....p.
0000:0070   7E 13 85 C9 A4 F0 00 F0-22 05 00 00 5F 5B 00 C0   ~......."..._[..
```

The address for the BIOS keyboard INT 09H is highlighted in boldface. The four hexadecimal values 23 19 B9 DF correspond to a memory address of DFB9:1923. This is the address of the routine that reads the keyboard. The address on your machine will probably be different, since it was not configured at boot time the same as the one used for this example. Recall from Figure 55.1 that the DIR command caused hardware in the disk drive to become activated through the use of BIOS and DOS interrupts. When the INT instruction issued by DOS is processed, the address of the interrupt service routine is loaded from the interrupt vector table. Thus, DOS does not have to know the exact address of the disk routine, just the number of the software interrupt that uses it. These DOS routines are simulated in the Windows environment, which provides support for DOS applications.

The entire interrupt vector table occupies memory from address 0:0 to address 0:3FF. Recall from Figure 55.3 that each vector requires four memory locations of storage. Thus, the address for a specific INT instruction equals four times the value of the interrupt number used in the instruction. For instance, the vector for INT 21H is located at address 0:84, since 21H (33 decimal) times 4 is 84H (132 decimal). To see only the INT 21H vector, use the following DEBUG command:

```
-D 0:84 L 4
```

HARDWARE INTERRUPT ASSIGNMENTS

The hardware interrupt circuitry on the motherboard is responsible for managing the interrupts generated by the various hardware devices used in the computer. These devices include the real-time clock, disk controllers, keyboard logic, and serial and parallel ports. A special chip called a *programmable interrupt controller* is used to do all the work. This chip is programmed during boot time so that it translates one of 16 interrupt requests (IRQ0 through IRQ15) into specific INT instructions. Table 55.3 shows the hardware interrupt request assignments for a typical PC.

Refer to Figure 55.2 once again. The interrupt signal generated by the keyboard logic when it has received a key code is IRQ1. The programmable interrupt controller translates this request into an INT 09H instruction, which forces the processor to suspend what it is doing and process the keyboard interrupt. The suspended main program will resume after the interrupt service routine does its job.

The hardware interrupt logic was extended to 16 request lines in the AT computer. A sample assignment of these interrupt request lines is shown in Figure 55.4. This information is the output of the MSD.EXE (Microsoft Diagnostics) program that comes with DOS. Other, similar utilities (such as the MFT.EXE program from Quarterdeck) allow system information to be examined. Even the system BIOS on new machines provides this type of information. It is useful to examine hardware settings to get an overall feel for the system hardware.

Often, when a new piece of hardware has just been installed, the setup software will examine the current IRQ assignments of the system and assign an IRQ that does not conflict with the others. This process is not foolproof, and sometimes the reason behind a mysterious hardware problem is the result of conflicting hardware interrupts.

TABLE 55.3 Interrupt assignments

IRQ Number	Purpose
0	System timer
1	Keyboard
2	Cascade from IRQ9
3	COM2, COM4
4	COM1, COM3
5	Parallel port 2
6	Floppy disk controller
7	Parallel port 1
8	Real-time clock
9	Sound card
10	Network card
11	Video card
12	Free
13	Coprocessor
14	Hard disk controller (primary)
15	Hard disk controller (secondary)

FIGURE 55.4 Sample output from MSD utility

```
IRQ  Address     Description        Detected            Handled by
---  ---------   ----------------   ------------------  ---------------
 0   0B41:0C46   Timer Click        Yes                 SCH
 1   DFB9:1923   Keyboard           Yes                 Block Device
 2   F000:EF6F   Second 8259A       Yes                 BIOS
 3   06C9:006F   COM2: COM4:        COM2:               Default Handlers
 4   E762:114D   COM1: COM3:        COM1: Serial Mouse  MOUSE.COM
 5   0B41:0E82   LPT2:              No                  SCH
 6   06C9:00B7   Floppy Disk        Yes                 Default Handlers
 7   D691:0896   LPT1:              Yes                 CTSOUND0
 8   06C9:0052   Real-Time Clock    Yes                 Default Handlers
 9   F000:ECF5   Redirected IRQ2    Yes                 BIOS
10   F000:EF6F   (Reserved)                             BIOS
11   15A4:0015   (Reserved)                             REDIR5
12   F000:EF6F   (Reserved)                             BIOS
13   F000:F0FC   Math Coprocessor   Yes                 BIOS
14   06C9:0117   Fixed Disk         Yes                 Default Handlers
15   F000:EF6F   (Reserved)                             BIOS
```

TROUBLESHOOTING TECHNIQUES

It pays to know the details of interrupt processing when troubleshooting a program. Many times, the fault of erratic execution in an 80x86-based system is a poorly written or incomplete interrupt handler. Understanding the basic principles can help eliminate some of the more obvious problems.

- A valid stack must exist to save all the information required to support the interrupt handler.
- A typical interrupt pushes the current flags and return address (CS:IP).
- Vector addresses are equal to four times the vector number.
- In the interrupt vector table, the handler address is stored in byte-swapped form, with IP as the first word and CS as the second.
- It may be necessary to save and restore registers (via PUSH/POP) in the interrupt handler.
- Use IRET (return from interrupt) to return from an interrupt handler. RET does not work properly with interrupts.
- For an interrupt to work, its vector must be loaded with the starting address of the handler, and the handler code must be in place as well.

These tips should come in handy if you try to write an interrupt handler of your own, or if you are looking at the code for someone else's handler, to determine how it works.

Windows 95/98 reports the current interrupt assignments when you right-click on the Computer icon in the Device Manager window. Figure 55.5(a) shows the initial set of interrupts and their assignments.

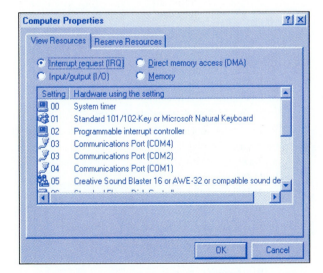

FIGURE 55.5 (a) Interrupt assignments in Windows 95/98 *(continued on the next page)*

693

FIGURE 55.5 (b) Windows NT IRQ Resources

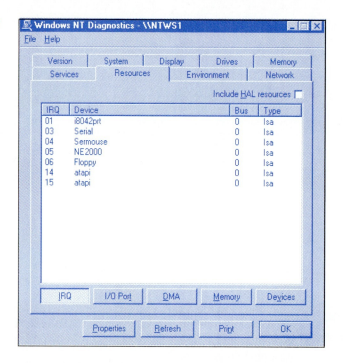

Figure 55.5(b) shows the Windows NT Diagnostics Resources display of interrupt resources used by the system. Note that I/O ports, DMA, and Memory are also considered resources used by the system. Their assignments are displayed in a fashion similar to the IRQ display.

SELF-TEST

This self-test is designed to help you check your understanding of the background information presented in this exercise.

True/False

Answer *true* or *false*.

1. Hardware and software operate independently in the PC.
2. The interrupt vector table is initialized during boot time.
3. The keyboard generates an interrupt once every second.
4. All devices use the same IRQ signal when they need service.

Multiple Choice

Select the best answer.

5. Software interrupts are used by
   a. BIOS only.
   b. Operating system only.
   c. Both BIOS and operating system.
6. An interrupt service routine is located at address 2000:1E00. The segment address of the routine is
   a. 2000.
   b. 1E00.
   c. 3E00.
   d. 20001E00.
7. Interrupt vector 0 is predefined by Intel for
   a. Out-of-memory errors.
   b. Divide-by-0 errors.
   c. Breakpoints.
   d. DOS and BIOS.

Completion

Fill in the blank or blanks with the best answers.

8. The keyboard logic generates a(n) _____ interrupt when a key is received.
9. _____ and _____ both use interrupts to control the personal computer hardware.
10. _____ is a utility that allows the user to display the contents of memory.
11. The device responsible for handling hardware interrupts is called a(n) _____ _____ _____.

FAMILIARIZATION ACTIVITY

1. Start the DEBUG program by entering

   ```
   C> DEBUG
   ```

2. Display the contents of the first portion of the interrupt vector table with

   ```
   -D 0:0
   ```

3. Are there any interrupt vectors that equal 0000:0000?
4. Find the interrupt vector location for INT 09H and read the associated service routine address.
5. Find the service routine address for the BIOS interrupts listed in Table 55.1.
6. Examine the rest of the interrupt vector table by using the D command three more times, as in

   ```
   -D
   ```

   You will get a new block of 128 bytes each time. Are there any vectors equal to 0000:0000? Are any of the vectors more popular than the others; that is, does one address appear more frequently than the others?
7. If you have MSD.EXE (or some other system utility program), display the configuration of the hardware interrupts for your system.

QUESTIONS/ACTIVITIES

1. DEBUG can be used to view the assembly language associated with an interrupt with its built-in U (unassemble) command. For example, if the address of the keyboard interrupt service routine is DFB9:1923, use the U command as follows:

   ```
   -U DFB9:1923
   ```

   You will get a display similar to this:

   ```
   DFB9:1923 9C              PUSHF
   DFB9:1924 FA              CLI
   DFB9:1925 2E              CS:
   DFB9:1926 833EB02300      CMP     WORD PTR [23B0],+00
   DFB9:192B 7510            JNZ     193D
   DFB9:192D 2E              CS:
   DFB9:192E FE06BA23        INC     BYTE PTR [23BA]
   DFB9:1932 2E              CS:
   DFB9:1933 FF1EC823        CALL    FAR [23C8]
   DFB9:1937 2E              CS:
   DFB9:1938 FE0EBA23        DEC     BYTE PTR [23BA]
   DFB9:193C CF              IRET
   DFB9:193D 50              PUSH    AX
   DFB9:193E 06              PUSH    ES
   DFB9:193F 2E              CS:
   DFB9:1940 8E062119        MOV     ES,[1921]
   ```

   Try this on your own computer.

2. Examine the hardware manual for any card plugged into the motherboard of your computer. What interrupt request line (or lines) does the card use, if any? What is it (are they) for?

REVIEW QUIZ

Within 15 minutes and with 100% accuracy,

1. Explain how hardware and software interact through interrupts.
2. Discuss the operation of the interrupt vector table.
3. Use DEBUG to view the interrupt vector table.
4. Find the address associated with DOS INT 21H.

56 The Advanced Intel Microprocessors

INTRODUCTION

Joe Tekk had an old case of floppies that contained programs written in an assembly language class many years ago. He picked one of the floppies at random and put it into the disk drive. A directory showed several files, one of which was an executable for a blackjack game. Joe typed in the name of the program and ran it. It worked just as it was designed to work, even though it was written for an original 8088 PC, and Joe was now running it on his Pentium-based office machine.

Don, his manager, walked up and asked Joe what he was doing.

"I'm running a program I wrote years ago, and it still works. Even though Intel has changed and improved the internal hardware architecture of the 80x86 family, all the processors still run code from the original 8086. Even the Pentium III still comes up in real mode at power-on."

"So all the advanced Intel processors initially act like superfast 8086 machines?"

"That's right," Joe replied. "Really fast 8086s. But the way the hardware does it has changed."

PERFORMANCE OBJECTIVES

Upon completion of this exercise, within 10 minutes and with 100% accuracy, you will be able to

1. Discuss the improvements offered by each new 80x86 processor.
2. Explain the basic operation of an instruction pipeline.

BACKGROUND INFORMATION

In this exercise we will survey the advanced Intel microprocessors. Most importantly, we will see that every processor in the 80x86 family runs programs written for the first machines in the series, the 8088 and 8086. Thus, even though new personal computers use more advanced Intel microprocessors than the original PC did, they are fully software-compatible with the older machines.

A SUMMARY OF THE 80186

Intel made great efforts to make future processors compatible with the 8086, while at the same time offering additional enhancements and features. The 80186 High-Integration 16-bit Microprocessor can be thought of as a super 8086. Its instruction set is compatible with the 8086, which allows programs written for the 8086 to run on the 80186. Ten additional instructions are included, some of which control the additional 80186 hardware features.

Unlike the 8086, which comes in a 40-pin Dual In-line Package (DIP), the 80186 contains 68 pins and comes housed in a variety of different packages. Plastic Leaded Chip Carrier (PLCC), Ceramic Pin Grid Array (PGA), and Ceramic Leadless Chip Carrier (LCC) are the three types of packages that the 80186 may be found in. Both the PLCC and LCC packages have 17 pins on each of their four sides. The pins on the PGA come out of the bottom of the package. These new types of packages occupy less space on printed circuit boards and are thus very popular with designers.

Many of the additional pins are needed for signals used and generated by the new hardware features of the 80186. A **programmable interrupt controller** has been added, which supervises the operation of five hardware interrupt lines, including a nonmaskable interrupt signal. These interrupt lines can be programmed for different modes of operation. In *fully nested* mode, the four maskable interrupts are used to generate internal interrupt vectors on a prioritized basis. In *cascade* mode, the four interrupt lines become handshaking signals for an external interrupt controller, greatly increasing the interrupt capability of the processor.

Three 16-bit **programmable timers** have also been added. Two of the timers interface with the outside world and can be programmed for many different operations, such as counting and timing external events and generating waveforms (e.g., square waves and pulses). The third timer is used for internal operations. The timers can be programmed to cause an interrupt when a certain count (called a *terminal count*) is reached. All timers are clocked at a frequency equal to one-quarter of the CPU's internal clock.

A **programmable DMA unit** connected to the internal processor bus allows two independent DMA channels to operate under processor control. The DMA channels are especially useful when transferring large blocks of data between the processor, memory, and I/O devices. Both 8-bit and 16-bit transfers are allowed, and may be transferred as fast as 2.5MB per second. Priorities may be assigned to the two DMA channels in two ways: with one channel having a higher priority than the other, or with both channels having the same priority. Each channel can be programmed to interrupt the processor when it has completed its data transfer.

The **bus interface unit** is similar to the one found in the 8086, with additional signals providing the use of synchronous and asynchronous bus transfers. Finally, a **chip-select unit** has been added, which contains programmable outputs that can be used to select banks of memory or I/O devices. This task was previously accomplished by additional chip-select logic outside the processor. Putting this logic inside, and making it programmable, further reduces the amount of hardware needed to operate the microcomputer system. Wait states can also be programmed to be automatically inserted into bus cycles to allow for slow memory or I/O devices.

From a software standpoint, the 80186 is compatible with the 8086, using the same register set. Many additional hardware registers have been added for the purpose of controlling the new services provided by the 80186. These registers are contained in a dedicated area called the *peripheral control block* (PCB). The PCB is automatically placed at the top of the processor's I/O space (port addresses FF00H to FFFFH) whenever the CPU is reset, but it can be moved to a different location by changing the contents of the processor's **relocation register.** The PCB contains interrupt control registers, timer control registers for all three timers, chip-select control registers, and DMA descriptors for both channels. Programming all the internal hardware enhancements is done through these registers.

Several new instructions have also been added. PUSHA and POPA deal with the stack and are used to push or pop *all* the 80186's registers (AX, BX, CX, etc.). The integer multiply instruction IMUL has been enhanced to allow immediate data as an operand. Also enhanced are the shift and rotate instructions, which now allow a count value to be included in the instruction. For I/O operations, two new instructions allow 8-bit and 16-bit data to be inputted or outputted between an I/O device and a memory location (instead of the usual use of the accumulator). These instructions are INS and OUTS. For byte operations, INSB and OUTSB are used, and INSW and OUTSW are used for word operations. A special form of the instruction allows a **string** of bytes or words to be transferred. Two other instructions are used to manipulate the stack area (with the help of the BP register). These instructions are ENTER and LEAVE, and have been included to assist in the implementation of procedure calls in high-level languages such as C. Finally, the BOUND instruction is included to help in the partitioning of memory for multiuser environments. BOUND checks the contents of a specified register against an allowable range and generates an interrupt if the register is "out of bounds."

All in all, the 80186 offers many significant improvements over the 8086, while still being compatible. For those designers still determined to use an 8-bit data bus, Intel offers the 80188. This processor is an exact internal copy of the 80186, but differs in its use of an external 8-bit data bus.

A SUMMARY OF THE 80286

The next improvement in Intel's line of microprocessors was the 80286 High-Performance Microprocessor with Memory Management and Protection. Unlike the 80186, the 80286 does not contain the internal DMA controllers, timers, and other enhancements. Instead, the 80286 concentrates on the features needed to implement **multitasking,** an operating system environment that allows many programs or tasks to run seemingly simultaneously. In fact, the 80286 was designed with this goal in mind. A 24-bit address bus gives the processor the capability of accessing 16MB of storage. The internal memory management feature increases the storage space to 1 **gigabyte** of virtual address space—more than *one billion* locations of virtual memory. **Virtual addressing** is a concept that has gained much popularity in the computing industry. Virtual memory allows a large program to execute in a smaller physical memory. For example, if a system using the 80286 contained 8MB of RAM, memory management and virtual addressing would permit the system to run a program containing 12MB of code and data, or even multiple programs in a multitasking environment, *all of which* could be larger than 8MB.

To implement the complicated addressing functions required by virtual addressing, the 80286 has an entire functional unit dedicated to address generation. This unit is called the *address unit.* It provides two modes of addressing: 8086 real address mode and protected virtual address mode. The 8086 real address mode is used whenever an 8086 program executes on the 80286. The 1MB addressing space of the 8086 is simulated in the 80286 by the use of the lower 20 address lines. Processor registers and instructions are totally compatible with the 8086.

Protected virtual address mode uses the full power of the 80286, providing memory management, additional instructions, and protection features, while at the same time retaining the ability to execute 8086 code. The processor switches from 8086 real address mode to protected mode when a special instruction sets the protection enable bit in the machine's status word. Addressing is more complicated in protected mode, and is accomplished through the use of **segment descriptors** stored in memory. The segment descriptor is the device that really makes it possible for an operating system to control and protect memory. Certain bits within the segment descriptor are used to grant or deny access to memory in certain ways. A section of memory may be write-protected, or made execute-only, by the setting of proper bits in the access rights byte of the descriptor. Other bits are used to control how the segment is mapped into virtual memory space and whether the descriptor is for a code segment or a data segment. Special descriptors, called *gate descriptors,* are used for other functions. Four types of gate descriptors are call gates, task gates, interrupt gates, and trap gates. They are used to change privilege levels (there are four), switch tasks, and specify interrupt service routines.

The instruction set of the 80286 is identical to that of the 80186, with an additional 16 instructions thrown in to handle the new features. Many of the instructions are used to load and store the different types of descriptors found in the 80286. Other instructions are used to manipulate task registers, change privilege levels, adjust the machine status word, and verify read/write accesses. Clearly, the 80286 differs greatly from the 80186 in the services it offers, while at the same time filling a great need for designers of operating systems.

A SUMMARY OF THE 80386

Intel continued its 8086-compatible trend with the introduction of the 80386 High Performance 32-bit CHMOS Microprocessor with Integrated Memory Management. Software written for the 8088, 8086, 80186, 80188, and 80286 will also run on the 386. A 132-pin Pin Grid Array (PGA) package houses the 80386, which offers a full 32-bit data bus and 32-bit address bus. The address bus is capable of accessing more than 4GB of physical memory. Virtual addressing pushes this to more than 64 *trillion* bytes of storage.

The register set of the 80386 is compatible with earlier models, including all general-purpose registers, plus the four segment registers. Although the general-purpose registers are 16 bits wide on all earlier machines, they can be extended to 32 bits on the 80386. Their new names are EAX, EBX, ECX, and so on. Two additional 16-bit segment registers, FS and GS, are included. These registers are illustrated in Figure 56.1. Note that the original 8086 registers may still be used as 8- or 16-bit registers (AH, AL, AX, for example).

Like the 80286, the 80386 has two modes of operation: real mode and protected mode. In real mode, segments have a maximum size of 64KB. In protected mode, a segment can be as large as the entire physical addressing space of 4GB. The new extended flags register contains status information concerning privilege levels, virtual mode operation, and other flags concerned with protected mode. The 80386 also contains three 32-bit control registers. The first, called the machine control register, contains the machine status word and additional bits dealing with the coprocessor, paging, and protected mode. The second, page fault linear address, is used to store the 32-bit address that caused the last page fault. In a virtual memory environment, physical memory is divided into several fixed-size **pages.** Each page will at some time be loaded with a portion of an executing program or other type of data. When the processor determines that a page it needs to use has not been loaded into memory, a **page fault** is generated. The page fault instructs the processor to load the missing page into memory. Ideally, a low page-fault rate is desired.

The third control register, page directory base address, stores the physical memory address of the beginning of the page directory table. This table is up to 4KB in length and may contain up to 1024-page directory entries, each of which points to another page table area, whose information is used to generate a physical address.

The segment descriptors used in the 80286 are also used in the 80386, as are the gate descriptors and the four levels of privilege. Thus, the 80386 functions much the same as the 80286, except for the increase in physical memory space and the enhancements involving page handling in the virtual environment.

FIGURE 56.1 Software model of the 80386

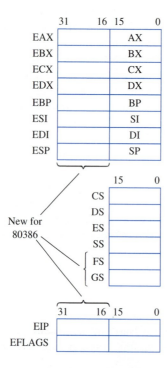

The computing power of each of the processors that have been presented can be augmented with the addition of a floating-point coprocessor. All sorts of mathematical operations can be performed with the coprocessors with 80-bit binary precision. The 8087 coprocessor is designed for use with the 8088 and 8086, the 80287 with the 80286, and the 80387 with the 80386.

A SUMMARY OF THE 80486

This processor is the next in Intel's line of compatible 80x86 architectures. Surprisingly, there are only a few differences between the 80486 and the 80386, but these differences create a significant improvement in performance.

Like the 80386, the 80486 is a 32-bit machine containing the same register set as the 80386 and all of the 80386's instruction set, with a few additional instructions. The 80486 has a similar 4GB addressing space using the same addressing features.

The first improvement over the 80386 is the addition of an 8KB **cache** memory. A cache is a very high speed memory, with an access time usually 10 times faster than that of conventional RAM used for external processor memory. The 80486's internal cache is used to store both instructions and data. Whenever the processor needs to access memory, it first looks for it in the cache. If the data is found in the cache, it is read out much faster than if it had to come from external RAM or EPROM. This is known as a *cache hit*. If the data is not found in the cache, the processor must then access the slower external memory. This is called a *cache miss*. The processor tries to keep the cache's hit ratio as high as possible. Consider the following example:

$$\text{RAM access time} = 70 \text{ ns}$$
$$\text{Cache access time} = 10 \text{ ns}$$
$$\text{Hit ratio} = 0.85$$
$$\text{Average memory access time} = 0.85 \times (10 \text{ ns}) \text{ Hit}$$
$$+ (1 - 0.85) \times (10 \text{ ns} + 70 \text{ ns}) \text{ Miss}$$
$$= 20.5 \text{ ns}$$

The average memory access time for a hit ratio of 0.85 is less than 21 ns. The reason for this is as follows: If data is found in the cache (85% of the time), the access time is only 10 ns. If data is not found (15% of the time), the access time is 80 ns (the cache access time plus the RAM access time), because the processor had to read the cache to find out that the data was *not* there.

If you consider that a large portion of a program (or even an *entire* program) might fit within the 8KB cache, you will agree that the program will execute very quickly, because most instruction fetches will be for code already in the cache. This architectural improvement significantly increases the processing speed of the 80486. Some of the new 80486 instructions are included to help maintain the cache.

The 80486 has two other improvements. Although it executes the same instruction set as the 80386, the 80486 does so with a redesigned internal architecture. This new design allows many 80486 instructions to execute with fewer clock cycles than required by the 80386. This reduction in clock cycles adds additional speed to the 80486's execution. Also, the 80486 comes with an on-chip coprocessor. You might recall that the 80386 can be connected to an external 80387 coprocessor to enhance performance. The 80486 has the equivalent of an 80387 built right into it. Moreover, since the coprocessor is closer to the CPU, data is transferred more quickly, which adds another performance boost.

Thus, although the 80386 and 80486 share many similarities, the 80486's differences create a much more powerful processor.

PIPELINING

Before we cover the Pentium processor, we will take a short look at the technique of **pipelining**. A pipelined processor executes instructions faster than a nonpipelined processor, as you can see in Figure 56.2. The nonpipelined processor, illustrated in Figure 56.2(a), is

FIGURE 56.2 Instruction execution in nonpipelined and pipelined processors

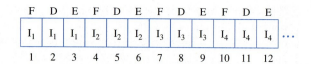

(a) Nonpipelined: 12 clock cycles required

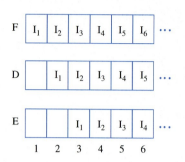

(b) Pipelined: six clock cycles required

designed to execute instructions in three clock cycles. The first clock cycle is the Fetch cycle (F), the second clock cycle is the Decode cycle (D), and the third clock cycle is the Execute cycle (E). A sequence of four instructions executed on this machine requires 12 clock cycles to execute. When the first instruction has been fetched, decoded, and executed, the second instruction begins, and so on. This is an inefficient way of executing instructions, since each stage of execution is idle for two clock cycles.

Figure 56.2(b) shows how pipelining reduces the number of clock cycles it takes to execute the same four instructions. The main difference is that in the pipelined processor the Fetch, Decode, and Execute operations *overlap*. For example, during the third clock cycle, I_3 is being fetched, I_2 is being decoded, and I_1 is executing. By the end of the sixth clock cycle, four instructions have made it through the Execute unit, and two more instructions have been partially completed. On a much larger scale, suppose that we execute 1000 instructions on the nonpipelined computer. This will require 3000 clock cycles of execution time. In contrast, the pipelined processor will need three clock cycles for the first instruction and *one* clock cycle for each of the remaining 999, for a total of 1002 clock cycles. This is a significant improvement over the nonpipelined result. In effect, an instruction pipeline is capable of executing one instruction every clock cycle.

Pipelining is made possible by the use of latches between consecutive pipeline stages, as shown in Figure 56.3. The latches make it possible for the three pipeline stages to operate in parallel. The result of one stage is latched for use by the next stage. This allows each stage to work on something different each clock cycle.

Intel uses pipelined computer architecture in its 80x86 family of microprocessors. With the design of the Pentium processor, the pipeline was given special attention to boost performance. In the next section you will see what Intel did with the Pentium's pipeline.

A SUMMARY OF THE PENTIUM

The newest, and fastest, chip in the Intel high-performance microprocessor line is the Pentium. As usual, upward compatibility has been maintained. The Pentium will run all

FIGURE 56.3 A three-stage pipeline

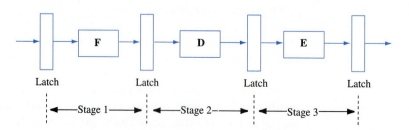

FIGURE 56.4 Architectural layout of the Pentium microprocessor

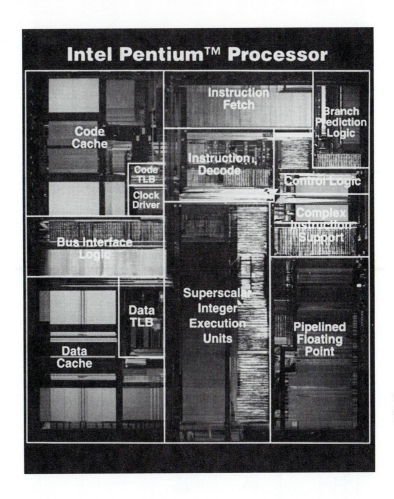

programs written for any machine in the 80x86 line, although it does so at speeds many times that of the fastest 80486. And the Pentium does so with a radically new architecture! Figure 56.4 shows a die shot of the Pentium.

There are two major computer architectures in use: CISC and RISC. CISC stands for Complex Instruction Set Computer. RISC stands for Reduced Instruction Set Computer. All the 80x86 machines prior to the Pentium can be considered CISC machines. The Pentium itself is a mixture of both CISC and RISC technologies. The CISC aspect of the Pentium provides for upward compatibility with the other 80x86 architectures. The RISC aspects lead to additional performance improvements. Some of these improvements are separate 8KB data and instruction caches, dual integer pipelines, and branch prediction.

As Figure 56.5 shows, the Pentium processor is a complex machine with many interlocking parts. At the heart of the processor are the two integer pipelines, the U pipeline and the V pipeline. These pipelines are responsible for executing 80x86 instructions. A floating-point unit is included on the chip to execute instructions previously handled by the external 80x87 math coprocessors. During execution, the U and V pipelines are capable of executing two integer instructions at the same time, under special conditions, or one floating-point instruction.

The Pentium communicates with the outside world via a 32-bit address bus and a 64-bit data bus. The bus unit is capable of performing *burst* reads and writes of 32 bytes to memory and, through bus cycle pipelining, allows two bus cycles to be in progress simultaneously.

An 8KB instruction cache is used to provide quick access to frequently used instructions. When an instruction is not found in the instruction cache, it is read from the external data bus and a copy, placed into the instruction cache for future reference. The branch target buffer and prefetch buffers work together with the instruction cache to fetch instructions as fast as possible. The prefetch buffers maintain a copy of the next 32 bytes of prefetched instruction code, and can be loaded from the cache in a single clock cycle, due to the 256-bit-wide data output of the instruction cache.

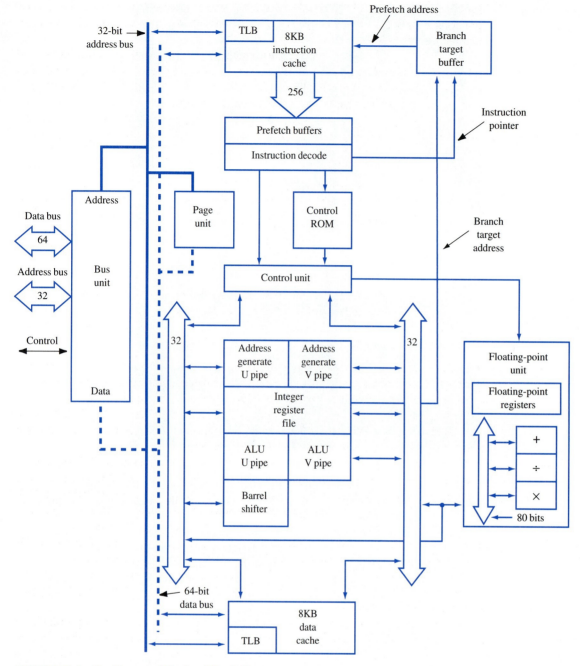

FIGURE 56.5 Pentium architecture block diagram

The Pentium uses a technique called *branch prediction* to maintain a steady flow of instructions into the pipelines. To support branch prediction, the branch target buffer maintains a copy of instructions in a different part of the program located at an address called the *branch target*. For example, the branch target of a CALL XYZ instruction is the address of the subroutine XYZ. Certain instructions, such as CALL, may cause the processor to jump to an entirely different program location, instead of simply proceeding to the instruction in the next location. So, just in case the code from the target address is needed, the branch target buffer maintains a copy of it and feeds it to the instruction cache.

A separate 8KB data cache stores a copy of the most frequently accessed memory data. Because memory accesses are significantly longer than processor clock cycles, it pays to keep a copy of memory data in a fast-reading cache. The data and instruction caches may both be enabled/disabled with hardware or software. Both also use a translation lookaside buffer, which converts logical addresses into physical addresses when virtual memory is used.

The floating-point unit of the Pentium maintains a set of floating-point registers and provides 80-bit precision when performing high-speed math operations. This unit has been completely redesigned from the one used inside the 80486 and is also pipelined. The floating-point unit uses hardware in the U and V pipelines to perform the initial work during a floating-point instruction (e.g., fetching a 64-bit operand), and then uses its own pipeline to complete the operation. Since both integer pipelines are used, only one floating-point instruction may be executed at a time.

Altogether, the Pentium processor includes many features designed to increase performance over earlier 80x86 machines. This was possible by blending CISC and RISC technologies together. The benefit to us as programmers lies in the fact that all Intel processors from the 8086 up, including the Pentium, run the same basic instruction set, but they do it faster and faster.

TROUBLESHOOTING TECHNIQUES

The advanced nature of the Pentium microprocessor requires us to think differently about the nature of computing. As we have already seen, exotic techniques such as branch prediction, pipelining, and superscalar processing have paved the way for improved performance. Let us take a quick look at some other improvements from Intel:

- Intel has added MMX technology to its line of Pentium processors (Pentium, Pentium Pro, Pentium II, and Pentium III). A total of 57 new instructions enhance the processor's ability to manipulate audio, graphic, and video data. Intel accomplished this major architectural addition by *reusing* the 80-bit floating-point registers in the FPU. Using a method called SIMD (single instruction multiple data), one MMX instruction is capable of operating on 64 bits of data stored in an FPU register.
- The Pentium Pro processor (and the Pentium II) use a technique called *speculative execution*. In this technique, multiple instructions are fetched and executed, possibly out of order, to keep the pipeline busy. The results of each instruction are speculative until the processor determines that they are needed (based on the result of branch instructions and other program variables). Overall, a high level of parallelism is maintained.
- First used in the Pentium Pro, a new bus technology called Dual Independent Bus architecture uses two data buses to transfer data between the processor and main memory (including the level-2 cache). One bus is for main memory, the second for the level-2 cache. The buses may be used independently or in parallel, significantly improving the bus performance over that of a single-bus machine.
- The five-stage Pentium pipeline was redesigned for the Pentium Pro into a *superpipelined* 14-stage pipeline. By adding more stages, less logic can be used in each stage, which allows the pipeline to be clocked at a higher speed. In addition, improvements in microchip manufacturing allow increases in operating frequency. This is easily verified by the more than 500-MHz processors currently available. Although there are drawbacks to superpipelining, such as bigger branch penalties during an incorrect prediction, its benefits are well worth the price.
- The Pentium III has several new features that make it a worthy successor to the Pentium II. Streaming SIMD extensions for multimedia applications, several low-power modes, and an Intel processor serial number are three new additions to the powerful Intel Pentium architecture. Examine the literature available on Intel's Web site for additional details.

Every aspect of computing must be studied in order to fully understand how to develop improved methods, software, and hardware for increased performance. Invest some time trying to think of an improvement of your own, as if you were designing a new microprocessor. Look through recent issues of computer architecture journals, or search the Web for information. You will find that a lot of other people are thinking about improvements, too.

SELF-TEST

This self-test is designed to help you check your understanding of the background information presented in this exercise.

True/False

Answer *true* or *false*.

1. A program written for the 8088 will run on the 80386.
2. A program written for the Pentium will run on the 80386.
3. All the 80x86 microprocessors support multitasking.
4. In real mode, an 80486 operates like a very fast 8086.
5. Branch prediction is a technique borrowed from RISC designs.

Multiple Choice

Select the best answer.

6. The 80286 accesses memory in protected mode through the use of
    a. An on-board coprocessor.
    b. Segment descriptors.
    c. 8086 instructions.
7. The extended registers on the 80386 are able to store
    a. 8 bits.
    b. 16 bits.
    c. 32 bits.
    d. 1 byte.
8. The 80486 uses a special high-speed memory called a
    a. Cache.
    b. Page directory.
    c. RAM.
9. A pipelined processor is capable of executing one instruction
    a. Every three clock cycles.
    b. Every other clock cycle.
    c. Every clock cycle.
10. CISC stands for
    a. Creative instruction scheduler.
    b. Complex instruction set computer.
    c. Computing index scalar.

Completion

Fill in the blanks with the best answers.

11. Running more than one program at a time is called _____.
12. Physical memory is divided into pages in a(n) _____ memory system.
13. The Pentium uses two integer _____.
14. Pipeline operation is improved through the use of branch _____.

FAMILIARIZATION ACTIVITY

1. If possible, prepare identical AUTOEXEC.BAT and CONFIG.SYS files for two different computers that have different processor speeds (33 MHz and 50 MHz, for example) or different processors (80386 and 80486). Run the same program (Windows, for example) on both machines, and measure the system response time for each. System response time can be gauged by how long it takes for the Windows desktop to appear, for each machine to spell check the same file within WordPerfect, or for some other similar type of operation.
2. Compare the two response times. Are they related to the clock speeds?
3. If possible, use a machine that has a **turbo** button on the front panel that switches the processor between two different execution speeds. Repeat the response-time

measurement described in step 1 and compare the results. Why would the lower processor speed be necessary once the faster one is available?

QUESTIONS/ACTIVITIES

1. Go to a library and search through back issues of *Computer Design, Electronic Design*, or *BYTE* magazine for articles on computer architecture. Read one article and share your thoughts with your instructor.

REVIEW QUIZ

Within 10 minutes and with 100% accuracy,

1. Discuss the improvements offered by each new 80x86 processor.
2. Explain the basic operation of an instruction pipeline.

57
A Detailed Look at the System BIOS

INTRODUCTION

Joe Tekk was working on a 386-based system. The BIOS setup parameters, necessary for proper operation of the computer, were constantly being lost, requiring them to be reentered every time the computer booted up.

Joe had seen this problem once before, at the college where he studied engineering. One of the computers in the department was always forgetting its BIOS parameters as well. His professor, Alonzo Dixon, told him that the battery in the CMOS RAM might be bad, causing it to lose data when the computer is turned off.

Joe took the CMOS RAM chip out of a spare motherboard and replaced the suspect chip in the 386 machine. Since he had no idea what the BIOS parameters were in the spare CMOS RAM, he once again set them to their proper values during power-on. To test the new RAM, he ran several power-down/power-up cycles, until he was convinced that the CMOS RAM was working properly.

Old Al knew what he was talking about, he said to himself.

PERFORMANCE OBJECTIVES

Upon completion of this exercise, within 10 minutes and with 100% accuracy, you will be able to

1. Describe the purpose of the system BIOS.
2. Explain why CMOS RAM replaced motherboard DIP switches for specifying system configuration parameters.

BACKGROUND INFORMATION

As we have seen in previous exercises, the system BIOS has a great deal to do with the overall operation of the personal computer. At power-on, the system BIOS program is responsible for testing all hardware in the computer, as well as starting the DOS boot sequence. In this exercise we will examine the operation of the BIOS setup program provided by AMIBIOS. Other BIOS programs for EISA systems (e.g., Award Modular or Phoenix Technology BIOS software) have similar features.

GETTING INTO THE SYSTEM BIOS

The AMIBIOS setup program is started when the machine is booting up if the user presses the Del key (or Ctrl-Alt-Esc for Award Modular BIOS). If the password option is enabled, the user must enter the correct password to gain access to the setup program.

The main menu of the AMIBIOS setup program looks like this:

>STANDARD CMOS SETUP
>ADVANCED CMOS SETUP
>ADVANCED CHIP SET SETUP
>AUTO CONFIGURATION WITH BIOS DEFAULTS
>AUTO CONFIGURATION WITH POWER-ON DEFAULTS
>CHANGE PASSWORD
>HARD DISK UTILITIES
>WRITE TO CMOS AND EXIT
>DO NOT WRITE TO CMOS AND EXIT

The arrow keys are used to highlight a menu item, which is selected when the user presses Enter. Notice the numerous references to CMOS.

THE CMOS RAM

Beginning with the AT model computer, the old way of configuring the system with motherboard-mounted DIP switches was eliminated in favor of a CMOS RAM that stored system parameters. There were simply too many options to be set with switches. The CMOS RAM stores 64 bytes of data and uses a battery backup circuit so that it retains its information when the computer is turned off. During the boot sequence, the CMOS RAM is read by the BIOS software to establish the required configuration. The BIOS setup program allows you to modify the CMOS RAM and thereby reconfigure your system.

STANDARD CMOS SETUP

This menu option allows you to change the system time and date, the types and numbers of hard and floppy drives, the type of display being used, and the keyboard test option.

When specifying the hard drive parameters, you must enter either a predefined type number (provided by the BIOS manufacturer) or your own set of parameters. For example, the parameters for drive C: may look like this:

	Cyln	Head	WPcom	LZone	Sect	Size
Hard Drive C:	5248	16	0	0	63	2.7GB

All the numbers shown may be altered by the user. Note that some BIOS setup programs have an *autodetect* feature that will interrogate the hard drive and read the setup parameters automatically.

In addition, this option also shows how much base and extended RAM is available in the system.

ADVANCED CMOS SETUP

This menu option allows you to alter the following settings:

- Keyboard typematic delay (milliseconds)
- Keyboard typematic rate (characters per second)
- Memory test sequence
- Boot sequence (C then A, or A then C)

- Internal/external cache (enable/disable)
- NUM-LOCK indicator at boot time (on/off)
- Password checking (enable/disable)
- Video ROM shadow (enable/disable)

The Video ROM *shadow* is a clever technique used to speed up the video ROM BIOS software. Newer display cards come equipped with their own video ROM BIOS chips that contain the software required to operate the display electronics. When a video ROM is shadowed, its contents are copied into RAM (which has faster access). The system then uses the RAM copy during video operations instead of the ROM copy. Thus, intensive video applications will run faster with ROM shadowing enabled.

ADVANCED CHIP SET SETUP

This menu option is used to set up the chip set that controls data transfers over the EISA bus. Recall that the EISA bus grew out of a need to expand the older ISA bus so that 32-bit processors could communicate with adapter cards. Thus, the ways in which 8-, 16-, and 32-bit transfers take place, and the clock speeds for the transfers, need to be specified. For example, an 80486DX2-66 is an 80486 processor running at 66 MHz *internally* and 33 MHZ *externally,* which means that the bus operates at a speed of 33 MHz (one-half the internal processor speed). The EISA chip set needs to be configured to operate under these speed conditions. Normally, only experienced technicians modify the chip set parameters. New chip sets that control operations on the local bus are configured here as well.

AUTO CONFIGURATION WITH BIOS DEFAULTS

This option sets all parameters to defaults stored in the system BIOS ROM. The default parameters are chosen to represent a broad range of typical EISA systems.

AUTO CONFIGURATION WITH POWER-ON DEFAULTS

This option is similar to the previous option, except that fewer parameters are enabled. This option is included to help diagnose hardware problems.

CHANGE PASSWORD

This option allows the BIOS password to be changed. In order to change the password, the current password ("AMI" by default) must be entered first.

HARD DISK UTILITIES

This option allows the hard disk to be low-level formatted (not a good idea for IDE and SCSI drives). Normally, the hard disk utilities available with DOS and Windows (and supplied by the hard drive manufacturer) would be used instead of this menu option.

WRITE TO CMOS AND EXIT

After the BIOS parameters have been modified, they can be saved in the CMOS RAM before the BIOS setup program is exited. This menu option is used to save the new BIOS parameters.

DO NOT WRITE TO CMOS AND EXIT

This option allows the user to exit the BIOS setup program without changing any of the BIOS parameters. This is a good menu option to have, especially if the user has forgotten which parameters have been changed. The changes made in the BIOS parameters during a setup session may be ignored when this option is used.

THE BIOS DATA AREA

Once the system BIOS has completed booting the system, a data table of BIOS parameters will have been written into system RAM, beginning at address 0040:0000. These BIOS parameters are used by the BIOS interrupt service routines to both control and reflect the status of all system hardware. Table 57.1 lists some of the predefined RAM locations and their associated parameters.

Experienced programmers are able to make use of the BIOS data area and control the system by manipulating selected parameters. For example, only a few instructions are needed to access and change the NUM-LOCK status bit stored in the byte at address 0040:0017.

EXAMINING THE BIOS DATA AREA

The DEBUG utility provides an easy way to examine the contents of the BIOS data area. Simply use the following command from inside DEBUG:

```
-D 40:0
```

The resulting display will be similar to the following:

```
0040:0000  F8 03 F8 02 00 00 00 00-78 03 00 00 00 00 00 00   ........x.......
0040:0010  63 C4 00 80 02 80 00 00-00 00 22 00 22 00 30 0B   c........."."0.
0040:0020  0D 1C 61 1E 74 14 61 1E-2E 34 74 14 78 2D 74 14   ..a.t.a..4t.x-t.
0040:0030  0D 1C 64 20 20 39 30 0B-34 05 30 0B 3A 27 00 00   ..d  90.4.0.:'..
0040:0040  D6 00 C0 00 00 00 00 00-00 03 50 00 00 10 00 00   ..........P.....
0040:0050  00 18 00 00 00 00 00 00-00 00 00 00 00 00 00 00   ................
0040:0060  0E 0D 00 D4 03 29 20 03-00 00 C0 20 C1 60 0C 00   .....) ....  .`..
0040:0070  00 00 00 00 00 02 00 00-14 14 14 34 01 01 01 01   ...........4....
```

Once again, only experienced programmers should attempt to work directly with BIOS data.

TABLE 57.1 Selected BIOS data area locations

Address	Data
0040:0000	I/O addresses for synchronous communication
0040:0008	I/O addresses for printers
0040:0010	Installed devices
0040:0013	Installed memory in KB
0040:0017	Keyboard shift flags
0040:001E	Keyboard buffer
0040:003F	Disk motor status
0040:0042	Disk controller status
0040:0049	Current video mode
0040:0072	Reset flag
0040:0076	Fixed disk control

TROUBLESHOOTING TECHNIQUES

The BIOS software is a very important part of the PC's operating system. New BIOS programs, called "Plug-and-Play" BIOS, work together with add-on peripherals (sound cards, modems, etc.) to automatically recognize new hardware when it is added to the machine. The user does not have to fool around with DIP switch settings or tiny option jumpers. Windows 95/98 contains built-in support for Plug-and-Play BIOS, and does a nice job of detecting and configuring new plug-and-play hardware.

SELF-TEST

This self-test is designed to help you check your understanding of the background information presented in this exercise.

True/False

Answer *true* or *false*.

1. The BIOS setup program controls only the video display and hard drive.
2. Password protection is provided with the BIOS setup program.
3. AMIBIOS is the only BIOS available for the PC.
4. The BIOS data area and the CMOS RAM data are stored in two different places.

Multiple Choice

Select the best answer.

5. The AMIBIOS setup program is started by
   a. Entering BIOS at the DOS prompt.
   b. Pressing Del when the machine is booting.
   c. Pressing both mouse buttons at once.
6. Using a video ROM shadow
   a. Slows down video BIOS routines.
   b. Increases the graphics resolution.
   c. Speeds up video BIOS routines.
7. The data stored in the CMOS RAM is
   a. Lost when the computer's power is turned off.
   b. Backed up by a battery.
   c. Loaded every time the machine boots.

Completion

Fill in the blank or blanks with the best answers.

8. Data transfers on the EISA bus are controlled by a(n) _____ _____.
9. The BIOS program is stored in a(n) _____.
10. The starting address of the BIOS RAM data is _____.

FAMILIARIZATION ACTIVITY

1. Reboot your computer by turning the power off and on. Enter the setup program as indicated by your machine.
2. If your system is equipped with two floppy drives, change the setting on the second drive (for drive B:) to NONE (or the equivalent). If your system has a single floppy, change the hard drive type to NONE (be sure to check with your instructor before doing this).
3. Save your changes and exit the BIOS program.
4. When DOS boots, attempt to get a directory of drive B: (or drive C: on the single-floppy system). What error message, if any, do you get?
5. Reboot the computer using Ctrl-Alt-Del. Are you able to enter the BIOS setup program from a warm boot?

6. Change the drive parameters back to their original settings, save your changes, and exit the setup program.
7. Repeat the DIR command. Is everything back to normal?

QUESTIONS/ACTIVITIES

1. Examine the motherboard of your computer. Can you find the BIOS ROMs? Write down the manufacturer's name and other information.
2. Use DEBUG to view the date of your BIOS ROM. This is done by displaying the contents of memory beginning at F000:FFF0. Use the following DEBUG command:

   ```
   -D F000:FFF0
   ```

   Your display will be similar to the following:

   ```
   F000:FFF0   EA F4 04 A6 02 30 34 2F-33 30 2F 39 30 00 FC EF    .....04/30/90...
   ```

   The bytes highlighted in boldface contain the date stamp of the installed BIOS.

REVIEW QUIZ

Within 10 minutes and with 100% accuracy,

1. Describe the purpose of the system BIOS.
2. Explain why CMOS RAM replaced motherboard DIP switches for specifying system configuration parameters.

58 Windows Internal Architecture

INTRODUCTION

Joe Tekk was browsing through the books in the computer section of a local bookstore. He found a book that illustrated the technical details of the internal operation of the Windows 98 operating system. The book was intended for developers writing application software. As Joe leafed through it, he found answers to many of the little questions he had in his mind about how Windows 98 did its job. For example, how does Windows 98 handle old hardware devices that are unsupported today?

Joe bought the book to put on his shelf as a reference.

PERFORMANCE OBJECTIVES

Upon completion of this exercise, within 10 minutes and with 100% accuracy, you will be able to

1. Discuss some of the architectural features of Windows 95/98 and Windows NT.
2. Explain why real-mode device drivers reduce performance.

BACKGROUND INFORMATION

The internal operation of Windows is a far more advanced topic than we are able to delve into here. But a short look at some of the major architectural components of Windows should provide you with a general picture of how things work in the operating system. You may never need to use this information during day-to-day computing, but it may be valuable when a problem crops up that you are unable to resolve using the standard help mechanism.

A LOOK INSIDE WINDOWS 95/98

Figure 58.1 illustrates a simplified view of the Windows 95/98 architecture. Three main components, the API layer (application programming interface), the system virtual machine, and the MS-DOS virtual machine all communicate with the base system through a set of protection rings (provided by 80x86 protected mode). Let us take a look at each component.

THE WINDOWS API LAYER

This layer provides system services to both 16-bit and 32-bit applications. The 16-bit API is used for old Windows 3.1 programs. The 32-bit API provides similar services, plus many more, for newer 32-bit applications.

FIGURE 58.1 Windows 95/98 internal architecture

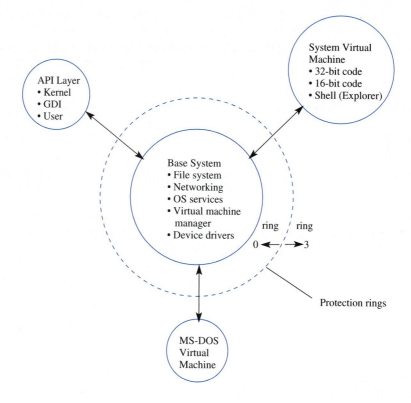

The API layer is composed of three components: the kernel, the GDI (graphical device interface), and the user portion. The kernel provides services such as memory management. The GDI controls what appears on the display screen, managing all graphical output. The user portion provides services to applications such as supplying icons for buttons.

THE SYSTEM VIRTUAL MACHINE

The system virtual machine is responsible for handling 32-bit and 16-bit applications. The 32-bit applications run in a preemptive multitasking mode, meaning they can be interrupted by another application and restarted at any time. The 16-bit applications (old Windows 3.1 programs) are executed in a different memory space, to simulate the Windows 3.1 environment and help protect the rest of Windows 95/98 from an out-of-control 16-bit application.

The shell portion of the system virtual machine is typically the Windows Explorer program. Users new to Windows 95 have the option of running the Program Manager from Windows 3.1 to maintain a sense of familiarity.

THE MS-DOS VIRTUAL MACHINE

Running a program in a DOS shell uses the MS-DOS virtual machine. This service is provided by the virtual-8086 mode of the advanced 80x86 processors. Each virtual machine contains 1MB of memory space, a copy of DOS (configured by CONFIG.SYS and AUTOEXEC.BAT at boot time), and operates as a unique, individual machine.

THE BASE SYSTEM

The base system contains several components. Unlike Windows 3.1, which ran on top of DOS and used DOS to manage files, Windows 95/98 contains its own file system, managed by the base system. Included is support for long file names and larger hard drives (up to 2GB partitions).

The base system also contains built-in networking support, allowing Windows 95/98 machines to be easily networked via network interface card, modem, or serial port. The base system also provides operating system support (e.g., DirectX graphical hardware control), manages any virtual machines created, and controls the various installed device drivers. There are two types of device drivers: virtual device drivers (files with .VXD extensions) and real-mode device drivers. Virtual device drivers run in protected mode. Real-mode device drivers (typically used for old or unsupported hardware) force Windows 95/98 to switch from protected mode to real mode and back whenever they are used. This could lead to a loss in performance for a heavily used device driver.

WINDOWS NT ARCHITECTURE

The Windows NT operating system internal architecture is much more complex than Windows 95/98 architecture because of the internal modifications necessary to achieve a more stable, reliable, and secure environment. As such, the Windows NT operating system can accommodate any type and size of organization. Figure 58.2 illustrates all the various components in both the *user* and *kernel* modes of Windows NT.

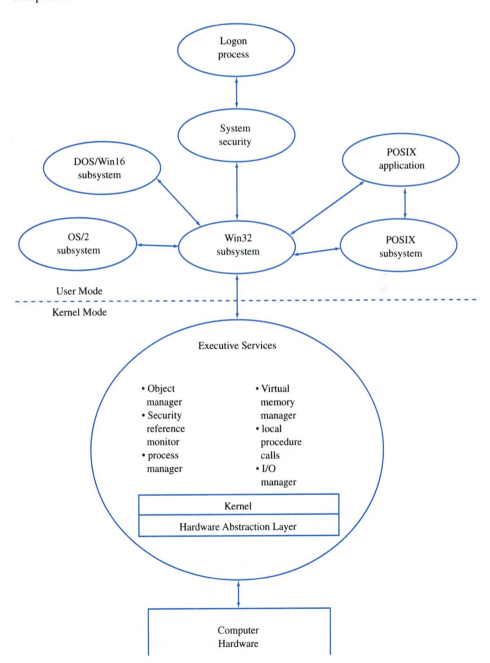

FIGURE 58.2 Windows NT system architecture

TABLE 58.1 Windows NT components

Component	Description
DOS, Win 16, and OS/2 subsystems	Support for applications written for earlier operating systems.
POSIX subsystem	Portable operating system interface for computing environments (POSIX) support.
Win32 subsystem	32-bit API support between application and operating system. Also manages keyboard, mouse, and display I/O for all subsystems.
Object manager	Responsible for creating, naming, protecting, and monitoring objects (operating system resources).
Security reference monitor	Manages security (access to objects).
Process manager	Manages processes and threads.
Virtual memory manager	Implements and manages virtual memory. All processes are allocated a 4GB virtual address space.
Local procedure calls	Message passing facility providing communication between client and server processes.
I/O manager	Interfaces with I/O drivers to provide all operating system I/O functions.
Kernel	Schedules tasks (processes of one or more threads) among multiple processors. Manages dispatcher objects and control objects.
Hardware abstraction layer	Provides a machine-independent interface by "virtualizing" the actual hardware of the base system.

Notice from Figure 58.2 that only the kernel mode has access to the physical hardware, thereby offering a high level of system protection. The executive services provided by the kernel are the foundation on which all other processing activities are performed in user mode. Table 58.1 shows a brief description for each of the components in each mode.

This modular approach to operating system design and development has allowed the Windows NT operating system to be ported to other hardware platforms, such as RISC-based microprocessors. You are encouraged to explore in more detail each of the internal components of the Windows NT operating system.

TROUBLESHOOTING TECHNIQUES

Knowing even a little bit of the internal operation of Windows may help you understand the reason behind a problem you encounter with an application, or serve as a starting point as you begin determining what is wrong. You are encouraged to learn more about Windows from books devoted to the subject, from information posted on the Web, and from other Windows users. The more information you have at your disposal, the better.

SELF-TEST

This self-test is designed to help you check your understanding of the background information presented in this exercise.

True/False

Answer *true* or *false*.

1. Windows 95/98 offers no type of protection mechanism.
2. The GDI is a component of the system virtual machine.
3. Only 32-bit code can be executed in the API layer.

4. You can use Program Manager to run applications under Windows 95/98.
5. All programs are executed using preemptive multitasking.
6. Windows NT runs on RISC-based microprocessors.

Multiple Choice

Select the best answer.

7. Device drivers are maintained by the
   a. API layer.
   b. Base system.
   c. System virtual machine.
8. GDI stands for
   a. General device indicator.
   b. Graphical device interface.
   c. Graphical data interface.
9. Windows NT Executive Services run in
   a. User mode.
   b. POSIX subsystem.
   c. Kernal mode.

Completion

Fill in the blank or blanks with the best answers.

10. The three components of the API layer are the _____, _____, and _____.
11. A 32-bit code is executed in the _____ virtual machine.
12. The typical shell used by the system virtual machine is _____.
13. The extension on a virtual device driver file is _____.
14. The _____ _____ _____ in Windows NT "virtualizes" the actual hardware of the base system.

FAMILIARIZATION ACTIVITY

1. Experiment with various options with CONFIG.SYS and AUTOEXEC.BAT. What does the DOS environment look like for each configuration?
2. Determine which RISC processors can run Windows NT.

QUESTIONS/ACTIVITIES

1. What real-mode device drivers are being used on your Windows 95/98 system, if any?
2. How many MS-DOS windows can you open?
3. Why is Windows NT considered more reliable than Windows 95/98?

REVIEW QUIZ

Within 10 minutes and with 100% accuracy,

1. Discuss some of the architectural features of Windows 95/98 and Windows NT.
2. Explain why real-mode device drivers reduce performance.

59 Computer Viruses

INTRODUCTION

Joe Tekk was puzzled. First, the time displayed on his Windows 98 taskbar was incorrect. Next, for no apparent reason, some applications would launch with a single-click instead of a double-click.

Then the printer began working erratically. Joe checked the cable, which was fine. He connected the printer to a different computer, where it also worked fine.

The problem must be inside my computer, he thought to himself. Deciding to look inside, he shut Windows 98 down and powered off. He took the case off and gave the motherboard, plug-in cards, and connecting cables a good visual. Seeing nothing out of the ordinary, he powered his machine back on and waited while it booted up.

A warning tone from his computer took Joe by surprise. One look at the screen told him what was wrong. His computer had a virus.

PERFORMANCE OBJECTIVES

Upon completion of this exercise, within 15 minutes and with 100% accuracy, you will be able to

1. Describe the operation of a typical file-infecting virus.
2. Discuss how viruses are transmitted between computers and files.
3. Use MSAV to scan a system for viruses.

BACKGROUND INFORMATION

There are literally thousands of programs available for the personal computer that provide meaningful and constructive service to the user. Unfortunately, there is a growing group of destructive programs, called *computer viruses,* as well. These virus programs are written by clever but dangerous programmers with the intent of doing some kind of damage to the computers of others. This damage can be as simple as a message on the display that reads "You are infected!" or as devastating as a destructive hard drive format. In this exercise, we will examine the operation of a computer virus, its method of infection and replication, its classification, and methods of preventing and eliminating virus infections.

OPERATION OF A TYPICAL FILE-INFECTING VIRUS

A virus is a computer program designed to place a copy of itself into another program. Programs are stored on floppy and hard disks as .COM and .EXE files. A virus program intercepts the .COM or .EXE file as DOS begins to load it into memory for execution. The virus checks the .COM or .EXE file to see if it already contains an infection. If the file is not infected, the virus inserts a copy of itself into the .COM or .EXE file and makes whatever other changes are necessary to the file so that the virus, and not the original .COM or .EXE program, executes first the next time the file is executed.

If the file is already infected, the virus does nothing, and allows the program to load normally. This is a clever way of avoiding detection, and it makes no sense to reinfect an already-infected file. This process is illustrated in Figure 59.1. Notice that control in the computer switches from DOS to the virus program, and back to DOS. This implies that the virus is already in memory, watching what is going on in the computer. How did the virus get there in the first place? The answer is given in the next section.

GETTING THE FIRST INFECTION

Virus infections come from a limited number of sources. First, you may get an infection by copying or running a program from someone else's floppy disk (or a swapped hard drive). If you copy an infected program (e.g., a game program, a popular hiding place for viruses), you must run the copied program to activate the virus. If you run an infected program, the virus code gets control first, and loads itself into memory to reinfect more programs later. The virus is active until you turn your computer off. It is not good enough to just reboot your

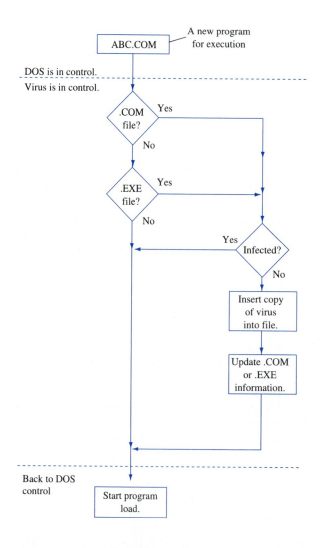

FIGURE 59.1 Typical virus infection sequence

FIGURE 59.2 How a virus spreads

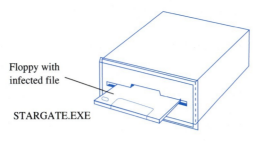

(a) User puts friend's floppy into disk drive.

Copy STARGATE.EXE C:\Games

(b) User copies the infected program.

CD\Games
STARGATE

(c) User runs the infected game program. The virus installs itself into memory.

CHESS

(d) User gets tired of game and starts a new one.

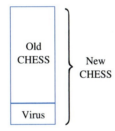

(e) Virus infects the CHESS program prior to execution.

machine using Ctrl-Alt-Del when you think you have an infection, because some viruses take over the reboot code and keep themselves in memory during a warm boot.

The second way to get an infection is to download an infected program from a computer bulletin board or network site. Most system operators in control of these types of installations already scan their software to help guarantee it is virus-free.

It is rare to find a virus hiding in a newly purchased software product or a box of preformatted disks, but anything is possible. It only takes one infected program to begin the spread of a virus, as you can see in Figure 59.2. The user in this example has unknowingly infected two programs. What happens the next time the STARGATE or CHESS program is executed? The virus will be back in business, resident in memory, waiting to infect any other programs that are run that day. So the user will spread the virus, once again unknowingly, to more and more files.

THE ANATOMY OF A VIRUS

For a self-replicating virus to be able to survive, it must be capable of the following:

- Operating as a memory-resident program
- Interfacing with the disk I/O routines
- Duplicating itself

FIGURE 59.3 Hooking an interrupt

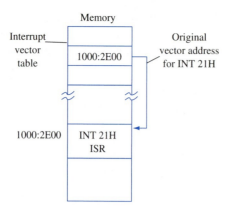

(a) Original vector

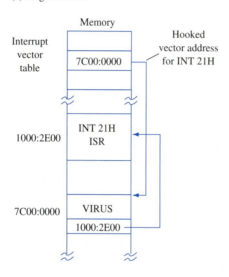

(b) Hooked vector. Virus gets control first.

You may think, from our brief introduction to assembly language, that the code for a virus that does all of this might be substantial. But the whole trick to writing a virus is to make it as small as possible, because large chunks of virus code might be more easily spotted. Many self-replicating viruses are written with fewer than 1000 bytes of machine code, but are still capable of mass destruction of system files.

For the virus to install itself as a memory-resident program, it requires a method of getting control of the system. This is usually done through the use of an interrupt *hook*. Recall that BIOS and DOS initialize the interrupt vector table at boot time with the addresses for each interrupt service routine. All the virus has to do is pick an interrupt frequently used by BIOS and DOS (e.g., the keyboard interrupt or INT 21H) and make the interrupt vector point to itself, instead of to the BIOS or DOS service routine. To give the impression that everything is fine, the virus always completes its job by running the original interrupt service routine. This concept of hooking an interrupt is illustrated in Figure 59.3. As you can see, the virus makes a copy of the original INT 21H service routine address, and uses it upon exit.

The process of hooking the interrupt and installation as a memory-resident program is called *initialization*.

When a virus detects that a file should be infected, it does the following:

1. Positions the file pointer at the end of the file. This is accomplished with calls to BIOS or DOS file I/O routines.
2. Writes a copy of itself (directly from memory) to the end of the file being infected.
3. Modifies the initial code of a .COM file (usually a JMP instruction), or the header information in an .EXE file, so that the virus code is executed first whenever the file loads.

4. Tells DOS to update the file information. This is done with another call to BIOS/DOS file routines.
5. Resumes the process of loading the file into memory, so that the user thinks everything is normal.

The duplication (or replication) phase of the virus continues either forever or until the virus has determined that it has reached a preset number of duplications. In the latter case, the virus takes whatever destructive action it was designed to take. Some viruses merely slow down the system response time when they are activated. For example, files that get sent to the printer take longer to print, the mouse gets sluggish, the display scrolls at a slower rate—anything annoying to the user. Other viruses are more troublesome, scrambling the internal key codes so that the keyboard becomes impossible to use (and Ctrl-Alt-Del does not work anymore). A destructive virus might swap numbers around in a spreadsheet file, encrypt a file so that it is impossible to decipher, format a few random tracks on a floppy disk, or delete important system files such as AUTOEXEC.BAT, CONFIG.SYS, and COMMAND.COM. Some viruses are mutated into *strains* or *variants* by other programmers who make a few changes in the virus code so that it does slightly different things when activated. File-infecting viruses are the most plentiful of the known viruses.

BOOT SECTOR VIRUSES

The viruses discussed so far infect actual files residing on floppy or hard disks. A **boot sector virus** is a virus that takes over the boot sector of a floppy or hard drive. Recall that the boot sector is the *first* piece of code read in from the disk and is in control of what happens next during the boot process. Boot sector viruses are very sophisticated and require good programming skills, which is why there are fewer boot sector viruses than file viruses.

The operation of a typical boot sector virus is detailed in Figure 59.4. The boot sector virus moves the original boot sector to a new sector on the disk and marks the new sector as bad so that DOS does not try to access it. The portion of the boot sector virus code that does not fit into the boot sector is stored in sectors marked bad also, to help avoid detection.

Any system booted from an infected disk begins running the virus immediately. Once again, the virus operates transparently to the user as a memory-resident program, possibly infecting the boot sector of any disk placed in the floppy drives.

WORMS

A **worm** is essentially a virus that does not replicate. When a worm-infected program is loaded into memory, the worm gets control first and does its damage (possibly transparently). Since it does not try to replicate, the worm is more difficult to discover by the user.

TROJAN HORSES

A **Trojan horse** is a virus disguised as a normal program. For example, a chess program with sophisticated 3-D graphics and sound could be a Trojan horse program containing a file-infecting virus. So, while the user is having fun playing chess, the virus is secretly running, too, examining directories on the hard drive for files to infect. The hard drive activity seems normal to the user because of the 3-D effects on the screen. But the effects are a mask for the file infection happening under the user's nose.

MACRO VIRUSES

Relative newcomers to the world of viruses are the *macro viruses*. Typically, a macro virus is written in a scripting language, such as Visual BASIC, and contains commands that can

FIGURE 59.4 Typical boot sector virus operation

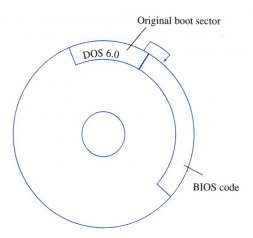

(a) Original floppy

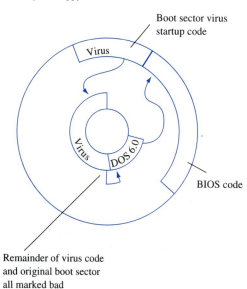

(b) Infected floppy

be processed from within a word-processing or spreadsheet application. These commands, stored as a macro for the particular application, can be very destructive, for instance by deleting files or randomly rearranging data within a spreadsheet when the macro is executed. Putting a simple macro virus together does not require the programming skills of an assembly language or C/C++ programmer, making them easier to write. Word-processing and spreadsheet documents should be included in the list of files routinely scanned by your virus scanner.

VIRUS DETECTION

Fortunately, there are methods that can easily detect the presence of a virus hiding in *any* file. One method involves the use of a file **checksum.** A checksum is a numerical sum based on every byte contained in a file. For example, an 8-bit checksum is made by adding all the bytes of the file, ignoring any carries out of the most significant bit. A file infected by a virus will practically always have a different checksum, thereby making detection a simple matter of comparing the current checksum of a file with a previous one.

A second method of virus detection involves searching a file for a known virus **signature.** A virus signature is an encoded string of characters that represents a portion of the actual virus code for a specific virus. A file is scanned for an entire collection of virus signatures to help guarantee that it is free of infection. The only problem with this technique is that a brand-new virus does not have a signature that the virus detection software will recognize, and it will escape detection.

There are many good virus scanning programs on the market. Many are also available as shareware, or through the computer centers at many colleges, which have an interest in keeping the number of virus infections on students' disks to a minimum. Students migrate from machine to machine on a large campus and will rapidly spread a virus before it is detected.

When DOS 6.0 was released, a virus scanning program called MSAV.EXE (Microsoft AntiVirus) was included; it runs under DOS and is very easy to use. MSAV contains a detailed list of all the viruses it scans for and is capable of removing a virus from an infected file. When MSAV is first started, it scans the current drive for all possible directories. A menu allows the user to choose one of the following operations:

1. Detect
2. Detect and clean
3. Select new drive
4. Options
5. Exit

Selections are made by using the arrows keys followed by Enter, by typing in specific letters, or by using the mouse. MSAV also contains an information menu, activated when F9 is pressed, which allows the user to view details of a specific virus chosen from MSAV's internal list of viruses.

VIRUS DETECTION IN WINDOWS

A number of companies make virus detection products for Windows. One popular virus scanner package is McAfee VirusScan. VirusScan's VShield runs at boot time, installing itself so that it can watch everything that is going on. VirusScan is used to scan entire drives for infected files. VShield scans files when they are accessed. When a floppy disk is inserted into the drive, it is automatically scanned for viruses. VirusScan is also able to automatically update its virus information database over the Web (for registered users). A screen shot of VirusScan's control window is shown in Figure 59.5. As shown, the hard drive C: has been scanned and no viruses were found. Several configuration options allow you to control what types of files are scanned and what to do when a virus is found. Figures 59.6 and 59.7 illustrate VShields Detection and Action option menus, respectively.

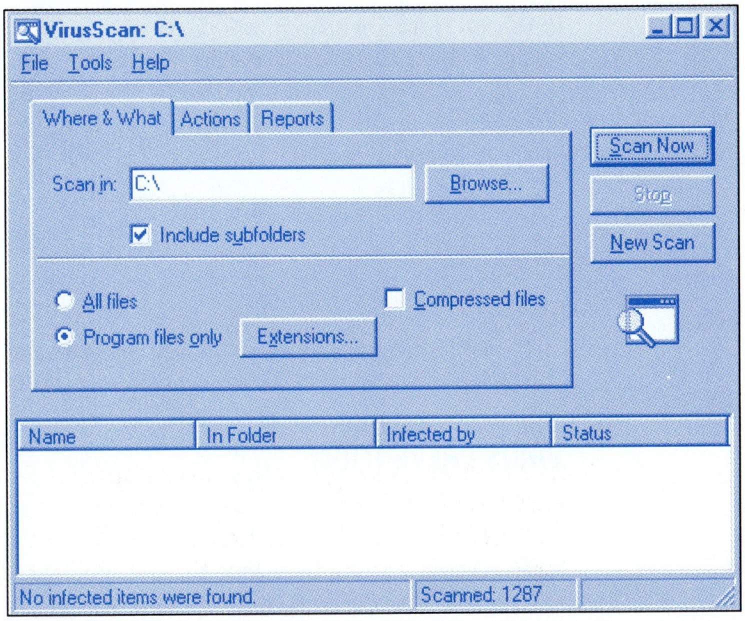

FIGURE 59.5 McAfee VirusScan control window

FIGURE 59.6 Detection menu

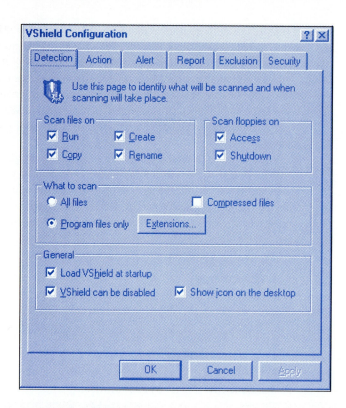

FIGURE 59.7 Action menu

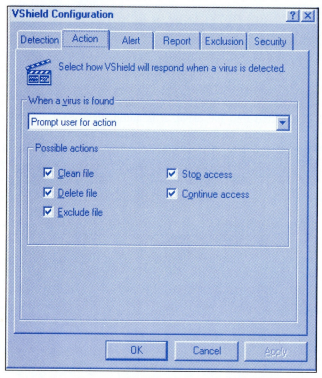

It is well worth the investment (money and scanning time) to use a good virus scanner. You will appreciate its value the first time it finds a virus.

VIRUS PREVENTION

The simple rules that follow should help eliminate the threat of infection.

- Never share your floppy disks with another person.
- Make sure your floppy disks are write-protected.

- Never copy or run software from another person's floppy before it has been scanned for viruses.
- Never execute downloaded software (from a computer bulletin board or network site) before it has been scanned for viruses.
- Never run a program from your disk on another person's computer.
- Run MSAV periodically on your computer to ensure that there are no spreading infections.

Keeping your disks and software to yourself is the best method of avoiding a virus infection.

TROUBLESHOOTING TECHNIQUES

Four simple words say it all: When in doubt, scan.

SELF-TEST

This self-test is designed to help you check your understanding of the background information presented in this exercise.

True/False

Answer *true* or *false*.

1. Only .COM files can be infected.
2. File infections come only from other infected files.
3. The hard drive is safe from infection.
4. Viruses are easy to find because they require huge amounts of code and cannot easily hide in a file.

Multiple Choice

Select the best answer.

5. In order to reproduce, a virus must be able to
    a. Install itself as a memory-resident program.
    b. Use BIOS/DOS disk I/O routines.
    c. Duplicate itself from its own image in memory.
    d. All of the above.
6. The best way to stop a virus from spreading is to
    a. Reboot the computer.
    b. Run as many programs as possible to tire the virus out.
    c. Eliminate the virus with a virus scanning program.
    d. Format the hard drive.
7. Viruses are written
    a. By beginning programmers.
    b. With the intent of doing damage to PCs.
    c. For amusement at computer trade shows.

Completion

Fill in the blanks with the best answers.

8. A virus disguised as a normal program is called a(n) _____ _____.
9. Viruses are detected by comparing _____ or _____.
10. A virus may also hide inside the _____ _____ of a floppy or hard disk.

FAMILIARIZATION ACTIVITY

1. Start up the MSAV program from hard drive C:.
2. When MSAV finishes checking directories, choose Detect from the menu. Detect will scan system RAM for memory-resident viruses and then begin scanning the hard drive. The current file (and its associated directory path) is displayed in the upper left corner

of the display. Warning! If MSAV finds a virus, the speaker will emit a short tone, and a message will be displayed such as:

Virus Friday 13th was found in: SHIP.COM

MSAV will give you the option of cleaning the file, continuing, or stopping the scan. Consult with your instructor if this happens.

You may also get a *verify* error for a particular file. This indicates that MSAV found a different time/date stamp on the file, or that the file's length or checksum has changed. MSAV allows you to update the file's information, or ignore the error and resume scanning. Consult with your instructor if this happens (in case you have discovered a brand-new virus).

3. Use the Select option to choose a new drive. Select the drive appropriate for the floppy disk you are going to scan.
4. Use Detect to scan your floppy disk. Once again, consult with your instructor if a virus is detected.
5. Press the F9 button to get into MSAV's information menu. An alphabetical list of viruses should appear, beginning with the As.
6. Enter the three characters "F," "R," and "I." MSAV should update the virus list so that the selected virus begins with the name "Friday." Use the arrow keys to select the Friday the 13th virus.
7. Press Tab until the Info box is highlighted. Then press Enter. You should get a screen of information on the virus that shows what types of files are infected, how long the virus code is, and what the side effects of the virus are.
8. Exit MSAV.
9. Restart MSAV with the following command line:

```
C> MSAV   C:\DOS
```

MSAV will now scan only the \DOS directory, not the entire hard drive. Individual files can also be scanned in this way by using their full path names after MSAV, as in:

```
C> MSAV C:\DOS\PRINT.EXE
```

QUESTIONS/ACTIVITIES

1. Use MSAV to make a list of 10 boot sector viruses. Compare their relative sizes and infection information. Include the Pakistani Brain, Michelangelo, and Alameda viruses in your list.
2. Repeat step 1 for the file-infecting viruses. Include the Friday the 13th and Columbus Day viruses in your list.

REVIEW QUIZ

Within 15 minutes and with 100% accuracy,

1. Describe the operation of a typical file-infecting virus.
2. Discuss how viruses are transmitted between computers and files.
3. Use MSAV to scan a system for viruses.

60 A Typical Computer Center

INTRODUCTION

Joe Tekk stopped by the computer center of the local community college to visit his friend Len Ore, a programmer analyst. Staffing problems force Len to perform many duties outside his basic job description, such as maintaining the college's Web server, e-mail accounts, and technical support. Joe waited patiently outside Len's office for 15 minutes while Len handled several urgent problems. A speaker phone allowed Len to talk to one person while working on three different computers simultaneously, his swivel chair constantly in motion as he switched from keyboard to keyboard.

"Hey, Joe," Len said, after entering a final command. "Sorry to keep you waiting."

Joe smiled. "Don't apologize. I know what it is like to work like that."

Len shrugged. "Sometimes I wish they would call someone else when they need help. But then I remind myself that they call me because I can fix their problems, and that makes me feel good."

PERFORMANCE OBJECTIVES

Upon completion of this exercise, within 10 minutes and with 100% accuracy, you will be able to

1. Describe the typical structure of a computer center.
2. Discuss the different positions available in a computer center.
3. List the important functions performed by computer center staff.

BACKGROUND INFORMATION

A computer center consists of the hardware, software, and personnel who are responsible for all the computing needs for an organization. The typical structure of a computer center can be either centralized or decentralized. A centralized computer center has most of the computer hardware in one central location. A central staff, consisting of administration, system analysts, programmers, operators, and technicians, is responsible for correct day-to-day operations. In contrast, a decentralized operation consists of computer hardware located in various areas of the organization connected using a LAN.

Figure 60.1 shows the floor plan of a typical computer center. Several server computers are surrounded by all the support equipment. Two of the servers physically share the same disk drives, making it possible to shut down one of the computers while continuing to allow

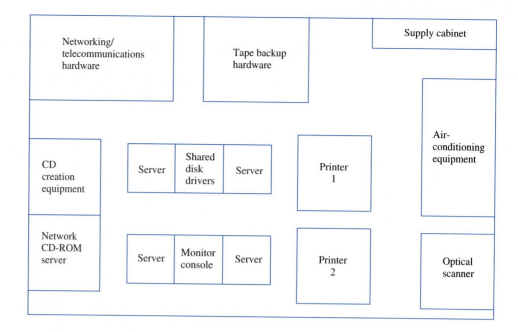

FIGURE 60.1 A typical computer center

access to the drives. Other server computers contain their own set of disks. The size of one disk is typically 9GB, but 18GB and larger are also becoming available. A centrally located monitor console is used to control each computer and view the output display from each on one central display.

Each of the computers is connected to the LAN and WAN using twisted-pair cables to a switch in the Networking and Telecommunications area. In an organization that maintains several buildings, the Networking and Telecommunications area will also be the place that the fiber optic cables coming from the other buildings are terminated.

The tape backup hardware area contains the devices used to back up all the information from each computer. A large number of disks require the use of top-end backup hardware to automatically switch the tapes as the backup progresses. Several tape drives (if available) are usually in operation simultaneously. The supply area contains new tape media, tape drive cleaning equipment, printer supplies, blank CD disks, and so on.

The printers in a computer center are usually capable of producing high-volume, high-quality output. It is becoming more and more common to replace the older impact printers with laser printers capable of printing on both sides of a form. The other equipment typically found in a computer center includes an optical scanner, CD-ROM servers, CD creation equipment, and other expensive computer hardware.

NETWORK EQUIPMENT

The actual network hardware of the organization may consist of hubs, switches, UTP and fiber cabling, and other telecommunication equipment. The organization may have an Internet presence, requiring a network administrator capable of managing the IP addresses assigned to the organization.

In addition to the hardware required to provide network access to all users is the equipment necessary to maintain the network. Test equipment (oscilloscopes, logic analyzers, network analyzers, cable testers), fabrication and repair tools (soldering iron, cable crimpers), and diagnostic software are all required to keep a large network running smoothly.

PERSONNEL

Many different types of positions are available in a computer center. In a large shop, responsibilities are broken down into small areas; in a small shop, one person wears multiple hats. In either case, many different functions need to be performed on a regular basis.

System Manager

The system manager for a particular organization manages the computer system. He or she is responsible for the day-to-day running of the computer. Duties for a system manager include operating system upgrades and maintenance, software installation, preventive maintenance and monitoring, and system security.

System Programmers

A system programmer is responsible for the proper operation of the computer system. Problems that affect all the users of a multiuser system, such as e-mail or group applications, fall under the domain of the system programmer. The system programmer works under the direction of the system manager, and may also perform some of the functions of the system manager such as installing software, monitoring computer resources, and offering support to the programmers.

System Analysts

A system analyst is responsible for the development of computer system software. The system analyst performs all necessary research to determine the best way to solve a particular problem requiring the use of a computer. A system analyst does not write the code to solve the problem, but provides all of the necessary information for the code to be written. The system analyst works very closely with a programmer or programmer/analyst to actually develop the necessary computer source code.

Programmer/Analysts

The programmer/analyst is a person with dual responsibility: first, to analyze the requirements for a system, and second, to actually write the computer program.

Programmers

A computer programmer writes computer code based on the information provided by an analyst. The programmer does not need to perform any analysis, as it has already been done. The programmer works from written specifications that precisely identify how the computer program should work.

Operators

Computer operators are responsible for the day-to-day activities required by the system. Some of the duties involve startup and shutdown of computer systems, performing backup operations, running the computer programs, verifying proper program execution, and producing any output generated by running a program.

Technicians

Computer technicians perform personal computer installations throughout an organization. This task usually involves installation of appropriate networking hardware and software for a computer to connect to the LAN. Computer technicians can perform hardware and software upgrades, printer installations, and offer troubleshooting to solve various computer problems.

Network Engineers

Network engineers are responsible for the design and development of the network used to connect all parts of an organization together. For a small organization, a network engineer might be responsible for a LAN.

Support Staff

The support staff in a computer center holds everything together. Receptionists, secretaries, and data entry staff perform all the essential functions necessary to maintain the center.

SUPPLIES

A quick look at the following list of supplies shows that running a computer center requires a significant amount of money:

- Paper (blank stock and preprinted forms).
- Ribbons.
- Ink cartridges.
- Floppy disks.
- High-capacity disks (e.g., Zip disks).
- Laser printer toner.
- Magnetic tape cartridges.
- Networking hardware (NICs, cable, hubs, etc.).
- Rewritable CD-ROM media.
- Spare PC hardware components (disk drives, keyboards, monitors).

MAINTENANCE AGREEMENTS

Along with expensive hardware come the expensive service agreements. Here are a few of the major ones:

- Hardware maintenance agreements to guarantee a quick response from the vendor to fix any type of hardware failure.
- Software license agreements to automatically receive the latest software updates and on-demand technical support directly from each vendor.
- Telecommunication expenses necessary to maintain external network connections (such as the cost of a T1 line to an ISP).

Although expensive, it is more cost-effective to maintain an agreement than deal with the problems associated with any type of hardware failure (disk crashes, power supply failures, etc.) or software bug that prevents people from getting their work done.

OTHER EXPENSES

Additional expenses for the typical computer center include:

- Salaries (full- and part-time).
- Utilities (heat, electricity, water, telephone).
- Travel and seminars.
- Miscellaneous expenses to maintain the mechanical, electrical, and physical environment (air-conditioning, humidification/dehumidification, etc.).

SERVICES PROVIDED

The services provided by a computer center fall into several categories:

- Administrative (payroll, human resources, budget, financial records, etc.).
- E-mail, Web page hosting, and Web applications.
- Help desk support for routine problems and questions.
- Software training and support.
- PPP dial-up access for off-site users.
- Data backup and recovery.
- Hardware upgrades.
- Software upgrades.
- Security administration.
- Resource monitoring and planning.
- Business-specific applications.

TROUBLESHOOTING TECHNIQUES

With all of the different types of positions in a computer center, a coordinated effort between all individuals is required to maintain high quality and standards throughout an organization. A partial list of the important issues that must be addressed includes the following:

- Monitor system activities to identify problems early.
- Create a centralized support center or help desk to provide fast access to necessary support personnel.
- Maintain adequate supplies for all supported equipment such as paper, toner, ink cartridges, ribbons, and tapes.
- Provide written documentation that explains how tasks are executed and what to do in the event of a failure.
- Offer training classes for employees within the organization to learn about new technologies, applications, and shared resources.
- Provide effective communication channels between all computer center personnel to allow for important information to be exchanged.
- Monitor system resources to plan for future hardware purchases.
- Implement a preventative maintenance schedule for all hardware devices, such as computers and printers.

Depending on the type of organization (bank, manufacturer, ISP, etc.), many other issues may also need to be considered.

SELF-TEST

This self-test is designed to help you check your understanding of the background information presented in this exercise.

True/False

Answer *true* or *false*.

1. A computer center can be easily run by three people.
2. Off-site users can connect to a central mainframe using PPP.
3. A programmer is responsible for fine-tuning all system software for peak performance.
4. Maintenance agreements are a luxury and not really necessary.
5. Only decentralized computer centers use LANs for communications.

Multiple Choice

Select the best answer.

6. Who is responsible for determining why a new PC cannot connect to the organization's network?
    a. System analyst.
    b. Technician.
    c. Network engineer.
7. Who is responsible for determining how to network PCs required for a new laboratory?
    a. System analyst.
    b. Technician.
    c. Network engineer.
8. Who is responsible for performing routine backups?
    a. Operator.
    b. System manager.
    c. Network engineer.
9. Problems that affect all of the users are handled by the
    a. Programmer.
    b. Programmer/analyst.
    c. System programmer.
10. Which is not a supply item?
    a. Maintenance agreement.
    b. A box of floppies.
    c. A laser printer cartridge.

Completion

Fill in the blanks with the best answers.

11. A computer center whose personnel are spread out geographically is called _____.
12. A system _____ is responsible for installing new software.
13. E-mail is a(n) _____ provided by the computer center.
14. A programmer relies on the expertise of the _____ analyst for guidance.
15. _____ are responsible for the day-to-day activities required by the system.

FAMILIARIZATION ACTIVITY

1. Visit the computer center at your school or work site. If one is not available, go to a local school or business. Inquire about the number of employees, their duties, and the services provided by the computer center.
2. Search the Web for hardware vendors that manufacture mainframe computers. How does a mainframe computer compare to an ordinary PC?
3. Estimate the hardware cost of setting up a computer center and LAN for 100 users. Assume that every piece of hardware and software must be purchased. For fun, assume that money is no object.
4. Using the Web, locate a job description of each position in a computer center.

QUESTIONS/ACTIVITIES

1. Which position in a computer center appeals to you? If you went to an interview, what would you say to the interviewer to demonstrate your ability to hold that position?
2. List some advantages and disadvantages of a centralized computer center.

REVIEW QUIZ

Within 10 minutes and with 100% accuracy,

1. Describe the typical structure of a computer center.
2. Discuss the different positions available in a computer center.
3. List the important functions performed by computer center staff.

Appendix A
ASCII Character Set

DECIMAL VALUE →		0	16	32	48	64	80	96	112	128	144	160	176	192	208	224	240
↓	HEXA-DECIMAL VALUE	0	1	2	3	4	5	6	7	8	9	A	B	C	D	E	F
0	0	BLANK (NULL)	►	BLANK (SPACE)	0	@	P	`	p	Ç	É	á				∝	≡
1	1	☺	◄	!	1	A	Q	a	q	ü	Æ	í				β	±
2	2	☻	↕	"	2	B	R	b	r	é	FE	ó				γ	≥
3	3	♥	‼	#	3	C	S	c	s	â	ô	ú				π	≤
4	4	♦	¶	$	4	D	T	d	t	ä	ö	ñ				Σ	∫
5	5	♣	§	%	5	E	U	e	u	à	ò	Ñ				σ	∫
6	6	♠	▬	&	6	F	V	f	v	å	û	ª				μ	÷
7	7	•	↨	'	7	G	W	g	w	ç	ù	º				τ	≈
8	8	◘	↑	(	8	H	X	h	x	ê	ÿ	¿				Φ	°
9	9	○	↓	)	9	I	Y	i	y	ë	Ö	⌐				Θ	•
10	A	◙	→	*	:	J	Z	j	z	è	Ü	¬				Ω	•
11	B	♂	←	+	;	K	[	k	{	ï	¢	½				δ	√
12	C	♀	∟	,	<	L	\	l	|	î	£	¼				∞	η
13	D	♪	↔	-	=	M	]	m	}	ì	¥	¡				∅	2
14	E	♫	▲	.	>	N	^	n	~	Ä	Pts	«				∈	■
15	F	☼	▼	/	?	O	_	o	△	Å	ƒ	»				∩	BLANK 'FF'

Answers to Odd-Numbered Self-Test Questions

EXERCISE 1

None.

EXERCISE 2

1. c 3. b 5. b 7. d
9. The nine safety rules presented in this exercise are: (1) no horseplay is allowed in the lab; (2) always get instructor approval; (3) immediately report any injuries; (4) use safety glasses; (5) use tools correctly; (6) use equipment correctly; (7) do not distract others; (8) use correct lifting techniques; and (9) remove jewelry.

EXERCISE 3

1. True 3. False 5. True 7. b 9. b 11. h, i 13. f 15. a
17. output 19. mouse

EXERCISE 4

1. True 3. False 5. False 7. a 9. b 11. b 13. d
15. 720K 17. FORMAT /S

EXERCISE 5

1. True 3. True 5. False 7. a 9. d 11. c, d 13. f 15. a
17. DIR 19. DIR/W

EXERCISE 6

1. True 3. True 5. True 7. b 9. d 11. a 13. path name 15. CD

EXERCISE 7

1. True   3. True   5. True   7. d   9. b   11. c   13. b   15. a   17. h
19. LASTDRIVE   21. AUTOEXEC.BAT   23. IF
25. It prints a file selected by the user to the printer.
27. The missing feature is a PAUSE reminding the user to turn on the printer and make sure it has paper.

EXERCISE 8

1. True   3. False   5. False   7. b   9. d   11. a   13. c   15. SORT

EXERCISE 9

1. False   3. True   5. b   7. protected   9. dialog box

EXERCISE 10

1. False   3. False   5. False   7. False   9. True   11. c   13. a   15. b
17. TCP/IP   19. Desktop   21. Registry   23. Shutdown

EXERCISE 11

1. True   3. False   5. False   7. b   9. b   11. Shutdown
13. Application   15. Diagnostics

EXERCISE 12

1. True   3. False   5. False   7. True   9. False   11. b   13. c   15. b   17. b
19. b   21. Taskbar   23. tiled   25. Briefcase   27. screen saver   29. Settings

EXERCISE 13

1. False   3. False   5. True   7. c   9. d   11. MPEG   13. Properties   15. WAV

EXERCISE 14

1. False   3. False   5. False   7. d   9. d
11. My Computer, Network Neighborhood, Recycle Bin
13. pull-down   15. Find Now

EXERCISE 15

1. False   3. False   5. False   7. b   9. c   11. b   13. dithering
15. local   17. default

EXERCISE 16

1. False   3. False   5. True   7. False   9. d   11. b   13. d
15. a   17. fragmented   19. WordPad

EXERCISE 17

1. False   3. True   5. False   7. a   9. a   11. Extended User Interface
13. Accessories   15. Internet service provider

EXERCISE 18

1. False   3. True   5. False   7. a   9. a   11. a   13. Installation
15. Application setting

EXERCISE 19

1. False   3. True   5. False   7. c   9. a   11. virtual device driver
13. interrupts   15. lower

EXERCISE 20

1. True   3. False   5. False   7. a   9. b   11. NDS
13. IPX   15. Time

EXERCISE 21

1. True   3. False   5. b   7. b   9. a   11. c
13. fully connected   15. data-link

EXERCISE 22

1. False   3. False   5. False   7. a   9. b   11. c   13. b   15. virtual
17. Carrier Sense Multiple Access, Collision Detection

EXERCISE 23

1. False   3. False   5. False   7. a   9. a   11. c   13. d   15. e
17. combo   19. five

EXERCISE 24

1. False   3. True   5. False   7. d   9. c   11. c   13. e   15. b
17. electronic mail (e-mail)   19. dotted-decimal notation

EXERCISE 25

1. False   3. True   5. True   7. c   9. b   11. NetBEUI   13. host
15. TRACERT

EXERCISE 26

1. False   3. True   5. False   7. c   9. a   11. HTML   13. active   15. HTTP

EXERCISE 27

1. False   3. True   5. False   7. c   9. c   11. d   13. a   15. e
17. workgroup   19. one

EXERCISE 28

1. True   3. True   5. True   7. b   9. a   11. T1
13. 13572468   15. cell

EXERCISE 29

1. True   3. False   5. c   7. c   9. b   11. UPS

EXERCISE 30

1. False 3. True 5. a 7. b 9. c 11. fan

EXERCISE 31

1. True 3. True 5. a 7. b 9. c 11. current

EXERCISE 32

1. True 3. False 5. False 7. a 9. d 11. b, c 13. a 15. d
17. disk drive 19. power supply 21. Answer depends on your system.
23. Answer depends on your system.
25. Yes, you must have an operating system (DOS or Windows).

EXERCISE 33

1. True 3. True 5. c 7. a 9. b 11. coprocessor 13. software

EXERCISE 34

1. True 3. False 5. d 7. b 9. b 11. c 13. RAM; virtual 15. level-2

EXERCISE 35

1. True 3. True 5. True 7. c 9. a 11. d 13. a, c 15. b 17. 1MB
19. switches
21. Interrupt request lines are used to temporarily get the attention of the microprocessor.
23. The EISA expansion slot is made compatible with the ISA expansion card by having two rows of connectors; the top row is compatible with the ISA card. Since the card isn't notched, it will go down only as far as the first row of connectors.
25. The distinguishing feature of micro-channel architecture is the use of smaller-sized expansion slots that will not accommodate earlier IBM PC, XT, or AT expansion cards but will accommodate 32-bit microprocessors such as the 80386 and up.

EXERCISE 36

1. True 3. False 5. b 7. c 9. a 11. math coprocessor

EXERCISE 37

1. False 3. True 5. True 7. d 9. b 11. unbootable
13. NLX 15. primary, secondary

EXERCISE 38

1. False 3. True 5. False 7. b 9. d 11. c
13. a 15. root 17. Small Computer Systems Interface

EXERCISE 39

1. False 3. False 5. True 7. d 9. b 11. a
13. b 15. BIOS 17. low-level

EXERCISE 40

1. False 3. True 5. False 7. d 9. a 11. b
13. sectors 15. FILE000N.REC

EXERCISE 41

1. False 3. False 5. True 7. a 9. d 11. c 13. d 15. b
17. resolution 19. 640×480
21. *Multisync* refers to a monitor capable of generating many different sync frequencies for various display resolutions.
23. 1280×1024
25. The monitor type is changed when you want to take advantage of the features of your new monitor.

EXERCISE 42

1. True 3. False 5. False 7. c 9. d 11. b 13. c
15. escape 17. portrait, landscape

EXERCISE 43

1. True 3. False 5. c 7. character 9. serial

EXERCISE 44

1. False 3. True 5. b 7. b 9. a 11. line 13. protocol

EXERCISE 45

1. True 3. True 5. False 7. b 9. c 11. 300KB/sec
13. advanced wave effects

EXERCISE 46

1. False 3. True 5. False 7. c 9. a 11. c 13. e 15. a
17. wireless 19. device driver

EXERCISE 47

1. True 3. False 5. True 7. b 9. a 11. Tools
13. view 15. grammatical

EXERCISE 48

1. True 3. True 5. True 7. c 9. a 11. selected
13. Wizard 15. A1

EXERCISE 49

1. True 3. False 5. False 7. c 9. b
11. Switchboard 13. relational 15. form

EXERCISE 50

1. False   3. True   5. True   7. a   9. b
11. slides   13. spell   15. templates

EXERCISE 51

1. False   3. False   5. True   7. c   9. a
11. unordered   13. sketch or drawing   15. .htm or .html

EXERCISE 52

1. True   3. False   5. True   7. b   9. c
11. promiscuous   13. colormaps   15. demo

EXERCISE 53

1. False   3. True   5. True   7. c   9. segment
11. byte-swapping

EXERCISE 54

1. False   3. False   5. False   7. d   9. b
11. object   13. compiler

EXERCISE 55

1. False   3. False   5. c   7. b   9. BIOS; the operating system
11. programmable interrupt controller

EXERCISE 56

1. True   3. False   5. True   7. c   9. c
11. multitasking   13. pipelines

EXERCISE 57

1. False   3. False   5. b   7. b   9. ROM

EXERCISE 58

1. False   3. False   5. True   7. b   9. c   11. system   13. .VXD

EXERCISE 59

1. False   3. False   5. d   7. b   9. checksums; signatures

EXERCISE 60

1. False   3. False   5. False   7. c   9. c   11. decentralized   13. service
15. Operators

Index

Accelerated graphics port (AGP), 432–434
   adapter, 509
Accessibility options, 122
   menu, 167
Accessories menus, 209–210
Adapters. *See* Video adapter card
Add New Hardware Wizard, 167
Add/Remove Programs, 168
Addresses. *See* IP addresses
Addressing modes, 80x86 microprocessor, 668–670
Address Resolution Protocol (ARP), 304
Address unit, 699
Administrative Tools, Windows NT, 222–224
   Disk Administrator, 456–457
Alphanumeric mode, 503
API layer, 715–716
Application layer, 268
Architecture
   80x86, 404, 659–670
   extended industry standard (EISA), 427–428
   micro-channel (MCA), 427
   Pentium, 703
   Windows 95/98, 715
   Windows NT, 717–718
   See also Topologies
Archive files, 94
Arithmetic instructions, 80x86 microprocessor, 666
ASCII code, 519–521
   character set, 737
   extended, 521
Aspect ratio, 502

Assembly language, vs. machine language, 674–675
Assembly procedures, 373
Asynchronous transfer mode (ATM), 349
AT expansion slots, 426–427
Attachments, 157
Attributes, 93–94
Audible error codes, 438–439
Auto configuration, 711
AUTOEXEC.BAT files, 79–80
   obsolescence of, 100

Background, desktop, 144–148
Backup
   schedules, 472
   using DOS, 480
   using PKZIP and PKUNZIP, 481–483
   Windows 95/98, 472–478
   Windows NT, 479
   See also Copying, Saving
Bar code readers, 562–563
Baseband system, 268
Based addressing, defined, 670
Based-indexed addressing, 670
Based-indexed with displacement addressing, 670
Base memory, 414–415
Base system, 716–717
Basic input/output system (BIOS), 411
   data area, 712
   defaults, auto configuration with, 711
   getting into, 710
   upgrades, 447

BASIC
   interpreter, 677
   sample program, 679–683
   statements, 680
   See also Visual BASIC, Visual C++
Batch files, 74–75
Batch parameters, 77
Batch processing, defined, 73
Baud rate, 538
   defined, 537
Binary numbers, 660–661
BIOS. *See* Basic input/output system (BIOS)
Bit manipulation instructions, 80x86 microprocessor, 666
Bitmap file, and desktop background, 145
Bits, 660
   defined, 411–412
Booting
   DOS, 25–27
   methods, 34
   sequence, 35
   See also Rebooting
Boot sector virus, 725
BOOTSTRAP program, 25–27
BREAK command, 68–69
Bridges, 290
Broadband system, 268
Broadcasting, defined, 269
Browser, defined, 211
Browse window, 150
Bubble-jet printer, 515
BUFFERS command, 69–70
Bus, 400, 402, 424
   local, 429–430

PC card, 430, 432–433
    peripheral component inter-
        connect (PCI), 430–431
Bus interface unit, 698
Bus mouse, 531
Bus network, 266, 275
    wiring, 276
Bytes, defined, 412, 664
    See also Gigabytes
Byte-swapping, 665

Cable connection, direct, 219
Cable modems, 292–294, 544
Cables
    Ethernet, 281–283
    printer, 515–516
Cable tester, 295
Cache, level-2, 413
Cache memory, 8KB, 701
Caching, 462
Calculator, 216
Calendar, 217
Cardfile, 217
Carrier-modulated signal, 267
Carrier Sense Multiple Access
        with Collision Detection
        (CSMA/CD), 269
CCITT standards, 543
C drive. *See* Hard disk drive,
        Primary partition
CD-ROM
    defined, 13
    installation, 552–553
    operation, 550
    photo, 552
    physical layout, 550–551
    Properties window, 554
    standards, 551–552
Cell, defined, 580
Central processing units
        (CPUs), 400
    compatible, 403–404
Character code, 529
Character map, 218
Character ROM, 502
Characters, defined, 681
Chart Wizard, 587–589
Chat servers, 315–316
Checksum, defined, 726
Chips, math, 404–405
Chip-select unit, 698
Chip sets, 447
    advanced, 711
Chip speed, 413
Circuits
    virtual, 274
    sample, Electronics Work-
        bench, 645–647
Circuit switching, 346–347

Clipboard viewer, 218
Clusters, 453–454, 467
CMOS RAM, 710
    do not write to, and exit, 712
    setups, 710–711
    write to, and exit, 711
Coaxial cables, 281–282
Code, Visual BASIC, writing, 653
Code page, defined, 98
Collision, defined, 269
Color composite monitor, 505
Color graphics array (CGA)
        adapter, 506
COMMAND.COM, 26–27
Common Gateway Interface
        (CGI), 325–327
Compact disk. *See* CD-ROM
Compiler, 677
    C++, 681
Compression agent, and Drive-
        Space, 216
Compression submenu, 388
Computer
    electrical environment, 361–366
    memory, 409–410
    physical environment, 356–361
    resetting, 36–37
    See also Central processing units
        (CPUs), Computer system
Computer center
    defined, 731
    services, 734
    supplies, 734
Computer network. *See* Network
Computer system
    basic, 20
    energy efficient, 501–502, 523
    overview, 371–372, 392
Concentrator. *See* Hub, Switches
Conditional, defined, 667
CONFIG.SYS, 68–73
Connectors, 283–284
Context-sensitive menus, 117
Control codes, 519
Control flags, 664
Control Panel, 165–166
    Windows NT, 133–134
Conventional memory, 414–416
Conversion, binary-to-
        decimal, 660–661
Cooperative multitasking, 121
Coprocessor, 404–405
    video, 428
COPY command, 48–49
    Windows Explorer, 188
Copying
    disk, 37
    group of files, 50
        from hard disk, 96–97

single file, 48
Corrosion, effect of, on
        computer, 360–361
COUNTRY command, 69, 70
Crossbar switch, 346–347
Cut feature, Windows Explorer, 188
Cylinders, 451–452

Data
    bits, 538
    elements, relationships
        between, 597–598
    migration, 256
    organization, 80x86 micro-
        processor, 664–665
    storage, 462
    types, C++, 681–683
Database
    creating, 600–602
    defined, 594
    design, 594
    existing, 594–597
    forms, 600
    internal structure, 598–600
    reporting and queries, 602–603
    software, 603
Data communications equipment
        (DCD), defined, 537
Data-link layer, 268
    IEEE standards, 308–309
Data terminal equipment (DTE),
        defined, 537
Data transfer instructions, 80x86
        microprocessor, 665
Date
    setting, 34, 36
    to limit search, Windows
        Explorer, 188–189
Date/Time Properties, 148–149
    menu, 168
DEBUG, viewing interrupt vector
        table with, 689–692
Decimal-to-binary conversion, 661
Defect mapping, defined, 483
Defragmentation, 212–213, 388,
        465–467
Desktop
    defined, 108
    getting to, from DOS, 158
    host, 318
    managing contents, 152–153
    Properties, changing, 150–152
    Web-style, with Windows 98, 159
    Windows, 140
        95, 116
        98, 122–123
        NT, 130, 133, 159
DEVICE command, 69, 70–71
Device drivers, 70

746

multimedia, 563
DHPC, 315
Diagnostics, Windows NT, 224
Dial-Up Networking, 219, 229–231
Digital cameras, 560–561
Digital data, methods of
 representing, 267
Digital signal processor, 554
Digital versatile disk (DVD), 562
DirecPC, 294–295
Direct addressing, defined, 669
Directories, 40, 42
 creating, 55–57
 deleting, non-empty, 61
 examples, 58–59
 getting in and out, 57–58
 removing, 60–61
Disassembly procedures, 372–373
Disk
 copying, 37
 storage capacity, 463–465
 structure, 463
 See also Floppy disks, Hard disk
Disk Administrator, Windows
 NT, 456–457
Disk caching, 462
Disk drive, 20–21
 controller, 389–390
 See also Floppy disk drive (FDD),
 Hard disk drive, Jaz drives,
 Zip drive
Disk Properties menu, Windows
 Explorer, 187
Display adapters. *See* Video
 adapter card
Display Properties menu, 169
Distance vector route discovery, 308
Domain clients, 336
Domain Name Service (DNS), 304
Domains, 334–336
 logging on, 336
 vs. workgroups, 333–334
 Windows NT, 133–134
DOS
 booting, 25–27, 34–35
 commands, internal and
  external, 37–39
 disk backup using, 480
 getting to desktop from, 158
 messages, 35–36
 prompt, 36
 changing, 80–81
 utilities, 88
 wild card characters, 49–50
 Windows 95 and, 120–121
DOS application, running, inside
 Windows, 113
DOS files, 26–27
 attribute, 93–94
 checking status, 95

creating, 51
defined, 39–40
displaying, 40–42
contents, 50
one, 49
editing, 53
erasing, 54
naming, 40
organizing, 54–55
redirecting, 50–51
renaming, 53–54
transferring, 48–49, 61
See also Archive files, Batch files,
 Hidden files, System files
DOS RECOVER command, 492–493
Dot-matrix printer, 514–515
Double words, defined, 665
Downward compatible, defined, 414
DRIVEPARM, 69, 71–72
DriveSpace, compression agent
 and, 216
Dual in-line memory module
 (DIMM), 413
Duplex mode, defined, 540–541
Dust, effect of, on computer, 359–360
Dynamic Host Configuration
 Protocol (DHCP), 309

Echo, 541
ECHO command, 75–76
Echo servers, 315–316
EDIT command, 52–53
Editing features, Windows
 Explorer, 188
Editor, C++, 681
80x86
 architecture, 404, 659–670
 microprocessor, 404, 659–670
 addressing modes, 668–670
 data organization, 664–665
 instruction types, 665–668
 real-mode software
  model, 662–663
80486, summary, 701
80186, summary, 698–699
80286, summary, 699–701
Electrical noise, 361–364
Electronics Workbench, 644–647
Electrostatic discharge (ESD), 366
E-mail, 157
Energy efficiency, 501–502, 523
Enhanced graphics adapter
 (EGA), 507
 monitor, 505
Enhanced Integrated Drive
 Electronics (EIDE)
 interface, 459–460
Enhanced Small Device Interface
 (ESDI), 459
Erasing, files, 54

See also Recycle bin
Errors, reported by System
 Properties, 176
Escape codes, printer, 522–523
Ethernet, 266, 268–270
 cabling, 281–283
 IEEE standards, 308–309
Event Viewer, 223
Expanded memory, 414–416
Expansion slots, 423–424
 design, 427–428
 makeup of, 424–427
 micro-channel, 428–429
Extended character set, 502–503
Extended data out DRAM (EDO
 DRAM), 413
Extended memory, 414–416
Extensions, 41

Familiarization activity, defined, 4
FAT32, 457–458
Fax/data modems, 544
FDISK tool, 453, 483
Fiber distributed data network
 (FDDI), 350
File allocation table (FAT), 454,
 456–458, 467–468
File names, long, 116–117
Files
 installing, 235–236
 restoring, 478–479, 481
 sharing, over network, 229
 See also DOS files
File system security, NetWare, 259
File Transfer Protocol (FTP), 306
FILES entry, 69, 72
Filters, packet capture, 643–644
FIND command, 93
 Windows Explorer, 188–190
Flag register, 664
Float numbers, 681, 683
Floppy disk drive (FDD), 12
 major components, 385–386
 operating sequence, 386–387
 operating system and, 387
 power cable, 390
 relationship of, to computer
  system, 392
 support system, 387
 troubleshooting, 393–394
Floppy disks, 21, 388
 formatting, 24–25
 labeling, 21–23
 storage capabilities, 23–24
 working with, 392
 write-protecting, 23–24
 See also SuperDisk, 100MB
Folder
 creating new, Windows
  Explorer, 183

747

deleting, Windows Explorer, 183–185
Folder Properties menu, Windows Explorer, 187
F1 key, 51
FORMAT command, 26–27
Formatting, floppy disks, 24–25
Formulas, for spreadsheets, 589–590
Fragmentation. *See* Defragmentation
Frame, 269
Frame relay, 348
Full duplex, defined, 540

Games, 210–211
Gate descriptors, 699
General-purpose registers, 663–664
GIF images, 620
Gigabytes, 459, 699
GOTO command, 78–79
  and IF, combining, 79
Go To menu, Windows Explorer, 191–192
Graphics
  adding, to Web page, 626–628
  creating, for Web page, 621–625
  Web page and, 620–621
Graphics accelerator adapters, 508–509
Graphics mode, 502
Graphs, 2- and 3-D, MATLAB, 648–650

Hard disk
  new, preparation of, 483–484
  optimizing performance, 468
  partitions, 452–456
  See also Platters
Hard disk drive
  backup, 472
  defined, 12
  interfaces, 458–461
  replacing, physical considerations, 490–492
Hard disk utilities, 711
Hardware
  adding new, 167
  installing, in Windows NT, 249–252
  new, detecting, 246–248
  printer, 516
Hardware interrupt, 688
  assignments, 692
Hayes compatible commands, 539
Heat, effect of, on computer, 356–357
Help
  Built-in printer, 199
  Windows 98, 158
  Windows 95, 118
Help menu, Windows Explorer, 193

Hexadecimal number system, 661–662, 677
Hidden files, 95
High Capacity Storage System (HCSS), 257
High Sierra format, 550
Host desktop, 318
Host name, 304
  See also IP addresses
HTML. *See* Hypertext markup language (HTML)
Hub, 275
  Ethernet, 289–290
  See also Switches
Hybrid networks, 276, 278
HyperTerminal, 220
Hypertext markup language (HTML), 323–325, 625
  software, 635
  HoTMetaL PRO, 625–626
Hypertext Transport Protocol (HTTP), 322–323

Icons, Windows Explorer, 181–182
IEEE 802 standards, 308–309
IF command, 77–78
  and GOTO, combining, 79
Images. *See* Graphics
Immediate addressing, defined, 668
Impact printers, 514–515
Inbox, 157
Include files, 681
Indexed addressing, defined, 669
Industry standard architecture. *See* Architecture, extended industry standard (EISA), ISA expansion slots
Information superhighway, 266
Ink-jet printers, 515
Input device, defined, 11
Instruments, Electronics Workbench, 645
Integers, 681, 683
Integrated Drive Electronics (IDE) drive
  adding second
    as master, 491–492
    as slave, 490
  interface, 459–461
  replacing old, 492
Integrated services data network (ISDN), 349
  modems, 543–544
Interactive processing, 73
Interconnection network. *See* Circuit switching
Interior Gateway Routing Protocol (IGRP), 308
Internet, 266

addresses. *See* IP addresses
  connecting to, 231
  organization, 322
  tools, 211
Internet Control Message Protocol (ICMP), 305
Internet Explorer, 322–323
Internet Protocol (IP), 302–303
  version 6, 303
  See also IP addresses, TCP/IP protocol suite
Internetwork Packet Exchange (IPX), 258, 300
Interrupt handler, defined, 688
Interrupt vector table, 688–690
  viewing, with DEBUG, 689–692
I/O (input/output) devices, 14
IP addresses, 277, 302–303
  See also Domain Name Service (DNS)
ISA expansion slots, 424–426
ISO–OSI Network Model, 299–300

Java, 327–329
Jaz drives, 391
Joystick, defined, 13
  Force-feedback, 124
JPG images, 620

Keyboard, 20, 528–531
  defined, 12
  servicing, 529–531
Keyboard Properties menu, 169–170
Keys, identifying, 529
Keywords, C++, 681

LanExplorer, 637–644
Languages
  assembly vs. machine, 674–675
  high-level, 676–679
  See also BASIC, Programming, Visual BASIC, Visual C++
LANs, 265
Laser hardware, 516–517
Laser printers, 515
LASTDRIVE, 69, 72
Level-2 cache, 413
Library files, 681
LINK program, 676
Linker, C++, 681
Links, adding, to Web page, 631–633
Link state routing, 308
Link Support Layer (LSL), 259
Linux operating system, 260–261
Little-endian format, 665
Local area networks. *See* LANs
Local echo, 541
Log file, defined, 480
Logging on, 336

Logical block addressing (LBA), 457
Login scripts, NetWare, 259
Low-level format, defined, 483

MAC address, 304
Machine language, vs. assembly language, 674–675
Macro Assembler, 675
Macro virus, 725–726
Magnetic fields, 361
Magneticware, 409
Main menu, Electronics Workbench, 644–645
Maintenance agreements, 734
Management, NetWare, 259–260
Map Network Drive, Windows Explorer, 190–191
Matched memory extension, 428–429
Math chips, 404–405
MATLAB, 647–650
Matrix operations, MATLAB, 650
Megabyte, defined, 668
Memory, 409–410
    allocation, 414–416
    map, 414–415
    organization, 414
    usage in Windows 95, 418–419
    virtual, 418
Menus
    Context-sensitive, 117
    Electronics Workbench, 644–645
    NetWare, 259
    See also Pull-down menus; individual types
Messaging, NetWare, 259
Meta files, MATLAB, 647–648
Micro-channel expansion slots, 428–429
Microcom Networking Protocol (MNP) standards, 542–543
Microcomputer system
    major parts, 12–13
    relationship of, to peripheral devices, 13
Microprocessor, 400–404
    80x86, 404, 659–670
Microsoft
    latest updates from, 237
    networking, 227–228
Microsoft Access. *See* Database
Microsoft Excel. *See* Spreadsheet
Microsoft Exchange, 122, 157
Microsoft PowerPoint, 607–608
Microsoft Word, 569–571
    creating new document, 571–573
    exiting, 573
    moving window, 574
    opening document in, 575–576
    printing documents, 576
    pull-down menus and toolbars, 574
    resizing window, 575
    saving document, 575
    scrolling through document, 575
Mobile communication, 351
MODE command, 97–100
Modem
    cable, 292–294, 544
    common problems, 545
    defined, 13, 536
    fax/data, 544
    installing, 245–246
    ISDN, 543–544
    setup, 537–538
    software, 538–539
    terminology, 539–541
Modems Properties menu, 170
Modes
    asynchronous transfer (ATM), 349
    duplex, 540–541
    graphics, 502
    monitor, 502–503
    multiplex, 450–451
    real and protected, 414
    scientific, 216
    standard, 216
    text, 502
    virtual 8086, 416
Modulation methods, 541–542
Monitor, 20, 497–498
    defined, 12
    modes, 502–503
    monochrome and color, 499–501
    servicing, 498–499
    types, 503–505
Monochrome composite monitor, 504
Monochrome display adapter (MDA), 505
Monochrome display monitor, 505
Monochrome graphics adapter (MGA), 506
MORE command, 92
Motherboard
    contents, 400–401
    defined, 399
    form factors, 446
    installing new, 448
    reasons to replace, 446
    removal process, 447–448
    See also Expansion slots
Mouse, 20
    defined, 13, 527
    serial and parallel, 531
Mouse Properties menu, 170–171
MS–DOS. *See* DOS
MS–DOS virtual machine, 716
Multicolor graphics array (MCGA) adapter, 508
Multimedia, 212
    devices, interfacing, 563
    miscellaneous applications, 563–564
Multimedia PC (MPC) compliant, defined, 549–550
Multimedia Properties menu, 170–171
Multiple Link Interface Driver (MLID), 258–259
Multiplex mode, defined, 540–541
Multiscan monitor, 505
Multistage switch, 347
Multitasking, preemptive, 121
Musical instrument digital interface (MIDI), 555
My Briefcase, 156–157
My Computer, 153–155
    contents, 192–193

Nanoseconds, 414
NetBEUI, 227–228, 300
Netscape Navigator, 322–323
NetWare, 255–260
    client software, 260
    installing/upgrading, 256
    network protocols, 257–259
NetWare Core Protocol (NCP), 258
Network
    administrator, 134
    bus, 266, 275
    defined, 265–266
    drives, Windows Explorer, 190–191
    equipment, 732
    engineers, 733
    fully connected, 266, 275
    hierarchy, 277
    hybrid, 276, 278
    layer, 268
    logging onto, 336
    menu, 171–172
    monitoring traffic, 638–640
    neighborhood, 155–156, 228
    printing, 205, 228
    protocols, NetWare, 257–259
    ring, 266, 276
    server, running, 336–338
    sharing files over, 229
    star, 266, 275
    token-ring, 270, 286–289
    topology, 266
    See also LANs, Satellite network system, WANs
Network Directory Service (NDS), 256–257
    security, 259
Networking
    Dial-Up, 219, 229–231
    Microsoft, 227–228

Windows 95, 119–120
Network interface card (NIC), 283, 285–287
   defined, 13
   and RFCs, 301–302
Network Link Services Protocol (NLSP), 308
Nodes, defined, 266
Nonimpact printers, 515
NOT command, 78
Notepad, 220
Numbers, integer and float, 681, 683
NUMOFF program, 675–676

Object-oriented programming, 683
OLE 2, support for, 122
OmniPage Pro, 561
Open Data-link Interface (ODI), 258–259
Open Shortest Path First (OSPF), 308
Operating system, FDD and, 387
Operators
   defined, 733
   redirection, 88–90
Optical character recognition (OCR), 561
Output
   and input, standard, 88
   serial and parallel, changing, 100
Output device, defined, 11

Packet capture, Lan-Explorer, 640–642
   filters, 643–644
Packet Internet groper (PING), 313–314
Packet switching, 348
Page fault, 700
Pages, defined, 700
Paint utility, 220–221
Parameter, defined, 77
Parity, 412, 538
Partitions, hard disk, 452–456
Parts bins, 647
Password
   change, 711
   Windows Explorer, 191
Passwords Properties menu, 172–173
Paste feature, Windows Explorer, 188
Path, setting, 68
PATH command, 468–469
Pathnames, 59–60
PAUSE command, 76–77
pcANYWHERE, 317–318
PCMCIA card. *See* Bus, PC card
Pentium microprocessor, 404
   summary, 702–705
Performance objectives, defined, 3
Peripheral control block (PCB), 698
Peripheral devices
   defined, 11
   relationship of, to microcomputer, 13

Personnel, in computer center, 732–733
Phase-shift keying, 267
Phone Dialer, 221
Photo CD, 552
Physical address, defined, 668
Physical layer, 268
Pipelining, 701–702
Pixels, 502
Planar. *See* Motherboard
Platters, 451–452
Point-to-Point Protocol (PPP), 306–308
Port addressing, defined, 670
Ports, for modem, 538
Power cable, FDD, 390
Power cycling, 356–359
Power-on defaults, auto configuration with, 711
Power-on self-test (POST), 411
   defined, 437–438
   error messages, 438–441
Power source problems, 363–366
Power supply
   characteristics, 378–381
   function, 377–378
Preemptive multitasking, 121
Presentation
   adding slides, 611–612
   creating, 614–616
   layer, 268
   opening existing, 608
   printing, 614
   saving, 613–614
   software, 616. *See also* Microsoft PowerPoint
   viewing, 609
Primary partition, 452
Primary storage, RAM and ROM, 410–411
Print devices, multifunction, 523
Printer, 20
   adding new, 202–203
   checking status, 204
   defined, 13
   drivers, 206
   features, 521–522
   setting, with MODE, 99–100
   technical considerations, 515–517
   testing, 517–518
   window, 172–173
Printer Properties menu, 197–202
Printer security, Windows NT, 206
Printing
   in Microsoft Word, 576
   network, 205, 228
   presentation, 614
Print job
   deleting, 205
   pausing and resuming, 204–205
Print services, NetWare, 260

Procedure, defined, 667
Processing, batch, and interactive, 73
Processor control instructions, 80x86 microprocessor, 667–668
Professional graphics adapter (PGA) monitor, 505
Program group, window, 109–110
Programmable DMA unit, 698
Programmable interrupt controller, defined, 698
Programmable timers, defined, 698
Programmer, defined, 733
Programmer/analysts, defined, 733
Programming
   Object-oriented, 683
   procedural vs. events-driven, 652
   See also Languages
Program transfer instructions, 80x86 microprocessor, 667
PROMPT command, 80–81
Properties
   checking/setting, Windows Explorer, 185–187
   setting, Visual BASIC, 652–653
Protected-mode device drivers, 248–249
Protected-mode operating system, 108
Protocol analyzer, 637
   See also LanExplorer
Protocols, 299–300
   communication, 267–268
   IPX/SPX, 258, 300
   modem, 544–545
   NetBEUI, 227–228, 300
   NetWare, 257–259
   routing, 308–309
   TCP/IP, 301
Pull-down menus
   Electronics Workbench, 645
   Microsoft Word, 574

Random access memory (RAM), 25, 410–411
Read-only memory (ROM), 25, 410–411
   character, 502
   files, 94
Read/write memory, 410
Real-mode device drivers, 248–249
Real-mode operating system, 108
Rebooting, 318
Recycle bin, 121, 155–156
Redirection operators, 88–90
Regional Settings menu, 173
Register
   addressing, defined, 668
   80x86 processor, 663–664
   indirect addressing, defined, 669
   relocation, 698
Registry, 118–119

Relocation register, 698
REM command, 75
Remote echo, 541
Remote procedure calls, 122
Renaming, files, 53–54
Repeaters, 276, 289
Request for Comments documents
(RFCs), 301–302
Resource meter, 213–214
Restoring files, 478–479
with DOS, 481
Reverse Address Resolution Protocol
(RARP), 304
Review quiz, defined, 4
RGB (red-green-blue) monitor, 505
Ring network, 266, 276
See also Token–ring network
Routers, 291–292
Routing Information Protocol
(RIP), 258, 308
Routing protocols, 308–309
RS-232 standard, 537

Satellite network system, 294–295
Saving
presentation, 613–614
Word document, 575
ScanDisk, 29–30, 213–214
Scanners, 523, 561
Scientific mode, 216
Screen editor, 680
Search. See FIND command
Secondary storage, 410
Sectors, 463
Security
NetWare, 259
printer, 206
Windows NT, 130–132, 340–341
Segment, defined, 668
Segment descriptors, 699
Segment registers, 663–664
Self-test. See Power-on self-test
(POST)
Sequenced Packet Exchange
(SPX), 258, 300
Serial Line Interface Protocol
(SLIP), 306–308
Serial and parallel outputs,
changing, 100
Serial port configuration, with
MODE, 100
Service Advertising Protocol
(SAP), 258
Session layer, 268
Settings menu, Windows 98, 158–159
Setup Wizard, 237–240
Shell, defined, 113
Shortcuts
creating
and using, 149–150
Windows Explorer, 185–186

missing, 218
Simple Mail Transport Protocol
(SMTP), 306
Simple Network Management
Protocol (SNMP), 307
Simplex, defined, 540
Single-in-line memory module
(SIMM), 412–413
Single instruction multiple data
(SIMD), 404
Slide, 607
inserting objects into, 612–613
Small Computer System Interface
(SCSI), 459, 461
adding drive, 492
Sockers, defined, 316
Software
database, 603
effect of, on hardware, 688
HTML, 635. See also HoTMetaL
PRO
installing new, from Windows
CD–ROM, 236
installing third-party, 237–240
NetWare client, 260
presentation, 616. See also
Microsoft PowerPoint
printer and, 519–521
removing, 240–241
science and technology, 654. See
also Electronics Workbench,
LanExplorer, MATLAB,
Visio, Visual BASIC,
Visual C++
spreadsheet, 590
Windows modem, 538–539
word processing, 576. See also
Microsoft Word
Software drivers, two types, 248–249
Software interrupt, 687
SORT command, 90–92
Sound card
installation, 555–556
operation, 554
Sounds Properties menu, 173–174
Source disk, defined, 480
Speakers, defined, 13
Speed. See Chip speed
Spool Settings menu, 200
Spreadsheet, 579–580
appearance, 590
building, 581–586
planning, 580–581
software, 590
viewing output from, 587–589
ST506 interface, 459
STACKS, 69, 73
Standard input and output, 88
Standard mode, 216
Star network, 266, 275
Start button, 140–144

Status flags, 664
Stop bits, 538
Storage
capacity, 463–465
floppy disk, 23–24
See also Data, storage; Memory;
Primary storage, RAM and
ROM; Secondary
storage
Store and forward switching, 290
String instructions, 80x86 micro-
processor, 666–667
Subdirectories, 56
Subnets, 277
Super VGA (SVGA)
adapter, 508
monitor, 505
SuperDisk, 100MB, 392
Switches, 277, 290–291
Switching. See Circuit switching,
Packet switching, Store
and forward switching
Switching power supply, 378–381
Synchronous DRAM (SDRAM), 413
Synchronous optical network
(SONET), 350
System. See Computer system
System analysts, defined, 733
System board. See Motherboard
System files, 95
configuration, 68–73
System information, 215
System manager, defined, 733
System monitor, 215
System programmers, defined, 733
System Properties menu, 174–175
error, 176
System settings area, 148–149
System tools, 212–216
System virtual machine, 716

Tables, adding, to Web page, 628–630
Tape backup unit, defined, 12
Target disk, defined, 480
Task Manager, Windows
NT, 131–132
Taskbar, 144–145
Windows 95, 116
Windows NT, 133
TCP/IP protocol suite, 301
Technicians, defined, 733
Television cards, 561–562
Telnet, 306–307
Terminal count, defined, 698
Text mode, 502
Thermal shock, 356–359
Time
setting, 34, 36
See also Date/Time Properties
Time-division multiplexing
(TDM), 346

Time domain reflectometer (TDR), 276
Token–ring network, 270, 286–289
    IEEE standards, 308–309
    See also Fiber distributed data network (FDDI)
Tokens, 681
Topologies
    computer network, 266
    physical versus logical, 274–275
TRACERT, 314–315
Trackballs, 531
Traffic, monitoring, on network, 638–640
Transceiver, 269, 289
Transmission Control Protocol (TCP), 304
    See also TCP/IP protocol suite
Transport layer, 268
Trojan horse, 725
Trust relationships, 335–336

UNDELETE command, 121
Undo feature, Windows Explorer, 188
Uninstall feature, 240–241
Universal Resource Locator (URL), 322
Unshielded twisted pair. See UTP cable
Upgrading
    NetWare, 256
    Windows applications, 236
Upper memory, 414
User Datagram Protocol (UDP), 304
User interface, designing, Visual BASIC, 652
User Manager, 223
User profiles, 338–340
Utilities, DOS, 88
UTP cable, 283

Video adapter card, 497–498, 505–509
Video controls, major, 502
Video coprocessor, 428
Video display, controlling, with MODE command, 98–99
Video Electronics Standards Association (VESA), 508
Video graphics array (VGA)
adapter, 507
monitor, 505
Video RAM (VRAM), 413
View, changing, Windows Explorer, 181–182
Virtual addressing, 699
Virtual device drivers. See Protected-mode device drivers
Virtual memory, 418
Virus, 721
    anatomy, 723–725
    boot sector, 725
    detection, 726–727
    in Windows, 727–728
    file–infecting, 722
    first infection, 722–723
    macro, 725–726
    prevention, 728–729
Virus signature, defined, 726–727
Visio, 650–651
Visual BASIC, 651–653
Visual C++, 653–654
    elements of, 681–683
    environment, 681
Volatile memory, 411
Volume labels, 28–29

WANs, 265–266
Web page
    creating, 625
    defined, 211
    final, 633–634
    maintaining, 634–635
    planning, 619–620
    service, reference, and technology-based, 329
Wide area networks. See WANs
Windows
    moving, 110–111
    resizing, 111–112
    scrolling through, 111
    virus detection in, 727–728
Windows CD–ROM, installing new software from, 236
Windows Explorer, 118–119
    desktop, 154
    Windows 98, 159, 179–180, 194–195
    Windows 95, 179–193
    Windows NT, 195
Windows modem software, 538–539
Windows 98
    desktop, 122–123
    Explorer, 179–180, 194–195
    Help, 158
    improvements, 122–125
    Settings menu, 158–159
    Web-style desktop, 159
Windows 95
    desktop, 116
    Explorer, 179–193
    memory usage, 418–419
    networking, 119–120
    new features, 122
    version B, 122
Windows 95/98
    architecture, 715
    backup, 472–478
    detecting new hardware, 246–249
    disk defragmenter, 212–213
Windows NT, 129–130
    Administrative Tools, 222–224
    architecture, 717–718
    backup, 479
    Control Panel, 133
    desktop evolution, 159–160
    diagnostics, 224
    Disk Administrator, 456–457
    domains. See Domains
    Explorer, 195
    installing hardware in, 249–252
    NTFS, 458
    operating system logon, 130
    printer ports menu, 200
    printer security, 206
    security menu, 130–132
    startup, 133–134
Windows 3.x
    leaving, 109
    running DOS application inside, 113
    starting, 108
    upgrading software, 236
WINIPCFG, 315–316
Wireless technology, 351
Wizards
    Add New Hardware, 167, 246–247
    Chart, 587–589
    Setup, 237–240
Word
    defined, 412, 664
    See also Microsoft Word
WordPad, 221–222
Word processing applications, 576
    See also Microsoft Word
Workgroup
    logging on, 336
    vs. domain, 333–334
World Wide Web, 322–323
    and Windows 98, 124
    See also Internet, Web page
Worm, 725
Write-protecting, 23–24

XCOPY command, 95–96

Zip drive, 20, 390–391
    defined, 12